BRAIN RESPONSES TO AUDITORY MISMATCH AND NOVELTY DETECTION

BRAIN RESPONSES TO AUDITORY MISMATCH AND NOVELTY DETECTION

Predictive Coding from Cocktail Parties to Auditory-Related Disorders

JOS J. EGGERMONT
Department of Physiology and Pharmacology, University of Calgary, Calgary, AB, Canada
Department of Psychology, University of Calgary, Calgary, AB, Canada

ELSEVIER

ACADEMIC PRESS
An imprint of Elsevier

Academic Press is an imprint of Elsevier
125 London Wall, London EC2Y 5AS, United Kingdom
525 B Street, Suite 1650, San Diego, CA 92101, United States
50 Hampshire Street, 5th Floor, Cambridge, MA 02139, United States
The Boulevard, Langford Lane, Kidlington, Oxford OX5 1GB, United Kingdom

Notices
Knowledge and best practice in this field are constantly changing. As new research and experience broaden our understanding, changes in research methods, professional practices, or medical treatment may become necessary.

Practitioners and researchers must always rely on their own experience and knowledge in evaluating and using any information, methods, compounds, or experiments described herein. In using such information or methods they should be mindful of their own safety and the safety of others, including parties for whom they have a professional responsibility.

To the fullest extent of the law, neither the Publisher nor the authors, contributors, or editors, assume any liability for any injury and/or damage to persons or property as a matter of products liability, negligence or otherwise, or from any use or operation of any methods, products, instructions, or ideas contained in the material herein.

ISBN 978-0-443-15548-2

For information on all Academic Press publications
visit our website at https://www.elsevier.com/books-and-journals

Publisher: Nikki P. Levy
Acquisitions Editor: Joslyn T. Chaiprasert-Paguio
Editorial Project Manager: Kristi L. Anderson
Production Project Manager: Swapna Srinivasan
Cover Designer: Greg Harris

Typeset by STRAIVE, India

Contents

Preface

Sensory perception has long been considered a bottom-up process, driven by environmental stimuli activating the sensory systems in the brain. The initially activated parts are coding the stimulus information in as rich a detail as possible. Along the pathway to the sensory cortices more and more derived aspects of the sensory information in a scene analysis take part. Ultimately, they may end up as neural group activity in the cortex that is tuned to faces, musical melodies, etc. However, already in the mid-1800s it was postulated by Helmholtz for vision that because retinal inputs are ambiguous, previous knowledge is required for perception in order to give sense to, and infer some properties of, the visual object. This is called top-down processing. In the early 1900s Gestalt psychology emerged in Austria and Germany as a theory of perception, which emphasized that organisms perceive entire patterns or configurations, not merely individual components. Thus, out of the fine discrimination in the peripheral sensory systems, a new whole was created. It was Valentino Braitenberg who noted that "the cortex is a machine that works on its own output," thereby combining the bottom-up with the top-down processing.

The notion of a processing hierarchy reflects the recognition that there is an interaction between forward and backward extrinsic connections in the brain. Mumford (1992) was among the first to put forward a hypothesis on the role of the reciprocal, topographic pathways between two cortical areas, one often a "higher" area dealing with more abstract information about the world, the other, "lower," area, dealing with sensory stimulus data. Mumford's ideas underlie the field of visual as well as auditory perceptual learning. The first application was to the visual system when Hochstein and Ahissar noted that processing along the feedforward hierarchy of areas, leading to increasingly complex representations, is automatic and implicit, whereas conscious perception begins at the hierarchy's top, gradually returning downward as needed. Thus an initial conscious percept—"vision at a glance"—matches a high-level generalized categorical scene interpretation, i.e., identifying the "forest before the trees." For later "vision with scrutiny," reverse hierarchy routines focus attention to specific, active, primary visual cortex units, incorporating into conscious perception detailed sensory information available there. This "Reverse Hierarchy Theory" dissociates between early explicit perception and implicit low-level vision, explaining various phenomena.

To phrase hierarchical learning slightly differently, how often is it that while reading your guess how the sentence proceeds often does not fit the actual words? Your brain had the idea how it should continue but was not matched with reality. In these cases, there is a mismatch between the brain's worldview and the actual sensory reflection of it. The brain then generates a mismatch response. These responses are so large that they can be recorded from the scalp using EEG or MEG, and intracranially using neuroimaging. Such mismatch responses were first recognized in the late 1970s by Risto Näätänen and colleagues, and subsequently studied in excruciating detail by his Finnish research group as the neural signs of sensory mismatches with the brain's internal model. This model can be seen as a short-duration memory of the recent past. This memory may be temporally built up by presenting a series of identical stimuli, e.g., tone pips of a fixed frequency, and then suddenly followed by an oddball tone pip of different frequency. The brain, in this case the auditory system, generates a so-called mismatch negativity (MMN), reflecting the mismatch in frequency. This can be extended to mismatches in sound duration, localization, and even sound omission.

In the early 2000s, Karl Friston generalized the idea that hierarchical learning is the most likely way the brain compares sensory stimuli with its own worldview by generating prediction errors in higher-level cortex that are sent down to lower levels and ultimately to primary cortex. There is then an interplay between the bottom-up prediction error signals and the top-down prediction signals. Thus the prediction errors result from a comparison between the sensory input and the top-down predictions. So even in the case of a sound omission there will be a difference with the prediction of an expected sound and the absence thereof. As we expect, the prediction errors could be reflected in the MMN, but other long-latency responses such as the P300 complex can also play a role. We are now faced with the question of what the standard event-related potentials (ERPs) that are generated in response to environmental stimuli mean. If we put a boundary at $\sim$100 ms, then we can state that ERPs with latencies <100 ms are reflecting prediction signals to higher sensory cortex. ERPs with latencies >100 ms can be considered candidates for prediction errors. Based on these principles, predictive coding was born, and it has taken cognitive science by storm. The theory appears to be universal, encompassing, e.g., MMN generation models based on repetition suppression of the frequent stimuli or on a comparison between memory and present. It appears that the MMN, either in the auditory, visual, or somatosensory domain, served as the ultimate prediction error. This applied particularly to local (in time) deviances, but also to global (in time) mismatches, in those cases followed by either a P3a or a P3b if a task was incorporated. This emphasizes that the MMN and P3a are preattentive responses albeit that they can be enhanced by paying attention to it.

The MMN paradigm has been applied to disorders that feature cognitive problems, such as developmental disorders—dyslexia, autism, and attention-deficit hyperactivity disorder—or age-related disorders—age-related hearing impairment, mild cognitive impairment, and Alzheimer's disease. Here it is appropriate to note that Näätänen and colleagues emphasized that "while not serving as a specific marker to any particular disorder, MMN may be useful for understanding factors of cognition in various disorders."

At this point it would be wise to ask whether this predictive coding only occurs in humans, in which cognition and its disorders play such a big role. Fortunately for studying the intricate neurophysiology underlying the interplay between prediction signals and prediction error, and identifying the responsible brain areas and connection networks, several of the human findings including the MMN have been identified in nonhuman primates. In addition, MMN-like signals have also been found in cat cortex and in rodent midbrain and cortex. These animal data have been dominantly found in the auditory system, with it well spaced and studied brainstem and midbrain structures. The putative equivalent of the MMN was identified and called stimulus-specific adaptation (SSA). The studies were largely based on local mismatches, and the role of adaptation in a forward-masking paradigm could describe most, but not all, of the findings. This gave rise to the idea that adaptation of the response to the frequent stimuli resulted in a smaller response in the last frequent before the response to the "fresh" deviant. Thus prediction error did not result from the SSA or at least could not account for all its effects.

In order to quantify the origin and effect of the prediction error, one has to identify putative sources along, say, the auditory pathway that contribute to the generation of the MMN. The putative cortical areas in humans are typically inferred from EEG/MEG recordings and their equivalent dipoles localized using various techniques. These source structures most frequently used in the dynamic causal modeling (DCM) are, for auditory MMN, the primary auditory cortex (A1), the superior temporal gyrus (STG), and the inferior frontal gyrus (IFG). DCM is an intrinsically straightforward procedure using for each of the identified structures a canonical cortical model that specifies the interaction between input and output pyramidal cells and their interaction with inhibitory neurons. The connection strengths between cortical areas, both bottom-up and top-down as well as interhemispheric, have to be estimated, as are the connections between the three types of cortical neurons. From all these estimated connections, resulting in a large number of possible networks, the optimal network in a Bayesian sense is then calculated such that it best predicts the recorded MMN in amplitude and latency. The network depends crucially on the putative areas selected; the A1-STG-IFG network can optionally be extended with a modulatory input from, e.g., prefrontal cortex (PFC) or orbitofrontal cortex (OFC), but other structures are possible.

Provided that the networks for people with normal cognition have been established, but still based on a putative set of sources in the brain, one may now apply predictive coding to people with various cognitive disorders, from developmental to aging related. It turns out that regardless of the disorder the same MMN network surfaces, albeit that changes in intrinsic and extrinsic connection strengths are generated by the DCM. This can provide insight into the underlying pathology. It is known that there are fairly specific network changes in each of the mentioned disorders. The purpose of this book is to link these disorder-specific changes with the "stable" disease-independent MMN networks.

The book comprises 12 chapters and an Appendix cataloging the myriad of ERPs used in the various EEG/MEG studies. Chapter 1 presents a fairly technical primer on predictive coding and the use of several computational models, e.g., dynamic causal modeling, to infer network connectivity and intrinsic activity of the various cortical network nodes. Chapter 2 gives a cursory overview of auditory stream segregation, deviance, and novelty detection, which is where predictive coding might shed light on the underlying mechanisms. Chapter 3 is again a fairly technical report on the role of adaptation in forward masking and amplitude modulation and the role it plays in explaining MMN as an alternative to the commonly proposed predictive coding network. Chapter 4 gives a detailed overview of animal studies of MMN like activity along the auditory pathway and its putative equivalent in stimulus-specific adaptation (SSA). A detailed overview of the role of neurotransmitters and modulators in SSA is followed by elucidating the role of animal auditory cortex in predictive coding, followed by the role of brain rhythms in separation prediction update and prediction error streams. Chapter 5 is devoted to human studies via the use of ERPs in deviance detection, local and global mismatch, novelty detection, and again the role of brain rhythms.

As a prelude to the use of predictive coding in developmental disorders, Chapter 6 reviews developmental and maturation aspects of ERPs, in particular the MMN. From newborns, infants, children to adolescents we describe the so-far-not-extensive use of predictive coding in these age groups. Chapter 7 further details the role of ERPs and brain rhythms in predictive coding, and the underlying MMN network with the role the auditory cortex and inferior frontal gyrus play therein. Chapter 8 reviews the use of auditory ERPs and especially the MMN and P300 in the following developmental disorders: autism spectrum disorder, attention-deficit/hyperactivity disorder, and dyslexia. Several questions arise: Do ERPs diagnose between ASD and ADHD? Does predictive coding help diagnose? Chapter 9 then has the focus on age-related disorders, among those the role that aging plays in tinnitus, mild cognitive impairment, and Alzheimer's disease. We discuss here in general the role of predictive coding in understanding cognitive impairment.

Chapter 10 takes an in-depth look at the sensory brain networks involved in deviance and novelty detection. We now include visual and somatosensory MMN and their generating networks and compare these with the auditory MMN. We pay particular attention to the role of the default mode network (DMN) and that of the inferior frontal gyrus therein. We also touch on networks underlying prediction error coding. In Chapter 11 we present a detailed study of the considerable role of predictive coding in the perception of music, speech, and language, and contrasts this with alternative prediction procedures. Chapter 12 reviews network changes underlying the previously mentioned neurological disorders and how these may affect the MMN-generating networks. Although it appears that the MMN networks are disorder independent, changes in the DMN that may modulate the canonical MMN network appear to be disorder specific. The contribution that predictive coding and especially in relation the dynamic causal modeling elucidates the intrinsic and extrinsic (connectivity) changes that ultimately may inspire effective clinical treatments.

I thank my wife Mary for listening and advising during the writing process at home. I also thank Kristi Anderson, Senior Editorial Project Manager at Academic Press, for prompt advice when I needed it, and for making writing this book a smooth process.

Reference

Mumford, D., 1992. On the computational architecture of the neocortex. II. The role of cortico-cortical loops. Biol. Cybern. 66, 241–251.

Abbreviations

A1	primary auditory cortex
AAF	anterior auditory field
ABR	auditory brain stem response
ACC	anterior cingulate cortex
ACx	auditory cortex
AD	Alzheimer's disease
ADHD	attention-deficit/hyperactivity disorder
AEP	auditory evoked potential
AG	angular gyrus
AI	anterior insula
AII	second auditory area
ALE	activation likelihood estimation
ALFF	amplitude of low-frequency fluctuations
AM	amplitude modulated
aMCI	amnestic MCI
aMMN	auditory MMN
AMPA	α-amino-3-hydroxy-5-methyl-4-isoxazole propionic acid
ANF	auditory nerve fiber
ApoE	apolipoprotein E
AS	Asperger syndrome
ASA	auditory scene analysis
ASD	autism spectrum disorder
ASD/+LI	ASD with language impairment
ASD/−LI	ASD without language impairment
ASD/MVNV	ASD with minimally verbal/nonverbal
BF	best frequency
BLA	basolateral amygdala
BOLD	blood oxygen level dependent
bvFTD	behavioral variant FTD
CAEP	cortical auditory evoked potential
CAMF	cortical auditory magnetic fields
CAP	compound action potential
CCEP	cortico-cortical evoked potential
CEN	central executive network
CF	characteristic frequency
CMC	canonical microcircuit
CN	cochlear nucleus
CNV	contingent negative variation
CON	cingulo-opercular network
DAN	dorsal attention network
DCM	dynamic causal modeling
DCN	dorsal CN
DD	developmental dyslexia

DEV	deviant
DLPFC	dorsolateral PFC
DMN	default mode network
DMPFC	dorsomedial PFC
DTI	diffusion tensor imaging
ECoG	electrocorticography
EEG	electroencephalography
EPSP	excitatory postsynaptic potential
ERAN	early latency right anterior negativity
ERN	error-related negativity
ERS/ERD	event-related synchronization/desynchronization
FA	fractional anisotropy
FC	functional connectivity
FEF	frontal eye fields
FFR	frequency-following response
fMRI	functional magnetic resonance imaging
fNIRS	functional near-infrared spectroscopy
FOP	frontal operculum
FPN	frontoparietal network
FRQ	frequency
FTD	frontotemporal dementia
GABA	gamma-aminobutyric acid
GC	Granger Causality
GM	gray matter
HFA	high-frequency activity; high functioning adults
HG	Heschl's gyrus
IC	inferior colliculus
ICA	independent component analysis
ICC	central nucleus of the IC
IFG	inferior frontal gyrus
IFJ	inferior frontal junction
IFOF	inferior fronto-occipital fasciculi
INT	intensity
IPC	inferior parietal cortex
IPD	interaural phase disparity
IPL	inferior parietal lobule
ISI	interstimulus interval
ITG	inferior temporal gyrus
kHz	kilo Hertz
LDN	late discriminative negativity
LFP	local field potential
LN	late negativity
LOFC	lateral OFC
MCI	mild cognitive impairment
MD	multiple demand system
MEG	magnetoencephalography
MGB	medial geniculate body

MLR	middle-latency response
MMN	mismatch negativity
MMNm	magnetic MMN
MMR	mismatch response
MMSE	mini-mental status exam
MNTB	medial nucleus of the trapezoid body
mPFC	medial PFC
mRAS	caudal mesencephalic RAS
MRI	magnetic resonance imaging
MRS	magnetic resonance spectroscopy
MTC	medial temporal cortex
MTG	middle temporal gyrus
MTL	medial temporal lobe
MU	multiunit
MUA	multiunit activity
NAc	nucleus accumbens
NMDA	N-methyl-D-aspartate
OCEP	obligatory cortical evoked potentials
OCN	occipital cortex network
OFC	orbitofrontal cortex
PAF	posterior auditory field
PC	predictive coding
PCC	posterior cingulate cortex
Pcu, pCUN	precuneus
PE	prediction error
PET	positron emission tomography
PFC	prefrontal cortex
PHC	parahippocampus
PLV	phase-locking values
PMC	posteromedial cortex
PP	planum polare
PPI	prepulse inhibition
PSP	postsynaptic potential
PT	planum temporale
PV	parvalbumin
pwPE	precision-weighted prediction errors
RAND	random
RAS	reticular activating system
RD	reading disorder
REG	regular
ReHo	regional homogeneity
ROI	region of interest
RON	reorienting negativity
RP	readiness potential
rs-FC	resting-state functional connectivity
rs-fMRI	resting-state fMRI
RSN	resting-state network

RT	reaction time
SAL	salience network
SAM	sinusoidally AM
SCD	subjective cognitive decline
SEM	structural equation modeling
SFG	superior frontal gyrus
SFR	spontaneous firing rate
SI	SSA index
SMA	supplementary motor area
SMA	supplementary motor area
sMMN	somatosensory MMN
SMN	sensorimotor network
SNHL	sensorineural hearing loss
SOA	stimulus onset asynchrony
SOC	superior olivary complex
SOM	somatostatin
SPL	sound pressure level
SRAF	suprarhinal auditory field
SRT	speech reception threshold
SSA	stimulus-specific adaptation
SSD	single-sided deafness
SSN	somatosensory network
STD	standard
STDP	spike time-dependent potentiation
STG	superior temporal gyrus
STS	superior temporal sulcus
STSD	short-term synaptic depression
TD	typical developing
TE	transfer entropy
TMS	transcranial magnetic stimulation
tMTF	temporal modulation transfer function
TPJ	temporo-parietal junction
TPN	task-positive network
TRN	thalamic reticular nucleus
VAF	ventral auditory field
VAN	ventral attention network
VBM	voxel-based morphometry
VCN	ventral CN
VLPFC	ventrolateral PFC
vMMN	visual MMN
VMPFC	ventromedial PFC
VOT	voice onset time
WM	white matter tracts

CHAPTER 1

A primer on predictive coding and network modeling

1.1 Introduction

At the core of predictive processing is the idea that the brain develops a generative model of the world that it uses to predict sensory input (Gregory, 1980). The comparison of predicted and actual sensory input then updates an internal representation of the world. This process is often described as a processing hierarchy. Although predictive processing is a contemporary framework in the context of brain function, the main concept was already described in the late 19th century, by Helmholtz (1962) as "unconscious inference." Helmholtz explained that, because retinal inputs are ambiguous, previous knowledge is required for perception in order to give sense to, and infer some properties of, the visual object (Pereira et al., 2019). I will first mention several working definitions of predictive coding (PC) and the "meaning" of the mismatch negativity (MMN), an endogenous event-related potential (ERP) that has a critical function in PC.

> Predictive coding theories posit that the perceptual system is structured as a hierarchically organized set of generative models in the brain which become increasingly general at higher levels. The difference between these model predictions and the actual sensory input, called prediction error, drives generative-model selection and adaptation processes that minimize this error. ERPs elicited by sensory deviance or mismatch reflect the processing of this prediction error at an intermediate level in the processing hierarchy. (Winkler and Czigler, 2012).

> Within the framework of predictive coding, prediction errors—reflecting the mismatch between incoming sensations and predictions established through experience—are minimized. The MMN amplitude and latency are measures of the prediction error. The predictive coding theory predicts that the MMN amplitude should decrease as the occurrence of a deviance becomes more predictable, e.g., during repetitive stimulation. (Lecaignard et al., 2015).

Brain Responses to Auditory Mismatch and Novelty Detection
https://doi.org/10.1016/B978-0-443-15548-2.00001-6

PC proposes that the brain constructs a hierarchical, generative model of the world that is capable of generating patterns of activity "from the top-down" that external stimuli would elicit "from the bottom-up." The brain continuously tries to "fit" such models by predicting the incoming sensory input. Bad-fits signal prediction errors resulting in increasingly accurate estimates, and following perceptual learning in a modified model. (Heilbron and Chait, 2018).

1.2 Hierarchical learning

The notion of a hierarchy depends upon the recognition that there is an interaction between forward and backward extrinsic connections in the brain. Mumford (1992) was among the first to put forward a hypothesis on the role of the reciprocal, topographic pathways between two cortical areas, one often a "higher" area dealing with more abstract information about the world, the other, "lower," area, dealing with sensory stimulus data. The higher area attempts to fit its abstractions to the sensory data it receives from lower areas by sending back a template reconstruction that best fits the lower-level view. In a sense, the higher areas modulate the sensory input layers. As it is likely that subcortical areas also contribute to this hierarchical processing, especially in the auditory system, this requires a multi-level corticofugal modulation of the auditory periphery (Fig. 1.1).

Mumford's ideas underlie the field of visual as well as auditory perceptual learning. The first application, however, was to the visual system when Hochstein and Ahissar (2002) noted that processing along the feedforward hierarchy of areas, leading to increasingly complex representations, is automatic and implicit, whereas conscious perception begins at the hierarchy's top, gradually returning downward as needed. Thus an initial conscious percept—"vision at a glance"—matches a high-level generalized categorical scene interpretation, i.e., identifying the "forest before the trees." For later "vision with scrutiny," reverse hierarchy routines focus attention to specific, active, primary visual cortex units, incorporating into conscious perception detailed sensory information available there. This "Reverse Hierarchy Theory" dissociates between early explicit perception and implicit low-level vision, explaining a variety of phenomena (Fig. 1.2 top). More recently, Wolff et al. (2022) noted that brain resting state studies show a hierarchy of intrinsic neural timescales with a shorter duration in unimodal regions (e.g., visual cortex and auditory cortex) and with a longer duration in transmodal regions (e.g., default mode network, DMN). This unimodal-transmodal hierarchy is present across acquisition modalities—electro/

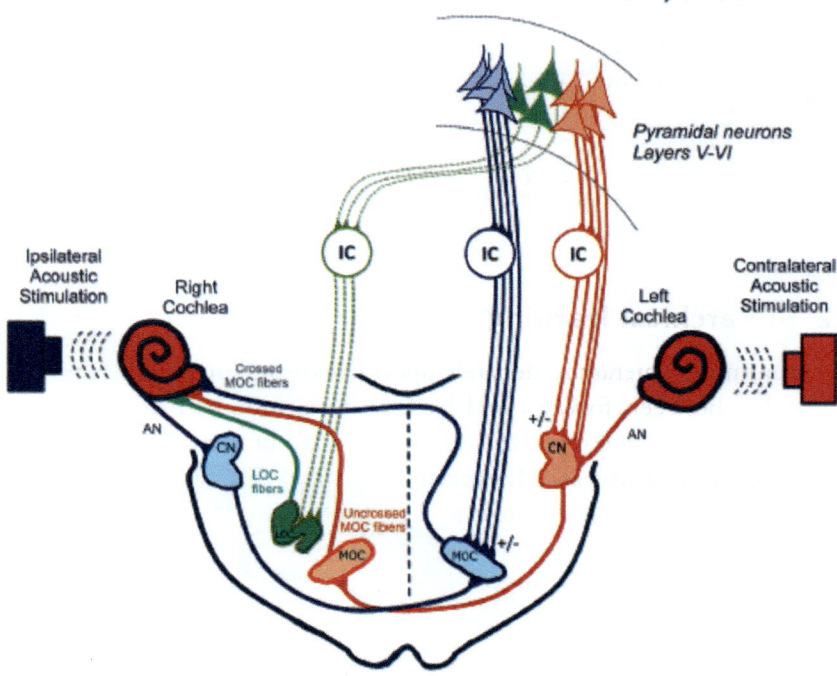

Brainstem

Fig. 1.1 The three pathways model for the cortico-collicular-olivocochlear and cochlear nucleus circuits. In order to simplify this model, the colliculo-thalamic-cortico-collicular loop has been omitted. In addition, only efferent pathways from the left auditory cortex to the right cochlea are presented. Three OC pathways are directed to the right cochlear receptor and auditory nerve, which are depicted in color *green*, *orange*, and *blue* corresponding to the: (i) right LOC fibers; (ii) right uncrossed MOC; and (iii) left crossed MOC neurons, respectively. Ipsilateral acoustic stimulation of the right cochlea activates right AN, right CN neurons that send projections to the contralateral MOC. In turn, left crossed MOC neurons modulate right cochlear responses *(blue brainstem pathways)*, constituting the ipsilateral OC reflex. On the other hand, contralateral acoustic stimulation of the left cochlea activates left AN, left CN neurons that send projections to the right uncrossed MOC fibers, which modulate right cochlear responses *(orange brainstem pathways)*, constituting the contralateral OC reflex that connects both ears. This model proposes that the descending pathways from the left auditory cortex directed to the left IC and to the left CN *(orange corticofugal pathways)* modulate the contralateral OC reflex, by regulating the activity of the left CN and right uncrossed MOC neurons. On the other hand, descending pathways directed to the left IC and left MOC *(blue corticofugal pathways)* regulate crossed MOC activity, which is involved in the ipsilateral OC reflex. Finally, corticofugal pathways to the contralateral IC *(green corticofugal pathways)* could regulate right LOC neurons, modulating the activity of right AN fibers. The +/− signs represent possible excitatory and inhibitory pathways. *AN*, auditory nerve; *CN*, cochlear nucleus; *IC*, inferior colliculus; *LOC*, lateral olivocochlear; *MOC*, medial olivocochlear. *(From Terreros, G., Delano, P.H., 2015. Corticofugal modulation of peripheral auditory responses. Front. Syst. Neurosci. 9, 134. https://doi. org/10.3389/fnsys.2015.00134. Open access.)*

magnetoencephalography, EEG/MEG, and functional magnetic resonance imaging, fMRI—and can be found during a variety of different task states. This suggested to them "that the hierarchy of intrinsic neural timescales is central to the temporal integration (combining successive stimuli) and segregation (separating successive stimuli) of external inputs from the environment, leading to temporal segmentation and prediction in perception and cognition (Fig. 1.2 bottom)."

Applying this strictly to the auditory system, Nahum et al. (2008) argued that the "increasing amount of evidence for an auditory processing hierarchy in which lower stations represent acoustic features of sounds, including

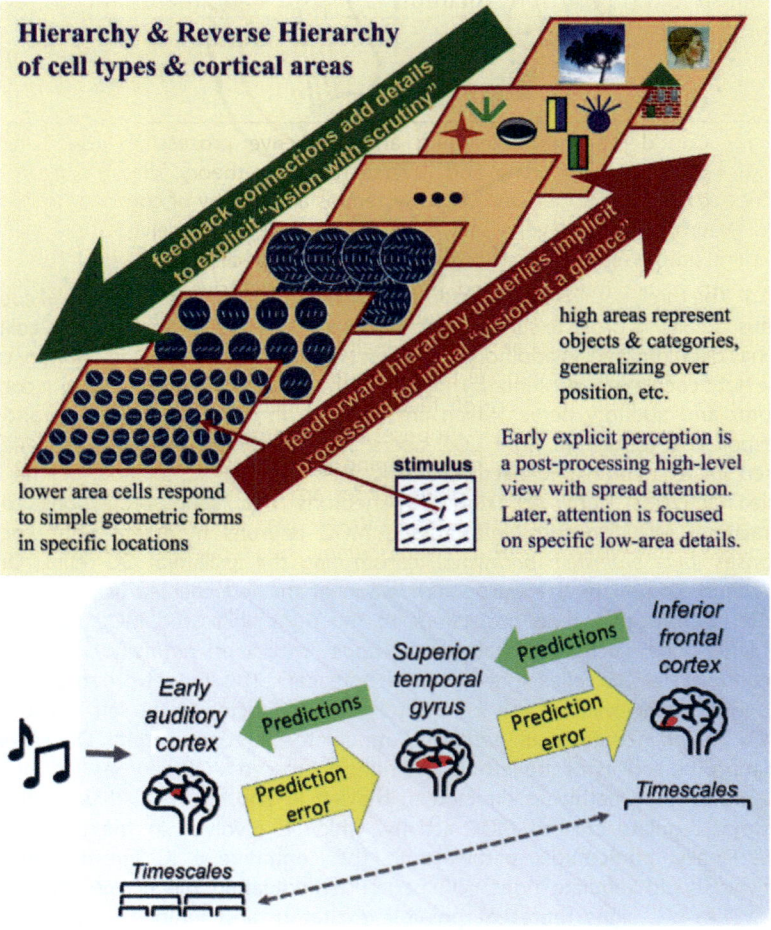

Fig. 1.2 See legend on opposite page.

precise frequency and level, while higher stations represent sounds more abstractly. Along this hierarchy, acoustic fidelity is presumably gradually replaced by ecologically relevant representations." Low-level representations—corresponding to the stages up to, and including, the inferior colliculus (IC; Fig. 1.1)—are determined by the physical, acoustic, nature of the stimulus. High-level representations, corresponding to cortical areas, converge across different low-level representations that denote the same objects or events. For example, cortical areas ventral ("belt") and posterior ("parabelt") to primary auditory cortex (A1), and portions of the superior temporal sulcus (STS), process temporal and spectral feature combinations that may be related to phoneme discrimination. This process was illustrated (Fig. 1.3) by Shamma (2008) in a comment on Nahum et al. (2008).

Fig. 1.2, Cont'd Hierarchical learning and predictive processing. Top. Schematic diagram of classical hierarchy and reverse hierarchy theory. Classical feedforward theory *(red arrow)* considers the visual system as a hierarchy of cortical areas and cell types. Neurons of low-level visual cortical areas (V1, V2) receive visual input and represent simple features such as lines or edges of specific orientation and location. Their outputs are integrated and processed by successive cortical levels (V3, V4, medial-temporal area MT), which gradually generalize over spatial parameters and specialize to represent global features. Finally, further levels (inferotemporal area IT, prefrontal area PF, etc.) integrate their outputs to represent abstract forms, objects, and categories. The function of feedback connections was unknown. Reverse Hierarchy Theory *(green arrow)* proposes that the above forward hierarchy acts implicitly, with explicit perception beginning at high-level cortex, representing the *gist of the scene* on the basis of a first-order approximate integration of low-level input. Later, explicit perception returns to lower areas via the feedback connections, to integrate into conscious *vision with scrutiny* the detailed information available there. Thus initial perception is based on spread attention (large receptive fields), guessing at details, and making binding or conjunction errors. Later vision incorporates details, overcoming such impaired vision. Bottom. Primary cortical areas (e.g., early auditory cortex) receive external sensory input (music notes). The number and repertoire of timescales in this primary sensory area is large. At the top of the hierarchy (inferior frontal cortex, IFG), higher-order areas have a smaller number and repertoire of timescales. In this "feedforward-feedback cascade," higher-order areas provide "top-down predictions" to lower-order areas. At the same time, these higher-order areas receive prediction error signals from lower-order areas. *(Top Panel: Reprinted from Hochstein, S., Ahissar, M., 2002. View from the top: hierarchies and reverse hierarchies in the visual system. Neuron 36 (5), 791–804; with permission from Elsevier. Bottom Panel: Reprinted from Wolff, W., Berberian, N., Golesorkhi, M., Gomez-Pilar, J., Zilio, F., Northoff, G., 2022. Intrinsic neural timescales: temporal integration and segregation. Trends Cogn. Sci. 26 (2), 159–173, with permission from Elsevier.)*

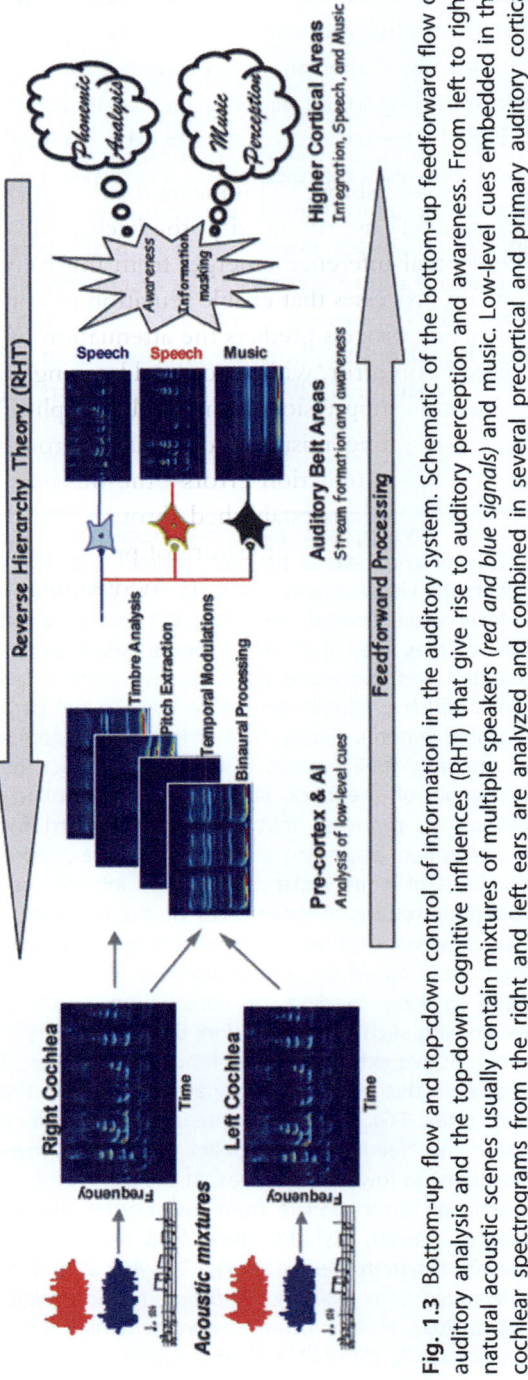

Fig. 1.3 Bottom-up flow and top-down control of information in the auditory system. Schematic of the bottom-up feedforward flow of auditory analysis and the top-down cognitive influences (RHT) that give rise to auditory perception and awareness. From left to right, natural acoustic scenes usually contain mixtures of multiple speakers *(red and blue signals)* and music. Low-level cues embedded in the cochlear spectrograms from the right and left ears are analyzed and combined in several precortical and primary auditory cortical (A1) stages. Neural correlates of consciously perceived streams of speech and music would emerge in the auditory belt areas beyond A1. In complex realistic scenes, ambiguous ("informationally masked") speech and musical streams are resolved through top-down influences described by the RHT. *(From Shamma, S., 2008. On the emergence and awareness of auditory objects. PLoS Biol. 6 (6), e155. Open Access.)*

1.3 From hierarchical learning to predictive coding

In the words of Friston (2005), "cortical responses can be seen as the brain's attempt to minimize the variability/uncertainty induced by a stimulus and thereby encode the most likely cause of that stimulus. [...] The use of hierarchical models enables the brain to construct prior expectations in a dynamic and context-sensitive fashion." This means that the causal structure of the world—a world model—is embodied in the backward connections (Fig. 1.2 bottom). Perceptual inference emerges from mutually informed top-down and bottom-up processes that enable sensation to constrain perception. This self-organizing process predicts the attenuation of peripheral responses, encoding prediction error, with perceptual learning and explains phenomena such as repetition suppression of the MMN amplitude. Within the framework of predictive coding, mismatch or deviance processing is part of an inference process where prediction errors—the mismatch between incoming sensations and predictions established through experience—are minimized. In this view, the MMN is a measure of prediction error (PE), which yields specific expectations regarding its modulations by various experimental factors (Lecaignard et al., 2015).

Chennu et al. (2013) tested these notions for auditory perception by independently manipulating top-down expectation and attention alongside bottom-up stimulus predictability. Their results support an integrative interpretation of ERPs such as MMN, P300, and contingent negative variation (CNV; see Appendix), as manifestations along successive levels of prediction errors. Early first-level processing—local (in time) deviations indexed by the MMN—is sensitive to stimulus predictability where attention enhanced early responses, but explicit top-down expectation diminishes it. This pattern contrasts with second-level processing—global (in time) deviations indexed by the P3b—while still sensitive to the degree of predictability and contingent on attention is sharpened by top-down expectation. At the highest processing level, the CNV functions as a marker of top-down expectation itself. Source reconstruction of high-density EEG and intracranial recordings implicate temporal and frontal brain regions that are differentially active at early and late levels. This suggests that the CNV might be involved in facilitating the consolidation of context-salient stimuli into conscious perception (Chennu et al., 2013).

Predictive coding also posits that this will be processed by two separate classes of neurons: (1) representational units, which process probabilistic representations (or predictions) about upcoming sensory input and (2) error units, which code prediction errors when there is a discrepancy between expected and actual sensory events (Friston, 2005). Within each level of the hierarchy, there is an exchange of information between the representational and error units, such that surprising events elicit a large, early response and locally update the prior probabilistic representations (i.e., they create a posterior probabilistic representation; cf. Fig. 1.6). Any unexplained surprise in the error units advances up the hierarchy to the representational units to evoke prediction errors. These representational units dynamically update prior predictions and project these again down the hierarchy. As such, they "explain away" error in the immediately preceding level of the hierarchy. Thus the system iteratively and rapidly minimizes surprise (or prediction errors) in sensory systems by updating probability distributions in the generative model, until the most probable cause of a sensory event is inferred (Apps and Tsakiris, 2014).

1.4 Predictive coding and Bayesian inference
1.4.1 Generative models and prediction levels

Predictive coding and generative models allow understanding the neuronal dynamics—firing patterns—in relation to perceptual categorization. This approach to perception involves adapting a putative internal model of the world to match sensory input (Mumford, 1992; Fig. 1.2). Recall that predictive coding models emphasize the role of backward connections in mediating the prediction, at lower or input levels, based on the activity of units in higher levels. The connection strengths between neurons, reflected in ERPs or fMRI activity, are changed so as to minimize the error between the predicted and observed inputs at any level (Friston, 2002). When there is a match, this means that the sensory input was successfully predicted and no updating of the internal prediction is required. However, if there is a mismatch, then the prediction failed to account for the incoming sensory input and the difference between prediction and input generates a prediction error. Prediction errors can be regarded as the information that remains to be explained in the input and serves to update subsequent predictions associated with the input (Mumford, 1992). De Ridder et al. (2014) noted that to reduce the prediction error, one of two things can happen: (1) the brain can either change its prediction or (2) change the way it gathers data from the environment. In order to do so efficiently, the brain will selectively sample the sensory inputs that it expects, to minimize surprise and minimize prediction errors, thereby also maximizing the sensory evidence for the

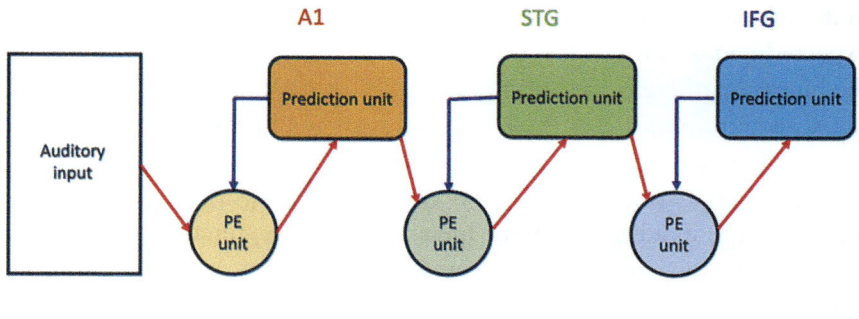

Fig. 1.4 Simplified scheme of a predictive processing model for auditory stimulation of the cortical hierarchy for perceptual inference. *A1*, primary auditory cortex; *IFG*, inferior frontal gyrus; *PE*, prediction error; *STG*, superior temporal gyrus. *(Based on Pereira, M.R., Barbosa, F., de Haan, M., Ferreira-Santos, F., 2019. Understanding the development of face and emotion processing under a predictive processing framework. Dev. Psychol. 55 (9), 1868–1881. https://doi.org/10.1037/dev0000706.)*

predicted stimulus' existence. Since the brain minimizes prediction error (Friston, 2009), both in the firing of neurons as in the wiring between them, De Ridder et al. (2014) proposed that changing connections between neurons is formally identical to Hebbian plasticity. By updating predictions through the incorporation of PEs, the system will minimize PE in the next round of comparisons, thus creating a better (predictive) model of the input.

Pereira et al. (2019) noted that at the first processing level, predictions are about immediate sensory input, but at the next level of the hierarchy, predictions are about the lower-level predictions. Furthermore, PEs at each level of the hierarchy reflect the mismatch between the prediction from the higher level and activity from the lower level. In this hierarchy, it is then possible to predict not only the immediate input (yielding rapid perceptual inference) but also more stable regularities of the environment (yielding slower timescale perceptual learning), in what may be considered a generative model of the causal structure of the world. This means that, early in the predictive hierarchy, there will be predictions and PEs about very fast occurring and local characteristics of the stimuli, but that, further along the hierarchy, predictions and PEs will, in turn, refer to long-lasting features of the stimuli and context (Pereira et al., 2019). A general illustration of this scheme is depicted in Fig. 1.4. The cortical regions for auditory processing are potentially, from lower to higher level, primary auditory cortex (A1), superior temporal gyrus (STG), inferior frontal gyrus (IFG), and orbitofrontal cortex (OFC).

1.4.2 Bayesian inference
1.4.2.1 Basics
Bayesian inference is a method in which Bayes' theorem is used to update the probability for a hypothesis as more evidence or information becomes available. Expanding on Bayesian inference theory, predictive coding describes adaptive and dynamic forward and backward processes in the brain, which involves dynamic adaptation of probability models due to continually changing environments. In predictive coding, backward connections convey predictions from "higher order" brain areas to earlier sensory areas. Incoming (bottom–up) sensory information is compared to these top–down predictions (Fig. 1.2). Feedforward connections from more peripheral areas convey the resulting prediction errors back to the higher-level areas. This adaptive cycle continues until, prediction errors are minimized through updating of the internal model. At the neuronal level, backward processes to subcortical structures primarily originate from deep cortical layers (layer V/VI; Fig. 1.1), whereas feedforward connections mainly originate from superficial cortical layers (layer II/III). In addition, backward connections not only inhibit firing, but also facilitate/drive firing rates of cells in lower-level brain areas. According to predictive coding theory, these top–down predictions should modulate early sensory areas to resolve conflicts between the incoming sensory input and the template model to create perception (Chan et al., 2016; Wacongne, 2016).

This interplay between bottom–up information from a sensory stimulus, $P(I)$, and the expectation or prior belief, $P(x)$, is graphically illustrated in Fig. 1.5, where $P(I,x)$ is the probability that the bottom–up sensory information, $P(I)$, and prior belief, $P(x)$ occur together. Then the actual percept, i.e., the revised belief, $P(x|I)$, is given by the Bayes' theorem:

$$P(x|I) = [P(I|x)P(x)]/P(I) \qquad (1.1)$$

In practice, most PC models involve a central prediction, which is compared to sensory input. When the two are mismatched, a "prediction error" occurs, which is used as a learning signal to modify the internal model. This scheme is consistent with numerous findings showing enhanced neural responses (e.g., the MMN) to unpredicted stimuli.

Asilador and Llano (2021) commented that most current theories in cognitive science rely on hierarchical predictive coding models involving a set of Bayesian priors generated by high-level brain regions (e.g., prefrontal cortex, PFC) that are used to influence processing at lower levels of the

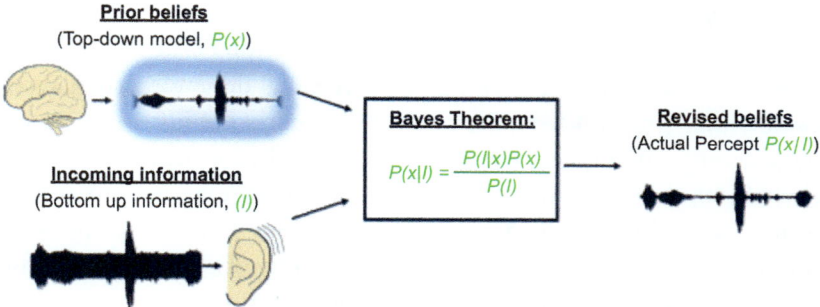

Fig. 1.5 Illustration of the principles of Bayesian inference in neuronal coding. Top-down preconceived notions about a sound (illustrated as a sound trace surrounded by a *blue haze*) are combined with noisy information from the periphery (illustrated as a noisy sound trace entering the ear). Bayes' theorem (in the box) combines the preconceived notions with the noisy sensory information to recover the original signal (here represented as the posterior probability or the actual percept). *(From Asilador, A., Llano, D.A., 2021. Top-down inference in the auditory system: potential roles for corticofugal projections. Front. Neural Circuits 14, 615259. https://doi.org/10.3389/fncir.2020.615259. Open access.)*

cortical sensory hierarchy (e.g., auditory cortex). As such, virtually all proposed models to explain top-down facilitation are focused on intracortical connections, and consequently, subcortical nuclei have scarcely been discussed in this context (in contrast to animal research, see Chapter 4). Asilador and Llano (2021) argue that corticofugal pathways (Fig. 1.1) contain the required circuitry to implement predictive coding mechanisms to facilitate perception of complex sounds. Consequently, top-down modulation at early (i.e., subcortical) stages of processing complements modulation at later (i.e., cortical) stages of processing. This has been conclusively shown for the inferior colliculus (IC) in animal studies (reviewed in Carbajal and Malmierca, 2018). In the words of Asilador and Llano (2021): "Most predictive coding schemes postulate that top-down predictions subtract from lower-level processors, leaving behind that which is not predicted—the prediction error. This scheme suggests that sub-cortical neurons are primarily responding to prediction errors—that which we do not predict. However, our behavior is just the opposite—we tend to ignore sensory data that do not fit into our predictions about the world. Thus, although predictive coding schemes that rely on the concept of prediction error can reproduce the responses of sub-cortical neurons, they do a poor job of explaining perception."

1.4.2.2 Implementation

"Under predictive coding, beliefs are represented using Gaussian probability distributions, or densities, over a given perceptual dimension such as stimulus intensity (Fig. 1.6 left). Precision is defined as the inverse variance of such a density; the expectation, or most likely belief, is defined as its mode. A belief is thus defined as both the prediction itself (expectation) as well as the level of confidence in said prediction (precision). Evidence is represented using a likelihood—i.e. the conditional probability of observing the sensory input given the prior belief—and has an expectation and precision of its own. Prediction error then becomes the precision-weighted difference between the prior belief and the evidence; the error itself thus also has an expectation and precision" (Hullfish et al., 2019).

Neural implementation of Bayesian inference is most often framed in terms of predictive coding. The brain's implementation of this strategy relies

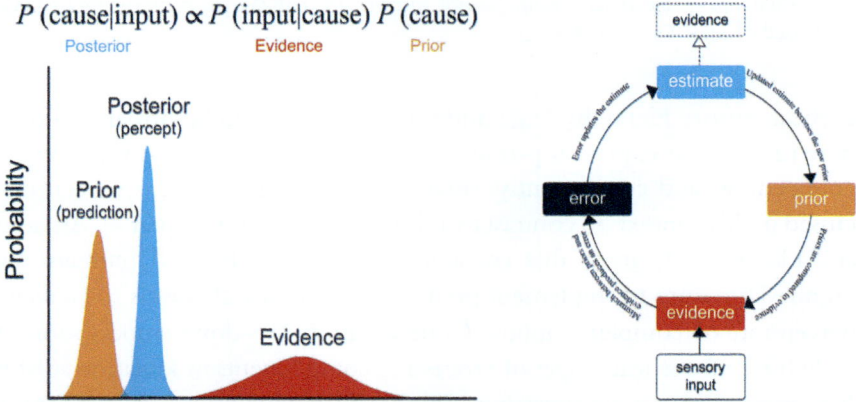

Fig. 1.6 Bayesian inference. Left. (top) The posterior belief (probability of the cause, given the input) is directly proportional to the product of the evidence (probability of the input, given the cause) and the prior belief (probability of the cause). The constant of proportionality is the probability of the input itself. (bottom) Illustration of Bayesian inference using probability density functions. Right. Diagram of hierarchical predictive coding via empirical Bayesian inference. The estimate serves as an empirical prior belief. The evidence is dependent on this empirical prior (top-down input) as well as sensory input (bottom-up input). Comparing the evidence and the prior produces a mismatch signal, i.e., the prediction error, which is used to update the original estimate. This updated estimate becomes the new empirical prior, and the process repeats until the prediction error is minimized. Extending this process over multiple hierarchical levels, the estimate at one level becomes the bottom-up input for the level above *(dotted lines). (Reprinted from Hullfish, J., Sedley, W., Vanneste, S., 2019. Prediction and perception: insights for (and from) tinnitus. Neurosci. Biobehav. Rev. 102, 1–12 with permission from Elsevier.)*

on a hierarchical generative model where bottom-up sensations are compared to top-down prior beliefs across multiple levels. Each level uses these bottom-up inputs as evidence with which to update its beliefs via Bayesian inference (Fig. 1.6 right). As we have seen (Fig. 1.4) the belief at one level serves as the prior for the level below. "The mismatch between evidence and priors, i.e. the prediction error, at that subordinate level then ascends and updates the original belief, at which point the process repeats. This empirical Bayesian approach enables the brain to estimate priors from data without needing to know them intrinsically, i.e. unsupervised learning. Because the causes of sensations cannot be directly observed, the brain must instead rely on minimizing prediction error as the major criterion for optimizing its generative model. This means that the brain's ability to recognize which prediction errors carry reliable information about changes in the environment is critical. Thus, each factor (prior belief and evidence) influences inference in proportion to its precision, meaning that only sufficiently precise prediction errors are allowed to significantly alter the generative model and thus affect perception" (Hullfish et al., 2019).

1.5 Brain areas and activity involved in predictive coding

1.5.1 Putative top-down prediction structures

Previously (Eggermont, 2022), I wrote: "The prefrontal cortex can modulate auditory activity via direct projections to auditory cortex (ACx), namely via the orbitofrontal cortex (OFC) to primary auditory cortex (A1; Winkowski et al., 2018), and via fronto-striatal circuits that show altered connections to the nucleus accumbens (NAc; Xu et al., 2019). In turn, the NAc core projects to the thalamic reticular nucleus (TRN; O'Donnell et al., 1997) and via this route the PFC can affect the input to ACx via the modulatory effect of the TRN on the medial geniculate body (MGB) output. The notion of the PFC as integration site is enhanced by its interconnections with hippocampus, basolateral amygdala (BLA), and NAc (Torres-García et al., 2012). Both the PFC and hippocampal formation send excitatory projections to the NAc and are interconnected with the amygdala."

These direct pathways between human auditory cortex and prefrontal cortex are illustrated in Medalla and Barbas (2014). From rhesus monkey brain analyses (Fig. 1.7), they noted that "pathways from auditory association cortices reach distinct sites in the lateral, orbital, and medial surfaces of the prefrontal cortex in rhesus monkeys. Among prefrontal areas, frontopolar area 10 [corresponding to Brodmann area 10, anterior prefrontal cortex,

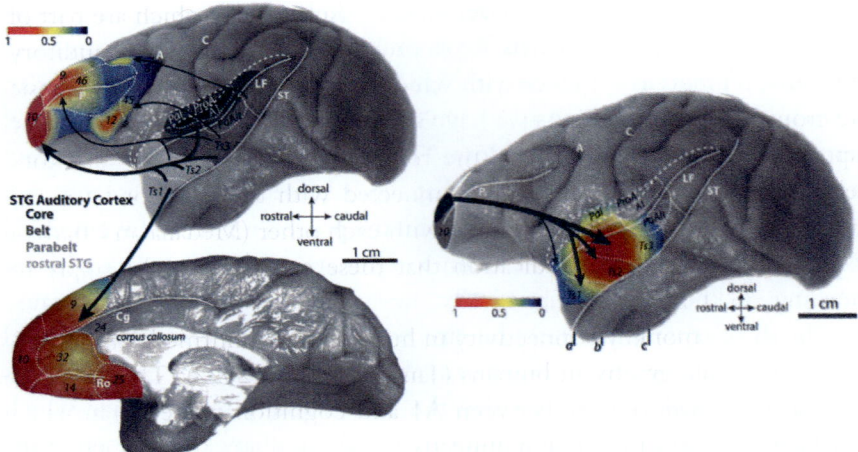

Fig. 1.7 Left. Gradient map of auditory input to the prefrontal cortex. Lateral (top) and medial (bottom) surfaces of the rhesus monkey brain show relative proportions of afferent pathways (projection neurons) from superior temporal gyrus (STG) to prefrontal cortex, represented by the weights of the arrows denoting the STG → prefrontal pathways. Long dashes demarcate banks of sulci schematically unfolded and short dashes delineate areal boundaries. Abbreviations for sulci: A, arcuate; C, central; Cg, cingulate; LF, lateral fissure; Ro, rostral; ST, superior temporal. Right. Topography and laminar terminals of pathways from area 10 in distinct auditory cortices. The gradient map shows the relative density of area 10 pathway terminations in distinct STG areas. Axon terminals were labeled after injection of anterograde tracers in dorsal area 10. Density is normalized to the highest in the set. *Long dashes* demarcate banks of sulci schematically unfolded; *short dashes* delineate areal boundaries. *(From Medalla, M., Barbas, H., 2014. Specialized prefrontal "auditory fields": organization of primate prefrontal-temporal pathways. Front. Neurosci. 8, 77. https://doi.org/10.3389/fnins.2014.00077. Open Access.)*

in humans] has the densest interconnections with auditory association areas, spanning a large antero–posterior extent of the superior temporal gyrus from the temporal pole to auditory parabelt and belt regions. Moreover, auditory pathways make up the largest component of the extrinsic connections of area 10, suggesting a special relationship with the auditory modality." Medalla and Barbas (2014) also reported that "frontopolar area 10 is indeed the main frontal "auditory field" as the major recipient of auditory input in the frontal lobe and chief source of output to auditory cortices. Area 10 is thought to be the functional node for the most complex cognitive tasks of multitasking and keeping track of information for future decisions. These patterns suggest that the auditory association links with area 10 are critical for complex cognition. [...] In addition to area 10, particularly strong auditory connections are seen for medial prefrontal areas 32 (dorsal) and 25 (subgenual) which are part of the anterior cingulate cortex (ACC). Importantly, these prefrontal auditory 'hotspots' are also robustly interconnected with each other through intrinsic pre-frontal pathways (Barbas et al., 2005; Medalla and Barbas, 2010)." Frontopolar area 10 and ACC areas, the rostral and medial frontal "auditory fields" that are most strongly interconnected with the superior temporal gyrus (STG), are also robustly linked with each other (Medalla and Barbas, 2014). There is a strong indication that these connections also apply to humans (Fig. 1.7).

The rhesus monkey connectivity in humans was confirmed using diffusion tensor tractography in humans (Jang and Choi, 2022). They assessed the structural connectivity between A1 and cognition-related brain areas illustrated in Fig. 1.8. This connectivity is based on streamlines (Yeh et al., 2019), which is a contiguous set of 3D points produced by tractography algorithms. Jang and Choi (2022) noted that A1 showed structural connectivity (over 50%) for a threshold of 5 streamlines with the following areas: the ventrolateral prefrontal cortex (VLPFC) (88.4%), OFC (81.4%), fornix (66.3%), hippocampus (55.8%), and parahippocampal cortex (53.5%); and

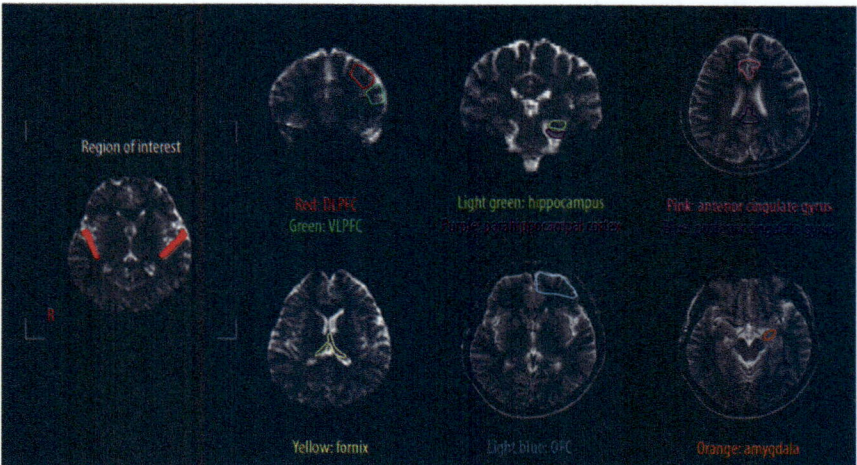

Fig. 1.8 Neural connectivity from primary auditory cortex (A1)—region of interest—to cognition-related brain areas: The dorsolateral prefrontal cortex (DLPFC), ventrolateral prefrontal cortex (VLPFC), anterior and posterior cingulate gyri, orbitofrontal cortex (OFC), amygdala, hippocampus, and parahippocampal cortex. *(From Jang, S.H., Choi, E.B., 2022. Evaluation of structural neural connectivity between the primary auditory cortex and cognition-related brain areas using diffusion tensor tractography in 43 normal adults. Med. Sci. Monit. 28, e936131. https://doi.org/10.12659/MSM.936131. Open access.)*

for a threshold of 15 streamlines: the VLPFC (82.6%), OFC (74.4%), and fornix (53.5%). Jang and Choi (2022) concluded that A1 showed a high degree of neural connectivity with brain areas associated with cognition and memory.

1.5.2 Event-related potentials as representing prediction errors

From a review of cellular recordings in animals, Carbajal and Malmierca (2018) detail how deviance detection and prediction error are generated throughout hierarchical levels of processing, following two pathways of increasing computational complexity and abstraction along the auditory system. The proportion of deviance detection accounted for by the prediction error increases from lemniscal (tonotopic) to nonlemniscal (nontonotopic) pathways and from the midbrain to the auditory cortex. The implication of areas beyond the auditory system is also expected, continuing the neural processing hierarchy in charge of extracting increasingly abstract relationships between stimuli, and allowing multisensory perceptions by sharing information with other neural systems. As local field potential (LFP) and ERP data from animal models have implicated prefrontal cortices in rat auditory deviance detection (Imada et al., 2013), and in the human MMN and the P3 (Dürschmid et al., 2016). This has also led to investigating various obligatory ERPs (e.g., N1, P2). From the predictive coding perspective, these evoked transient responses are as predicted if recurrent message passing was involved in resolving prediction errors—and suppressing the activity of error units (usually associated with cortical superficial pyramidal cells, putatively underlying the MMN). Considering that excitatory and inhibitory neurons form tight recurrent networks in ACx, Carbajal and Malmierca (2018) hypothesize that adapting interneurons can amplify deviance detection in excitatory neurons through differential postsynaptic integration by excitatory neurons (Natan et al., 2015). Consequently, inhibitory interneurons in ACx may play a prominent role regulating cortical deviance detection (details in Chapter 4).

1.6 Modeling networks underlying predictive coding

I will introduce here three well-established computational models that provide the needed effective connectivity estimates that predictive coding requires, i.e., isolate the forward and backward connectivity between cortical (and putatively subcortical) areas, and elucidate intrinsic inhibitory changes aimed at reducing prediction error.

1.6.1 Granger causality and transfer entropy

Granger Causality (GC) is based on the estimation of causal statistical influences between simultaneously recorded neural time series, and, if available, in the context of task performance. Causality in the Granger sense is based on the statistical predictability of one time series that derives from knowledge of one or more others (Bressler and Seth, 2011). The core idea behind GC is that X causes Y if X contains information that helps predict the future of Y better than information already in the past of Y (and in the past of other "conditioning" variables). The most common implementation of GC is via linear autoregressive modeling of time series data, enabling both statistical significance testing and estimation of GC magnitudes. However, GC is not limited to this implementation; it can use nonlinear, time-varying, and nonparametric models (Friston et al., 2013).

Transfer entropy (TE; Gourévitch and Eggermont, 2007) represents an information-theoretic generalization of GC that is model free. "More precisely, the transfer entropy estimates the part of activity of a neuron that is not dependent on its own past but dependent on the past activity of another neuron. In a nutshell, it estimates the information transferred between two neurons in both directions. [...] To a certain extent, this tool is able to distinguish information resulting from common history and exclude it by appropriate conditioning of the entropy. Transfer entropy also detects asymmetry in neural relations, allowing studies of possible feedback in neural circuits, a topic that recently gained considerable interest. Finally, but not unimportantly, transfer entropy takes into account linear and nonlinear flows and thus may represent a very general way to define the causality strength between two spikes trains. In particular, the window size for which maximum information is transferred may be useful to study neural integrative properties" (Gourévitch and Eggermont, 2007). In the words of Friston et al. (2013): "Specifically, the TE from X to Y is zero if, and only if, Y is conditionally independent of X's past, given its own past. Importantly, for Gaussian data, TE is equivalent to GC, furnishing a useful interpretation of GC in terms of information transfer in 'bits'."

In neuroimaging and human electrophysiology, Granger causality is the most popular of these two techniques, because it is simple to estimate, given (stationary stochastic) time series. As Friston et al. (2013) put it: "GC has some useful properties including a decomposition of causal influence by frequency and formulation in an 'ensemble' form, allowing evaluation of GC between multivariate sets of responses. GC has provided useful descriptions of directed functional connectivity in many electrophysiological studies. GC can also be applied to standard EEG or MEG signals, either at the source or sensor level (following spatial filtering to reduce the impact of volume conduction). [...] In the analysis of electrophysiological timeseries, GC is widely accepted because there is no temporal lag between the responses recorded and their underlying (neuronal) causes and because the data can be sampled at fast timescales."

From a methodological perspective, it is advised to combine exploratory (TE or GC) and confirmatory (structural equation modeling, Section 1.6.2, or dynamic causal modeling, Section 1.6.3) approaches to the analysis of network activity (Chan et al., 2021). Exploratory and confirmatory approaches have distinct and complementary ambitions that are usefully considered in relation to the detection of functional connectivity and the identification of models of effective connectivity. Thus exploratory approaches model dependency among observed responses, while confirmatory approaches model coupling among the hidden states generating observations (Friston et al., 2013).

1.6.2 Structural equation modeling

From neural imaging and structural equation modeling (SEM), insights can be gained from examining covariances (C_{ij}), in low-frequency fluctuations in the blood oxygen level-dependent (BOLD) response measured in fMRI activity or in ERPs based on time-integrated activity measures of ensembles or brain regions i and j. Formally, covariances are related to correlations, the latter being normalized—by the product of the standard deviations (σ_i and σ_j) or auto-covariances of the two ensemble activities—covariances such that the correlation coefficients are between -1 and $+1$.

$$r_{ij} = C_{ij}/\sigma_i \, \sigma_j \qquad (1.2)$$

where r_{ij} is the correlation coefficient (Eggermont, 1990). Expressed in terms of neural systems, a measure of (cross)covariance represents the degree to which the activities of two regions (or two neurons) are related to one another or how they vary together. A high covariance between areas A and B means that if area A increases its activity, so too will B (in the case of a positive covariance). The dependent variables (regional activity of brain

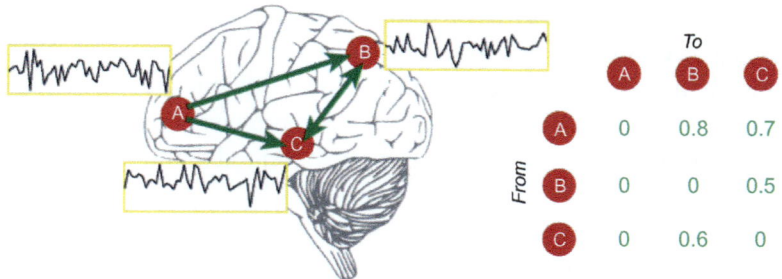

Table (correlation matrix):

		To		
		A	B	C
From	A	0	0.8	0.7
	B	0	0	0.5
	C	0	0.6	0

- Define network *nodes* (spatial coordinates or regions of interest)
- Identify a timeseries associated with each node
- Estimate the *edge strengths*, or connections between the nodes
  - For example, correlate each timeseries with every other timeseries
  - If the data (and method for estimating edges) permits the estimation of causality, the edges maybe uni-directional, resulting in an asymmetric network matrix

Fig. 1.9 The main steps in using resting-state functional MRI data (with a fluctuating activity time series at every voxel in the brain) are identifying network nodes, and then estimating the network edges from the internodal correlation strengths. These are shown in the correlation matrix. Note that in this example the connectivity from A to B or C is unilateral, whereas that from B to C it is bilateral but with different connection strengths, resulting in a nonsymmetrical matrix. *MRI, magnetic resonance imaging. (Reprinted from Smith, S.M., Vidaurre, D., Beckmann, C.F., Glasser, M.F., Jenkinson, M., Miller, K.L., et al., 2013. Functional connectomics from resting-state fMRI. Trends Cogn. Sci. 17 (12), 666–682, with permission from Elsevier.)*

areas) may be anatomically connected to one another, if not the connectivity is called functional. Functional interactions between neural elements can be detected by examining the covariances of measured activity within the central nervous system (Fig. 1.9).

This can be expressed (McIntosh and Gonzalez-Lima, 1994) as in Eq. (1.3), which represents effects of brain areas X and Z on the regional activity of area Y.

$$Y = \alpha + \beta_{yx}X + \beta_{yz}Z + \psi \tag{1.3}$$

The β weights are the degree to which these two areas influence the variance, or activity, of area Y, i.e., β_{yx} reflects the influence of activity in area X on that in area Y. Eq. (1.3) represents multiple linear regression. If expressed as matrices (Eq. 1.4), this can define the interconnections of an entire network. The matrix representation is the foundation for structural equation models.

$$
\begin{bmatrix} X \\ Z \\ Y \end{bmatrix} =
\begin{bmatrix} 0 & 0 & 0 \\ \beta_{zx} & 0 & 0 \\ \beta_{yx} & \beta_{yz} & 0 \end{bmatrix}
\begin{bmatrix} X \\ Z \\ Y \end{bmatrix} +
\begin{bmatrix} \psi_x \\ \psi_z \\ \psi_y \end{bmatrix} \tag{1.4}
$$

In Eq. (1.4), the variances of all three regions, X, Y, and Z, are represented as a vector that is determined by the weighted influence of the other regions (matrix of β weights) plus the residual influences contained in the vector ψ. The zero values in the matrix β represent connections that do not exist in the model. Eq. (1.4) can be represented graphically, as shown in Fig. 1.10.

An extension of Eq. (1.4) can examine experimental effects on both regional activity and interregional covariance.

$$Y = \alpha + \beta_{y/x}X + \beta_{y/a}A + \beta_{y/xa}XA + \psi \qquad (1.5)$$

Eq. (1.5) represents the effect of area X on Y, the experimental effect of A on Y, and the interaction of the impact of brain area X and the experimental effect A (XA). The interaction assesses whether the relationship between X and Y changes because of the experimental effect. The β weights represent the degree to which these three factors influence the variance, or activity, of Y. The influence of area X on Y, independent of the experimental effect, is evaluated by testing the statistical significance of $\beta_{y/x}$. The significance of experimentally related regional differences in the activity of Y, independent of the influences of region X, is assessed by evaluating the significance of the slope $\beta_{y/xa}$. This indicates whether the influence of region X changes depending on the experimental condition, i.e., whether the covariance of X and Y changes because of the experimental manipulation. All structural

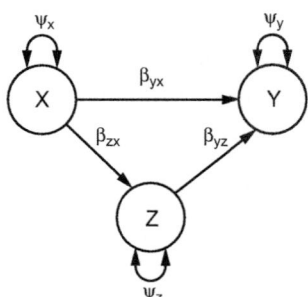

Fig. 1.10 Graphic representation of structural equation model from Eq. (1.3). *Circles represent the measured variance from regions X, Y, and Z. Unidirectional arrows represent the path for the influences of these sources of variance on each other, the weighting of the influence given by β. Curved bidirectional arrows represent residual influences whose size is indicated by ψ. (From McIntosh, A.R., Gonzalez-Lima, F., 1994. Structural equation modeling and its application to network analysis in functional brain imaging. Hum. Brain Mapp. 2, 2–22, ©1994 Wiley-Liss, Inc. with permission from John Wiley & Sons, Inc.)*

equation models are derived from covariance matrices and from a putative causal structure resulting from explorative modeling (McIntosh and Gonzalez-Lima, 1994).

1.6.3 Dynamic causal modeling

The goal of dynamic causal modeling (DCM) of neuroimaging and neuro-physiological data is to study experimentally induced changes in functional integration among brain regions. This requires (1) biophysically plausible and physiologically interpretable models of neuronal network dynamics that can predict distributed brain responses to experimental stimuli and (2) efficient statistical methods for parameter estimation and model comparison. We follow the outline provided by Daunizeau et al. (2011).

DCMs are based on so-called generative models, i.e., a quantitative description of the mechanisms by which observed data are generated. Typically, both hemodynamic (fMRI) and electromagnetic (EEG/MEG) signals arise from a network of brain regions or neuronal populations. This network can be thought of as a directed graph. Graph theory represents networks by way of nodes and edges. In brain networks the nodes can be individual neurons, or they can be EEG or MEG activity at particular electrodes or field sensor locations, or voxels in fMRI studies of resting-state networks (RSNs; Fig. 1.9). The sensor locations and voxels represent large collections of neurons, depending on voxel size or spatial resolution as in EEG and MEG. A standard fMRI voxel has about the size of $10\,\text{mm}^3$ and contains on the order of 10^5 neurons and 10^9 synapses (Honey et al., 2009). The network edges in brain studies represent axonal connections, cross-covariances, or cross-correlation coefficients between fMRI, EEG, or MEG time series, or spike trains. There is a modeling and a statistical component in the DCM. With respect to the modeling component it is noted that DCMs are causal in at least two senses (Daunizeau et al., 2011):

(1) DCMs describe how experimental manipulations (u) influence the dynamics of hidden (neuronal) states of the system (x), using ordinary differential equations (so-called evolution equations):

$$\partial x/\partial t = f(x, u, \theta) \tag{1.6}$$

where $\partial x/\partial t$ is the rate of change of the system's states x, f comprises the biophysical mechanisms underlying the temporal evolution of x under u, and θ is a set of unknown evolution parameters. The structure of the evolution function f determines both the presence/ absence of edges in the graph and how these influence the dynamics of the system's states.

(2) DCMs map the system's hidden states (x) to experimental data (y). This is typically written as a static observation equation:

$$y = g(x, \varphi) \tag{1.7}$$

where g is the instantaneous mapping from system states to observations and φ is a set of unknown observation parameters.

To be biophysically plausible, DCM for evoked responses utilizes canonical microcircuits (CMCs; Douglas and Martin, 1991; Fig. 1.11) as

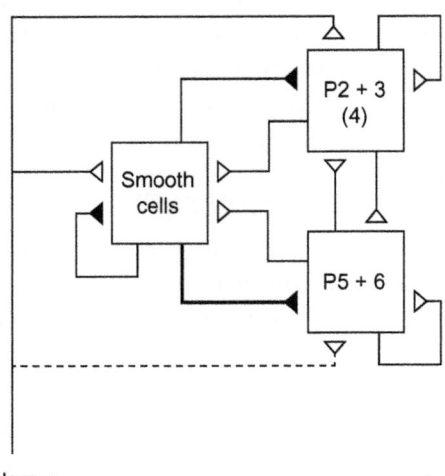

Thalamus

Fig. 1.11 Block diagram of circuit that successfully models the intracellular responses of cortical neurons to stimulation of thalamic afferents. Three populations of neurons interact with one another: one population is inhibitory (smooth cells, *filled synapses*), and two are excitatory *(open synapses)*, representing superficial (P2+3) and deep (P5 +6) layer pyramidal neurons. The layer 4 spiny stellate cells (4) are incorporated with the superficial group of pyramidal cells. Some neurons within each population receive excitatory input from the thalamus. *Continuous versus dashed lines* indicate that thalamic drive to the superficial group is stronger. The inhibitory inputs activate both GABA$_A$ and GABA$_B$ receptors on pyramidal cells. The *thick continuous line* connecting smooth cells to P5+6 indicates that the inhibitory input to the deep pyramidal population is relatively greater than that to the superficial population. However, the increased inhibition is due to enhanced GABAA drive only. The GABAB inputs to P5+6 are similar to those applied to P2+3. *(From Douglas, R.J., Martin, K.A., 1991. A functional microcircuit for cat visual cortex. J. Physiol. 440, 735–769, with permission from John Wiley & Sons, Inc.)*

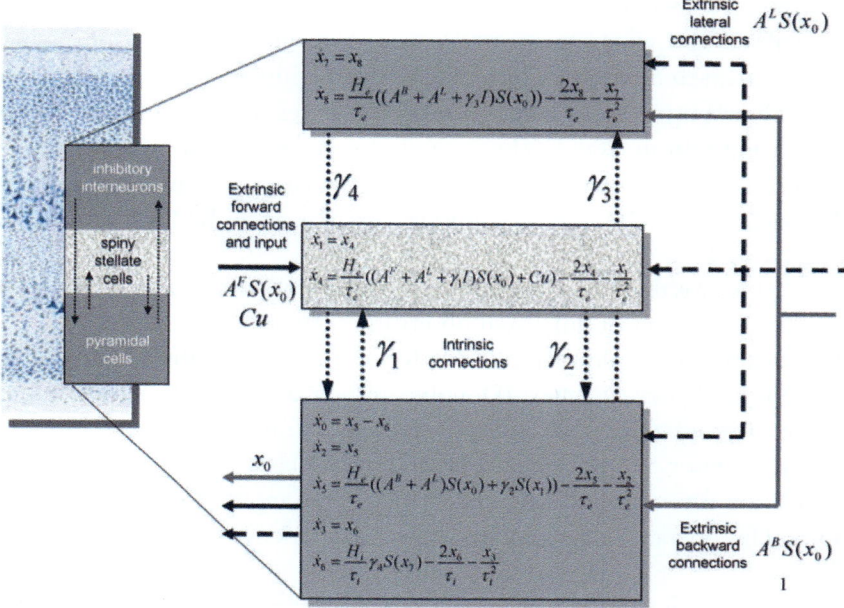

Fig. 1.12 Neuronal state equations. A source consists of three neuronal subpopulations, which are connected by four intrinsic connections with weights $\gamma_{1,2,3,4}$. Mean firing rates (Eq. 1.3) from other sources arrive via forward AF, backward AB, and lateral connections AL. Similarly, exogenous input Cu enters receiving sources. The output of each subpopulation is its transmembrane potential (Eq. 1.2). *(Reprinted from Kiebel, S.J., Garrido, M.I., Friston, K.J., 2007. Dynamic causal modelling of evoked responses: the role of intrinsic connections. Neuroimage 36 (2), 332–345, with permission from Elsevier.)*

generative models for each of the cortical regions involved in the modeling. The generative model presented here (Fig. 1.12) is based on this canonical microcircuit (Kiebel et al., 2007; Bastos et al., 2012), which includes—for each involved cortical area (cf. Fig. 1.4)—four populations of neurons: spiny stellate cells, inhibitory interneurons, and superficial and deep pyramidal cells. Within a cortical region or node, intrinsic communication between these populations results in overall inhibition. Between regions, extrinsic excitatory forward connections project from superficial pyramidal cells in lower cortical areas to spiny cells and deep pyramidal cells in higher cortical areas, whereas extrinsic inhibitory backward connections project from deep pyramidal cells in higher areas to superficial pyramidal cells and inhibitory interneurons in lower areas. Note that some excitatory connections result in inhibition of their targets, mediated by intermediate inhibitory synapses not modeled explicitly.

The state equations are ordinary second-order differential equations and are derived from the behavior of the three neuronal subpopulations, which operate as linear damped oscillators. The integration of the differential equations pertaining to each subpopulation can be expressed as a convolution (David and Friston, 2003). This convolution transforms the average density of its presynaptic inputs into an average postsynaptic membrane potential (Kiebel et al., 2007). The convolution kernel is given by Eq. (1.8).

$$p(t)_e = \frac{H_e}{\tau_e} t \, exp\left(-t/\tau_e\right) t \geq 0, \quad p(t)_e = 0 \; t < 0 \tag{1.8}$$

where subscript "e" stands for "excitatory," alternatively the subscript "i" is used for inhibitory synapses. $H_{e,i}$ controls the maximum excitatory/inhibitory postsynaptic potential and $\tau_{e,i}$ represents a lumped rate constant.

An operator S transforms the compound synaptic potential of each subpopulation into a mean firing rate, which is the input to other subpopulations. This operator is assumed to be an instantaneous sigmoid nonlinearity

$$S(x) = \frac{1}{1 + exp(-\rho_1(x-\rho_2))} - \frac{1}{1 + exp(\rho_1\rho_2)} \tag{1.9}$$

where the free parameters ρ_1 and ρ_2 determine its form (slope and translation). Interactions, among the subpopulations, depend on internal coupling constants $\gamma_{1,2,3,4}$, which control the strength of intrinsic connections and reflect the total number of synapses expressed by each subpopulation (Fig. 1.12). The integration of this model, to form predicted responses, rests on formulating these two operators (Eqs. 1.8 and 1.9) in terms of a set of differential equations as shown in Fig. 1.12.

The DCM is thus specified in terms of its state equations and output equations (cf. Eqs. 1.6, 1.7 and 1.8, 1.9). The state equations are ordinary second-order differential equations and are derived from the behavior of the three neuronal subpopulations, which operate as linear damped oscillators. The integration of this model, to form predicted responses, rests on formulating this in terms of a set of differential equations as shown in the shaded boxes in Fig. 1.12 (Kiebel et al., 2007). Note that in the equations in these boxes, x-dot is the Newtonian form of the more common Leibnitz form, $\partial x/\partial t$, for differentiation. This set of equations is used for each of the putative sources of the error signal, e.g., the MMN, connected as suggested in Fig. 1.4.

The need for neurobiological plausibility, as reflected in Fig. 1.12, can sometimes make DCMs fairly complex, at least compared to conventional regression-based models of effective connectivity, such as structural equation modeling (McIntosh and Gonzalez-Lima, 1994). This complexity, with potential nonidentifiability problems, requires sophisticated model inversion techniques, which are typically cast in the statistical component of DCM within a Bayesian framework (Daunizeau et al., 2011):

- Using statistical assumptions about residual errors in the observation process, Eqs. (1.6) and (1.7) are combined to derive a likelihood function $p(y|\vartheta,m)$. This specifies how likely it is to observe a particular set of observations y, given parameters $\vartheta \equiv (\theta, \varphi)$ of model m.
- One then defines priors $p(\vartheta|m)$ on the model parameters ϑ which reflect knowledge about their likely range of values. Such priors can be (i) principled (e.g., certain parameters cannot have negative values), (ii) conservative (e.g., "shrinkage" priors that express the assumption that coupling parameters are zero), or (iii) empirical (based on previous, independent measurements).
- Combining the priors and the likelihood function allows one, via Bayes' theorem, to derive the marginal likelihood of the model (the so-called model evidence)

$$p(y|m) = \int p(y|\vartheta, m)p(\vartheta|m)d\vartheta \qquad (1.10)$$

The model evidence is used for model comparison (e.g., different network structures embedded in different evolution functions f). The posterior density, $p(\vartheta|y,m)$, is used for inference on model parameters (e.g., context-dependent modulation of effective connectivity). Eq. (1.10) is the "statistical" (or model inversion) component of DCM (Daunizeau et al., 2011). Harris et al. (2018) describe how this method can be adapted for EEG/MEG using the Statistical Parametric Mapping (SPM) software package for MATLAB.

1.6.4 Application to auditory networks

Using an auditory oddball paradigm, (Friston, 2005) elicited ERPs that exhibited a strong modulation of late components. Auditory stimuli of between 1000- and 2000-Hz tones were presented binaurally. The tones were presented for 15 min, every 2 s in a pseudo-random sequence with 2000-Hz tones on 20% of occasions (oddballs) and 1000-Hz tones for 80% of the time (standards). The subject was instructed to keep a mental record of the number of 2000-Hz tones. Data were acquired using 128

EEG electrodes with 1000-Hz sample frequency. Before averaging, data were referenced to mean earlobe activity and band-pass filtered between 1 and 30 Hz. Six sources—left and right A1, right STG, left and right OFC, and PCC—were identified using procedures described in David et al. (2006) and were used to construct DCMs (Fig. 1.13).

To establish evidence for changes in backward and lateral connections beyond changes in forward connections, Friston et al. (2005) employed a Bayesian model selection procedure. This comprised the specification of four model variations that allowed for changes in forward, backward, forward and backward directions, and changes in all connections including those between hemispheres. The models were compared using the prediction error as an approximation to the log evidence for each model, shown in a lower-right insert in Fig. 1.13. The model with the highest evidence (by a margin of 27.9) is a DCM that allows for changes in forward, backward, and lateral connections between cortical areas. These results require—given the putative sources—that changes in backward and lateral connections are needed to explain the observed differences in auditory cortical responses. Note that animal studies (e.g., Medalla and Barbas, 2014; Fig. 1.7) have implicated OFC and PFC—both nodes in the default mode network—as candidate areas for providing backward information to A1.

1.7 Predictive coding networks and cognition

Garrido et al. (2009) reviewed studies that focus on neuronal mechanisms underlying the MMN generation, and proposed PC as a general framework to unify hypotheses where current inputs are predicted from past inputs. In the case of a prediction error, i.e., when there is a mismatch between the predicted and the actual sensory input, the neural system implementing the model must be adjusted (for example, by short-term synaptic plasticity). During the repetition of subsequent events, that adjustment is reflected in the suppression of prediction error, i.e., ideally the disappearance of the MMN but in practice a strong reduction thereof. The underlying network as found by DCM often includes A1, bilateral STG and right IFG, a node in the DMN (Fig. 1.14).

Why is the "MNN network" shown in Fig. 1.14 incorporating the right IFG, different from that in Fig. 1.13 as used by Friston (2005) based on OFC and PCC? The reason is that each model is defined by its postulated number of sources—typically resulting from previous ERP studies localizing sources for the MMN (Chapter 7)—and their connectivity. Given a particular

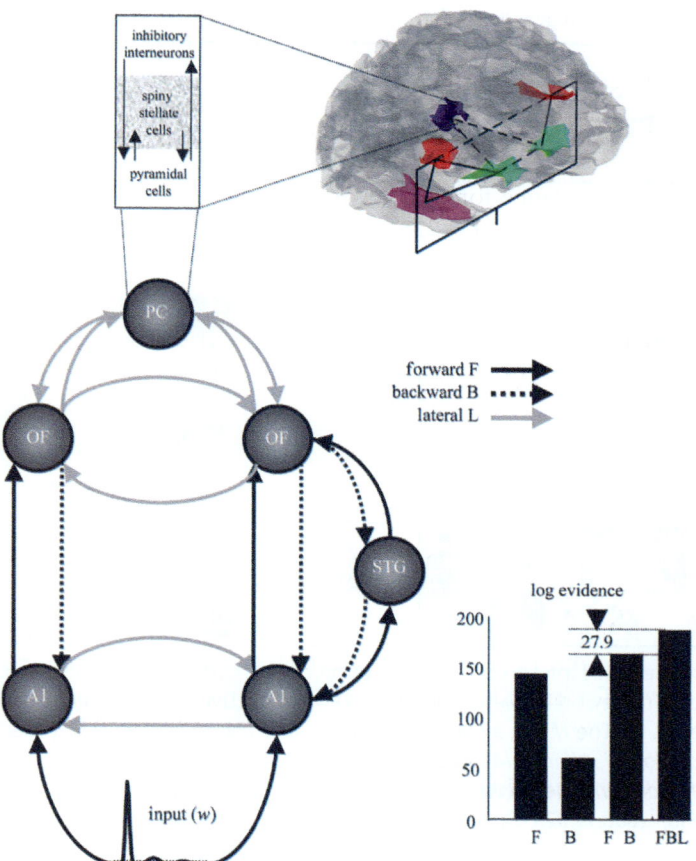

Fig. 1.13 Upper right: Transparent views of the cortical surface showing localized sources that entered the DCM. A bilateral extrinsic input acts on primary auditory cortices *(red)*, which project reciprocally to orbito-frontal regions *(green)*. In the right hemisphere, an indirect pathway was specified via a relay in the superior temporal gyrus *(magenta)*. At the highest level, orbitofrontal and left posterior cingulate *(blue)* cortices were assumed to be laterally and reciprocally connected *(broken lines)*. Lower left: Schematic showing the extrinsic connectivity architecture of the DCM used to explain empirical data. Sources were coupled with extrinsic cortico-cortical connections. *A1*, primary auditory cortex; *OF*, orbitofrontal cortex; *PC*, posterior cingulate cortex; *STG*, superior temporal gyrus (right is on the right and left on the left). The free parameters of this model included extrinsic connection strengths that were adjusted to best explain the observed ERPs. Critically, these parameters allowed for differences in connections between the standard and oddball trials. Lower right: The results of a Bayesian model selection are shown in terms of the log evidence for models allowing changes in forward (F); backward (B); forward and backward (FB); and forward, backward, and lateral connections (FBL). There is very strong evidence that both backward and lateral connections change with perceptual learning as predicted theoretically. *(Reprinted from Friston, K., 2005. A theory of cortical responses. Philos. Trans. R. Soc. B 360, 815–836. ©The Royal Society (U.K.) with permission.)*

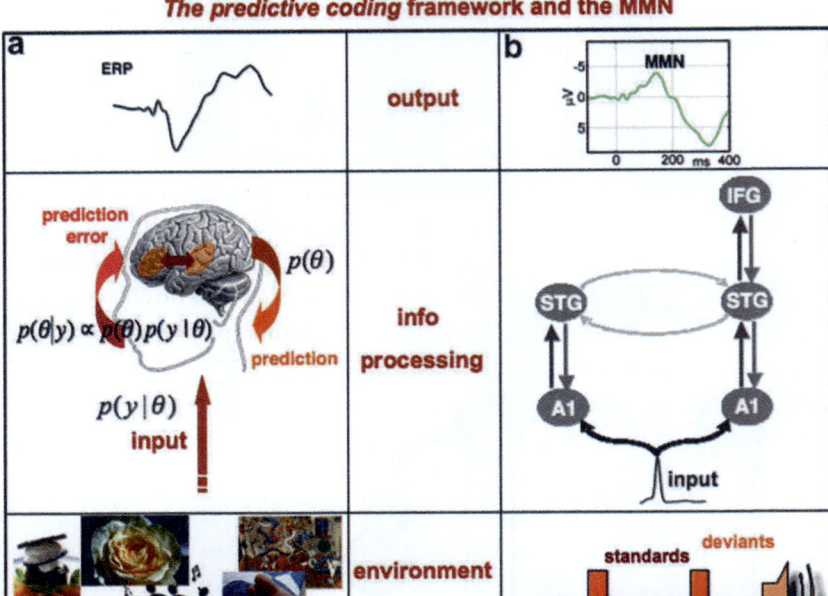

Fig. 1.14 The MMN interpreted in terms of predictive coding. (A) Illustrative scheme of the general framework of hierarchical Bayes and predictive coding as an explanation for ERP emerge. (B) The MMN, a concrete example and plausible underlying mechanisms. *(From Garrido, M.I., Kilner, J.M., Stephan, K.E., Friston, K.J., 2009. The mismatch negativity: a review of underlying mechanisms. Clin. Neurophysiol. 120 (3), 453–463. https://doi.org/10. 1016/j.clinph.2008.11.029. Open Access.)*

model, DCM will optimize the parameters of that model; however, there is no universal optimal model. Garrido et al. (2007) also explain that DCM models dynamic responses or transients that are continuous in time. This means that the DCM is not an explanation for a particular response component (e.g., the MMN) but the compound response over all peristimulus times; it is likely that this paradigm includes ERP N1-component effects (due to its difference in response to standard and oddball tones) and, at least phenomenologically, a P300-like component (although the analysis presented by Garrido et al. (2007) only went up to 250 ms). However, all these components could be explained by differences in a simple network model of interacting neuronal populations. Indeed, there may be other plausible models. A common factor in the two models is that IFG, OFC, and PCC are all nodes in the DMN, which is involved in introspection, mind-wandering, active episodic memory, and becomes deactivated during specific goal-directed behavior (Tessitore et al., 2019). Based on MEG recordings, Phillips et al. (2015) found evidence for using the A1-STG-IFG network, but supplemented this—in case of temporal structure violations—with an "expectancy" input from the PFC located at the highest level of the model's hierarchy (cf. Fig. 1.16 right and Fig. 7.6).

Human auditory perception is thought to be realized by a network of neurons that maintain a model of and predict future stimuli. Much of the evidence for this comes from experiments where a stimulus unexpectedly differs from previous ones, which generates a well-known "mismatch response." But what happens when a stimulus is unexpectedly omitted altogether? By measuring the brain's electromagnetic activity, it is shown that it also generates an "omission response" that is contingent on the presence of attention. Chennu et al. (2016) applied DCM of evoked electromagnetic responses recorded by EEG and MEG to an auditory paradigm in which the presence versus absence of "bottom-up" stimuli with the presence was crossed with absence of "top-down" attention to either auditory or visual stimuli. Model comparison revealed that both mismatch and omission responses were mediated by increased forward and backward connections, differing primarily in the driving input. In both responses, modeling results suggested that the presence of attention selectively modulated backward "prediction" connections (Fig. 1.15).

1.8 Is predictive coding complete?

Heilbron and Chait (2018) listed a number of key assumptions of predictive coding that may be empirically tested: "(1) Sensory cortex implements a hierarchical, generative model of the world: neurons at higher processing stages generate predictions that bias processing at lower levels. (2) Population responses (i.e., gross activity measured with MEG, EEG or BOLD responses reflect (at least in part) 'transient expressions of prediction error'—therefore, neural responses should be shaped by (hierarchically nested) expectations. (3) Prediction-generation and error-detection are implemented by separate neural subpopulations that reside in different cortical layers—as a consequence, prediction and error computations should have distinct laminar profiles. (4) Attention is the weighting of sensory input by its reliability—accordingly, the gain on upward projections should reflect (estimated) sensory precision. (5) In standard predictive coding, top-down predictions and bottom-up errors have distinct oscillatory profiles: predictions are conveyed via lower frequencies (beta and delta) and prediction

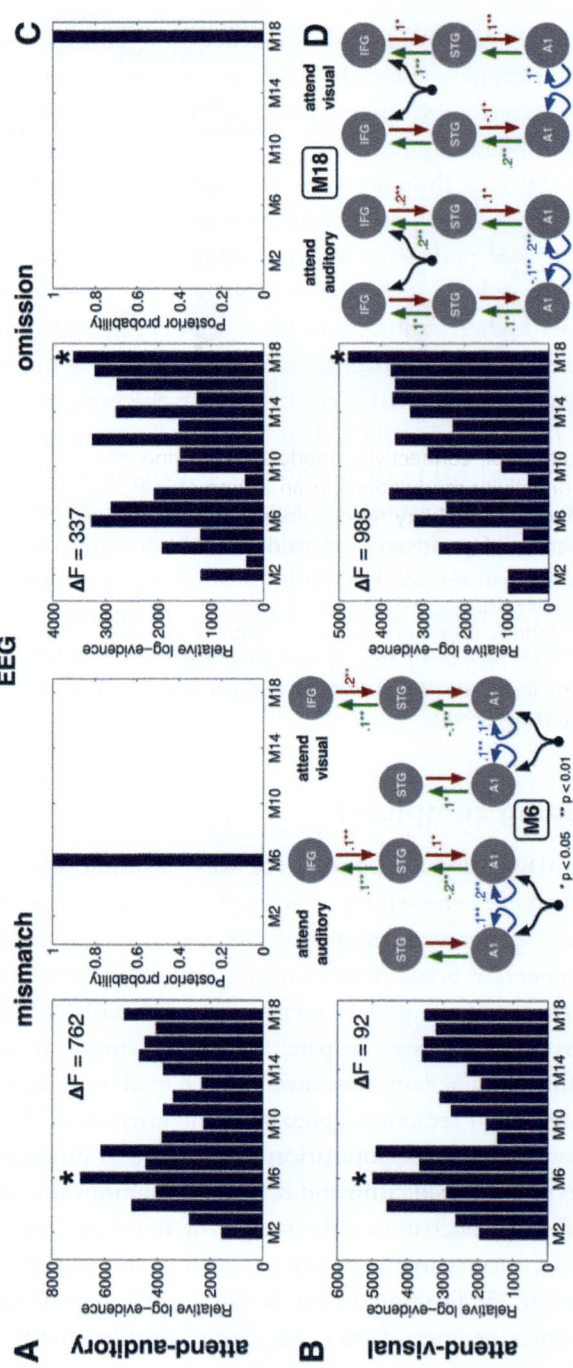

Fig. 1.15 Model evidence for the mismatch and omission effects. (A and B) Left, Relative log evidence and results of FFX BMS over the DCMs—labeled M1–M18—in their ability to model the mismatch effect ERP contrast in the attend-auditory and attend-visual conditions, respectively (left panels). *The model with the highest log evidence. The difference between log evidence of models with highest and second-highest evidence (Δ*F*) is shown in each case. A Δ*F* of 5 is equivalent to a Bayes factor of 150 in favor of the winning model. (C and D) Left, Log evidence of the same DCMs, but for modeling the omission effect ERP contrasts in the attend-auditory and attend-visual conditions. Among the DCMs instantiated and tested, model M6 (B, right) was the winning model of the mismatch effect (A, right), whereas model M18 (D, right) was the winning model of the omission effect (C, right), in both attention conditions. For connections in the winning models that were significantly modulated by these effects in each attention condition, the posterior expectations of the strength of this modulation calculated with BMA are indicated alongside. (*From Chennu, S., Noreika, V., Gueorguiev, D., Shtyrov, Y., Bekinschtein, T.A., Henson, R., 2016. Silent expectations: dynamic causal modeling of cortical prediction and attention to sounds that weren't. J. Neurosci. 36 (32), 8305–8316.*)

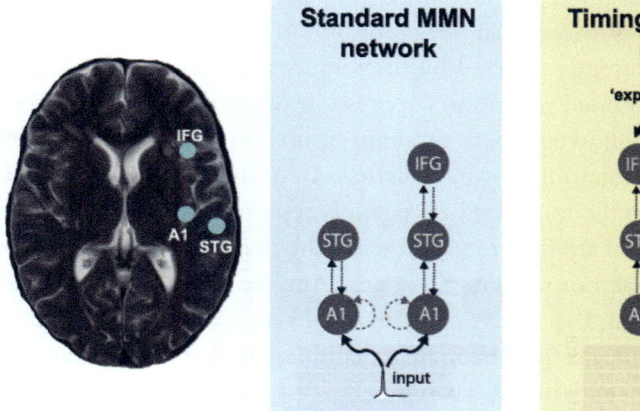

Fig. 1.16 Graphical specification of connectivity models underlying the MMN as suggested by DCM. Left: Connectivity modulations in an asymmetric frontotemporal network, combined with neuronal excitability modulations in A1, was shown to best explain the MMN across a variety of paradigms and modalities (Garrido et al., 2007, 2009; Phillips et al., 2015; Chennu et al., 2016). Right: Connectivity model including left IFG and frontal "expectancy inputs" which was found to best explain MMN responses to temporal irregularities (duration and silent gap) or omissions (Phillips et al., 2015; Chennu et al., 2016). *(Reprinted from Heilbron, M., Chait, M., 2018. Great expectations: is there evidence for predictive coding in auditory cortex? Neuroscience 389, 54–73, with permission from Elsevier.)*

errors via higher frequencies (gamma)." Critically, Heilbron and Chait (2018) noted that "even if we fully accept the network modulations suggested by DCM (Fig. 1.16), this doesn't mean that these changes necessarily reflect predictive coding, or even a single underlying mechanism. Indeed, it is difficult to see why changes in A1 excitability and STG/IFG connectivity should be uniquely characteristic of predictive coding—as shown by a comparison of the models illustrated in Figs. 1.13 and 1.14. This problem is reinforced by the fact that the discussed studies have mostly used designs in which expectation and adaptation are confounded, which makes arbitrating between predictive and nonpredictive interpretations difficult. As such, while the discussed studies constitute exciting methodological developments in the analysis of non-invasive electrophysiological data, their strength as empirical support for predictive coding theory seems limited."

Denham and Winkler (2020) also posed some general challenges to the predictive coding framework, among others a precise definition of what is meant by prediction is needed (see also Hogendoorn and Burkitt (2019)).

One answer to this question may come from the work of Costa-Faidella et al. (2011) who showed that the auditory N1 and P2 event-related responses were differently affected by the temporal regularity of repeating tones; N1 only exhibited repetition suppression when the tones were isochronous, while P2 showed repetition suppression for regular and randomly timed sequences, suggesting that information about the content (what) and timing (when) of predictions may be separately represented in audition. But if this is the case, what are the prerequisites for the two representations to be utilized in a conjoined or separate manner? "Answering this question may lead to understanding the equivocal pattern of results obtained regarding the role of patterns in auditory stream segregation: perhaps when predictions refer to both content and timing, a pattern effect emerges, but when only the content is predicted, the presence of patterns is not effective." It is clear from Denham and Winkler's (2020) review that there is a lack of consensus in the field regarding the role of predictions in auditory perception: (1) from a basic theoretical perspective, there is no agreement on the need for object representations to be predictive; (2) models of sequence processing (sequential auditory stream segregation, MMN) differ in their implementation specifically with regard to the need for explicit vs implicit predictions; and (3) behavioral experiments demonstrate sensitivity to patterns, but evidence regarding their role in decomposing the auditory scene is inconclusive.

In this book we will look into these issues in depth. Ultimately (Chapter 12) we conclude that: "There appears to be limited connections between disorder-specific network changes in the DMN and those in the IFG and STG that play such a dominant role in the MMN network. Consequently, the use of predictive coding networks based on auditory MMN as error signal only emphasizes the communality of cognitive decrease— potentially differentiated between temporal and frequency mismatches— however does not discriminate nor is of further diagnostic value. On the other hand, dynamic causal modeling elucidates the underlying intrinsic, particularly inhibitory changes that are affected by neurological disorders, in the various network nodes that affect the intra-node functional connectivity (FCs). This may be the most relevant contribution of predictive coding for therapeutics."

1.9 Summary

Generalizing the concept of hierarchical learning to predictive coding, based on the difference in expectation from a central world model and the actual sensory one, resulted in the assumption of prediction errors that are represented in feedforward signals, event-related potentials (ERPs), to higher-level prediction neurons. These higher-level cortical neurons produce prediction signals that are returned (feedback) to the lower cortical, and even subcortical, levels where they are compared to their sensory evoked activity. In short-latency ERPs this feedback does not produce changes, but at longer latencies, results in a mismatch signal that reflects the difference between the expected feedback information and the sensory evoked one. The nervous system aims to minimize this mismatch by adjusting the processing gain in the cortical prediction structures by changing inhibitory interactions within and between the various prediction levels. Estimating the relevant prediction structures is usually based on estimating putative sources from ERP or fMRI studies, whereas the forward and backward connectivities are adjusted using Bayesian inference to decide between the often very numerous potential networks. The often-used "standard" mismatch networks are, at least for the auditory modality, based on bilateral primary auditory cortex, bilateral superior temporal gyrus, and right inferior frontal gyrus, sometimes modulated by prefrontal cortex activity.

References

Apps, M.A.J., Tsakiris, M., 2014. The free-energy self: a predictive coding account of self-recognition. Neurosci. Biobehav. Rev. 41, 85–97.

Asilador, A., Llano, D.A., 2021. Top-down inference in the auditory system: potential roles for corticofugal projections. Front. Neural Circuits 14, 615259. https://doi.org/10.3389/fncir.2020.615259.

Barbas, H., Medalla, M., Alade, O., Suski, J., Zikopoulos, B., Lera, P., 2005. Relationship of prefrontal connections to inhibitory systems in superior temporal areas in the rhesus monkey. Cereb. Cortex 15, 1356–1370. https://doi.org/10.1093/cer-cor/bhi018.

Bastos, A.M., Usrey, W.M., Adams, R.A., Mangun, G.R., Fries, P., Friston, K.J., 2012. Canonical microcircuits for predictive coding. Neuron 76, 695–711.

Bressler, S.L., Seth, A.K., 2011. Wiener–Granger causality: a well established methodology. Neuroimage 58, 323–329.

Carbajal, G.V., Malmierca, M.S., 2018. The neuronal basis of predictive coding along the auditory pathway: from the subcortical roots to cortical deviance detection. Trends Hear. 22, 1–33. https://doi.org/10.1177/2331216518784822.

Chan, J.S., Langer, A., Kaiser, J., 2016. Temporal integration of multisensory stimuli in autism spectrum disorder: a predictive coding perspective. J. Neural Transm. 123, 917–923. https://doi.org/10.1007/s00702-016-1587-5.

Chan, J.S., Wibral, M., Stawowsky, C., Brandl, M., Helbling, S., Naumer, M.J., et al., 2021. Predictive coding over the lifespan: increased reliance on perceptual priors in older adults—a magnetoencephalography and dynamic causal modeling study. Front. Aging Neurosci. 13, 631599. https://doi.org/10.3389/fnagi.2021.631599.

Chennu, S., Noreika, V., Gueorguiev, D., Blenkmann, A., Kochen, S., Ibáñez, A., et al., 2013. Expectation and attention in hierarchical auditory prediction. J. Neurosci. 33 (27), 11194–11205.

Chennu, S., Noreika, V., Gueorguiev, D., Shtyrov, Y., Bekinschtein, T.A., Henson, R., 2016. Silent expectations: dynamic causal modeling of cortical prediction and attention to sounds that weren't. J. Neurosci. 36 (32), 8305–8316.

Costa-Faidella, J., Grimm, S., Slabu, L., Diaz-Santaella, F., Escera, C., 2011. Multiple time scales of adaptation in the auditory system as revealed by human evoked potentials. Psychophysiology 48, 774–783.

Daunizeau, J., David, O., Stephan, K.E., 2011. Dynamic causal modelling: a critical review of the biophysical and statistical foundations. Neuroimage 58, 312–322.

David, O., Friston, K.J., 2003. A neural mass model for MEG/EEG: coupling and neuronal dynamics. Neuroimage 20, 1743–1755.

David, O., Kiebel, S.J., Harrison, L.M., Mattout, J., Kilner, J.M., Friston, K.J., 2006. Dynamic causal modeling of evoked responses in EEG and MEG. Neuroimage 30, 1255–1272.

De Ridder, D., Vanneste, S., Freeman, W., 2014. The Bayesian brain: phantom percepts resolve sensory uncertainty. Neurosci. Biobehav. Rev. 44, 4–15.

Denham, S.L., Winkler, I., 2020. Predictive coding in auditory perception: challenges and unresolved questions. Eur. J. Neurosci. 51, 1151–1160.

Douglas, R.J., Martin, K.A., 1991. A functional microcircuit for cat visual cortex. J. Physiol. 440, 735–769. https://doi.org/10.1113/jphysiol.1991.sp018733.

Dürschmid, S., Edwards, E., Reichert, C., Dewar, C., Hinrichs, H., Heinze, H.J., et al., 2016. Hierarchy of prediction errors for auditory events in human temporal and frontal cortex. Proc. Natl. Acad. Sci. 113 (24), 6755–6760. https://doi.org/10.1073/pnas.1525030113.

Eggermont, J.J., 1990. The Correlative Brain; Theory and Experiment in Neural Interaction. Springer Verlag, Berlin.

Eggermont, J.J., 2022. Tinnitus and Hyperacusis. Facts, Theories and Clinical Implications. Academic Press, London.

Friston, K.J., 2002. Functional integration and inference in the brain. Prog. Neurobiol. 68, 113–143.

Friston, K., 2005. A theory of cortical responses. Philos. Trans. R. Soc. B 360, 815–836. https://doi.org/10.1098/rstb.2005.1622.

Friston, K., 2009. The free-energy principle: a rough guide to the brain? Trends Cogn. Sci. 13 (7), 293–301.

Friston, K., Moran, R., Seth, A.K., 2013. Analysing connectivity with Granger causality and dynamic causal modelling. Curr. Opin. Neurobiol. 23, 172–178.

Garrido, M.I., Kilner, J.M., Kiebel, S.J., Stephan, K.E., Friston, K.J., 2007. Dynamic causal modelling of evoked potentials: a reproducibility study. Neuroimage 36, 571–580.

Garrido, M.I., Kilner, J.M., Stephan, K.E., Friston, K.J., 2009. The mismatch negativity: a review of underlying mechanisms. Clin. Neurophysiol. 120 (3), 453–463. https://doi.org/10.1016/j.clinph.2008.11.029.

Gourévitch, B., Eggermont, J.J., 2007. Evaluating information transfer between auditory cortical neurons. J. Neurophysiol. 97, 2533–2543.

Gregory, R.L., 1980. Perceptions as hypotheses. Philos. Trans. R. Soc. Lond. B 290, 181–197.

Harris, C.D., Rowe, E.G., Randeniya, R., Garrido, M.I., 2018. Bayesian model selection maps for group studies using M/EEG data. Front. Neurosci. 12, 598. https://doi.org/10.3389/fnins.2018.00598.

Heilbron, M., Chait, M., 2018. Great expectations: is there evidence for predictive coding in auditory cortex? Neuroscience 389, 54–73.

Helmholtz, H., 1962. Concerning the perceptions in general. Treatise on Physiological Optics. Dover, New York, NY.

Hochstein, S., Ahissar, M., 2002. View from the top: hierarchies and reverse hierarchies in the visual system. Neuron 36 (5), 791–804. https://doi.org/10.1016/s0896-6273(02) 01091-7.

Hogendoorn, H., Burkitt, A.N., 2019. Predictive coding with neural transmission delays: a real-time temporal alignment hypothesis. eNeuro 6 (211), 1–12. https://doi.org/ 10.1523/ENEURO.0412-18.2019.

Honey, C.J., Sporns, O., Cammoun, L., Gigandet, X., Thiran, J.P., Meuli, R., 2009. Predicting human resting-state functional connectivity from structural connectivity. Proc. Natl. Acad. Sci. U. S. A. 106 (6), 2035–2040.

Hullfish, J., Sedley, W., Vanneste, S., 2019. Prediction and perception: insights for (and from) tinnitus. Neurosci. Biobehav. Rev. 102, 1–12.

Imada, A., Morris, A., Wiest, M.C., 2013. Deviance detection by a P3-like response in rat posterior parietal cortex. Front. Integr. Neurosci. 6, 127. https://doi.org/10.3389/ fnint.2012.00127.

Jang, S.H., Choi, E.B., 2022. Evaluation of structural neural connectivity between the primary auditory cortex and cognition-related brain areas using diffusion tensor tractography in 43 normal adults. Med. Sci. Monit. 28, e936131. https://doi.org/10.12659/ MSM.936131.

Kiebel, S.J., Garrido, M.I., Friston, K.J., 2007. Dynamic causal modelling of evoked responses: the role of intrinsic connections. Neuroimage 36 (2), 332–345. https://doi. org/10.1016/j.neuroimage.2007.02.046.

Lecaignard, F., Bertrand, O., Gimenez, G., Mattout, J., Caclin, A., 2015. Implicit learning of predictable sound sequences modulates human brain responses at different levels of the auditory hierarchy. Front. Hum. Neurosci. 9, 505. https://doi.org/10.3389/ fnhum.2015.00505.

McIntosh, A.R., Gonzalez-Lima, F., 1994. Structural equation modeling and its application to network analysis in functional brain imaging. Hum. Brain Mapp. 2, 2–22.

Medalla, M., Barbas, H., 2010. Anterior cingulate synapses in prefrontal areas 10 and 46 suggest differential influence in cognitive control. J. Neurosci. 30, 16068–16081.

Medalla, M., Barbas, H., 2014. Specialized prefrontal "auditory fields": organization of primate prefrontal-temporal pathways. Front. Neurosci. 8, 77. https://doi.org/10.3389/ fnins.2014.00077.

Mumford, D., 1992. On the computational architecture of the neocortex. II. The role of cortico-cortical loops. Biol. Cybern. 66, 241–251.

Nahum, M., Nelken, I., Ahissar, M., 2008. Low-level information and high-level perception: the case of speech in noise. PLoS Biol. 6 (5), e126. 978–991.

Natan, R.G., Briguglio, J.J., Mwilambwe-Tshilobo, L., Jones, S.I., Aizenberg, M., Goldberg, E.M., Geffen, M.N., 2015. Complementary control of sensory adaptation by two types of cortical interneurons. eLife 4, 1–27. https://doi.org/10.7554/ eLife.09868.

O'Donnell, P., Lavin, A., Enquist, L.W., Grace, A.A., Card, J.P., 1997. Interconnected parallel circuits between rat nucleus accumbens and thalamus revealed by retrograde transynaptic transport of pseudorabies virus. J. Neurosci. 17 (6), 2143–2167.

Pereira, M.R., Barbosa, F., de Haan, M., Ferreira-Santos, F., 2019. Understanding the development of face and emotion processing under a predictive processing framework. Dev. Psychol. 55 (9), 1868–1881. https://doi.org/10.1037/dev0000706.

Phillips, H.N., Blenkmann, A., Hughes, L.E., Bekinschtein, T.A., Rowe, J.B., 2015. Hierarchical organization of frontotemporal networks for the prediction of stimuli across multiple dimensions. J. Neurosci. 35 (25), 9255–9264.

Shamma, S., 2008. On the emergence and awareness of auditory objects. PLoS Biol. 6 (6), e155.

Tessitore, A., Cirillo, M., De Miccoa, R., 2019. Functional connectivity signatures of Parkinson's disease. J. Parkinsons Dis. 9, 637–652. https://doi.org/10.3233/JPD-191592.

Torres-García, M.E., Solis, O., Patricio, A., Rodríguez-Moreno, A., Camacho-Abrego, I., Limón, I.D., et al., 2012. Dendritic morphology changes in neurons from the prefrontal cortex, hippocampus and nucleus accumbens in rats after lesion of the thalamic reticular nucleus. Neuroscience 223, 429–438.

Wacongne, C., 2016. A predictive coding account of MMN reduction in schizophrenia. Biol. Psychol. 116, 68–74.

Winkler, I., Czigler, I., 2012. Evidence from auditory and visual event-related potential (ERP) studies of deviance detection (MMN and vMMN) linking predictive coding theories and perceptual object representations. Int. J. Psychophysiol. 83, 132–143.

Winkowski, D.E., Nagode, D.A., Donaldson, K.J., Yin, P., Shamma, S.A., Fritz, J.B., et al., 2018. Orbitofrontal cortex neurons respond to sound and activate primary auditory cortex neurons. Cereb. Cortex 28 (3), 868–879.

Wolff, W., Berberian, N., Golesorkhi, M., Gomez-Pilar, J., Zilio, F., Northoff, G., 2022. Intrinsic neural timescales: temporal integration and segregation. Trends Cogn. Sci. 26 (2), 159–173. https://doi.org/10.1016/j.tics.2021.11.007.

Xu, J.-J., Cui, J., Feng, Y., Yong, W., Chen, H., Chen, Y.-C., et al., 2019. Chronic tinnitus exhibits bidirectional functional dysconnectivity in frontostriatal circuit. Front. Neurosci. 13, 1299. https://doi.org/10.3389/fnins.2019.01299.

Yeh, C.-H., Smith, R.E., Dhollander, T., Calamante, F., Connelly, A., 2019. Connectomes from streamlines tractography: assigning streamlines to brain parcellations is not trivial but highly consequential. Neuroimage 199, 160–171.

CHAPTER 2

Setting the stage: Cocktail parties, auditory streaming, mismatch negativity, and stimulus-specific adaptation

2.1 Introduction

In a noisy room with multiple people talking, listeners with normal hearing can still recognize and understand attended speech and simultaneously ignore background noise and irrelevant speech. Cherry (1953) coined the term "cocktail party problem" for this phenomenon. There are two conceptually distinct challenges for a listener in a "cocktail party" situation. The first is the problem of sound segregation. To assess source locations, the auditory system needs to derive the properties of individual sounds from the mixture entering the ears, i.e., through a binaural filter. The second is directing attention to the sound source of interest while ignoring the others, i.e., through an attentional filter, and switching attention between sources when intermittently following two conversations. This assumes that the binaural filter is preattentional (McDermott, 2009). Listeners are also better at comprehending speech in a cocktail party setting if the attended speech is in the listener's native language, as personal experience testifies. This suggests the importance of expectation and/or memory. Because speech is first and foremost a temporal stimulus, the timing of events within a speech stream is critical for its decoding, attribution to the correct source, and ultimately to whether it will be included in the attentional "spotlight" or forced to the background of perception (Giraud and Poeppel, 2012; Shamma et al., 2011).

2.2 Auditory stream segregation

Auditory stream segregation—a generalization of the cocktail party setting—involves linking temporally separate acoustic events into one or more coherent sequences. Auditory stream segregation can be demonstrated by presenting a sequence of low (A) and high (B) frequency tones in an

Brain Responses to Auditory Mismatch and Novelty Detection
https://doi.org/10.1016/B978-0-443-15548-2.00002-8

alternating pattern, ABAB. When the tone presentation rate—between A and B—is low, or the frequency separation (ΔF) between the tones is small (<10%), a connected alternating sequence ABAB is perceived. However, when ΔF is large the alternating sequence perceptually splits into two parallel auditory streams—one composed of interrupted "A" tones and the other of interrupted "B" tones (Micheyl et al., 2007; Fig. 2.1).

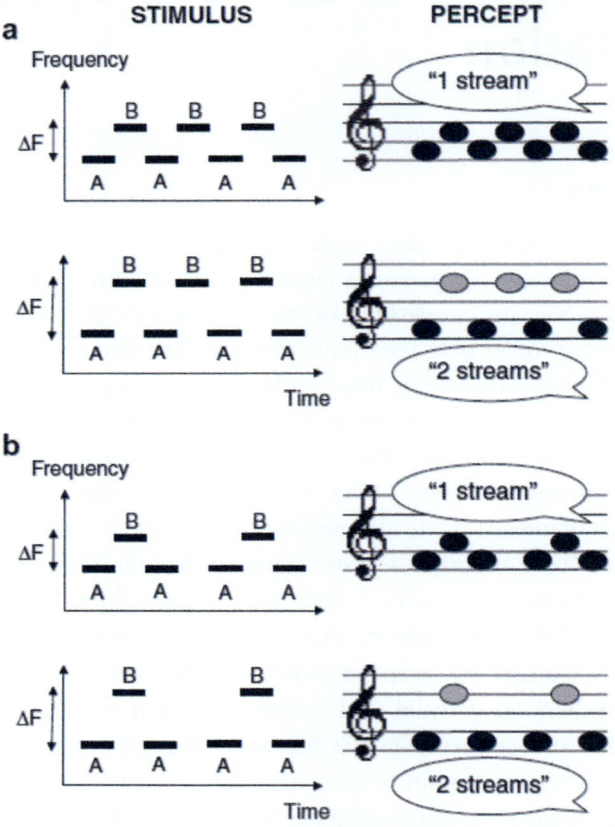

Fig. 2.1 Schematic representation of the stimuli commonly used to study sequential auditory streaming and of the corresponding auditory percepts. The stimulus (left) is a temporal sequence of pure tones alternating between two frequencies, represented here and in the text by the letters A and B. The A–B frequency difference (ΔF) is either small (top left panel) or large (bottom left panel). In the former case, the percept is that of a single, coherent stream of tones alternating in pitch. In the latter case, the percept is that of two separate streams of tones; since the tones in each stream have a constant frequency, the sense of pitch alternation is lost. (*Reprinted from Micheyl, C., Carlyon, R.P., Gutschalk, A., Melcher, J.R., Oxenham, A.J., Rauschecker, J.P., et al., 2007. The role of auditory cortex in the formation of auditory streams. Hear. Res. 229, 116–131, with permission from Elsevier.*)

As reiterated by Micheyl and Oxenham (2010), sounds that have common properties are more likely to be integrated by the auditory system. When properties differ enough between sound elements, they probably arise from different sources and are more likely to be assigned to different auditory objects or streams. Some of the acoustic properties of sound that play important roles in auditory scene analysis (ASA) include fundamental frequency and harmonic relationships among spectral components, temporal onsets/offsets, timbre, and patterns of amplitude modulation (reviewed in Zion Golumbic et al., 2013). Differences in the perceived location of sound sources, i.e., sound parts that have consistent interaural differences—suggesting a stable location—are a major basis for stream segregation, particularly in a cocktail party setting.

2.3 Human electrophysiology and neural imaging

2.3.1 The mismatch negativity

A specific class of cortical auditory evoked potentials (CAEPs) is formed by those responses that can only be obtained following an unexpected sound—and then called an event-related potential (ERP)—such as an infrequent (10%–15% of the time) tone of 1000 Hz among a series of more frequent (85%–90%) tones of 1100 Hz. Under passive listening conditions the difference between the responses to the standard and the unexpected, also called deviant or oddball, sound obtained is the mismatch negativity (MMN; Fig. 2.2).

The MMN reflects neural information in the brain that will allow behavioral detection of a difference, or discrimination of the two sounds, and was discovered by Näätänen et al. (1982) and named by Mäntysalo and Näätänen (1987). Haenschel et al. (2005) investigated changes in ERPs to different numbers of repetitions of standards, delivered in a roving-stimulus paradigm in which the frequency of the standard stimulus changed randomly between stimulus trains. Healthy volunteers were engaged in two experimental conditions: during passive listening and while actively discriminating changes in tone frequency. With increasing number of standards the oddball becomes more and more unpredictable and the MMN increases (Fig. 2.3). If during the presentation of a series of standard and deviant sounds, the subject is required to press a button or count the deviant tones an additional positive peak with a latency of 300 ms (P3b, see Chapter 5 and Appendix) appears. A still later response component, the N400, can be elicited when a word at the end of a sentence is perceived as semantically wrong.

The Mismatch Negativity (MMN)

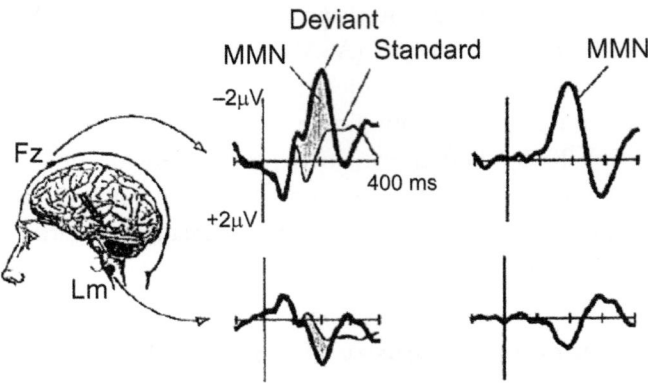

Fig. 2.2 A schematic illustration of the mismatch negativity (MMN). The MMN is usually maximal at frontocentral areas (Fz; upper panel, in the middle column) and reverses polarity at the mastoids (lower panel). The *thin line* represents brain response elicited by the standard stimulus and the *thick line* that to a deviant stimulus. The *shaded area* is the MMN. The MMN can be more clearly seen in subtraction waves (right column). The activity source of the MMN is indicated by the *arrow* in the auditory areas. *(Reprinted from Kujala, T., Näätänen, R., 2001. The mismatch negativity in evaluating central auditory dysfunction and dyslexia. Neurosci. Biobehav. Rev. 25, 535–543, with permission from Elsevier.)*

In a review of MMN use, Näätänen et al. (2014) emphasized that "while not serving as a specific marker to any particular disorder, MMN may be useful for understanding factors of cognition in various disorders." They also illustrate the mismatch response, which is of positive polarity, in young infants (Fig. 2.4). The changes in the mismatch response from a positive to negative wave during early maturation and with a potential explanation have been detailed in Eggermont and Moore (2012) and will be taken up again in Chapter 6.

Ponton et al. (2000b) described the maturation of the changes in the MMN compared to the CAEPs from 5 to 20 years of age and in response to left-ear stimulation in 118 subjects tested across 137 sessions. They used an interstimulus interval (ISI) of 700 ms. The age-related changes in the CAEPs were recorded over the hemisphere ipsilateral and contralateral to the stimulated left ear. They noted that in many respects the MMN appears to mature relatively early, particularly when compared to the obligatory N1

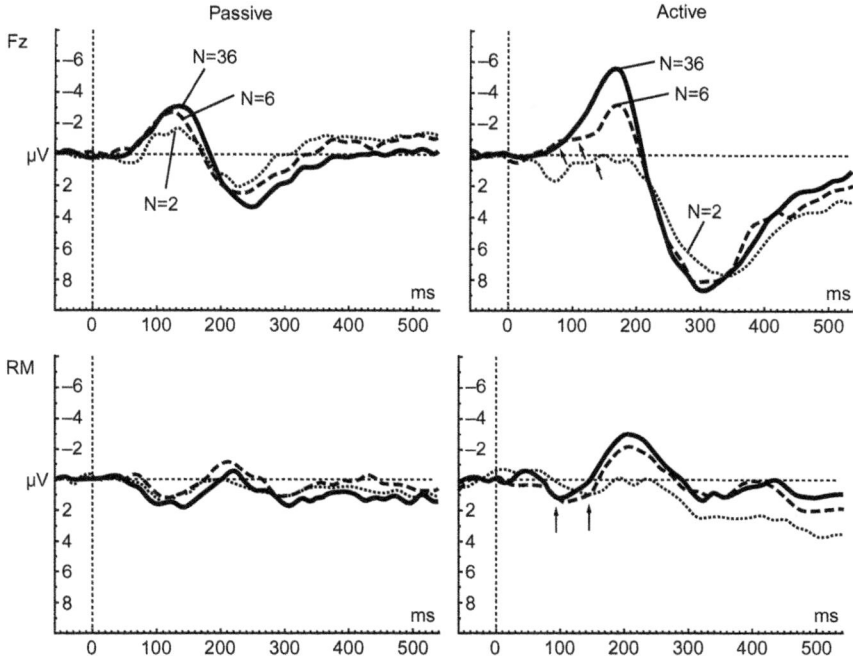

Fig. 2.3 MMN difference waves. The MMN difference waves for 2 *(dotted lines)*, 6 *(dashed lines)*, and 36 *(black lines)* standard repetitions in the passive (left) and the active (right) conditions are shown for central frontal and the right mastoid electrode. The *arrows* indicate the shortening of the onset latency into the P50 latency range. *RM*, right mastoid. *(From Haenschel, C., Vernon, D.J., Dwivedi, P., Gruzelier, J.H., Baldeweg, T., 2005. Event-related brain potential correlates of human auditory sensory memory-trace formation J. Neurosci. 25 (45), 10494–10501, Open access.)*

peak of the CAEP (Ponton et al., 2000a). The MMN data combined with the CAEP maturation data are illustrated in Fig. 2.5. The CAEP data are presented as grand-mean waveforms for 14 age-related groups. For this analysis, CAEPs from 5- and 6-year-olds—the youngest cohort—are averaged together as one group. Similarly, the CAEPs from 19- and 20-year-olds are also averaged together. Depending on age, each grand-mean waveform represents the CAEPs recorded from 7 to 16 subjects per group. The CAEPs are presented as surface plots, with latency on the *x*-axis, CAEP amplitude on the *y*-axis, and age on the *z*-axis. The surface plots provide visual continuity in assessing the maturational changes in the CAEPs. In detail, Fig. 2.5A contains a surface plot of the grand-mean CAEPs recorded from normal-hearing children. The CAEPs of younger children are dominated by a large positivity with a peak latency slightly earlier than that of the adult N1. This large

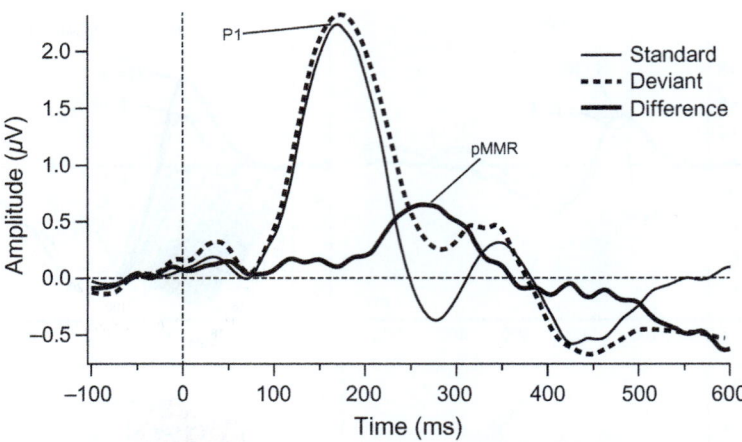

Fig. 2.4 Grand-mean event-related potentials (ERPs) evoked by a standard vowel sound *(solid thin line)* and deviant vowel sound *(dashed line)* in a group of infants ages 3–47 months of age. The difference waveform (deviant-minus standard, *thick solid line*) delineates the mismatch response (MMR) which has a positive polarity (labeled pMMR) in this age group. The P1 component of the ERPs is also labeled. The displayed waveforms are recorded from a Geodesic net site (Geodesic 3), which is approximately 1 cm in front of and to the right of Fz. *(From Näätänen, R., Sussman, E.S., Salisbury, D., Shafer, V.L., 2014. Mismatch negativity (MMN) as an Index of cognitive dysfunction. Brain Topogr. 27, 451–466. Fig. 2. Permission granted by Springer Nature.)*

positivity, which is labeled P1, is followed by a negative trough at about 180 ms suggesting that this negativity corresponds with the adult N2, although with a somewhat shorter latency. For the used ISI, the N1 peak is not consistently present in the CAEP until about 9–10 years of age— we evaluate this in detail in Chapter 6. In the surface plot, N1 appears as a deepening trough that separates the P1 peak from a second positivity that has the latency of the adult P2. As N1 becomes increasingly more negative, P1 decreases in amplitude. By 12–13 years of age, the CAEP waveforms assume an adult-like morphology. Significant latency decreases continue for the P1 and N1 peaks until adult-like values are attained at 15–16 years of age. Age-related changes in CAEP amplitude follow a similar maturation time course as latency, becoming adult-like by the age of 15–16 years (Ponton et al., 2000a). Fig. 2.5B shows the much earlier appearance of the MMN than the N1, thereby indicating that the MMN is not based on, but maybe causes, an enhancement of the N1 in the deviant stimulus condition, as for instance suggested by May and Tiitinen (2010) and discussed in Chapter 3.

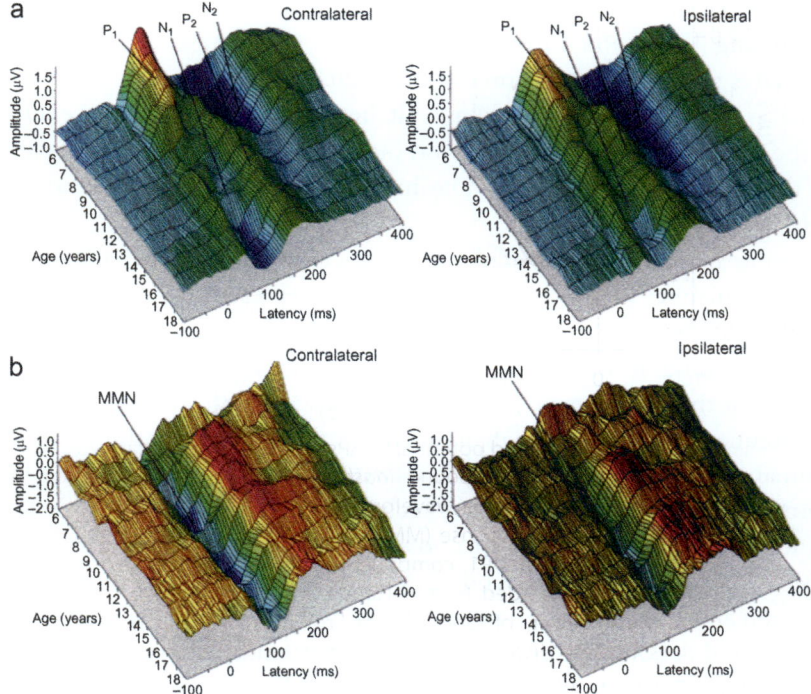

Fig. 2.5 Surface plots of the grand-mean waveforms for normal-hearing individuals between 5 and 20 years of age recorded at electrode sites located over the hemisphere contralateral (C4) and ipsilateral (C3) to the stimulated left ear. Each grand-mean waveform represents AEPs recorded from at least 7 subjects. (A) Standard CAEP grand-mean waveforms. (B) Difference waveforms containing the MMN and some P300 from age 5 (contralaterally) and age 8 (ipsilaterally) on. *(From Ponton, C.W., Eggermont, J.J., Don, M., Waring, M.D., Kwong, B., Cunningham, J., Trautwein, P., 2000, Maturation of mismatch negativity: effects of profound deafness and cochlear implant use. Audiol. Neuro-Otol. 5, 167–185. With permission from S. Karger AG, Basel.)*

2.3.2 Electrophysiological and neuroimaging of auditory streaming in human cortex

In preattentive stream segregation, MMN studies are mostly concerned with the role of spectral and temporal separation between the tones. Tiitinen et al. (1994) note: "As the occurrence of MMN is not usually affected by the direction of attentions, MMN reflects the operation of automatic sensory (echoic) memory the earliest memory system that builds traces of the acoustic environment against which new stimuli can be compared." Consequently, Sussman et al. (1999) used the MMN to probe whether the segregation associated with the streaming effect occurs passively. Alternating high- and low-frequency tones were presented at fast and slow rates while subjects ignored the stimuli. At the slow rate, tones were heard as alternating high and low pitches, and no MMN was elicited. At the fast pace a streaming effect was induced and an MMN was observed for the low-frequency stream. The high-frequency deviant did not elicit an MMN.

Using a cocktail party paradigm, Zion Golumbic et al. (2013) investigated the manner in which speech streams are represented in brain activity, and the way that selective attention affects the representation of speech from direct recordings from the auditory cortex in surgical epilepsy patients. They found that this brain activity dynamically tracks speech streams using both low-frequency phase and high-frequency amplitude fluctuations and that optimal encoding likely combines the two (Fig. 2.6; see also Fig. 7.1). Furthermore, their data "suggest that speech tracking in these two bands—low-frequency phase and high-gamma power—reflects distinct neuronal mechanisms for auditory speech encoding, since they differed in their spatial distribution and response time course. Sites with significant tracking effects are either "modulation" sites showing significant tracking of both talkers, albeit biased toward the attended one, or "selection" sites showing significant tracking of the attended talker only. Amplitude modulation sites were found in and near the superior temporal gyrus (STG) and in higher-order language processing and attentional control regions such as inferior frontal cortex (IFG), anterior and inferior temporal cortex, and inferior parietal lobule" (IPL; Zion Golumbic et al., 2013).

To investigate auditory stream segregation in a sound localization task, Zündorf et al. (2013) presented five different natural sounds from five virtual spatial locations (Fig. 2.7 left). Using functional magnetic resonance imaging (fMRI), auditory stream segregation activity was revealed in posterior STG bilaterally, anterior insula, supplementary motor area, and a frontoparietal network (Fig. 2.7 right). For orienting spatial attention to the target sound, they found critical roles of left planum temporale for extracting the sound of interest among acoustical distracters, and the precuneus which seems to be crucial for accurately determining locations of auditory targets in an acoustically complex scene of multiple sound sources. The IFG (Fig. 2.7 right) is involved in a cortical network for sound identification and spatial analysis of realistic sound sources providing spectrotemporal localization cues. As Chapter 1 illustrates, the bilateral STG and right IFG are part of auditory MMN networks involved in predictive coding.

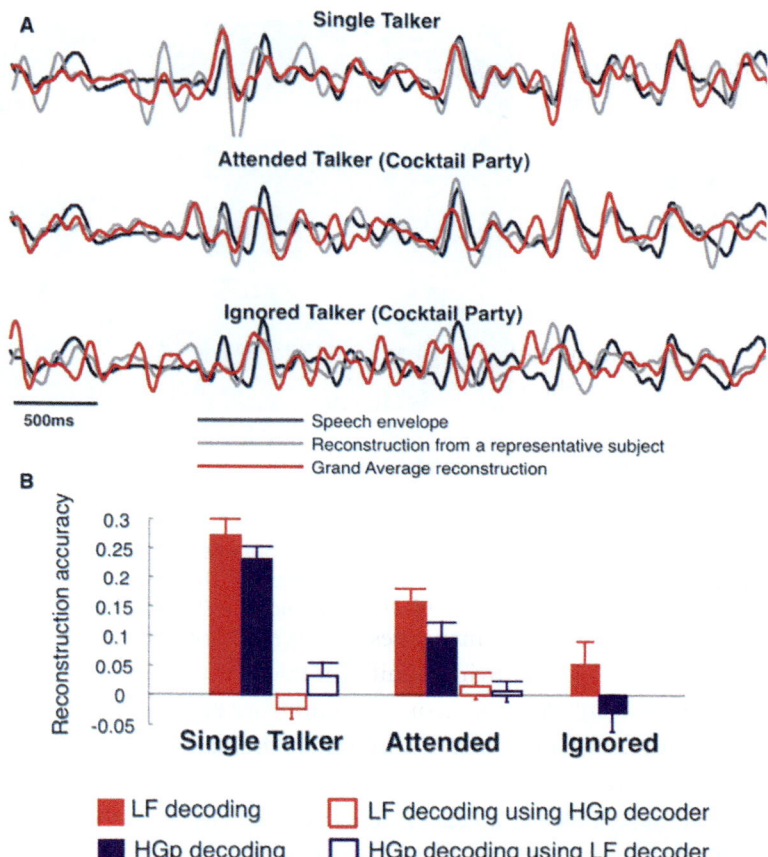

Fig. 2.6 Reconstruction of the Speech Envelope from the Cortical Activity. (A) A segment of the original speech envelope *(black)* compared with the reconstruction achieved using the low-frequency (LF) signal from one participant *(light gray)* and from all participants *(red)*. Reconstruction examples are shown for the Single Talker condition (top) as well as for the attended (middle) and ignored stimuli (bottom) in the cocktail party condition. (B). Full bars: Grand averaged of the reconstruction accuracy (i.e., the correlation coefficients between the actual and reconstructed time courses) across all participants using LF *(gray)* or HGp (high-gamma power; *black*). The Single Talker and the Attended Talker in the cocktail party condition could be reliably reconstructed using either measure, and in both cases, significantly better than the ignored speaker. Empty bars: Envelope reconstruction accuracy obtained by applying each of the single-band decoders to data in the other band. As shown here, decoders constructed using either band performed poorly when applied to data in the other band. This implies that the two single-band decoders have systematically different features. *(Reprinted from Zion Golumbic, E.M., Ding, N., Bickel, S., Lakatos, P., Schevon, C.A., McKhann, G.M., et al., 2013. Mechanisms underlying selective neuronal tracking of attended speech at a "Cocktail Party". Neuron 77, 980–991, with permission from Elsevier.)*

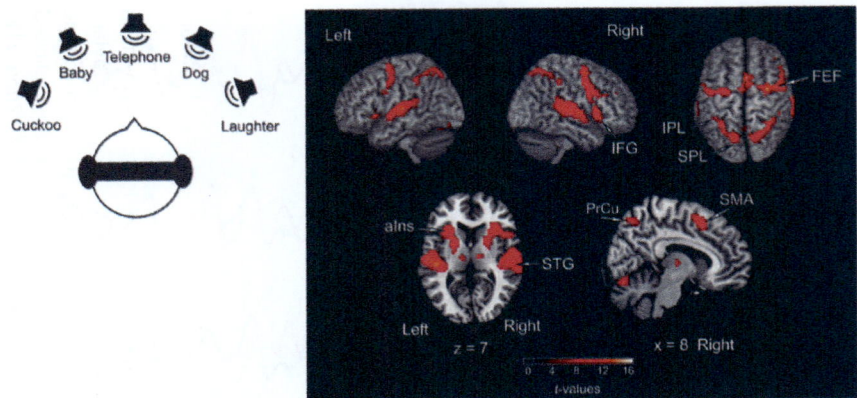

Fig. 2.7 Left. Example of one virtual auditory scene used for the "cocktail" and "passive" conditions. Each sound was presented as coming from a different location. Right. Activations of brain regions as revealed by the contrast of "cocktail" condition versus rest ($P < .05$). *aIns*, anterior insula; *FEF*, frontal eye fields; *IFG*, inferior frontal gyrus; *IPL*, inferior parietal lobule; *PrCu*, precuneus; *SMA*, supplementary motor area; *SPL*, superior parietal lobule; *STG*, superior temporal gyrus. The color code refers to *t*-values (see bar). *(From Zündorf, I.C., Lewald, J., Karnath, H-O., 2013. Neural correlates of sound localization in complex acoustic environments. PLoS One 8, e64259. Open Access.)*

Symonds et al. (2020) used a three-stream paradigm with four acoustic features uniquely defining each sound stream (frequency, envelope shape, spatial location, tone quality). The task load was manipulated by combining a difficult auditory task and an easy movie-viewing task with the same set of sounds in separate conditions. The MMN was measured to evaluate sound processing in both conditions. They found no effect of task demands on unattended sound processing, i.e., MMNs were elicited by unattended deviants during both low- and high-load task conditions. In the auditory task, however, the P3b component demonstrated a two-stage process of target evaluation and detection.

Fig. 2.8 displays the deviant-minus-standard difference waveforms and the voltage distribution maps of the P3b component elicited by the single and double deviants. Neither P3a (which is largest at Fz) nor P3b (which is largest at Pz) was elicited by envelope deviants in unattended low or middle streams in either attend-visual or attend-auditory conditions. P3b was elicited by the high stream single (nontarget) deviants in the attend-auditory condition and had a typical P3b scalp distribution (Fig. 2.8A, pink trace, 600–800 ms range). In contrast, when participants ignored the sounds and watched a movie, and when the double deviant was not designated as a target for a button press response, P3b was not elicited by single or double deviants

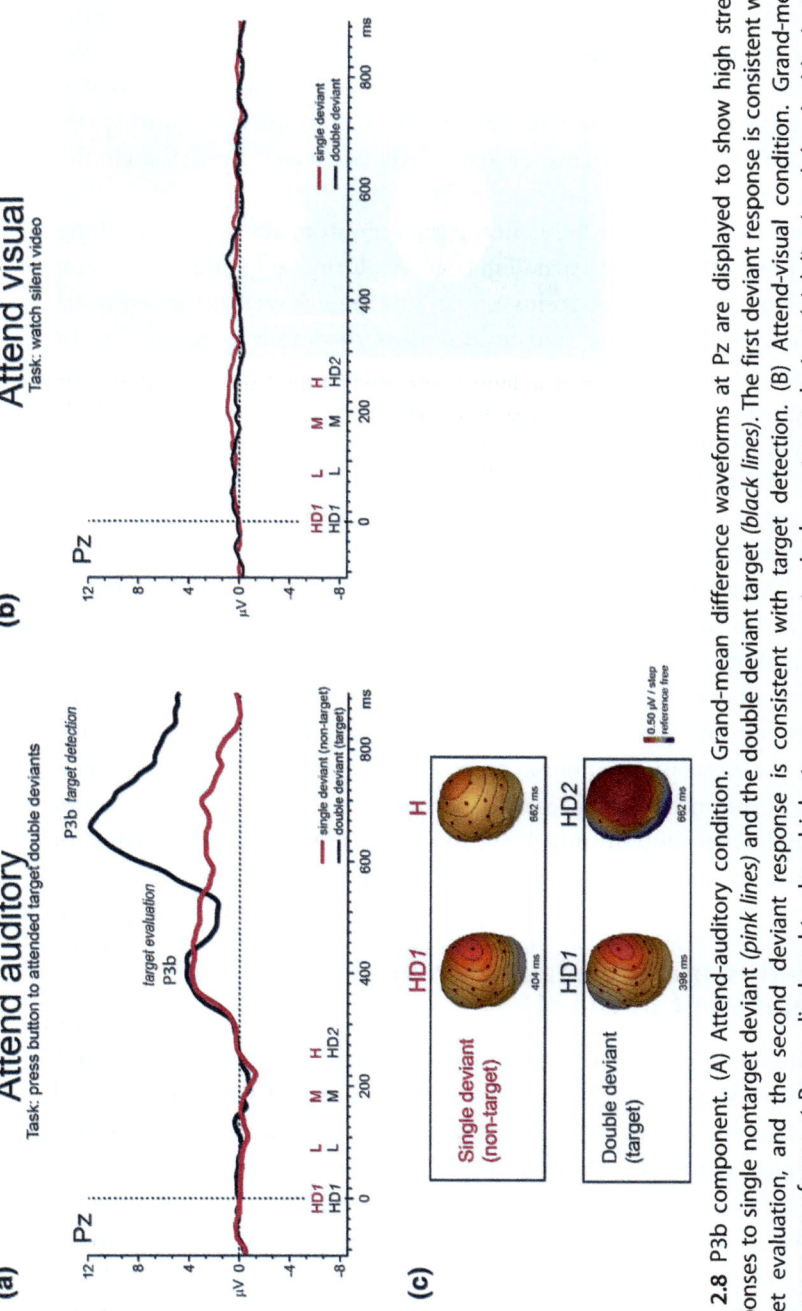

Fig. 2.8 P3b component. (A) Attend-auditory condition. Grand-mean difference waveforms at Pz are displayed to show high stream responses to single nontarget deviant (*pink lines*) and the double deviant target (*black lines*). The first deviant response is consistent with target evaluation, and the second deviant response is consistent with target detection. (B) Attend-visual condition. Grand-mean difference waveforms at Pz are displayed to show high stream responses to single nontarget deviant (*pink lines*) and the double deviant target (*black lines*). Tones were irrelevant to the task of watching a movie, and deviants did not elicit P3b components. (C) Scalp voltage topography. P3bs elicited by the single and double deviant in the attend-auditory condition are displayed. Similar topography elicited by single and double deviants suggests different chronological phases of the target detection process. (*From Symonds, R.M., Zhou, J.W., Cole, S.L., Brace, K.M., Sussman, E.S., 2020. Cognitive resources are distributed among the entire auditory landscape in auditory scene analysis. Psychophysiology 57 (2), e13487. © 2019 Society for Psychophysiological Research. With permission from John Wiley and Sons.*)

(attend–visual condition, Fig. 2.8B). Thus P3bs were elicited by the first and by the second of the target double high-frequency deviants, showing a temporal progression of events time locked to the deviants. Early on, there was an evaluation period at the occurrence of the first deviant, and later there was detection of the target at the second deviant that then required a button press response. P3b amplitude was smaller to the early target evaluation than to the later target detection.

Denham and Winkler (2020) noted that auditory streaming has long been a favored experimental paradigm for exploring sequential grouping in ASA because it allows one to investigate the cues as well as the temporal dynamics of grouping. The two-tone auditory-streaming stimulus most commonly used consists of a sequence of pure tones with alternating low (A) and high (B) frequencies (cf. Fig. 2.1), arranged either as a simple alternating sequence, BABABA. . ., or as series of triplets separated by a silent interval (_), BAB_BAB_BAB_. Due to the ambiguous nature of this stimulus, listening to long segments of the stimulus leads to perception switching back and forth between alternative sound organizations (Denham and Winkler, 2006; Pressnitzer and Hupe, 2006; Winkler and Schröger, 2015). Denham and Winkler (2020) then described two computational models of auditory streaming—those of Barniv and Nelken (2015) and Mill et al. (2013)—that may be used to illustrate contrasting views on the need for explicit predictions and the nature of the representations necessary for explaining the perceptual phenomenon. Both models build and maintain alternative (nondominant) representations in parallel, consistent with electrophysiological results that suggest that representations of alternative sound organizations are maintained in the brain (Sussman et al., 2014).

2.4 Stimulus-specific adaptation in animal models as an equivalent of the MMN

We saw that under passive listening conditions a difference waveform between the CAEP to the unexpected (also called deviant or oddball) sound and to the frequent sound occurs. This MMN was interpreted as a reflection of neural information processing in the brain that would allow behavioral detection of a difference in, or discrimination of, the two sounds (Chapter 1). At the single-neuron level in animal primary auditory cortex (A1) this phenomenon was named "stimulus-specific adaptation" (SSA; Fig. 2.9) and considered a microscopic correlate of the neural population-based MMN (Ulanovsky et al., 2003; Dean et al., 2005, 2008). It is noted

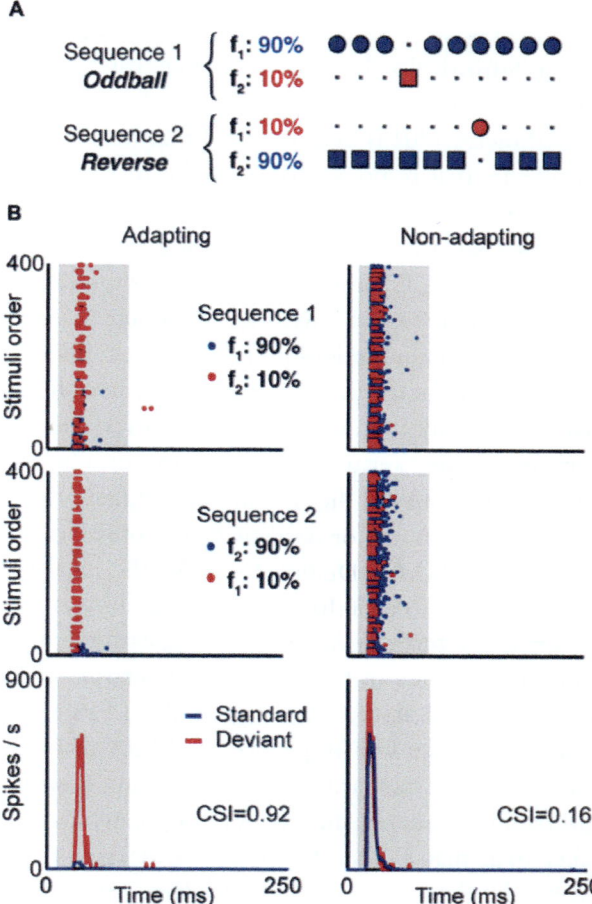

Fig. 2.9 (A) In the oddball paradigm, a low probability stimulus (f_2, *red*, "deviant") is embedded in a train of high probability stimuli (f_1, *blue*, "standard"). To compensate the responses to the different physical stimuli f_1 and f_2, a second sequence is presented where the probability of each stimulus is reversed. Examples of the responses of two neurons in the inferior colliculus recorded using this paradigm are shown in (B), an IC neuron showing SSA (adapting, left) and another not showing SSA (nonadapting, right). Here, f_1 and f_2 are pure tones of different frequencies, and the frequency difference is the same for both neurons. The top and middle panels in (B) show the dot raster in response to sequence 1 (top) and sequence 2 (middle), where the *blue dots* represent spikes in response to the standard and the *red dots* the response to the deviant. In the adapting neuron, the response to the standard stimulus decays after the first presentations, while the response to the deviant stimulus remains constant, as a typical example of stimulus-specific adaptation. The bottom panels show the poststimulus time histograms for the responses to the standard and deviant stimuli, combining the spikes for both stimuli at the same probability. The value of the common SSA index (CSI) is shown for each neuron; CSI values close to one indicate strong SSA while values close to zero indicate weak SSA. IC, inferior colliculus. *(From Pérez-González, D., Malmierca, M.S., 2014. Adaptation in the auditory system: an overview. Front. Integr. Neurosci. 8, 19. https://doi.org/10.3389/fnint.2014.00019. Open Access.)*

that stimulus deviance/novelty detection in animals can be demonstrated in diverse subcortical auditory structures (Chapter 4). Importantly, Pérez-González and Malmierca (2014) demonstrated that neurons in the inferior colliculus (IC) reduce their responses to a stimulus that is presented repeatedly, but when a novel sound is presented, the same neurons are able to overcome the adaptation and respond quickly and vigorously (Fig. 2.9).

Back to auditory cortex, Ulanovsky et al. (2004) showed that SSA in cat A1 could be described concurrently with several timescales, spanning many orders of magnitude, from hundreds of milliseconds to tens of seconds. Similar timescales are known for the auditory memory span of humans, as measured both psychophysically and using CAEPs. A simple model, with linear dependence on both short-term and long-term stimulus history—SSA resulting from the suppression of the responses (firing rate or CAEP amplitude) to the standard—provided a good fit to responses in A1. The same group (Taaseh et al., 2011) found that in rat A1 the local field potential (LFP) or multiunit activity (MUA) to a deviant stimulus was at least partially due to the change it represented relative to the regularity set by the standard tone, indicating the presence of true deviance detection (Fig. 2.10). The MUA responses measured at the same site as the LFPs are displayed in Fig. 2.10C. Similarly, to the LFP in part B, the MUA response to each of the tones was smaller when standard than when deviant. Remarkably, while in the equal-probability sequence the MUA response evoked by f2 was substantially weaker than that evoked by f1, in the deviant f2 sequence the MUA evoked by the deviant f2 was actually larger than that evoked by the standard f1, the opposite of what one would predict from the frequency selectivity of this site and what is reflected in the LFPs.

Pienkowski and Eggermont (2009) found that SSA features in the context dependence of spectrotemporal receptive fields in cat A1. Responses to sound stimulus frequencies close to the neuron's best frequency (on-BF bands) adapt with an average time constant of approximately 7 s (Gourévitch and Eggermont, 2008). In contrast, responses away from the best frequency (off-BF bands) do not adapt, but slightly increased over the 30-s observation window. Such stimulus-specific adaptation could function in enhancing stimulus discrimination and in maximizing neural information transmission by reducing redundancy.

2.5 Novelty detection

I emphasize that, although often used interchangeably, mismatch detection is distinct from novelty detection. Mismatch detection occurs, for instance,

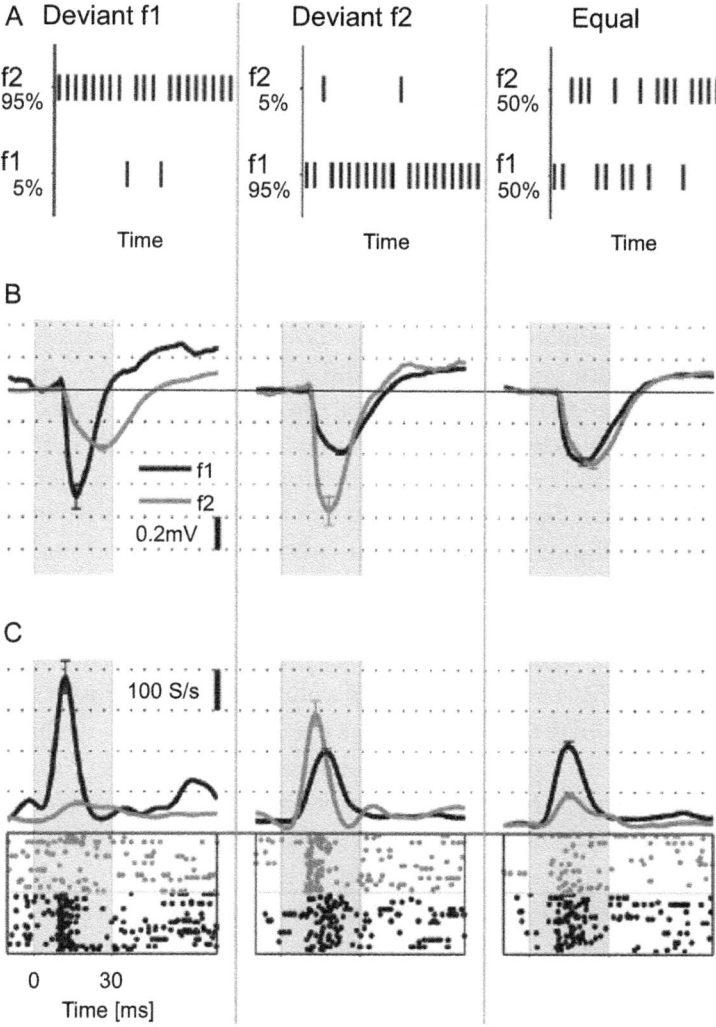

Fig. 2.10 The oddball paradigm. (A) A schematic spectrographic representation of the three basic sequences used in this study. In each trial, either f1 or f2 is presented pseudorandomly according to their probability of occurrence. (B) The average local field potential responses in a typical recording site to the two frequencies of the paradigm in each of the sequences (f1 = 13.3 kHz, *black*, and f2 = 19.2 kHz, *gray*). The level was 30-dB attenuation (~70 dB SPL). Error bars: ± s.e.m., shaded interval: stimulus. (C) Multiunit responses at the same site. The raster plots show 25 presentations for each of the 2 frequencies, corresponding to 5% of the 500 tone presentations in the sequence. For the standard and equal-probability conditions, which had more than 25 presentations in a sequence, the 25 presentations were selected so that they represent the spike count distribution of all responses in the time window shown. The line graphs represent poststimulus time histograms smoothed by a 10-ms Hamming window. *(From Taaseh, N., Yaron, A., Nelken, I., 2011. Stimulus-specific adaptation and deviance detection in the rat auditory cortex. PLoS One 6 (8), e23369. https://doi.org/10.1371/journal.pone.0023369. Open Access.)*

by disrupting a sequence of sounds with a high pitch followed after a random number in the sequence by a lower pitched sound. Varying the stimulus-onset intervals suggests that this depends on sensory memory with a duration of ~10 s. The neural responses reflecting this mismatch have latencies <150 ms. Novelty detection, which does not rely on interrupted repetition, takes longer and relies dominantly on working memory, with response latencies that are much longer than that of the MMN. The ability to detect such unexpected stimuli in the acoustic environment and determine their behavioral relevance to plan an appropriate reaction is critical for survival. Nevertheless, the brain response to auditory novelty comprises two main EEG components: an early MMN and a late P3b. Whereas the former has been proposed to reflect a prediction error (Chapter 1), the latter is often associated with working memory updating. Auditory novelty detection has thus been associated with two different cognitive processes. Bekinschtein et al. (2009) developed an experimental paradigm for separating these two processes (Fig. 2.11). It relies on two levels of auditory novelty that are independently manipulated: a change in pitch within series of five sounds (local novelty) gives rise to an MMN, whereas a rare change in series of five sounds in a fixed context (global novelty) triggers a P3b. The same group (Pegado et al., 2010) used high-density scalp event-related potentials during an active version of the auditory oddball paradigm to explore the lifetimes of these processes by varying the stimulus onset asynchrony (SOA). They observed that early MMN (90–160 ms) decreased when the SOA increased, confirming the time limitation of this echoic memory system. Subsequent neural events, including late MMN (160–220 ms) and P3a/P3b components of the P3 complex (240–500 ms), did not decay with SOA but showed a systematic delay effect supporting a two-stage model of accumulation of evidence. Pegado et al. (2010) proposed "a distinction within the MMN complex of two events: an early, pre-attentive and fast-decaying MMN associated with generators located within superior temporal gyri and frontal cortex, and a late MMN more resistant to SOA, corresponding to the activation of a distributed cortical network including fronto-parietal regions."

Again, the same group (El Karoui et al., 2015), recording from epileptic patients with intracranial electrodes, noted that local and global novelty has mostly been studied using ERPs. However, underlying these ERPs is local spiking activity as indexed by gamma (60–120 Hz) power and interactions between brain regions as indexed by modulations in beta-band (13–25 Hz) power and functional connectivity (as reviewed in Eggermont, 2021). El Karoui et al. (2015) found that local novelty triggered an early response

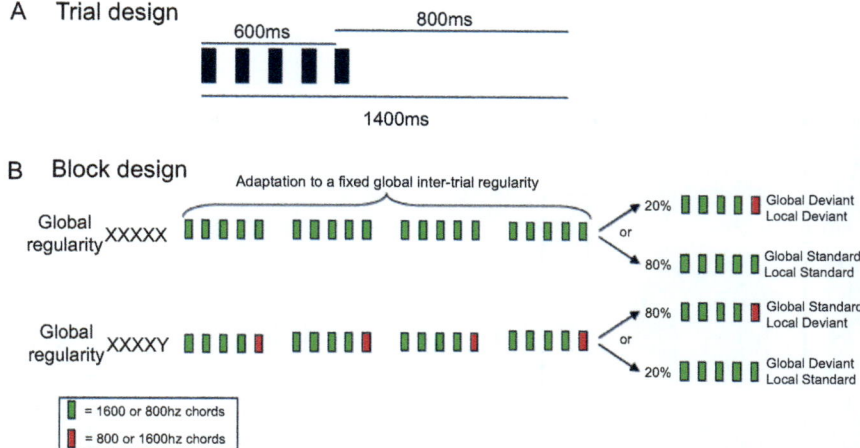

Fig. 2.11 Experimental design. (A) On each trial, 5 complex sounds of 50-ms duration each were presented with a fixed stimulus onset asynchrony of 150 ms between sounds. Four different types of series of sounds were used, the first 2 were prepared using the same 5 sounds (AAAAA or BBBBB), and the second 2 series of sounds were either AAAAB or BBBBA. (B) Each block started with 20–30 frequent series of sounds to establish the global regularity before delivering the first infrequent global deviant stimulus. *(From Bekinschtein, T.A., Dehaene, S., Rohaut, B., Tadela, F., Cohen, L., Naccache, L., 2009. Neural signature of the conscious processing of auditory regularities. PNAS 106 (5), 1672–1677. No permission needed.)*

observed as an intracranial MMN contemporary with a strong power increase in the gamma band and an increase in connectivity in the beta band. Importantly, all these responses were strictly confined to the auditory cortex. In contrast, global novelty gave rise to a late ERP response distributed across brain areas, contemporary with a sustained power decrease in the beta band (13–25 Hz) and an increase in connectivity in the alpha band (8–13 Hz) within the frontal lobe (Fig. 2.12). The authors concluded that "local novelty is associated with an early increase in functional connectivity in the beta band, mostly in pairs of temporal electrodes. [...] In contrast, global novelty was associated with a late increase in alpha-band functional connectivity, suggesting that this increase is related to genuine synchronization between brain areas."

2.6 Mismatch responses and predictive coding

"Predictive coding is possibly one of the most influential, comprehensive, and controversial theories of neural function. While proponents praise its

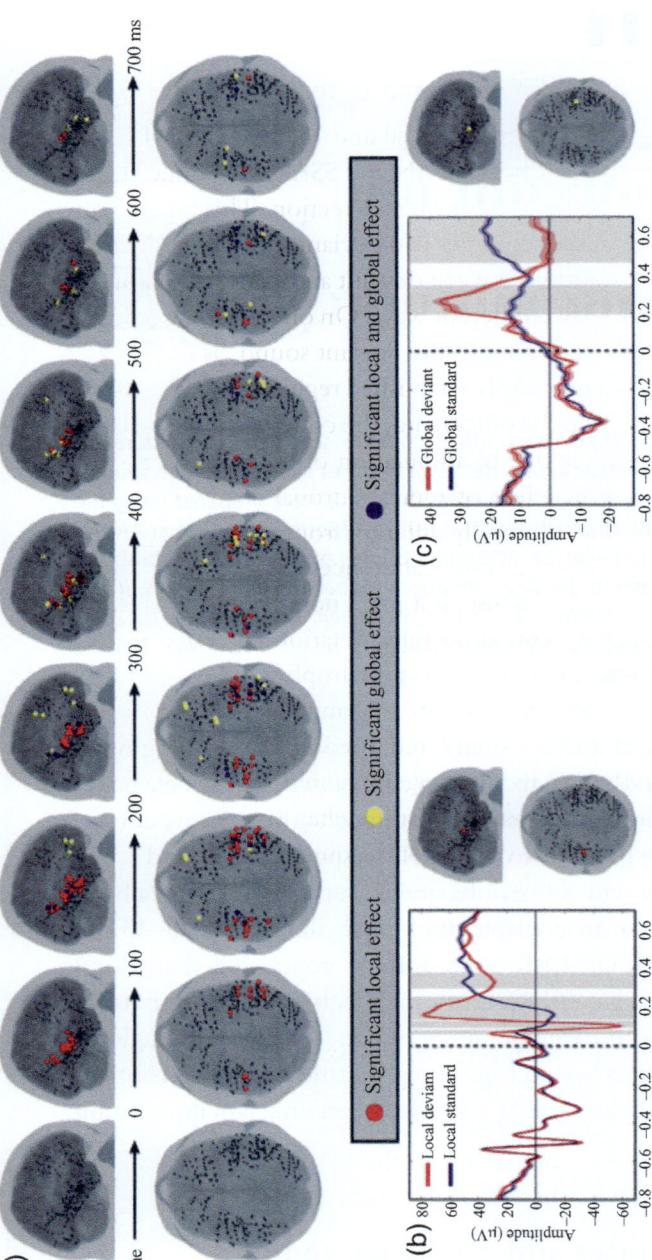

Fig. 2.12 Event-related potentials in response to local and global novelty. (A) Localization of significant differences between standard and deviant stimuli. Each *black dot* represents an electrode. *Red, yellow, and blue dots* represent electrodes showing a significant local, global, and both local and global effect, respectively, in the time window indicated at the extremities of the *black arrows*. (B) Example of a temporal electrode showing significant differences between local deviant (*in red*) and local standard (*in blue*) stimuli. Four components, peaking at 63, 105, 165, and 330ms, can be identified. *Red and blue shadings* represent standard error of the mean (SEM). *Gray shading* represents significant differences between conditions ($P_{corr} < .05$). The precise localization of this electrode (*red*) is shown on the right. (C) Example of a temporal electrode showing differences between global deviant (*in red*) and global standard (*in blue*) stimuli. Two components can be identified: the first one shows a peak at 260ms, and the second one starts at 462 ms. *Red and blue shadings* represent SEM. *Gray shading* represents significant differences between conditions ($P_{corr} < .05$). The precise localization of this electrode (*yellow*) is shown on the right. (*From El Karoui, I., King, J.-R., Sitt, J., Meyniel, F., Van Gaal, S., Hasboun, D., et al., 2015. Event-related potential, time-frequency, and functional connectivity facets of local and global auditory novelty processing: an intracranial study in humans. Cereb. Cortex 25, 4203–4212. By permission from Oxford University Press.*)

explanatory potential, critics object that key tenets of the theory are untested or even untestable" (Heilbron and Chait, 2018).

2.6.1 Role of MMN and SSA in deviance detection

I follow here the impressive review by Carbajal and Malmierca (2018). In the context of an oddball paradigm, both MMN and SSA can be understood as indices of automatic—preattentive—deviance detection. This is based on the overall difference between the responses to a deviant stimulus compared to a standard stimulus. This contrast between deviant and standard responses may be accounted for in at least two different ways. On one hand, it could be due to an enhancement in the response to the deviant sound, as its appearance represents a violation of a previously established regularity. Carbajal and Malmierca (2018) note that: "On the other hand, the contrast between deviant and standard could be simply due to attenuation of the response to the standard repetitive sound, as an effect of mere neuronal adaptation. The appearance of the deviant sound, physically different from the standard stimuli, would elicit the response of other novel afferences. The deviant sound would not produce an enhanced response, but just a non-adapted one (May and Tiitinen, 2010; Chapter 3). This latter interpretation conforms to the adaptation hypothesis that is favored in most neurophysiological studies about SSA. The definition of SSA in terms of probabilities is rather similar to the classic conceptualization of MMN recorded in the human scalp (Näätänen et al., 1978). In spite of its advantageous and versatile simplicity, the adaptation hypothesis was considered somewhat deficient in fully explaining deviance detection, and could not account for all the aspects of the MMN (Winkler et al., 2009)," but see Chapter 3 for convincing explanatory simulations using the adaptation model (May, 2021).

2.6.2 Some insights from a predictive coding perspective

One may explore the connection between SSA and MMN from a predictive coding standpoint. SSA takes place in the same cortical regions of the brain where (early) MMN sources are located, i.e., in auditory cortex, furthermore the time course of SSA is comparable to the MMN. Moreover, both SSA and MMN are affected by the manipulation of N-methyl-D-aspartate (NMDA) receptors, suggesting their involvement in the generation of SSA and MMN. This may imply that SSA and MMN are correlates of the same deviance-detection process (Heilbron and Chait, 2018). However, SSA and MMN are only theoretically comparable in the context of an oddball paradigm, as the definition of SSA (Ulanovsky et al., 2003) describes the effects of representing a stimulus in the system and establishing the simple prediction that the following stimulus will be similar to the previous encoded, thereby suppressing the response to it. This does not work for explaining MMN signals generated in contexts where the repetition rule does not occur. SSA is an index for the violation of only that particular representation. The physiological mechanism is deviance detection as defined by predictive coding. Prediction error and repetition suppression effects are confounded in the deviance-detection signal elicited by the oddball sequence. These components can be disentangled by the use of a control sequence that does not feature the standard-repetition rule, while generating a similar state of refractoriness in the system than the oddball (Heilbron and Chait, 2018).

These control sequences are the "many-standards" (Bekinschtein et al., 2009) and "cascade" (Parras et al., 2017) controls which make it possible to explain SSA in terms of predictive coding empirically. The application of control sequences has unveiled a predictive processing hierarchy in the auditory system, which roots originate in the cortices of the inferior colliculus (Carbajal and Malmierca, 2018). The proportion of deviance detection accounted for by prediction error in the auditory system increases from lemniscal to nonlemniscal pathways, and from the midbrain to the auditory cortex (Chapter 1). Therefore predictive activity in the auditory modality is not exclusively cortical. Within the auditory pathway, the high heterogeneity in the functioning of single neurons exposed to deviant stimulation hints at the great importance of local excitatory/inhibitory interactions in predictive processing. Along with the fact that deactivating the auditory cortex does not eliminate deviance detection on the IC, it follows that predictive activity must be emerging to a certain extent de novo at subcortical levels instead of being just passed down from the cortex (Carbajal and Malmierca, 2018).

2.7 Summary

Predictive coding, one would expect, could be a mechanism that allows to segregate an intended speaker's voice from background noise or a melee of other voices. Expectation of what to hear in a particular language or dialect requires memory but first of all the speaker should be localized, which is typically preattentional. This is then followed by paying attention to the localized source. The brain activity that accompanies this shows this stream segregation in the amplitude fluctuations of high-gamma-band activity and the phase of lower-frequency oscillations. In preattentive stream segregation, MMN studies are mostly concerned with the role of spatial separation between the sounds. Animal studies have identified stimulus-specific adaptation (SSA), partly based on repetition suppression of the frequent sound and partly based on sound novelty in deviance detection. SSA has been interpreted as a form of MMN, but this depends on the brain area generating it with deviance detection in the midbrain based on adaptation of the frequent sound in auditory cortex whereas novelty also impacts the response. For separating these two processes in humans, an experimental paradigm has been developed that relies on two levels of auditory deviance: one local (in time) that gives rise to an early MMN, and one global in time that triggers a late MMN and an P3b response.

References

Barniv, D., Nelken, I., 2015. Auditory streaming as an online classification process with evidence accumulation. PLoS One 10, e0144788.

Bekinschtein, T.A., Dehaene, S., Rohaut, B., Tadela, F., Cohen, L., Naccache, L., 2009. Neural signature of the conscious processing of auditory regularities. PNAS 106 (5), 1672–1677.

Carbajal, G.V., Malmierca, M.S., 2018. The neuronal basis of predictive coding along the auditory pathway: from the subcortical roots to cortical deviance detection. Trends Hear. 22, 1–33. https://doi.org/10.1177/2331216518784822.

Cherry, E.C., 1953. Some experiments on the recognition of speech, with one and two ears. J. Acoust. Soc. Am. 25, 975–979.

Dean, I., Harper, N.S., McAlpine, D., 2005. Neural population coding of sound level adapts to stimulus statistics. Nat. Neurosci. 8, 1684–1689.

Dean, I., Robinson, B.L., Harper, N.S., McAlpine, D., 2008. Rapid neural adaptation to sound level statistics. J. Neurosci. 28 (25), 6430–6438.

Denham, S.L., Winkler, I., 2006. The role of predictive models in the formation of auditory streams. J. Physiol. Paris 100, 154–170.

Denham, S.L., Winkler, I., 2020. Predictive coding in auditory perception: challenges and unresolved questions. Eur. J. Neurosci. 51, 1151–1160.

Eggermont, J.J., 2021. Brain Oscillations, Synchrony and Plasticity. Basic Principles and Application to Auditory-Related Disorders. Academic Press, London, ISBN: 978-0-12-819818-6, pp. 1–250.

Eggermont, J.J., Moore, J.K., 2012. Morphological and functional development of the auditory nervous system. In: Werner, L.A., et al. (Eds.), Human Auditory Development. Springer Handbook of Auditory Research, vol. 42. Springer Science+Business Media, New York, pp. 61–105.

El Karoui, I., King, J.-R., Sitt, J., Meyniel, F., Van Gaal, S., Hasboun, D., et al., 2015. Event-related potential, time-frequency, and functional connectivity facets of local and global auditory novelty processing: an intracranial study in humans. Cereb. Cortex. 25, 4203–4212.

Giraud, A.-L., Poeppel, D., 2012. Cortical oscillations and speech processing: emerging computational principles and operations. Nat. Neurosci. 15, 511–517.

Gourévitch, B., Eggermont, J.J., 2008. Spectrotemporal sound density dependent long-term adaptation in cat primary auditory cortex. Eur. J. Neurosci. 27, 3310–3321.

Haenschel, C., Vernon, D.J., Dwivedi, P., Gruzelier, J.H., Baldeweg, T., 2005. Event-related brain potential correlates of human auditory sensory memory-trace formation. J. Neurosci. 25 (45), 10494–10501.

Heilbron, M., Chait, M., 2018. Great expectations: is there evidence for predictive coding in auditory cortex? Neuroscience 389, 54–73.

Mäntysalo, S., Näätänen, R., 1987. The duration of a neuronal trace of an auditory stimulus as indicated by event-related potentials. Biol. Psychol. 24, 183–195.

May, P.J.C., 2021. The adaptation model offers a challenge for the predictive coding account of mismatch negativity. Front. Hum. Neurosci. 15, 721574. https://doi.org/10.3389/fnhum.2021.721574.

May, P.J.C., Tiitinen, H., 2010. Mismatch negativity (MMN), the deviance-elicited auditory deflection, explained. Psychophysiology 47, 66–122.

McDermott, J.H., 2009. The cocktail party problem. Curr. Biol. 19, R1024–R1027.

Micheyl, C., Oxenham, A.J., 2010. Objective and subjective psychophysical measures of auditory stream integration and segregation. J. Assoc. Res. Otolaryngol. 11, 709–724.

Micheyl, C., Carlyon, R.P., Gutschalk, A., Melcher, J.R., Oxenham, A.J., Rauschecker, J.-P., et al., 2007. The role of auditory cortex in the formation of auditory streams. Hear. Res. 229, 116–131.

Mill, R.W., Bohm, T.M., Bendixen, A., Winkler, I., Denham, S.L., 2013. Modelling the emergence and dynamics of perceptual organisation in auditory streaming. PLoS Comput. Biol. 9, e1002925.

Näätänen, R., Gaillard, A.W.K., Mäntysalo, S., 1978. Early selective-attention effect on evoked potential reinterpreted. Acta Psychol. (Amst) 42, 313–329.

Näätänen, R., Simpson, M., Loveless, N.E., 1982. Stimulus deviance and evoked potentials. Biol. Psychol. 14, 53–98.

Näätänen, R., Sussman, E.S., Salisbury, D., Shafer, V.L., 2014. Mismatch negativity (MMN) as an index of cognitive dysfunction. Brain Topogr. 27, 451–466.

Parras, G.G., Nieto-Diego, J., Carbajal, G.V., Valdés-Baizabal, C., Escera, C., Malmierca, M.S., 2017. Neurons along the auditory pathway exhibit a hierarchical organization of prediction error. Nat. Commun. 8, 2148. https://doi.org/10.1038/s41467-017-02038-6.

Pegado, F., Bekinschtein, T., Chausson, N., Dehaene, S., Cohen, L., Naccache, L., 2010. Probing the lifetimes of auditory novelty detection processes. Neuropsychologia 48, 3145–3154.

Pérez-González, D., Malmierca, M.S., 2014. Adaptation in the auditory system: an overview. Front. Integr. Neurosci. 8, 19. https://doi.org/10.3389/fnint.2014.00019.

Pienkowski, M., Eggermont, J.J., 2009. Effects of adaptation on spectrotemporal receptive fields in primary auditory cortex. Neuroreport 20, 1198–1203.

Ponton, C.W., Eggermont, J.J., Kwong, B., Don, M., 2000a. Maturation of human central auditory system activity: evidence from multi-channel evoked potentials. Clin. Neurophysiol. 111, 220–236.

Ponton, C.W., Eggermont, J.J., Don, M., Waring, M.D., Kwong, B., Cunningham, J., Trautwein, P., 2000b. Maturation of mismatch negativity: effects of profound deafness and cochlear implant use. Audiol. Neurootol. 5, 167–185.

Pressnitzer, D., Hupe, J.M., 2006. Temporal dynamics of auditory and visual bistability reveal common principles of perceptual organization. Curr. Biol. 16, 1351–1357.

Shamma, S.A., Elhilali, M., Micheyl, C., 2011. Temporal coherence and attention in auditory scene analysis. Trends Neurosci. 34, 114–123.

Sussman, E., Ritter, W., Vaughan Jr., H.G., 1999. An investigation of the auditory streaming effect using event-related brain potentials. Psychophysiology 36, 22–34.

Sussman, E.S., Bregman, A.S., Lee, W.W., 2014. Effects of task-switching on neural representations of ambiguous sound input. Neuropsychologia 64, 218–229.

Symonds, R.M., Zhou, J.W., Cole, S.L., Brace, K.M., Sussman, E.S., 2020. Cognitive resources are distributed among the entire auditory landscape in auditory scene analysis. Psychophysiology 57 (2), e13487. https://doi.org/10.1111/psyp.13487.

Taaseh, N., Yaron, A., Nelken, I., 2011. Stimulus-specific adaptation and deviance detection in the rat auditory cortex. PLoS One 6 (8), e23369. https://doi.org/10.1371/journal.pone.0023369.

Tiitinen, H., May, P., Reinikainen, K., Näätänen, R., 1994. Attentive novelty detection in humans is governed by pre-attentive sensory memory. Nature 372 (6501), 90–92.

Ulanovsky, N., Las, L., Nelken, I., 2003. Processing of low-probability sounds by cortical neurons. Nat. Neurosci. 6 (4), 391–398.

Ulanovsky, N., Las, L., Farkas, D., Nelken, I., 2004. Multiple time scales of adaptation in auditory cortex neurons. J. Neurosci. 24 (46), 10440–10453.

Winkler, I., Schröger, E., 2015. Auditory perceptual objects as generative models: setting the stage for communication by sound. Brain Lang. 148, 1–22.

Winkler, I., Denham, S.L., Nelken, I., 2009. Modeling the auditory scene: predictive regularity representations and perceptual objects. Trends Cogn. Sci. 13 (12), 532–540.

Zion Golumbic, E.M., Ding, N., Bickel, S., Lakatos, P., Schevon, C.A., McKhann, G.M., et al., 2013. Mechanisms underlying selective neuronal tracking of attended speech at a "Cocktail Party". Neuron 77, 980–991.

Zündorf, I.C., Lewald, J., Karnath, H.-O., 2013. Neural correlates of sound localization in complex acoustic environments. PLoS One 8, e64259.

CHAPTER 3

Neural adaptation and forward masking along the auditory pathway

3.1 Introduction

For studying the mismatch negativity (MMN), the effect of previous stimuli, i.e., forward masking, is important as a putative cause. Here we describe in some detail the effects of auditory adaptation from auditory nerve to cortex, and as reflected in forward masking and temporal modulation transfer functions.

Historically, in auditory neuroscience the name adaptation has been given to many phenomena. The most common was initially for the per-stimulatory decrease of single-unit firing rate over the first few hundred milliseconds of a long sound burst (Galambos and Davis, 1943). Per-stimulatory firing rate adaptation in the auditory periphery can be the result of time-dependent changes in the mechano-transduction, in the release of transmitter in the inner hair cells, and in the postsynaptic receptor mechanisms of auditory nerve fibers. Based largely on my descriptions in Eggermont (2015a,b), we will explore this in subcortical structures and auditory cortex.

One of the first electrophysiological recordings of equilibration (adaptation) in the neurophonic response—the compound phase-locked neural responses to the tone frequency—of the cat's auditory nerve was provided by Derbyshire and Davis (1935): "During continued stimulation, the [compound] action potentials of the nerve show a progressive diminution in size (equilibration). [...] There is a 'fast' equilibration complete within 2 seconds or less. [...] Recovery requires about 30 seconds." In a subsequent paper they showed results from single units, presumably from auditory nerve fibers but likely originating from cell bodies in the cochlear nucleus, Galambos and Davis (1943, 1948) and used the name [firing] rate adaptation. In response to tone bursts of constant sound level, the auditory-nerve neurophonic as well as the firing rate from cochlear nucleus units was maximum at onset, followed

Brain Responses to Auditory Mismatch and Novelty Detection
https://doi.org/10.1016/B978-0-443-15548-2.00003-X

by an adaptation to a quasisteady value within about 150 ms. This may involve several processes with different rates of adaptation (Harris and Dallos, 1979).

3.2 Electrophysiology of adaptation and forward masking

3.2.1 Auditory nerve

Forward masking of an auditory nerve fiber's (ANF's) response is generally defined as a reduction in the magnitude of the probe-evoked firing rate caused by a preceding masking stimulus. Young and Sachs (1973) measured the discharge rate of cat single ANFs to tone bursts after exposure to long-duration tones. Both the exposure tone and the test-tone bursts were at the fiber's characteristic frequency (CF). Following exposure, the discharge rate to the test-tone bursts was transiently depressed. For moderate exposures (180 s or less in duration and intensity less than 80-dB SPL), the recovery of discharge rate to the preexposure level was well described by a single decaying exponential. The time constant of the recovery increased as the level or duration of the exposure increased, and decreased as the level of the probe-tone bursts was increased. For all exposure and test conditions employed, time constants were in the range of 1–30 s.

In response to short tone-pip trains presented at 65-dB SPL, the amplitude and latency of the round window-recorded compound action potential (CAP) in the guinea pig reaches a steady-state value after about 5 stimuli (Eggermont and Spoor, 1973a). This steady-state value, expressed in percent of the compound action potential (CAP) amplitude to the first tone pip, depends on the stimulus level, tone-pip duration, and interstimulus interval (ISI). The steady-state value is generally reached in an exponential way. For ISIs larger than 100 ms there is typically no decrement for tone-pip durations <40 ms. The adaptation properties of the CAP (i.e., those depending on the ISI) measured in this way are identical to those for single-nerve-fiber firings. Goldstein and Kiang (1958) proposed that the CAP reflects the properties of the nerve-fiber population by means of a weighted sum of the single-nerve-fiber contributions.

Eggermont and Odenthal (1974) compared the dependency of the amplitude of the CAP on the ISI between successive tone pips for humans and guinea pigs. Recordings in humans were made from the promontory of the cochlea with a needle electrode inserted through the eardrum (under local anesthesia) and using an identical stimulation paradigm as in guinea pigs (Eggermont and Spoor, 1973b). Surprisingly, the ISI value for which depression of the CAP amplitude no longer occurred was a factor 4 larger than in the guinea pig (Fig. 3.1). Because peripheral adaptation is likely based on both pre– and postsynaptic mechanisms, the rather large difference in the recovery time constants of the adaptation process between human and guinea pig must, therefore, be related to structural or reaction-kinetic differences or both in their cochlear synapses and/or differences in neural synchronization that underlie the CAP.

3.2.2 A peripheral adaptation model

Early on (Eggermont, 1975, 1985) and updated in Eggermont (2015b), I proposed a model for auditory adaptation based on linear stochastic Markov models that are widely studied for birth and death processes and used this as a vehicle to review single–unit and CAP studies as well as various models

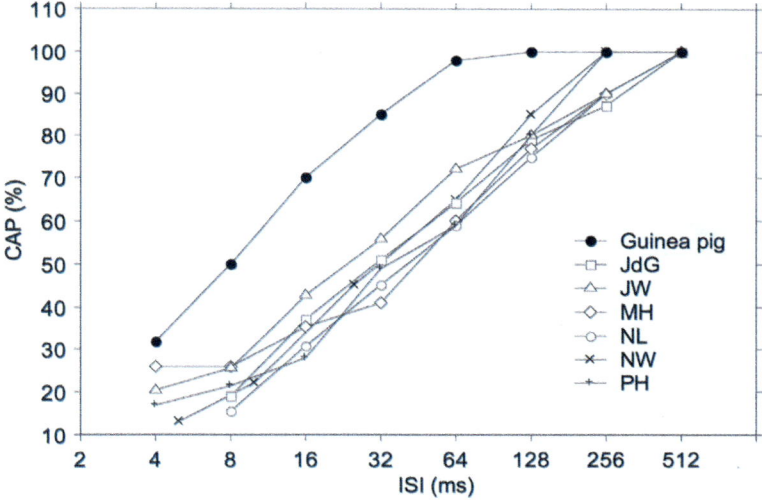

Fig. 3.1 The steady-state value of the CAP amplitude in tone-burst train stimulation as a function of the ISI value at stimulus intensities of 65–70 dB SPL in humans as compared with the guinea pig. The mean data for guinea pigs were obtained for 6-kHz tone bursts. The human data are shown for 6 individual normal-hearing participants for either 4-kHz (MH, NL) or 8-kHz (JdG, JW, NW, and PH) stimulation. The mean CAP amplitude, dependence on the ISI, calculated for the human subjects, differs essentially from the values found in the guinea pig at the same sound level above threshold. The half-value time (50% relative amplitude) is about 4 times longer in humans than in the guinea pig. CAP, compound action potential; ISI, interstimulus interval. *(Based on data from Eggermont, J.J., Odenthal, D.W., 1974. Electrophysiological investigation of the human cochlea recruitment, masking and adaptation. Audiology 13, 1–22. Reprinted from Eggermont, J.J., 2015. Animal models of auditory temporal processing. Int. J. Psychophysiol. 95, 202–215. ©2014, with permission from Elsevier.)*

for adaptation and forward masking. It appears that such a model inherently incorporates a relation between the per-stimulatory adaptation time constant; the poststimulatory recovery time constant; and the ratio between adapted, steady-state, firing rate and onset firing rate. The knowledge of any two of these parameters determines the third one. Although onset firing rates depend on a nonlinear way upon stimulus level, the above-mentioned time constants and the adapted rate to onset rate ratio are largely intensity independent. The per-stimulatory adaptation time constant in guinea pigs was ~30 ms, and the recovery time constant ~50–60 ms, in agreement with the model. Our model (Eggermont, 1975, 1985) in its simplest form is described in Fig. 3.2 left. It is formally equivalent to the model of Buran et al. (2010) shown at the right, which is based on presynaptic changes. In our model, free postsynaptic receptor sites are activated at a very fast rate and then converted with a slower rate λ to an occupied receptor state. The occupied states are liberated with rate μ determined by an enzymatic process. Consider the number of receptor sites that can be occupied by the maximum amount of transmitter that can be released by a stimulus of given intensity to be a, which is a nonlinear, slightly saturating function of stimulus intensity.

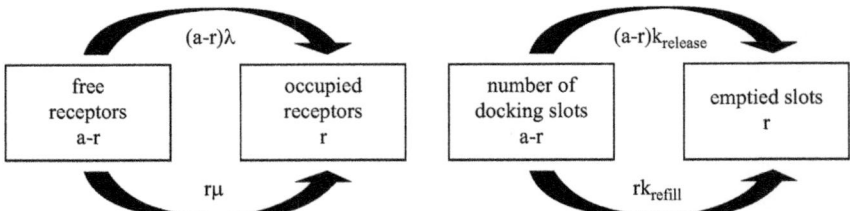

Fig. 3.2 The basic adaptation model in two forms. Left, model from Eggermont (1985): We assume that transmitter released from the hair cell is capable of activating receptor sites at the neural dendrite very fast; they are then converted to an occupied state with a slower rate λ_n which depends on the number of receptors that can be occupied as well as on the number that is still occupied: $\lambda_r = (a - r)\lambda$. Occupied receptor sites are by enzymatic action again liberated with a rate of μ per receptor sites: $\mu_r = r\mu$. In this model it is assumed that transmitter combines with receptor molecules that then undergo a conformational change such as to open Na^+ channels. After enzymatic breaking of the bond it is assumed that the receptor is in an insensitive state from which it recovers by rate μ. Right, model based on data from Buran et al. (2010): They assumed that there are docking slots in the ribbon synapse, when r slots are emptied, then $a - r$ are available for immediate release. The number of releasable docking slots is refilled in proportion to the number of empty slots r. The release rate $k_{release}$ is equivalent to λ, and the refill rate k_{refill} is equivalent to μ. *(Reprinted from Eggermont, J.J., 2015. Auditory Temporal Processing and Its Disorders. Oxford University Press, Oxford.)*

Let r be the number of actually occupied, inactive receptor sites. Then the rate at which the remaining $(a - r)$ free receptor sites are activated is proportional to $(a - r)$ and the rate at which they become occupied is as follows:

$$\lambda_r = (a - r)\,\lambda \tag{3.1}$$

Assume that the liberation rate is proportional to the number of occupied sites r.

$$\mu_r = r\mu \tag{3.2}$$

The dynamics of the occupation and liberation of receptor sites are considered to depend only upon the slower rates λ and μ, but not on the very fast transmitter release rate (which in the auditory system allows phase locking up to $5\,\mathrm{kHz}$):

$$\frac{dP_r(t)}{dt} = \lambda_{r-1}P_{r-1}(t) - (\lambda_r + \mu_r)P_r(t) + \mu_{r+1}P_{r+1}(t); r \geq 1 \tag{3.3}$$

where $P_r(t)$ is the probability that r receptor sites out of a are occupied. This is a forward Chapman-Kolmogorov differential equation (Feller, 1966). A unique solution is given by a binomial distribution:

$$P_r(t) = \binom{a}{r} p_0^r (1 - p_0)^{a-r} \tag{3.4}$$

where

$$p_0 = \frac{\lambda}{\mu + \lambda}\left(1 - e^{-(\lambda+\mu)}\right) \tag{3.5}$$

for large t this results in $p_0 = \lambda/(\mu + \lambda)$. The steady-state distribution is therefore determined by (i) the maximum number of receptor sites that can be occupied, a, which is stimulus dependent; (ii) the occupation rate constant λ for one transmitter quantum and one receptor; and (iii) the liberation rate constant μ for one occupied receptor site. The expected number of occupied receptor sites is presented as follows:

$$M(t) = \sum_{r=0}^{a} rP_r(t) = ap_0 = \frac{a\lambda}{\mu + \lambda}\left(1 - e^{-(\lambda+\mu)t}\right) \tag{3.6}$$

The number of receptor sites that still can be occupied is therefore $a - M(t)$. If we now consider that the size of an excitatory postsynaptic potential (EPSP) is linearly related to the number of occupied receptor sites, and in addition that the probability of firing is linearly related to EPSP size, then we can derive an expression for the firing rate as a function of time after stimulus onset. The expected firing rate will then, in first approximation, be proportional to the expected number of receptor sites that can be occupied at time t after stimulus onset. The maximum firing rate for a given stimulus intensity is determined by a, the relative mean firing rate $R(t)$ will thus be given by:

$$R(t) = 1 - \frac{M(t)}{a} \tag{3.7}$$

and according to Eq. (3.6):

$$R(t) = 1 - \frac{\lambda}{\mu + \lambda}\left(1 - e^{-(\lambda+\mu)t}\right) \tag{3.8}$$

This can be rewritten as follows:

$$R(t) = (1 - R_{ss})\exp(-t/\tau_A) + R_{ss} \tag{3.9}$$

in which the relative steady-state firing rate R_{ss} is given by:

$$R_{ss} = 1 - \frac{\lambda}{\mu + \lambda} = \frac{\mu}{\mu + \lambda} \tag{3.10}$$

and the adaptation time constant by

$$\tau_A = (\lambda + \mu)^{-1} \tag{3.11}$$

Our short-term adaptation model is very similar to that of Smith (1977) but does not take rapid adaptation into account (Yates et al., 1985; Westerman and Smith, 1988). In cat ANFs, Chimento and Schreiner (1991) found that the sum of two exponentials with time constants of 5.5 ms and ~94 ms described the per-stimulatory adaptation. Relkin et al. (1995) equated the forward–masking recovery times of 5-, 10-, and 40-dB tone-evoked CAPs to a 70-dB SPL-masking tone of 167 ms, 240 ms, and 154 ms, respectively, with that of low-SFR (high threshold) single units. In human subjects, long recovery times for the 93-dB SPL masked CAP of ~1.4 s were found (Murnane et al., 1998) in agreement with single-unit data in chinchilla auditory nerve (Relkin et al., 1995).

3.2.3 Brainstem and midbrain
3.2.3.1 Auditory brainstem
Boettcher et al. (1990) used a forward-masking paradigm and found nearly complete recovery at ~250 ms after the masker for primary-like,

primary-like-notch, and chopper units in the cochlear nucleus (CN), similar to ANF data. Results for pauser/buildup- and on-units were more varied. Kaltenbach et al. (1993) characterized forward-masking patterns in the dorsal cochlear nucleus (DCN). They found two patterns of suppression. In the first pattern, the suppression of the probe response became evident immediately following offset of the masker. In the second class, the suppression of the probe response did not become evident until well after offset of the masker. Ingham et al. (2016) examined whether the enhancement of suppression can occur in the ventral cochlear nucleus (VCN). They compared these responses with those from the central nucleus of the inferior colliculus (ICC) using the same preparation. In both nuclei, onset-type neurons showed the greatest amounts of suppression and, in the VCN, these recovered with the fastest time constants (14.1–19.9 ms). Neurons with sustained discharge demonstrated reduced masking with recovery time constants of 27.2–55.6 ms. Ingham et al. (2016) found that "the threshold elevations recorded for most unit types are insufficient to account for the magnitude of forward masking as measured behaviorally, however, onset responders, in both the cochlear nucleus and inferior colliculus demonstrate a wide dynamic range of suppression, similar to that observed in human psychophysics."

In single-unit recordings from superior olivary complex (SOC) neurons in anesthetized Long-Evans rats, Finlayson and Adam (1997) observed rapid exponential decreases with time constants of less than 20 ms in discharge rate during adapting tones. In a typical example, the time constants for excitatory and inhibitory responses were identical at ~14 ms. The recovery from adaptation was exponential with time constant of 106 ± 20.0 ms.

Gao and Berrebi (2016) recorded from single unit in the medial nucleus of the trapezoid body (MNTB) of rats. Using a forward-masking paradigm they showed that effects were greater when masker level and masker duration were increased, when the masker frequency approached the MNTB unit's characteristic frequency, and as the masker-to-probe delay was shortened (Fig. 3.3). Gao et al. (2017) characterized the offset response of single units in the superior para-olivary nucleus—a GABA-ergic cell group located in the SOC of the brainstem—to a forward-masking paradigm. They observed inhibited as well as facilitated response types to forward-masked stimuli. These forward-masking results do not follow the features observed in previously studied auditory nuclei, including the auditory nerve, MNTB, and ICC.

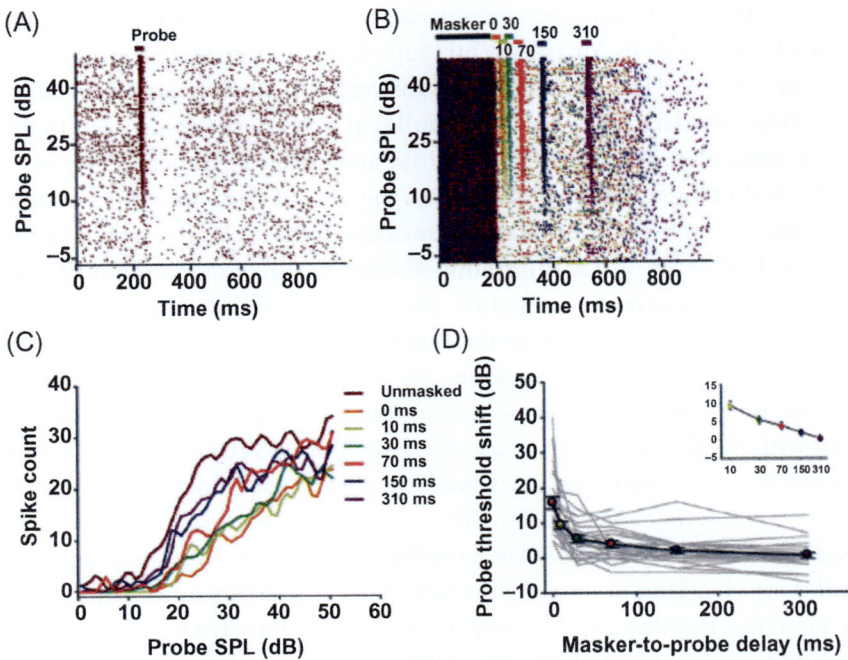

Fig. 3.3 (A) Unmasked responses of a typical MNTB unit (neuron 12-173-3: CF = 9 kHz, threshold = 12.5-dB SPL). (B) Masked responses of the same unit, with the masker preceding the probe by different masker-to-probe delays, represented by specific colors. (C) Probe rate–level functions for this unit with masker-to-probe delays ranging from 0 to 310 ms. (D) Threshold shift plotted over masker-to-probe delays for the sample of 29 recorded neurons. Individual units are represented by *gray lines;* the average (mean ± SEM) probe threshold shift across the population is depicted by the *black line. Colored circles* indicate the various masker-to-probe delays. Inset the recovery of the threshold to the baseline value followed a linear decay in log time with increasing masker-to-probe delays. *(From Gao, F., Berrebi, A.S., 2016. Forward masking in the medial nucleus of the trapezoid body of the rat. Brain Struct. Funct. 221, 2303–2317. With permission from Springer Nature.)*

3.2.3.2 *The auditory midbrain*

By recording local field potentials (LFPs) from the ICC in awake chinchillas, Arehole et al. (1987) found that the average forward-masking time constants ranged from 50 to 90 ms. Using a stimulus paradigm designed to exclude the influence of adaptation below the level of binaural integration in the ICC of urethane-anesthetized guinea pigs Ingham and McAlpine (2004) investigated spike-frequency adaptation of neurons sensitive to interaural phase disparities (IPDs). They fitted exponential decay functions to the response to best IPD steps and found an average (± SD) adaptation time constant of 52.9 ± 26.4 ms. Recovery from adaptation to best IPD steps showed an average time constant of 225.5 ± 210.2 ms. Recovery time constants were not correlated with adaptation time constants. The mean adaptation time constant at stimulus onset (at worst IPD) was 34.8 ± 19.7 ms, similar to the 38.4 ± 22.1 ms recorded to contralateral stimulation alone.

These data suggest that, unlike for auditory nerve fibers, forward masking in the brainstem and midbrain is not a direct result of peripheral neural adaptation; the central pathway apparently modifies the representation in a way that further attenuates the input's response to short probe signals. Nelson et al. (2009) show that much of this transformation is complete by the level of the IC. Probe threshold shifts for single units were not usually caused by a persistent excitatory response to the masker; instead may involve a wide-dynamic range inhibitory mechanism locked to sound offset to explain several key aspects of the data.

3.2.4 Thalamus and auditory cortex

3.2.4.1 Animal data

In cat primary auditory cortex (A1), Brosch and Schreiner (1997) induced forward inhibition at the shortest stimulus onset asynchrony between masker and probe. Recovery from forward inhibition occurred first at the lower- and higher-frequency borders of the masking tuning curve and lasted the longest for frequencies close to the neuron's characteristic frequency. The maximal duration of forward inhibition was measured as the longest period over which reduction of probe responses was observed. It was in the range of 53–430 ms, with an average of 143 ± 71 ms.

Eggermont (1998) investigated the degree of similarity of three different auditory cortical areas in the cat with respect to the coding of various periodic stimuli. Simultaneous local field potentials (LFPs), single- and multiunit recordings in response to these stimuli, were made from primary auditory cortex (A1), anterior auditory field (AAF), and second auditory area (AII) in the cat to address the following questions: (1) Is there, within each cortical area, a difference in the temporal coding of periodic click trains, amplitude-modulated (AM) noise bursts, and AM-tone bursts? (2) Is there a difference in this coding between the three cortical fields? I found that: (1) AM stimuli produced much higher limiting rates than periodic click trains. (2) For periodic click trains and AM noise, the limiting rates were not significantly different for the three areas (Fig. 3.4). However, for AM tones the limiting rates were about a factor 2 lower in AAF compared with the other areas. The BMFs were correlated positively with characteristic frequency in AAF. Thus

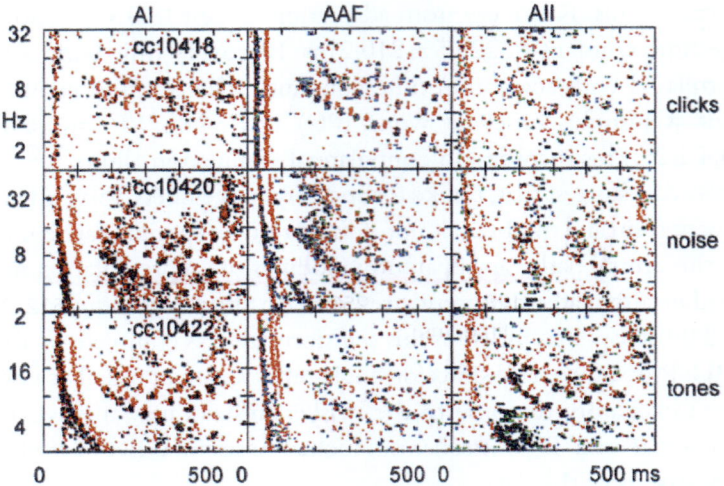

Fig. 3.4 Dot displays for simultaneous LFP and activity in A1, AAF, and AII in response to periodic click trains, and amplitude-modulated noise and tone bursts (2.5 kHz). LFP triggers are indicated by the *orange dots*, single-unit spikes—sorted by spike waveform—by *blue, green, or black dots*. Click-repetition rate (1–32 Hz) and the AM frequencies (2–64 Hz) are represented on a logarithmic axis. First 600 ms after stimulus (click train or AM burst) is shown. Characteristic frequencies (CFs) of the recording sites were 5.2 kHz (A1), 6 kHz (AII), and 6 kHz (AAF). *(From Eggermont, J.J., 1998. Representation of spectral and temporal sound features in three cortical fields of the cat. Similarities outweigh differences. J. Neurophysiol. 80, 2743–2764.)*

the coding of periodic stimuli in these areas was fairly similar with the exception of the very poor representation of AM tones in AAF. This suggests a strong parallel processing organization in these three cortical areas of the cat.

Using in vivo patch-clamp recordings, Wehr and Zador (2005) studied the synaptic mechanisms for forward masking in rat A1. They found that postsynaptic inhibition only contributed for the first 50–100 ms after a stimulus, and because the recovery typically lasted up to 400 ms, synaptic depression must play a dominant role. Thalamic neurons—in the medial geniculate body (MGB)—recovered from forward masking more quickly than cortical neurons. This would allow coding of periodic stimuli up to higher repetition rates or modulation frequencies (Fig. 3.5).

Neuronal stimulus selectivity is shaped by feedforward and recurrent excitatory-inhibitory interactions. In the auditory cortex (ACx), parvalbumin- (PV) and somatostatin-positive (SOM) inhibitory interneurons differentially modulate frequency-dependent responses of excitatory neurons. Responsiveness of neurons in the ACx to sound is also dependent

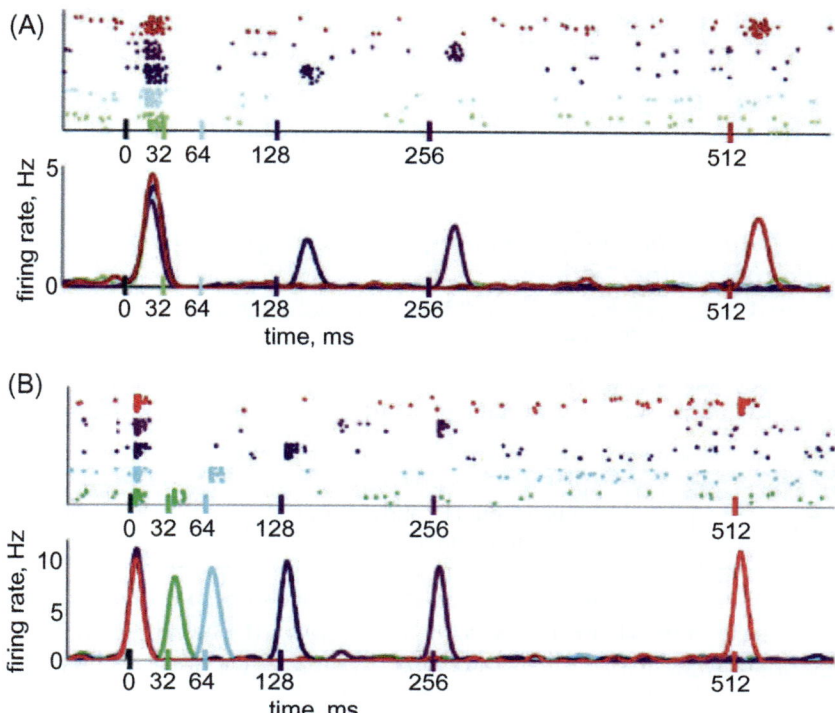

Fig. 3.5 (A) Forward suppression (in auditory cortex) of spiking responses lasts hundreds of milliseconds. Example of single-unit activity recorded in cell-attached mode showing responses to randomly interleaved click pairs (top: rasters, bottom: firing rates, stimuli indicated by ticks on abscissa, *colors* indicate different intervals). The response to the second click was completely suppressed for intervals shorter than 128 ms and progressively recovered for longer intervals. (B) Thalamic neurons recover from forward suppression more quickly than cortical neurons. (A) Example of single-unit activity recorded in cell-attached mode from a thalamic neuron, which responded strongly at an interval of 32 ms (an interval at which cortical neurons rarely showed any response at all, cf. A). *(Reprinted from Wehr, M., Zador, A.M., 2005. Synaptic mechanisms of forward suppression in rat auditory cortex. Neuron 47, 437–445. © 2005, with permission from Elsevier.)*

on stimulus history. Natan et al. (2017) found "that the inhibitory effects of SOMs and PVs diverged as a function of adaptation to temporal repetition of tones. Prior to adaptation, suppressing either SOM or PV inhibition drove both increases and decreases in excitatory spiking activity. After adaptation, suppressing SOM activity caused predominantly disinhibitory effects, whereas suppressing PV activity still evoked bi-directional changes. SOM, but not PV-driven inhibition, dynamically modulated frequency tuning with adaptation. Unlike PV-driven inhibition, SOM-driven inhibition elicited gain-like increases in frequency tuning reflective of adaptation." Natan et al. (2017) suggested that distinct cortical interneurons differentially shape tuning to sensory stimuli across the neuronal receptive field, altering frequency selectivity of excitatory neurons during adaptation.

3.2.4.2 Adaptation reflected in responses to amplitude modulation along the auditory pathway

As stated in Eggermont (2015a), "The temporal representation of sound in the auditory nervous system manifests itself in two forms; the first is based on the phase locking of neuronal firings to the period of pure tones or to the fine structure of complex sounds. The second form is the locking to the, slower, amplitude-modulation in a complex stimulus; a phenomenon that I will refer to as envelope locking. The degree of envelope locking can be expressed in the temporal modulation transfer function (tMTF). In psychoacoustic studies, the tMTF is generally expressed as a detection threshold for modulation depth, whereas in electrophysiological studies it is given generally as number of synchronized spikes, or normalized as vector strength, as a function of modulation frequency. The tMTF is a measure of the temporal acuity of the auditory system up to the point of measurement."

The relation between the rate of adaptation and tMTFs for auditory cortex units has been derived in Eggermont (1999). For repetitive stimulation, generally with short-duration stimuli such as clicks, the cumulative effects of incomplete per-stimulatory adaptation and incomplete recovery have to be considered. Under the assumption that the adaptation and recovery after the second click in a train are scaled versions of those after the first click, i.e., the adaptation starts from the level given by Eq. (3.9) (see Section 3.2.2) instead of the unadapted value 1, the onset firing rate for click 3 is given by:

$$r_{on}(2\Delta t) = r_{on}(\Delta t)^2 \tag{3.12}$$

Thus one can write this cumulative effect for stimulation with a click train with interstimulus interval $= \Delta t$ and consisting of $N+1$ (≥ 2) clicks (N depends on the click-repetition rate for fixed duration click trains), as:

$$r_{on}(N\Delta t) = [1 - d \; exp\left(-\Delta t / \tau_{recov}\right)]^N \tag{3.13}$$

where $\tau_{recov} = \mu^{-1}$

This is an adaptation in which the effect of subsequent clicks becomes progressively less. Facilitation can be introduced into this model. It has been considered additive in modeling adaptation for visual cortex cells

(Varela et al., 1997), but I found a multiplicative update, analogous to that for depression, to provide much better results (Eggermont, 1999).

For neurons in the ventral cochlear nucleus. Sayles et al. (2013) presented unmodulated tones to reflect the firing rate adaptation, and sinusoidal amplitude-modulated tones with systematically varying modulation depth from 0% to 100% at a range of modulation frequencies, to anesthetized guinea pig to assess the tMTF. Neurons showed either low-pass or band-pass tMTFs, with the proportion of band-pass responses increasing with increasing sound level. Fig. 3.6 illustrates the correspondence between the adaptation functions and the tMTFs.

Malone et al. (2015) recorded responses to pairs of sinusoidally amplitude modulated (SAM) tones in the auditory cortex of awake squirrel monkeys and showed that the prior presentation of the SAM masker elicited persistent and tuned suppression of the firing rate to subsequent SAM signals. Population averages of these effects are compatible with adaptation in broadly tuned modulation channels. In contrast, modulation context had little effect on the synchrony of the cortical representation of the second SAM stimuli and the tuning of such effects did not match that observed for firing rate. This suggests that the temporal representation of modulated signals is more robust to changes in stimulus context than representations based on average firing rate.

In Fig. 3.7, I compiled tMTFs from auditory nerve to auditory cortex in anesthetized animals. The tMTFs are expressed as modulation gain of the frequency response function of a putative temporal filter. It appears that tMTFs are very much like those for ANFs up to the lateral superior olive in the localization pathway—VCN bushy cells, superior olivary complex, medial nucleus of the trapezoid body, dorsal nucleus of the lateral lemniscus, but may increasingly become tuned to lower BMFs for the identification pathway (stellate cells and octopus cells in the VCN, ventral nucleus of the trapezoid body, intermediate and ventral nuclei of the lateral lemniscus) and all converging on the central IC. Enhancements of modulation gain are found in the CN stellate cells and in the ICC, whereas an overall reduction is seen in auditory cortex (A1 and AAF).

3.2.4.3 Human studies

Using magnetoencephalography (MEG) recordings in humans, Nishimura et al. (2003) found nonmonotonic recovery of the N1m component in auditory cortex following a masker; the minimum N1m amplitude was observed at a signal delay of 40 ms. As the signal delay decreased from 40 ms, the N1m

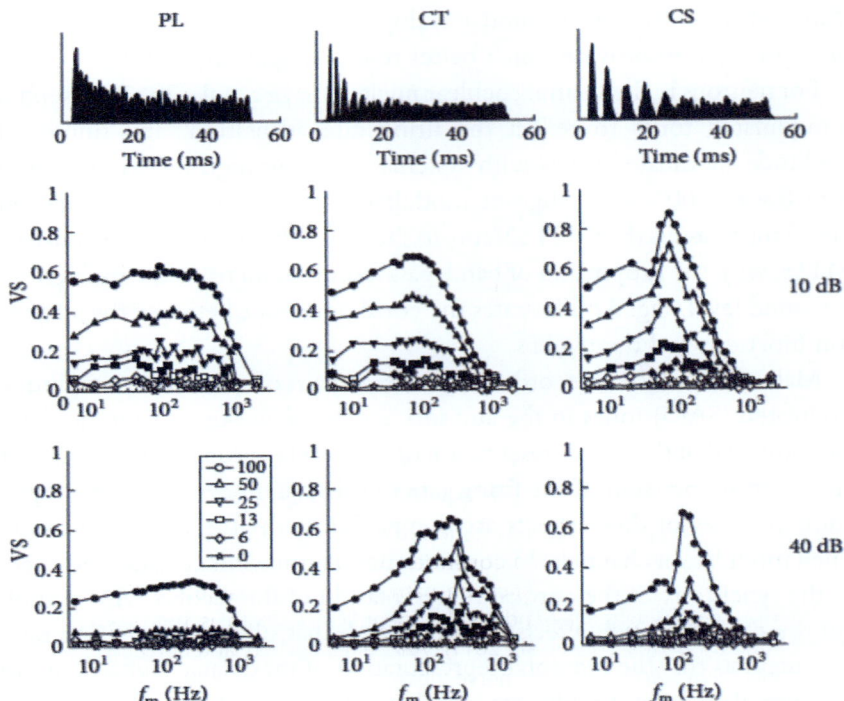

Fig. 3.6 tMTFs as a function of modulation depth (m), sound level, and unit type. The PL unit has low-pass MTFs. The CT unit becomes more band-pass at the higher sound level. The CS unit is band-pass tuned at both low and high sound levels. BMF does not change with modulation depth. Top row, peri-stimulus time histograms in response to 50-ms duration unmodulated BF tones. Middle row, tMTFs at 10 dB above pure-tone threshold. Bottom row, tMTFs at 40 dB above pure-tone threshold. *fm*, modulation frequency; *VS*, vector strength. Left column, PL unit (BF = 7.87 kHz). Middle column, CT unit (BF = 7.51 kHz). Right column, CS unit (BF = 8.15 kHz). Modulation depth (m, %) is shown in the inset. *Filled symbols* are significant (P < .001, Rayleigh statistic > 13.8), and *open symbols* are nonsignificant. *CS*, sustained chopper neuron; *CT*, transient chopper neuron; *PL*, primary-like neuron. (From Sayles, M., Füllgrabe, C., Winter, I.M., 2013. Neurometric amplitude-modulation detection threshold in the guinea-pig ventral cochlear nucleus. J. Physiol. (Lond.) 591, 3401–3419. © 2013 The Physiological Society, with permission from John Wiley and Sons.)

amplitude increased although the masking increased. This suggested to them that the growth of the N1m amplitude largely depends on temporal integration at signal delays below 40 ms.

Lanting et al. (2013) investigated the temporal properties of the long-latency CAEPs in humans. They found that the P2 component of the CAEP was more strongly affected by adaptation than the N1, suggesting that the

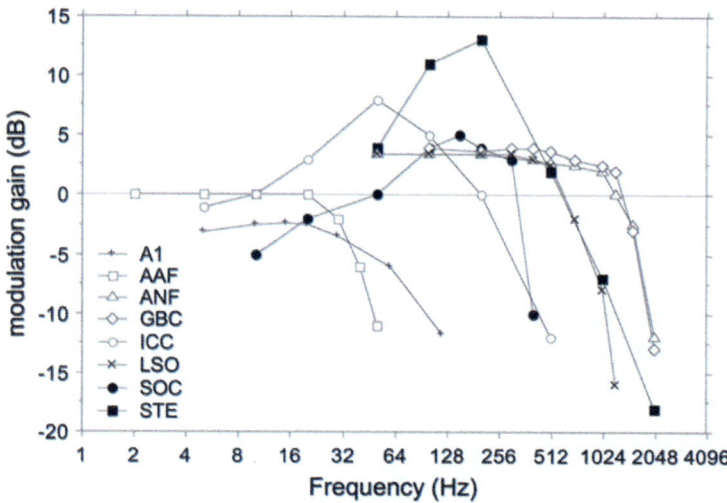

Fig. 3.7 tMTFs obtained from various stations along the auditory pathway. Limitations in the capacity to follow high AM frequencies start to occur for the LSO units in the localization pathway. In the identification pathway the gradual loss of sensitivity to high AM frequencies when ascending the identification path is noted. *A1*, awake primary auditory cortex (Yin et al., 2011); *AAF*, anesthetized anterior auditory field (Eggermont, 1998); *ANF*, auditory nerve fiber (Joris and Yin, 1992); *GBC*, globular bushy cell (Joris and Yin, 1998); *ICC*, central nucleus of the inferior colliculus (Rees and Møller, 1983); *LSO*, lateral superior olive (Joris and Yin, 1998); *SOC*, superior olivary complex, sustained cells (Kuwada and Batra, 1999); *STE*, cochlear nucleus stellate cells (Frisina et al., 1985). *(From Eggermont, J.J., 2015. Animal models of auditory temporal processing. Int. J. Psychophysiol. 95, 202–215. © 2014, with permission from Elsevier.)*

two deflections originate from different cortical generators (see also Ponton et al., 2002). They described that: "In contrast to our results, Eggermont (2000) found that adaptation in single- and multiunit spiking activity in cat primary auditory cortex decreased with increasing SOA (for short adapter durations) and ISI (for long durations) but depended little on adapter duration. This suggested a strong contribution to adaptation by the transient adapter." This led Lanting et al. (2013) to conclude that the difference between their results and ours showing persistence of adaptation may be due to the use of anesthesia in the cat. They analyzed their data using our adaptation model (Fig. 3.2) extended for auditory cortex with respect to the values of the transmitter depletion rate λ and the recovery rate μ. Lanting et al. (2013) found that the exponential fits of the adaptation functions with ISI indicated that, in human auditory cortex, the adaptation time constant was $(\mu + \lambda)^{-1} = 125.9$ ms and recovered with a time constant of $\mu^{-1} = 1271$ ms. These values are much longer than those observed in our anesthetized cats. Note that this was also the case in our electrochleographic studies of CAPs in humans compared to the CAPs recorded in anesthetized guinea pigs (Fig. 3.1). However, anesthesia is not considered to affect auditory nerve responses, suggesting a pronounced species difference.

3.3 A putative role of neural adaptation in explaining the mismatch negativity

3.3.1 An argument for adaptation

May and Tiitinen (2010) studied the possibility that the MMN elicited by stimulus change might be generated by so-called unadapted neuronal activity. This possibility contrasts with accounts relying on a memory-based explanation for MMN. May and Tiitinen (2010) proposed that "the MMN is, in essence, a latency- and amplitude-modulated expression of the auditory N1 response, generated by fresh-afferent activity of cortical neurons that are under nonuniform levels of adaptation." They contrast this to a memory-based explanation (Fig. 3.8). They continued: "the physiologically based adaptation model aimed at answering the following three questions that are not explained by the memory-based interpretation: 1) What is the physiological manifestation of the sensory memory trace underlying MMN? 2) How does the comparison process work? 3) How is the MMN generated? Furthermore, the rationale for extracting the MMN through subtraction of the response to the frequent from the one to the odd-ball is unfounded. Notably, there is no conclusive way to separate the MMN from the 'exogenous' N1 component since it may also have changed in response to the odd-ball. In addition, there is no convincing neurophysiological evidence for MMN generation according to the memory-based interpretation." May and Tiitinen (2010) conclude that "the memory-based interpretation, therefore, rules out an explanation of the MMN based on the known properties of auditory cortex, namely, adaptation and spatial representation of sound structure. Importantly, the rejection of the adaptation model and the validity of the memory-based model are based on circular argumentation." Furthermore, the adaptation ('fresh-afferents') model suggests that the MMN is part of a modulated N1 response and offers a physiological explanation for the memory trace and comparison process underlying MMN.

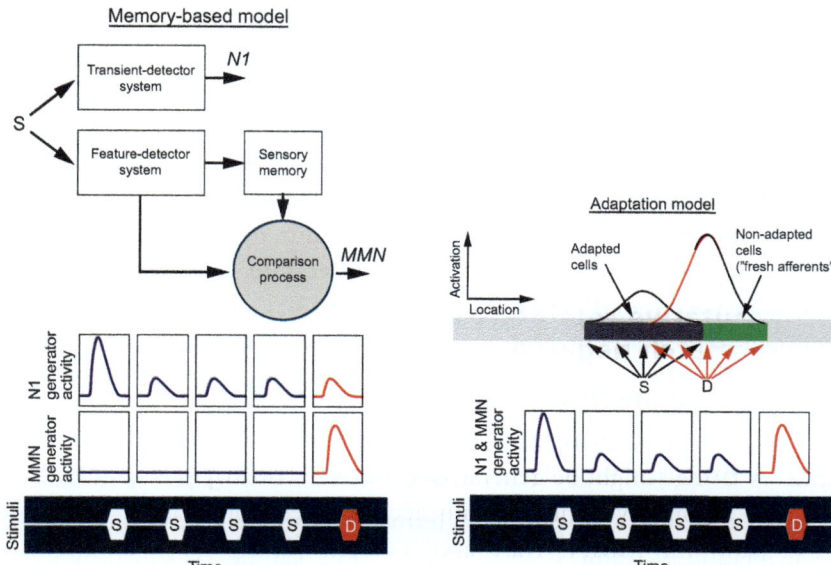

Fig. 3.8 Schematic representations of the memory-based and adaptation model of the MMN. Left: In the memory-based model, a stimulus is analyzed by an N1-generating transient-detector system and a separate, MMN-generating system that first analyzes the stimulus for its features (frequency, intensity, duration, etc.). The result is deposited in sensory memory. A comparison process compares the features of incoming stimuli with representations of past stimuli in the sensory memory store, and, when the two differ, an MMN response is generated. Also shown is the beginning of a stimulus sequence, four standards (S) followed by a deviant (D), and the event-related responses produced by the separate N1 and MMN generators to the stimuli. The N1 is largest for the first standard. In contrast, the MMN generator reacts only when the deviant follows an already established memory trace for the standards *(red curve)*. Therefore it produces no response to any of the standards, including the first stimulus in the sequence. Right: In the adaptation model, the standards and deviants activate overlapping neural populations. The repetitive standard leads to cells tuned to the standard to become adapted. When the deviant is presented, nonadapted cells—"fresh afferents"—contribute to an enhanced response. Being in a nonadapted state, the MMN generator responds vigorously to the first standard of the sequence. It also produces attenuated responses to the subsequent standards. In this model, the N1 and MMN are generated by the same neural populations, and the MMN is, essentially, an enhanced N1 response. *(From May, P.J.C., Tiitinen, H., 2010. Mismatch negativity (MMN), the deviance-elicited auditory deflection, explained. Psychophysiology 47, 66–122. © 2009 Society for Psychophysiological Research, with permission from John Wiley and Sons.)*

May and Tiitinen (2013) then introduced a computational model that structurally copies the gross anatomy of the auditory cortex and where the synapses are modulated by short-term synaptic depression (STSD). In further simulations (May et al., 2015), the model replicated single-unit forward masking and SSA (Ulanovsky et al., 2003, 2004; Chapter 4) as well as forward facilitation (Brosch et al., 1999; Brosch and Schreiner, 2000). Further, the model reproduced repetition suppression of the N1 as well as several types of MMN. The model was able to recreate such a wide variety of phenomena was found to be a consequence of STSD. Removing STSD also abolished SSA, masking, facilitation, combination sensitivity, N1 adaptation, and the MMN.

3.3.2 A broader look from a predictive coding perspective

Kiebel et al. (2007) showed that in terms of causal architectures (Fig. 1.12), the adaptation of feature-selective populations can be explained by purely intrinsic neuronal mechanisms, whereas hierarchical inference using predictive coding invokes recurrent interactions that are mediated by extrinsic forward and backward connections. Under the predictive coding hypothesis, differences between responses to predictable and unpredictable stimuli would be manifest as changes in both intrinsic and extrinsic coupling. In Fig. 3.9, the posterior means of the coupling gains of the two best models (Adaptation and Forward-Backward Inhibition, FBI, combining intrinsic in A1 and extrinsic mechanisms) are shown for changes that have a posterior probability of 95% or more. For the FBI model, this probability is more than 99% for the two intrinsic modulations, two of the forward connections, and the left backward connection. For the Adaptation model, this is true for the intrinsic adaptation of four sources (both A1 and STG). Kiebel et al. (2007) found that models based solely on changes in extrinsic connectivity are not the best. Rather, one should augment these models by intrinsic modulations as done for the first two A1 areas. Kiebel et al. (2007) also found that a pure 'local adaptation' model attained the highest model evidence. The local adaptation model explains the data well, so models with more (e.g., extrinsic) parameters must counter their increased complexity with a better fit. In the case of the combined FBI model, the increased complexity and accuracy were balanced so that there was no real difference between the simple adaptation and the more complicated (FBI) model.

Lieder et al. (2013) introduced a mathematical framework wherein competing ideas about the computational quantities indexed by MMN responses

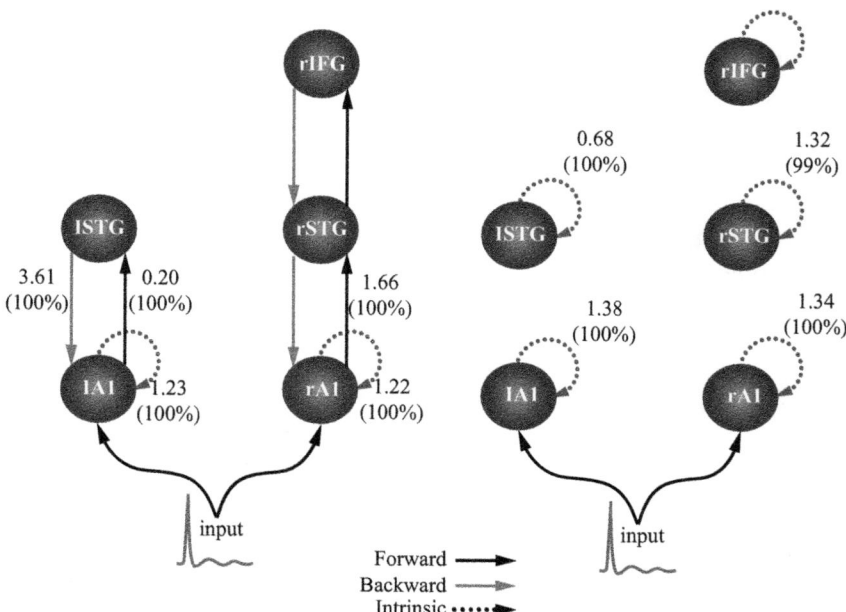

Fig. 3.9 Mismatch data: These two graphs show the posterior means of gains with a conditional probability of greater than 95% of being present. Left: FBI model. Right: adaptation model (A). *(Reprinted from Kiebel, S.J., Garrido, M.I., Friston, K.J., 2007. Dynamic causal modelling of evoked responses: the role of intrinsic connections. Neuroimage 36 (2), 332–345. © 2007, with permission from Elsevier.)*

can be formalized and tested against single-trial EEG data. This framework was applied to five major theories of the MMN, comparing their ability to explain trial-by-trial changes in MMN amplitude. In some detail, these five major hypotheses are: (1) *Change Detection Hypothesis*: The MMN reflects the detection of a local physical change in the sensory input (Schröger and Winkler, 1995). (2) *Adaptation Hypothesis*: The MMN reflects the difference in stimulus-evoked activity between adapted and nonadapted sensory neurons (May and Tiitinen, 2010). (3) *Model Adjustment Hypothesis*: The auditory cortex maintains a model of the acoustic environment, and stimulus-induced updates of this model are indexed by the MMN (Winkler et al., 1996). (4) *Novelty Detection Hypothesis*: The MMN reflects the degree to which the current event is surprising (novel; Tiitinen et al., 1994). An event is surprising, if its occurrence violates a (probabilistic) prediction. Surprise is different from change: when a change occurs predictably in a given context, its absence will be more surprising than its presence. Surprise is an undirected quantity; this distinguishes it from prediction error.

(5) *Prediction Error Hypothesis*: The cortex implements approximate Bayesian inference using predictive coding. The MMN reflects the neural activity encoding the prediction errors that drive this process, i.e., differences between actual and predicted inputs (Friston, 2005; Garrido et al., 2009; Chapter 1). In contrast to surprise, a prediction error indicates the direction in which the event deviated from the brain's prediction. Models based on the free-energy principle (Friston, 2009; Chapter 1) link the MMN to the neuronal encoding of posterior beliefs that is postulated by the Bayesian brain hypothesis. According to this hypothesis, the brain represents probabilistic beliefs, and updates them in an (approximately) Bayesian fashion (Lieder et al., 2013).

Three of these theories (predictive coding, model adjustment, and novelty detection) were formalized by linking the MMN to different manifestations of the same computational mechanism: approximate Bayesian inference based on minimizing prediction errors (Chapter 1). Lieder et al. (2013) thereby propose a unifying view on three distinct theories of the MMN. The relative plausibility of each theory was assessed against empirical single-trial MMN amplitudes acquired from eight healthy volunteers in a roving oddball experiment. Lieder et al. (2013) suggested that models based on minimizing prediction errors provided more plausible explanations of trial-by-trial changes in MMN amplitude than models representing the two more traditional theories (change detection and adaptation).

3.3.3 A rebuttal

In a follow-up study, May (2021) provided a reminder that the issue of MMN generation is far from settled, and that an alternative model in terms of adaptation continues to be relevant. Simulations of auditory cortex using the May and Tiitinen (2010) adaptation model show that locally operating STDS accounts both for adaptation due to stimulus repetition and for MMN responses. This happens even in cases where adaptation has been ruled out as an explanation of the MMN (e.g., in the stimulus omission paradigm and the multistandard control paradigm; Chapter 2). Results are presented in Fig. 3.10, which shows that the adaptation model produces a wide variety of MMNs which have been used as arguments against the adaptation hypothesis. It is beyond the current scope to explore in detail what is generating the MMN in each experiment. As explained in May et al. (2015), SSA on the single-unit level is only part of the explanation, with tuning to stimulus features also playing a major role. Omission responses are to

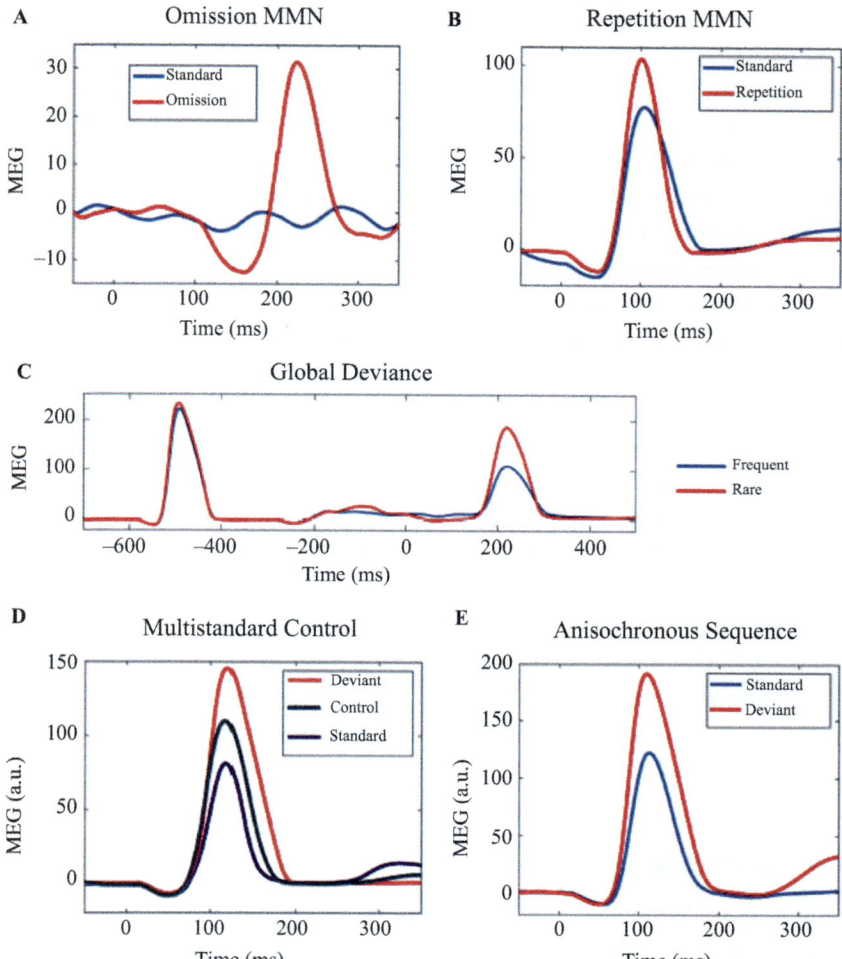

Fig. 3.10 Simulation results of the May and Tiitinen (2010) model. (A) Standard stimuli presented at a fast rate *(blue curve)* elicit no discernible response, whereas the occasional stimulus omission *(red curve)* results in a prominent MMN. (B) Occasionally repeating a tone *(red)* in a sequence of alternating tones *(blue)* results in an MMN. (C) The *blue curve* is the response to a sequence xxxxX of five tones presented as a global standard, and the *red curve* is the response elicited by the same xxxxX as an infrequent global deviant. When the sequence is a global deviant, the ending of the sequence elicits a much stronger response than when it is a global standard. Zero time indicates the onset of the fifth tone. (D) In the classic oddball paradigm, frequency deviants *(red)* elicit a stronger response than the standards *(blue)*. The response to the deviants is also stronger than the response elicited by the same deviants when these are presented as part of a random sequence of tones, in the so-called multistandard control condition *(black)*. (E) Standards *(blue)* and deviants *(red)* were presented as a series of anisochronous stimuli where the SOI varied randomly. *(From May, P.J.C., 2021. The adaptation model offers a challenge for the predictive coding account of mismatch negativity. Front. Hum. Neurosci. 15, 721574. doi: https://doi.org/10.3389/fnhum.2021.721574. Open Access.)*

be expected as resonance effects, given that interacting excitatory and inhibitory neural populations are dynamically equivalent to driven oscillators with damping (May and Tiitinen, 2010; Hajizadeh et al., 2019, 2021). In addition, the omission response could be enhanced or even caused by high-pass filtering acting on the sudden, omission-related drop in the sustained activity which is elicited by fast-rate stimulation (May and Tiitinen, 2010). As for the multistandard control results, these arise from the cortical columns being interconnected rather than acting as isolated frequency channels. Therefore the response of each column depends not only on the stimulation rate (which would be required for the multistandard control condition to be valid), but it is also modulated by lateral connections and the pattern of synaptic depression over the entire network, as established by the previous stimulation (May and Tiitinen, 2010).

May (2021) concluded: "It is too early to discard the adaptation model as an explanation of deviance detection as revealed in the MMN. Its modern version is able to reproduce a wide variety of MMN responses as well as intracortical results. PC as currently formulated provides a mostly conceptual explanation, and therefore it is difficult to contrast the relative successes of these models. Whilst the adaptation model is incomplete and it lacks the normative power and elegance of predictive coding, there are challenges ahead before the PC can match the adaptation model on a mechanistic level."

3.4 Summary

Early on, the MMN had been explained as resulting from adaptation (repetition suppression) of the N1 response to repeated frequents and leaving the unadapted neurons to respond to the oddball. Several alternative explanations have been based on unfulfilled expectations and were formalized in predictive coding. However, convincing arguments continue to be presented for a role of adaptation. Therefore we review the adaptation mechanism, the role it plays in forward masking, and repetition suppression—as also manifested in temporal modulation transfer functions. Dynamic causal modeling based on independent intrinsic changes in various cortical areas (A1, STG, and IFG) and compared to the standard MMN generation model—including bidirectional and lateral excitatory connections coupled with intrinsic inhibitory ones—showed comparable predictive success. This was once more illustrated by recent simulations based on the May and Tiitinen (2010) model.

References

Arehole, S., Salvi, R.J., Saunders, S.S., Hamernik, R.P., 1987. Evoked response 'forward masking' functions in chinchillas. Hear. Res. 30, 23–32.

Boettcher, F.A., Salvi, R.J., Saunders, S.S., 1990. Recovery from short-term adaptation in single neurons in the cochlear nucleus. Hear. Res. 48, 125–144.

Brosch, M., Schreiner, C.E., 1997. Time course of forward masking tuning curves in cat primary auditory cortex. J. Neurophysiol. 77, 923–943.

Brosch, M., Schreiner, C.E., 2000. Sequence sensitivity of neurons in cat primary auditory cortex. Cereb. Cortex 10, 1155–1167. https://doi.org/10.1093/cercor/10.12.1155.

Brosch, M., Schulz, A., Scheich, H., 1999. Processing of sound sequences in macaque auditory cortex: response enhancement. J. Neurophysiol. 82, 1542–1559. https://doi.org/10.1152/jn.1999.82.3.1542.

Buran, B.N., Strenzke, N., Neef, A., Gundelfinger, E.D., Moser, T., Liberman, M.C., 2010. Onset coding is degraded in auditory nerve fibers from mutant mice lacking synaptic ribbons. J. Neurosci. 30, 7587–7597.

Chimento, T.C., Schreiner, C.E., 1991. Adaptation and recovery from adaptation in single fiber responses of the cat auditory nerve. J. Acoust. Soc. Am. 90 (1), 263–273.

Derbyshire, A.J., Davis, H., 1935. The action potentials of the auditory nerve. Am. J. Physiol. 113, 476–504.

Eggermont, J.J., 1975. Cochlear adaptation: a theoretical description. Biol. Cybern. 19, 181–189.

Eggermont, J.J., 1985. Peripheral auditory adaptation and fatigue: a model oriented review. Hear. Res. 18, 57–71.

Eggermont, J.J., 1998. Representation of spectral and temporal sound features in three cortical fields of the cat. Similarities outweigh differences. J. Neurophysiol. 80, 2743–2764.

Eggermont, J.J., 1999. The magnitude and phase of temporal modulation transfer functions in cat auditory cortex. J. Neurosci. 19, 2780–2788.

Eggermont, J.J., 2000. Neural responses in primary auditory cortex mimic psychophysical, across frequency-channel, gap-detection thresholds. J. Neurophysiol. 84, 1453–1463.

Eggermont, J.J., 2015a. Animal models of auditory temporal processing. Int. J. Psychophysiol. 95, 202–215.

Eggermont, J.J., 2015b. Auditory Temporal Processing and Its Disorders. Oxford University Press, Oxford.

Eggermont, J.J., Odenthal, D.W., 1974. Electrophysiological investigation of the human cochlea recruitment, masking and adaptation. Audiology 13, 1–22.

Eggermont, J.J., Spoor, A., 1973a. Cochlear adaptation in guinea pigs. Int. J. Audiol. 12 (4), 193–220.

Eggermont, J.J., Spoor, A., 1973b. Masking of action potentials in the guinea pig cochlea, its relation to adaptation. Int. J. Audiol. 12 (4), 221–241.

Feller, W., 1966. An Introduction to Probability Theory and its Applications. Wiley, New York.

Finlayson, P.G., Adam, T.J., 1997. Excitatory and inhibitory response adaptation in the superior olive complex affects binaural acoustic processing. Hear. Res. 103, 1–18.

Frisina, R.D., Smith, R.L., Chamberlain, S.C., 1985. Differential encoding of rapid changes in sound amplitude by second-order auditory neurons. Exp. Brain Res. 60, 417–442.

Friston, K., 2005. A theory of cortical responses. Philos. Trans. R. Soc. B 360, 815–836. https://doi.org/10.1098/rstb.2005.1622.

Friston, K., 2009. The free-energy principle: a rough guide to the brain? Trends Cogn. Sci. 13 (7), 293–301.

Galambos, R., Davis, H., 1943. The response of single auditory-nerve fibers to acoustic simulation. J. Neurophysiol. 6, 39–57.

Galambos, R., Davis, H., 1948. Action potentials from single auditory-nerve fibers? Science 108, 513.

Gao, F., Berrebi, A.S., 2016. Forward masking in the medial nucleus of the trapezoid body of the rat. Brain Struct. Funct. 221, 2303–2317.

Gao, F., Kadner, A., Felix II, R.A., Chen, L., Berrebi, A.S., 2017. Forward masking in the superior paraolivary nucleus of the rat. Brain Struct. Funct. 222, 365–379.

Garrido, M.I., Kilner, J.M., Stephan, K.E., Friston, K.J., 2009. The mismatch negativity: a review of underlying mechanisms. Clin. Neurophysiol. 120 (3), 453–463. https://doi.org/10.1016/j.clinph.2008.11.029.

Goldstein, M.H., Kiang, N.Y.S., 1958. Synchrony of neural activity in electric responses evoked by transient acoustic stimuli. J. Acoust. Soc. Am. 30, 107–114.

Hajizadeh, A., Matysiak, A., May, P., König, R., 2019. Explaining event-related fields by a mechanistic model encapsulating the anatomical structure of auditory cortex. Biol. Cybern. 113, 321–345. https://doi.org/10.1007/s00422-019-00795-9.

Hajizadeh, A., Matysiak, A., Brechmann, A., König, R., May, P., 2021. Why do humans have unique auditory event-related fields? Evidence from computational modeling and MEG experiments. Psychophysiology 58, e13769. https://doi.org/10.1111/psyp.13769.

Harris, D.M., Dallos, P., 1979. Forward masking of auditory nerve fiber responses. J. Neurophysiol. 42 (4), 1083–1107.

Ingham, N.J., McAlpine, D., 2004. Spike-frequency adaptation in the inferior colliculus. J. Neurophysiol. 91, 632–645.

Ingham, N.J., Itatani, N., Bleeck, S., Winter, I.M., 2016. Enhancement of forward suppression begins in the ventral cochlear nucleus. Brain Res. 1639, 13–27.

Joris, P.X., Yin, T.C.T., 1992. Responses to amplitude-modulated tones in the auditory nerve of the cat. J. Acoust. Soc. Am. 91, 215–232.

Joris, P.X., Yin, T.C.T., 1998. Envelope coding in the lateral superior olive. III. Comparison with afferent pathways. J. Neurophysiol. 79, 253–269.

Kaltenbach, J.A., Meleca, R.J., Falzarano, P.R., Myers, S.F., Simpson, T.H., 1993. Forward masking properties of neurons in the dorsal cochlear nucleus: possible role in the process of echo suppression. Hear. Res. 67, 35–44.

Kiebel, S.J., Garrido, M.I., Friston, K.J., 2007. Dynamic causal modelling of evoked responses: the role of intrinsic connections. Neuroimage 36 (2), 332–345. https://doi.org/10.1016/j.neuroimage.2007.02.046.

Kuwada, S., Batra, R., 1999. Coding of sound envelopes by inhibitory rebound in neurons of the superior olivary complex in the unanesthetized rabbit. J. Neurosci. 19, 2273–2287.

Lanting, C.P., Briley, P.M., Sumner, C.J., Krumbholz, K., 2013. Mechanisms of adaptation in human auditory cortex. J. Neurophysiol. 110, 973–983.

Lieder, F., Daunizeau, J., Garrido, M.I., Friston, K.J., Stephan, K.E., 2013. Modelling trial-by-trial changes in the mismatch negativity. PLoS Comput. Biol. 9 (2), e1002911. https://doi.org/10.1371/journal.pcbi.1002911.

Malone, B.J., Beitel, R.E., Vollmer, M., Heiser, M.A., Schreiner, C.E., 2015. Modulation-frequency-specific adaptation in awake auditory cortex. J. Neurosci. 35 (15), 5904–5916.

May, P.J.C., 2021. The adaptation model offers a challenge for the predictive coding account of mismatch negativity. Front. Hum. Neurosci. 15, 721574. https://doi.org/10.3389/fnhum.2021.721574.

May, P.J.C., Tiitinen, H., 2010. Mismatch negativity (MMN), the deviance-elicited auditory deflection, explained. Psychophysiology 47, 66–122.

May, P.J.C., Tiitinen, H., 2013. Temporal binding of sound emerges out of anatomical structure and synaptic dynamics of auditory cortex. Front. Comput. Neurosci. 7, 152. https://doi.org/10.3389/fncom.2013.00152.

May, P.J.C., Westö, J., Tiitinen, H., 2015. Computational modelling suggests that temporal integration results from synaptic adaptation in auditory cortex. Eur. J. Neurosci. 41, 615–630. https://doi.org/10.1111/ejn.12820.

Murnane, O.D., Prieve, B.A., Relkin, E.M., 1998. Recovery of the human compound action potential following prior stimulation. Hear. Res. 124, 182–189.

Natan, R.G., Rao, W., Geffen, M.N., 2017. Cortical interneurons differentially shape frequency tuning following adaptation. Cell Rep. 21, 878–890.

Nelson, P.C., Smith, Z.M., Young, E.D., 2009. Wide-dynamic-range forward suppression in marmoset inferior colliculus neurons is generated centrally and accounts for perceptual masking. J. Neurosci. 29 (8), 2553–2562.

Nishimura, T., Nakagawa, S., Sakaguchi, T., Hiroshi Hosoi, H., Tono, M., 2003. Effect of a forward masker on the N1m amplitude: varying the signal delay. Neuroreport 14, 891–893.

Ponton, C.W., Eggermont, J.J., Khosla, D., Kwong, B., Don, M., 2002. Maturation of human central auditory system activity: separating auditory evoked potentials by dipole source modeling. Clin. Neurophysiol. 113, 407–420.

Rees, A., Møller, A.R., 1983. Responses of neurons in the inferior colliculus of the rat to AM and FM tones. Hear. Res. 10, 310–330.

Relkin, E.M., Doucet, J.R., Sterns, A., 1995. Recovery of the compound action potential following prior stimulation: evidence for a slow component that reflects recovery of low spontaneous-rate auditory neurons. Hear. Res. 83, 183–189.

Sayles, M., Füllgrabe, C., Winter, I.M., 2013. Neurometric amplitude-modulation detection threshold in the guinea-pig ventral cochlear nucleus. J. Physiol. (Lond.) 591, 3401–3419.

Schröger, E., Winkler, I., 1995. Presentation rate and magnitude of stimulus deviance effects on human pre-attentive change detection. Neurosci. Lett. 193, 185–188.

Smith, R.L., 1977. Short-term adaptation in single auditory nerve fibers: some poststimulatory effects. J. Neurophysiol. 40, 1098–1112.

Tiitinen, H., May, P., Reinikainen, K., Näätänen, R., 1994. Attentive novelty detection in humans is governed by pre-attentive sensory memory. Nature 372, 90–92.

Ulanovsky, N., Las, L., Nelken, I., 2003. Processing of low-probability sounds by cortical neurons. Nat. Neurosci. 6, 391–398. https://doi.org/10.1038/nn1032.

Ulanovsky, N., Las, L., Farkas, D., Nelken, I., 2004. Multiple time scales of adaptation in auditory cortex neurons. J. Neurosci. 24, 10440–10453. https://doi.org/10.1523/JNEUROSCI.1905-04.2004.

Varela, J.A., Sen, K., Gibson, J., Fost, J., Abbott, L.F., Nelson, S.B., 1997. A quantitative description of short-term plasticity at excitatory synapses in layer 2/3 of rat primary visual cortex. J. Neurosci. 17, 7926–7940.

Wehr, M., Zador, A.M., 2005. Synaptic mechanisms of forward suppression in rat auditory cortex. Neuron 47, 437–445.

Westerman, L.A., Smith, R.L., 1988. A diffusion model of the transient response of the cochlear inner hair cell synapse. J. Acoust. Soc. Am. 83 (6), 2266–2276.

Winkler, I., Karmos, G., Näätänen, R., 1996. Adaptive modeling of the unattended acoustic environment reflected in the mismatch negativity event-related potential. Brain Res. 742, 239–252.

Yates, G.K., Robertson, D., Johnstone, B.M., 1985. Very rapid adaptation in the guinea pig auditory nerve. Hear. Res. 17, 1–12.

Yin, P., Johnson, J.S., O'Connor, K.N., Sutter, M.L., 2011. Coding of amplitude modulation in primary auditory cortex. J. Neurophysiol. 105, 582–600.

Young, E.D., Sachs, M.B., 1973. Recovery from sound exposure in auditory nerve fibers. J. Acoust. Soc. Am. 54, 1535–1543.

CHAPTER 4

Animal studies of deviance detection along the auditory pathway

In humans, deviant auditory stimuli elicit an event-related potential (ERP) component that reflects the operation of cortical neurons on low-probability stimulus change. Traditionally, deviance detection is measured by the mismatch negativity (MMN) in studies of cortical evoked potentials. At the single-unit level in animal auditory cortex this phenomenon was named "stimulus specific adaptation" (SSA) and proposed as a neural correlate of the MMN (Chapter 2).

4.1 MMN in animals?

Several studies have observed MMN-like responses in animals. I will first review results from nonhuman primates as these were among the first obtained and relate most closely to findings in humans. Thereafter numerous studies in cats and rodents are presented.

4.1.1 Findings in nonhuman primates

Javitt et al. (1992) recorded epidural auditory ERPs from three monkeys in response to soft and loud clicks. Low-probability—also called "oddball" or "deviant"—loud or soft stimuli elicited a long-duration frontocentral negativity, peaking at approximately 85 ms, and was superimposed upon obligatory cortical auditory evoked potentials (CAEPs; Appendix). Subsequently, Javitt et al. (1994) demonstrated a significant contribution of primary auditory cortex (A1) to scalp-recorded MMN in the monkey, as reflected by greater response of A1 to loud or soft clicks presented as deviants than to the same stimuli presented as repetitive standards. The MMN-like activity was localized primarily in supragranular laminae within A1—particularly layers I and II—where responses were significantly larger to deviant stimuli than to standards.

Brain Responses to Auditory Mismatch and Novelty Detection
https://doi.org/10.1016/B978-0-443-15548-2.00004-1

Fishman and Steinschneider (2012) then investigated possible homologs of the MMN in macaque A1 using a frequency-oddball paradigm in which rare "deviant" tones are randomly interspersed among frequent "standard" tones. Standards and deviants had frequencies equal to the best frequency of the recorded neural population or to a frequency that evoked a response half the amplitude of the best-frequency response. Early and late field potentials, current source density components, multiunit activity, and induced high-gamma band responses were larger when elicited by deviants than by standards of the same frequency (Fig. 4.1). Differences between deviant and standard responses were more prominent in later activity, thus suggesting cortical amplification of initial responses driven by thalamocortical inputs. However, unlike the human MMN, larger deviant responses were characterized by the enhancement of "obligatory" responses (P1, N1, P2, N2) rather than the introduction of new components. Furthermore, a control condition wherein deviants were interspersed among many tones of variable frequency replicated the larger responses to deviants under the oddball condition. Fishman and Steinschneider (2012) suggest that differential responses under the oddball condition in macaque A1 reflect SSA rather than deviance detection per se and concluded that neural mechanisms of deviance detection likely reside in cortical areas outside of A1.

More recently, Camalier et al. (2019) recorded single-neuron activity from the auditory cortex (ACx), dorsolateral prefrontal cortex (DLPFC), and basolateral amygdala (BLA) of two macaque monkeys during an auditory oddball paradigm, which was modeled after that used in humans. Consistent with the predictive coding hypothesis (Friston, 2005; Chapter 1), novelty signals in DLPFC had longer latencies than in ACx and resulted from stimulus-specific effects seen in auditory cortex. Responses in BLA were comparable in magnitude and timing to those in DLPFC, and both were generally much weaker than those in ACx. These observations place constraints on putative generators of the auditory oddball-based MMN and also indicate that there are subcortical areas, such as the amygdala, that may be involved in novelty detection in an auditory oddball paradigm.

Lakatos et al. (2020) used simultaneous thalamic and A1 laminar recordings in seven macaques to evaluate the relative contributions of lemniscal and nonlemniscal thalamic afferents to MMN generation. They demonstrated deviance-related activity mainly in nonlemniscal—dorsal and medial—regions of the auditory thalamus and found MMN generators most prominent in layer 1 of A1—suggesting a nonlemniscal source, likely the reticular activating system (Chapter 5)—as opposed to lemniscal sensory responses

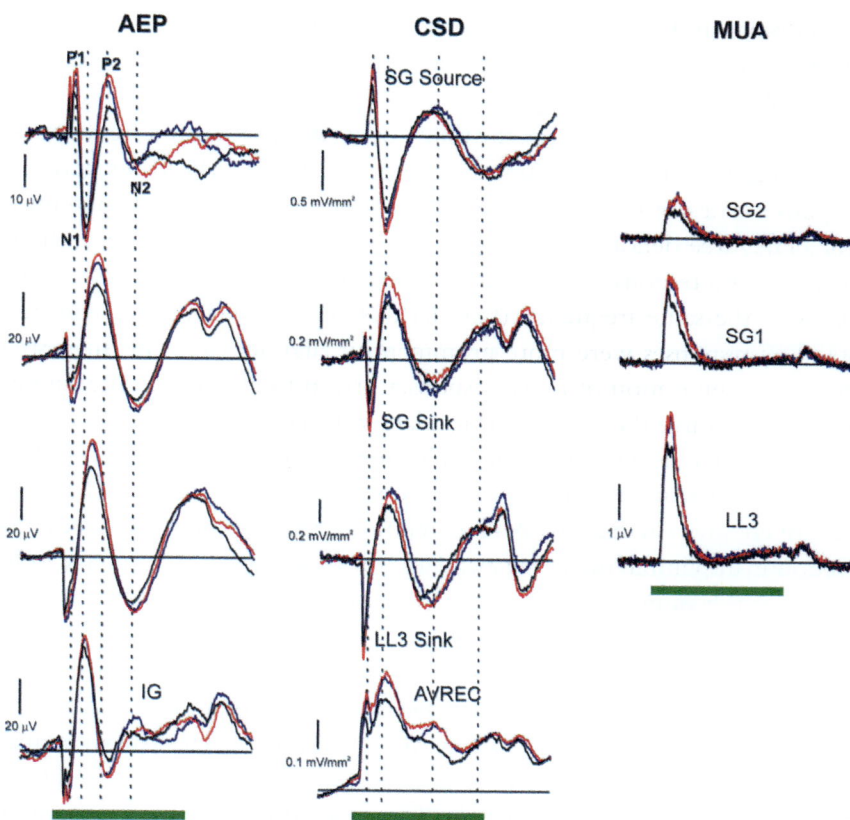

Fig. 4.1 Average AEP (left column), current source density (CSD; center column), and multiunit activity (MUA; right column) elicited in A1 by BF tones in the oddball and many-standards control conditions ($n = 30$ electrode penetrations). Stimulus duration (200 ms) is represented by the *horizontal green bars*. Responses to the BF tone when it was presented as a standard and as a deviant in the oddball condition are plotted in *black* and *red*, respectively. Responses to the BF tone under the control condition are plotted in *blue*. Mean AEP and CSD waveforms at laminar depths corresponding to the LL3 sink, supragranular (SG) sink, and SG source are plotted in separate rows, as indicated. The mean AEP recorded 600 μm below lower lamina 3 (LL3) sink in inferior granular layers (IG) is also included. The average rectified current source density (AVREC), quantifying total net extracellular current flow, is shown in the plot at the bottom of the center column. The MUA recorded in lower lamina 3 and in two adjacent supragranular channels are labeled LL3, SG1, and SG2, respectively. Major deflections in the superficial AEP are labeled (P1, N1, P2, and N2), and their peak latencies marked by *dashed vertical lines* superimposed on the waveforms. Peak latencies of major CSD components in the corresponding superficial CSD channel are also indicated by *dashed vertical lines*. Note different amplitudes of responses recorded at different depths, as indicated by the *vertical calibration bars*. (From Fishman, Y.I., Steinschneider, M., 2012. *Searching for the mismatch negativity in primary auditory cortex of the awake monkey: deviance detection or stimulus specific adaptation? J. Neurosci. 32 (45), 15747–15758.*)

that activate layer 3/4 first and sequentially all cortical layers. The MMN was elicited independent of the frequency tuning of A1 neuronal ensembles (Fig. 4.2).

O'Reilly (2021) compared the utility of deviance-detection and sensory-processing theories for describing epidural-recorded MMN-like auditory responses of a common marmoset monkey during roving oddball stimulation. They observed an enlarged auditory response to deviant stimuli, which decreased exponentially with stimulus repetition, characteristic of sensory gating. A slow positive deflection was viewed over approximately 300–800 ms after the deviant stimulus, which is more difficult to ascribe to afferent sensory mechanisms. When split into ascending and descending frequency transitions, the resulting difference waveforms were disproportionally influenced by descending frequency deviant stimuli. This asymmetry is inconsistent with the general deviance-detection theory of MMN and tentatively suggests that MMN-like responses from common marmosets are predominantly influenced by rapid sensory adaptation and frequency preference of the auditory cortex, while deviance detection may play a role in long-latency activity.

Jiang et al. (2022) investigated the hierarchical depth of predictive auditory processing by combining fMRI and high-density whole-brain electrocorticography (ECoG) in marmoset monkeys during an auditory local-global paradigm in which the temporal regularities of the stimuli were designed at two hierarchical levels. The prediction errors and prediction updates were examined as neural responses to auditory mismatches and omissions. Using fMRI, they identified a hierarchical gradient along the auditory pathway: midbrain and sensory regions represented local, shorter-timescale predictive processing followed by associative auditory regions, whereas anterior temporal and prefrontal areas represented global, longer-timescale sequence processing (see also Wolff et al., 2022; Chapter 1). The combined fMRI and ECoG recordings confirmed the activations at cortical surface areas and further differentiated the signals of prediction error and prediction update, which were transmitted via putative bottom-up gamma and top-down beta oscillations, respectively. Jiang et al. (2022) also found that omission responses caused by absence of input, reflecting solely the two levels of prediction signals that are unique to the hierarchical predictive coding framework, demonstrated the hierarchical top-down process of predictions in the auditory, temporal, and prefrontal areas.

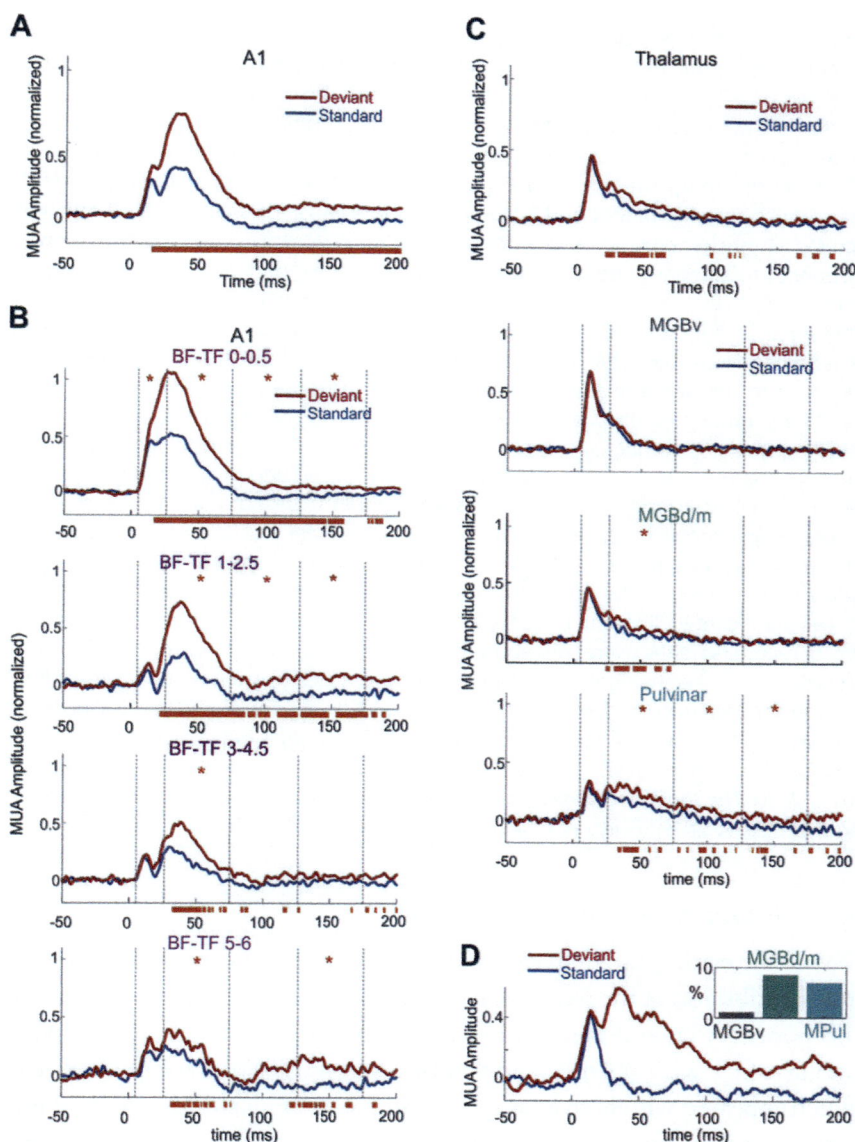

Fig. 4.2 Deviant tone–specific multiunit activity (MUA) responses in the auditory thalamocortical system. (A) MUA averaged across all A1 layers and experiments in response to rarely occurring frequency deviant *(red)* and frequently repeated standard *(blue)* tones (mean [SD] frequency difference = 20% [10%]). Note that significant response amplitude difference (Wilcoxon signed rank, $P < .05$, Bonferroni corrected), denoted by the *orange strip* below the x-axis, occurs almost immediately after response onset, at 16 ms. (B) Frequency dependence of early (5–25 ms) vs late (>25 ms) MUA responses to deviant tones. The MU-tone frequency (BF-TF) difference

(Continued)

Fig. 4.2, Cont'd (see *magenta labels*, on top of each graph). *Asterisks* denote significant response amplitude differences in the timeframes demarcated by the gray dotted lines: 5–25 ms, 25–75 ms, 75–125 ms, and 125–175 ms (Wilcoxon signed rank, $P < .05$, Bonferroni corrected). Note that while the early response amplitude difference is observable only in the first group, when the frequency of presented tones and the BF of the recording site are a close match, the later response amplitude differences are frequency independent and occur in all four groups. (C) Thalamic MUA averaged across all nuclei on top and nucleus-specific MUA traces on bottom showing responses to standard and deviant tones. Similar to the previous panels, the orange strip below the *x*-axis denotes time-resolved significant response amplitude differences, while the orange asterisks denote differences in the four different timeframes. Note that while neuronal firing in the ventral nucleus is not increased in response to deviant stimuli, there is a significant response amplitude increase in the 25–75-ms timeframe in the nonlemniscal nuclei of the medial geniculate body (MGB) and the pulvinar (Wilcoxon signed rank, $P < .05$, Bonferroni corrected). The pulvinar also displays a long-lasting response amplitude increase to frequency deviants. (D) Thalamic MUA responses to standard and deviant tones averaged across single electrode recording sites (up to 21 sites per recording corresponding to the individual electrodes of the linear electrode array during a trial block) where single-trial standard tone vs deviant tone–related response amplitudes were found to be significantly different in the 25–50-ms timeframe. Bar plot inset shows how these recording sites were distributed across the different thalamic nuclei (% of recording sites with significant deviant tone–specific response, $n = 8$ for MGBv, $n = 138$ for MGBd/m, and $n = 55$ for pulvinar). *MGBd/m*, dorsal and medial nuclei of the MGB; *MGBv*, ventral nuclei of the MGB; *MPul*, medial pulvinar. (*Reprinted from Lakatos, P., O'Connell, M.N., Barczak, A., McGinnis, T., Neymotin, S., Schroeder, C.E., et al., 2020. The thalamocortical circuit of auditory mismatch negativity. Biol. Psychiatry 87, 770–780. © 2019 Society of Biological Psychiatry, with permission from Elsevier.*)

4.1.2 Findings in cats

One of the earlier studies in cats (Pincze et al., 2001) investigated the amplitude distribution of the frequency–oddball MMN and that of the P1 and N1 components to reveal their sources in the auditory cortical areas. The mean latency of P1 was ~14 ms, followed by a second positive component (P2), which appeared with highest amplitude in the same location as the P1 and its mean latency value was 25.5 ms. The first negative component (N1) succeeding the early positive ones appeared with a peak latency between 45 and 60 ms. Its amplitude distribution was nearly the same as that of the P1, with a maximum in the middle part of the auditory cortex. The components described before could be recorded in responses elicited both by the standard and deviant stimuli. In the responses elicited by the deviant stimuli an additional negative deflection followed the N1. This deflection corresponded in many aspects to the human frequency–oddball MMN. The peak

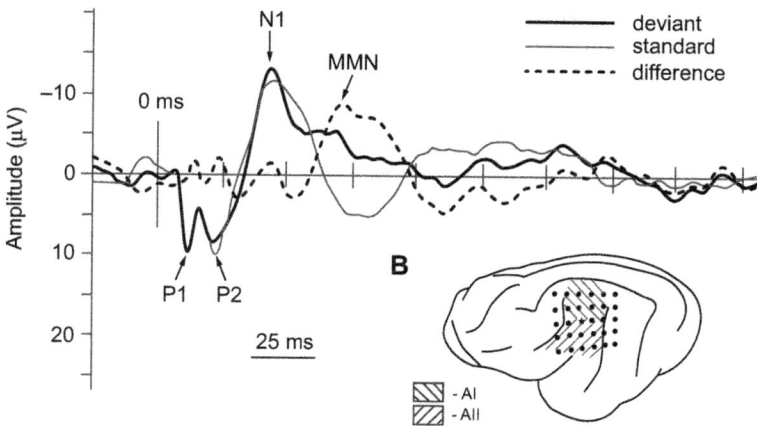

Fig. 4.3 (A) Typical auditory ERP waveforms elicited by identical frequency standard and deviant tones and their difference curve recorded from Cat 01 in the case of the 20% deviance. The *arrows* show the studied components. (B) Location of the electrode matrix above the auditory areas (AI and AII). The *asterisk* (centrally in the electrode assembly) shows the electrode the ERPs are recorded from. *(Reprinted from Pincze, Z., Lakatos, P., Rajkai, C., Ulbert, I., Karmos, G., 2001. Separation of mismatch negativity and the N1 wave in the auditory cortex of the cat: a topographic study. Clin. Neurophysiol. 112, 778–784. © 2001 Elsevier Science Ireland Ltd., with permission from Elsevier.)*

latency of this putative MMN varied between 62 and 74 ms, depending on the degree of frequency deviance (Fig. 4.3). The P1 and N1 components had the highest amplitude on the middle ectosylvian gyrus, which corresponds to the transition region between the primary and secondary auditory areas. However, the amplitude maximum of the MMN was ventral and rostral to them on the secondary auditory cortex area, hence well separated from the sources of the P1 and N1 components.

Ulanovsky et al. (2003) identified a form of adaptation that was present in single neurons in cat A1 but not in all major subdivisions of the medial geniculate body (MGB). The adaptation was stimulus specific, very rapid (started to develop within one trial), highly sensitive to stimulus statistics, and had a long latency, suggesting that intracortical processing contributes to this stimulus-specific adaptation. A comparison with MMN suggested that SSA in single A1 neurons may underlie cortical MMN as recorded from the scalp (see also Chapter 2).

4.1.3 Findings in rodents

In a seminal study, Kraus et al. (1994) investigated the role of the thalamus in the generation of a tone-evoked MMN in guinea pigs. Electrodes were placed in the caudomedial (nonlemniscal) and ventral (lemniscal) subdivisions of the auditory thalamus. Surface epidural electrodes were placed at the midline and over the temporal lobe. The MMN was elicited by a deviant stimulus (2450-Hz tone burst) embedded in a sequence of standard stimuli (2300-Hz tone bursts). They found that a tone-evoked MMN was present in nonlemniscal thalamus but was absent in the lemniscal thalamus. Surface-recorded MMNs were obtained at the midline electrode but not over the temporal lobe.

Srivastava and Bandyopadhyay (2020) showed that adaptation properties of single neurons in mouse orbitofrontal cortex (OFC) behaved drastically different from auditory cortex. OFC neurons show extremely long history dependence and hence adaptation lasting >10s as compared to the ventral A1 responses for a variety of interstimulus intervals (ISIs). The OFC neurons show pure oddball detection during oddball stimulus streams, while ceasing to respond after the first instant of repetition of the standard stimulus. Auditory responses in OFC are shaped by both lemniscal and nonlemniscal thalamic sources. The latter underlies a large component of the observed oddball detection and additionally controls persistent activity in the OFC through the amygdala. Note the OFC, as well as the prefrontal cortex (PFC), was featured in one of the earliest suggested MMN networks (Fig. 1.13; Friston, 2005). Thus there are at least two frontal-cortex sources underlying automatic deviance detection. The detection starts the auditory cortex, which encodes spectral regularities and reports frequency-specific deviances. Then, more abstract representations in the PFC and OFC allow to detect contextual changes of potential behavioral relevance. This was further elucidated by Casado-Román et al. (2020) who presented auditory oddball paradigms along with "no-repetition" controls to record mismatch responses (MMR) in neuronal spiking activity and LFPs at the rat medial PFC. Whereas MMRs in the auditory system are mainly induced by stimulus–dependent—repetition suppression—effects, the influence of frequency-specific effects in medial PFC neurons was almost irrelevant. In addition, auditory responsiveness in the PFC was driven by unpredictability, yielding context-dependent, comparatively delayed, more robust and longer-lasting MMRs mostly comprised of prediction error signaling. This characteristically different composition shows that mismatch responses in the PFC could not be inherited or amplified downstream from the auditory system. Thus it is more plausible for the PFC to exert top–down influences on the ACx, since the PFC exhibits flexible and potent predictive processing, capable of suppressing redundant input more efficiently than the ACx.

Nieto-Diego and Malmierca (2016) mapped the entire rat auditory cortex with multiunit activity (MUA) recordings, using an oddball paradigm. They demonstrated that SSA occurs outside A1 and differs between primary and nonprimary cortical fields. In particular, SSA is much stronger and develops faster in the nonprimary than in the primary fields, paralleling the organization of subcortical SSA (Section 4.2.2). Importantly, strong SSA is present in the nonprimary auditory cortex within the MUA latency range of the MMN in the rat and correlates with an MMN-like difference wave in the simultaneously recorded local field potentials (LFP). This links SSA at the cellular level to the MMN. Extending these findings, the same group (Parras et al., 2021) investigated functional specialization of the rat ACx fields (Fig. 4.4) in repetition suppression and prediction error by combining a tone frequency oddball paradigm with two different control sequences (many standards and cascade). They found that single-unit recordings show by far the largest prediction error effects for the posterior auditory field (PAF), whereas the A1, the anterior auditory field (AAF), the ventral auditory field (VAF), and the suprarhinal auditory field (SRAF) were dominated by repetition-suppression effects. Statistically significant repetition-suppression effects occurred in all ACx fields, whereas prediction error effects were less robust in the lemniscal areas A1 and AAF. Results indicate that the nonlemniscal PAF is more engaged in context-dependent processing underlying deviance detection than the other ACx fields, which are more sensitive to stimulus-dependent effects underlying differential degrees of neural adaptation.

Again, the Malmierca group (Pérez-González et al., 2021) recorded from fast-spiking—putative inhibitory—and regular-spiking—putative excitatory—units in ACx (Atencio and Schreiner, 2008). In response to an oddball paradigm, they found that both types of units showed similar amounts of deviance detection. When considering each ACx field separately, only in A1 fast-spiking neurons showed higher deviance-detection levels than regular-spiking neurons, while in the rest of the fields there was no such distinction. Interpreting these responses in the context of the predictive coding framework, Pérez-González et al. (2021) concluded that the responses of both types of units reflect mainly prediction error signaling (i.e., genuine deviance detection) rather than repetition suppression.

O'Reilly and Conway (2021) recorded epidural field potentials from urethane-anesthetized and awake mice during oddball and many-standards control paradigms (Fig. 4.5) with stimuli varying in duration, frequency, intensity, and interstimulus interval. They observed that stimulus duration

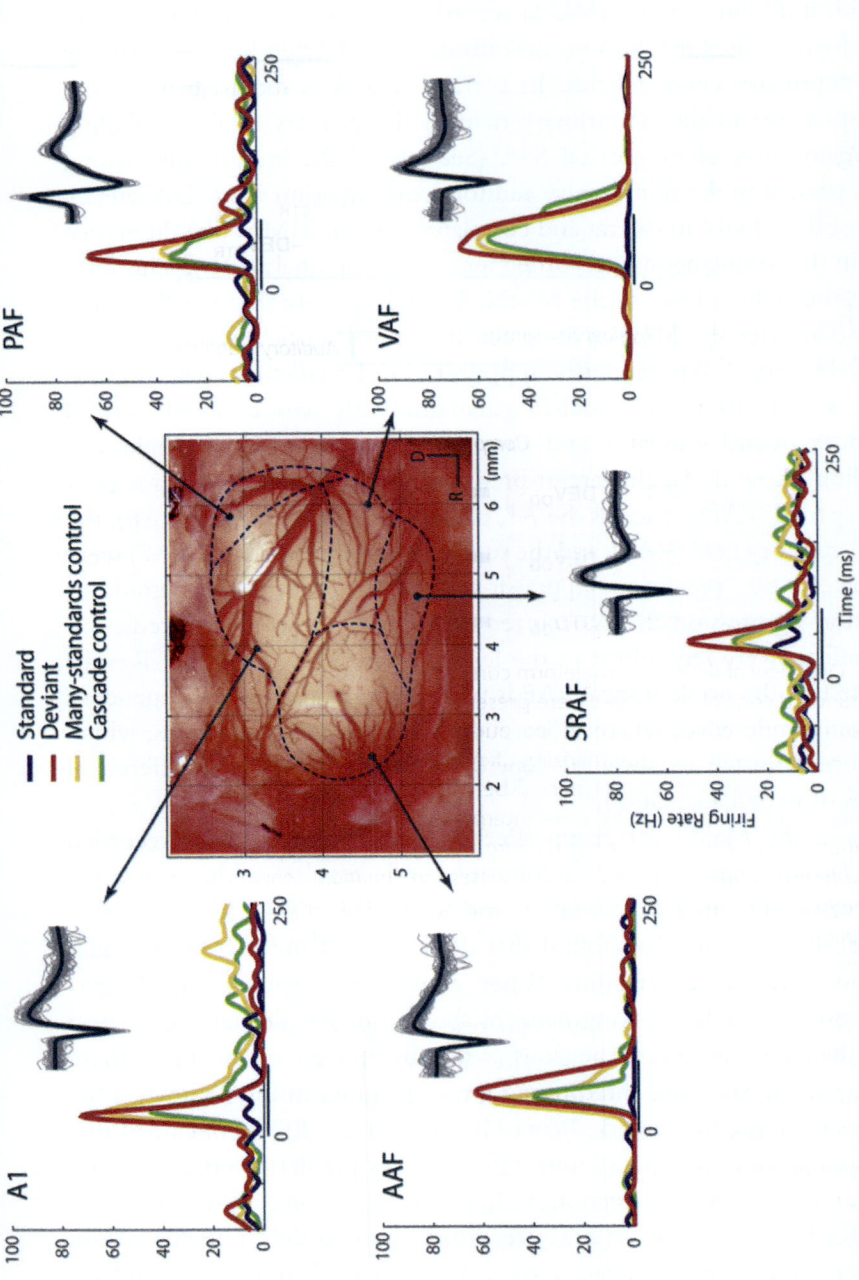

Fig. 4.4 Example of individual neurons from each AC field. Photography of a sample case craniotomy, cortical field boundaries in millimeters from bregma, and five representative examples of single-unit responses. Peristimulus time histograms showing the firing rate elicited by a particular frequency tone for all tested conditions and the five AC fields (A1, primary auditory cortex; AAF, anterior auditory field; PAF, posterior auditory field; SRAF, suprarhinal auditory field; VAF, ventral auditory field). (Reprinted from Parras, G.G., Casado-Román, L., Schröger, E., Malmierca, M.S., 2021. The posterior auditory field is the chief generator of prediction error signals in the auditory cortex. Neuroimage 242, 118446. © 2021 The Authors, with permission.)

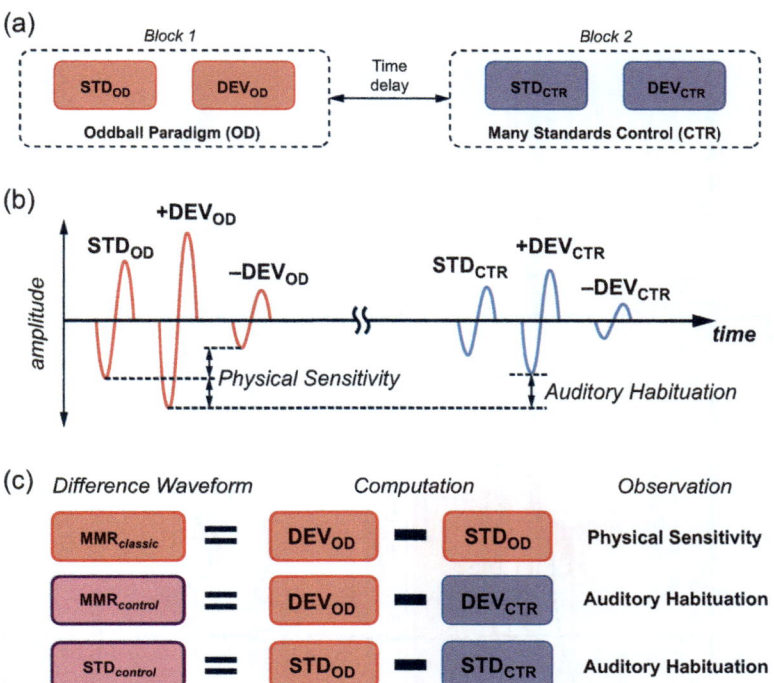

Fig. 4.5 Overview of difference waveform computations. (A) The oddball paradigm and many-standard control sequences were presented in consecutive blocks. (B) Simplified illustration of the effects observed on auditory evoked potentials. (C) Difference waveform computations, summarizing key observations. These analyses were designed to identify any differences between standard, oddball, and control conditions; however, the results were interpreted to reflect physical sensitivity and habituation of the auditory response. *(From O'Reilly, J.A., Conway, B.A., 2021. Classical and controlled auditory mismatch responses to multiple physical deviances in anaesthetised and conscious mice. Eur. J. Neurosci. 53, 1839–1854. © 2020 Federation of European Neuroscience Societies and John Wiley & Sons Ltd., with permission from John Wiley and Sons.)*

oddballs correlated with stimulus–off response peak latency, whereas frequency, intensity, and interstimulus interval deviants correlated with stimulus–onset N1 and P1 (awake only) peak amplitudes (Fig. 4.6). Controlled MMR waveforms appeared to exhibit habituation to auditory stimulation over time, which was equally observed in response to oddball and standard stimuli. Overall, this study suggests that difference waveforms in mice are fundamentally influenced by physical sensitivity and habituation of the auditory response. These findings, using a many-standards control sequences paradigm, are inconsistent with the mechanisms thought to underlie human MMN.

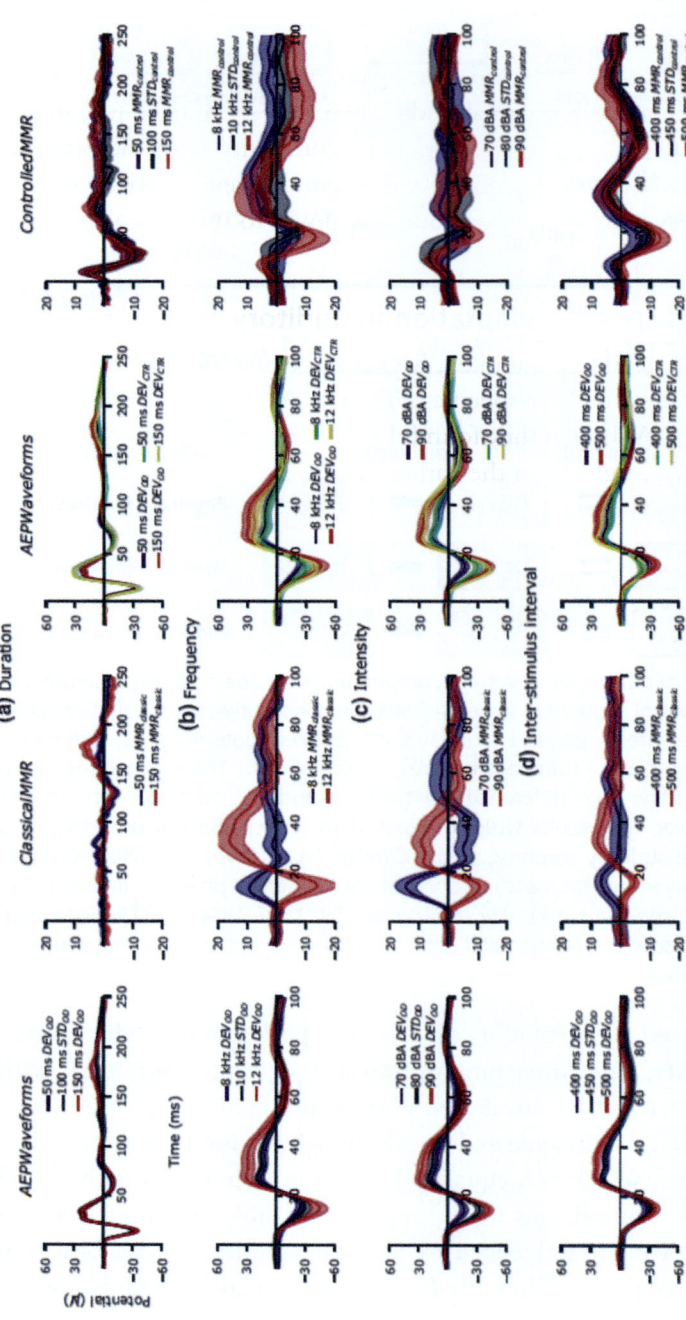

Fig. 4.6 AEP, classical MMR, and controlled MMR waveforms from conscious mice in response to duration (A), frequency (B), intensity (C), and ISI (D) deviance. Standard (STDOD), oddball (DEVOD), and control (DEVCTR and STDCTR) stimuli are labeled in the respective keys of each plot. None of these difference waveforms achieved the threshold for statistical significance with FDR correction. Note the different x-axis scales used for AEP and MMR waveforms; different y-axis scales for duration-varying waveforms in (A); shaded regions represent SEM. *(From O'Reilly, J.A., Conway, B.A., 2021. Classical and controlled auditory mismatch responses to multiple physical deviances in anaesthetised and conscious mice. Eur. J. Neurosci. 53, 1839–1854. © 2020 Federation of European Neuroscience Societies and John Wiley & Sons Ltd., with permission from John Wiley and Sons.)*

4.2 Stimulus-specific adaptation in animals

Most of the relevant studies on stimulus-specific adaptation originate(d) from Malmierca's group in Salamanca, Spain, and from Nelken and colleagues in Jerusalem, Israel. Animal studies on SSA were initially performed in the auditory midbrain, but to keep some continuity with the putative MNN sources in auditory cortex, we will begin our review with findings in animal auditory cortex and work our way down to subcortical areas.

4.2.1 Stimulus-specific adaptation in auditory cortex

Just as Javitt et al. (1994) in monkey, Szymanski et al. (2009) used current source density analysis of layer-specific LFPs in rat A1 in response to standard and oddball tones. Although they found that SSA can be observed throughout all layers of A1, right from the earliest part of the response, there were nevertheless significant differences between layers, with SSA becoming significantly stronger as stimulus-related activity passes from the main thalamo-recipient layers 3 and 4 to layer 5. This is different from the putative origin of the MMN in supragranular layers of primates (Section 4.1.1).

Farley et al. (2010) examined psychophysical and pharmacological properties of multiunit activity in the auditory cortex of awake rodents and found evidence dissociating SSA from novelty responses and the MMN. First, during an oddball paradigm with frequency deviants, neuronal responses showed clear SSA but failed to encode novelty in a manner analogous to the human MMN. Second, oddball paradigms using intensity or duration deviants revealed a pattern of unit responses that showed sensory adaptation, but again without any measurable deviance correlates aligning to the human MMN. Finally, NMDA antagonists, which are known to disrupt the MMN (Javitt et al., 1996), suppressed the magnitude of multiunit responses in a nonspecific manner, leaving the process of SSA intact. Consequently, Farley et al. (2010) suggest that auditory deviance detection as indexed by the MMN is dissociable from SSA at the level of activity encoded by auditory cortex neurons. Recording LFPs and MU activity in the auditory cortex of halothane-anesthetized rats, Taaseh et al. (2011) showed that responses to tones of the same frequency were larger when they were deviant than when they occurred as standard, even with interstimulus time intervals of almost 2 s (cf. Fig. 2.10). SSA was present even when the frequency difference between deviants and standards was as small as 10%, substantially smaller than the typical width of cortical tuning curves, revealing hyperresolution in frequency. Thus SSA is present and strong in rat auditory cortex.

Chen et al. (2015) obtained whole-cell recordings from pyramidal cells, and somatostatin (SST+) and parvalbumin-positive (PV+) GABAergic interneurons in layer 2/3 of mouse auditory cortex and measured tone-evoked membrane potential responses. All cell types displayed SSA of fast ("early") subthreshold and suprathreshold membrane responses— underlying the generation of LFPs and ERPs—with oddball tones eliciting enlarged responses compared with adapted standards. In addition, they identified a slower "late" response (200–400 ms after tone onset), most clearly in excitatory and PV+ neurons, which also displayed SSA. For excitatory neurons, this late component reflected genuine deviance detection (Fig. 4.7). Moreover, intracellular blockade of NMDA receptors reduced early and late responses in excitatory but not PV+ neurons. The late component in excitatory neurons thus shares time course, deviance detection, and pharmacological features with the deviant-evoked MMN and provides a potential link between neuronal SSA and MMN. In summary, Chen et al. (2015) suggested a two-phase cortical activation upon oddball stimulation, with oddball tones first reactivating the adapted auditory cortex circuitry and subsequently triggering delayed reverberating network activity.

Using implanted 32-channel drivable microelectrode arrays and telemetry, Polterovich et al. (2018) recorded LFPs and MU activity from A1 of awake, freely moving rats. They observed highly significant SSA in the awake state. Moreover, the responses to a tone when deviant were significantly larger than the responses to the same tone in the control condition. These results establish the presence of true deviance sensitivity in A1 in awake rats. Polterovich et al. (2018) also compared SSA in A1 and ventral auditory field. While SSA was present in both areas, overall adaptation was stronger in VAF than in A1. Remarkably, the contrast between standards and deviants was larger in A1 than in VAF. This is consistent with findings in awake mice (Parras et al., 2017), where SSA was larger in the lemniscal auditory cortex, compared to nonlemniscal auditory cortex. The stronger SSA that Polterovich et al. (2018) found in A1 than in VAF is, however, inconsistent with the findings of Nieto-Diego and Malmierca (2016), which showed a progressive increase in SSA from primary to nonprimary auditory fields (see also Parras et al., 2021). Polterovich et al. (2018) suggested that these differences are due to the use of anesthetized rats by Nieto-Diego and Malmierca (2016).

4.2.2 The auditory thalamus

To explore the presence of thalamic SSA, Anderson et al. (2009) recorded extracellularly from single units and multiunits in the ventral, medial, and

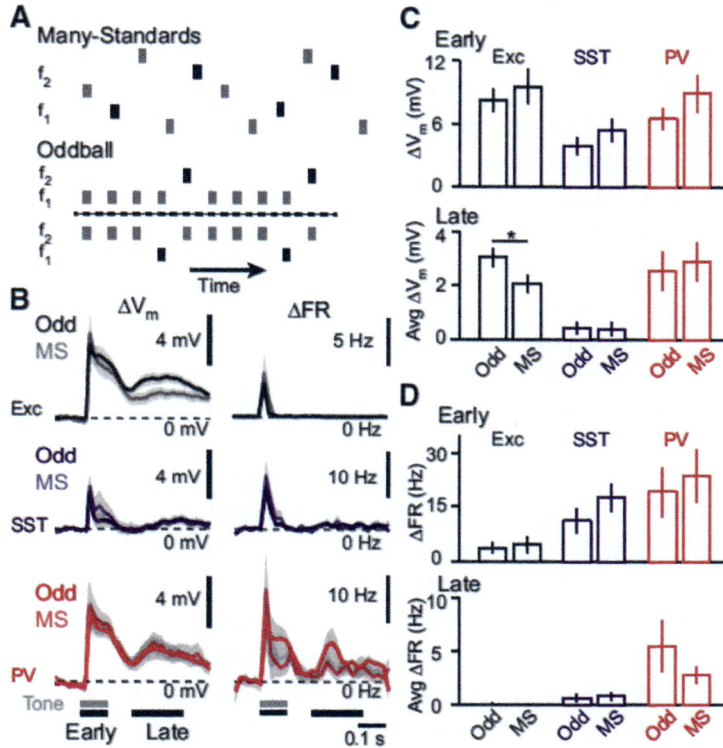

Fig. 4.7 Subthreshold late oddball-response component carries signs of deviance detection. (A) Schematic illustration of the relation between many-standards (MS) and oddball paradigm. In the MS condition, tones of frequencies *f*1 and *f*2 are presented within a pseudo-random sequence of tones of 15 different frequencies. Compared with the oddball condition, no regularity is established. Increased activity to oddballs compared with the same frequency tones in MS condition (MS tones) could indicate genuine deviance detection. (B) Grand-average V_m amplitudes and FR changes (both relative to baseline activity before each tone) for oddball (Odd; POdd = 0.1) and MS tones. Average responses are shown as *colored lines* and the SEM across each subpopulation as *gray shades*. Tone duration (0.1 s for both oddball and MS tones) is indicated as *gray bars* below traces and early and late time windows as *black bars*. (C and D) Comparisons of peak V_m amplitudes (C) and FR changes (D) evoked by oddball and MS tones within early and late time windows. Bar graph indicates the averages and error bars indicate 1 SEM. *$P < .05$. (From Chen, I.W., Helmchen, F., Lütcke, H., 2015. Specific early and late oddball-evoked responses in excitatory and inhibitory neurons of mouse auditory cortex. J. Neurosci. 35 (36), 12560–12573.)

dorsal subdivisions of the mouse MGB, while presenting the anesthetized animals with sequences of standard and deviant tones at interstimulus intervals of 400, 500, and 800 ms. They found SSA in the auditory thalamus at all three stimulus presentation rates, primarily in the nonlemniscal medial subdivision but to a lesser degree also in the lemniscal ventral MGB. These results demonstrate that SSA is present at subcortical levels, primarily in but not restricted to the nonlemniscal auditory pathway (as later also found by Rui et al., 2018). The same group (Antunes et al., 2010) demonstrated that a significant percentage of neurons in rat MGB adapt in a stimulus–specific manner. Neurons in the medial and dorsal subdivisions showed the strongest SSA, linking this property to the nonlemniscal pathway. However, common SSA index (SI) values—defined as follows:

$$SI = \{[d(f1) + df(2)] - [s(f1) + s(f2)]\} / \{[d(f1) + df(2)] + [s(f1) + s(f2)]\},$$

where $d(f1)$ and $s(f1)$ are responses to frequency $f1$ when it was deviant and standard, respectively, and similarly for $f2$ (Ulanovsky et al., 2003, 2004)—reported in A1 of the cat are far in excess of the values measured in cat MGB, presumably in the ventral division (Ulanovsky et al., 2003) or in rat ventral MGB. Thus Antunes et al. (2010) suggested that A1 is the first lemniscal station in which SSA is widespread and strong. Antunes and Malmierca (2011) then reported that SSA in the MGB of the rat remains intact when the ACx is deactivated by cooling, thus demonstrating that the ACx is not necessary for the generation of SSA in the thalamus. The ACx does, however, modulate the responses of MGB neurons in a way that strongly indicates a gain modulation mechanism. The changes imposed by the ACx in thalamic neurons depend on the level of SSA that they exhibit.

Bäuerle et al. (2011) also showed that pharmacological inactivation (using Muscimol) of the auditory cortex demonstrated that SSA in the ventral MGB is modulated by the corticofugal system. Duque et al. (2014) examined what role MGB GABAergic circuits play in the generation and/or modulation of SSA. Microiontophoretic activation of $GABA_A$ receptors ($GABA_ARs$) with GABA or with the selective $GABA_AR$ agonist gaboxadol significantly increased SSA by decreasing responses to common stimuli while having a lesser effect on responses to novel stimuli. In contrast, $GABA_AR$ blockade using gabazine resulted in a significant decrease in SSA. In all cases, decreases in SSA during gabazine application were accompanied by an increase in firing rate to the stimulus paradigm. The present findings suggest that $GABA_A$-mediated inhibition does not generate the SSA response but can regulate the level of SSA sensitivity in a gain control manner.

4.2.3 The thalamic reticular nucleus

The activity of the GABAergic neurons of the thalamic reticular nucleus (TRN) modulates the flow of information between thalamus and cortex.

Yu et al. (2009) found that TRN neurons responded more strongly to pure-tone stimuli presented as deviant stimuli than those presented as standard stimuli. MGB neurons also showed deviance detection in this procedure, albeit to a smaller extent. TRN neuron deviance detection either enhanced (14 neurons) or suppressed (27 neurons) MGB neuronal responses to a probe stimulus (Fig. 4.8). Both effects disappeared after inactivation of the auditory TRN.

Subsequently, Xu et al. (2017) also found that TRN neurons exhibited stronger responses to a tone when it was presented rarely as opposed to frequently at a certain spatial location. They also reported a significant correlation between the amount of adaptation and deviant discriminability. Jia et al. (2021) performed intracellular recordings in the MGB of anesthetized guinea pigs and examined whether and how the TRN modulates the SSA of MGB neurons with inhibitory inputs. For that purpose, they used microinjection of lidocaine to inactivate the neural activity of the TRN. Jia et al. (2021) found that (1) MGB neurons with hyperpolarized membrane potentials exhibited SSA at both the spiking and subthreshold levels; (2) SSA of MGB neurons depends on the interstimulus interval (ISI), where a shorter ISI results in stronger SSA; and (3) the long-lasting hyperpolarization of MGB neurons decreased after the burst firing of the TRN was inactivated. As a result, SSA of these MGB neurons was diminished after inactivation of the TRN (Fig. 4.9).

4.2.4 The auditory midbrain

4.2.4.1 Electrophysiology

Unquestionably, Malmierca's group dominates the SSA studies in the auditory midbrain. Pérez-González et al. (2005) compared SSA across different populations of neurons in the inferior colliculus (IC) of the rat and showed that a subclass of neurons with rapid and pronounced SSA respond selectively to novel sounds. These neurons, located in the dorsal and external cortex of the IC, regions that receive input from nonlemniscal ACx, fail to respond to multiple repetitions of a sound but briefly recover their excitability when some stimulus parameter is changed. The finding of neurons that respond selectively to novel stimuli in the mammalian auditory midbrain suggests that they may contribute to a rapid subcortical pathway for directing attention and/or orienting responses to novel sounds. To compare these findings across the IC with the studies in cat cortex, Malmierca et al. (2009) used an oddball stimulus paradigm, similar to that used by Ulanovsky et al. (2003) in cortex. IC neurons that expressed SSA responded with a variety of discharge patterns. However, those neurons that had a high degree of

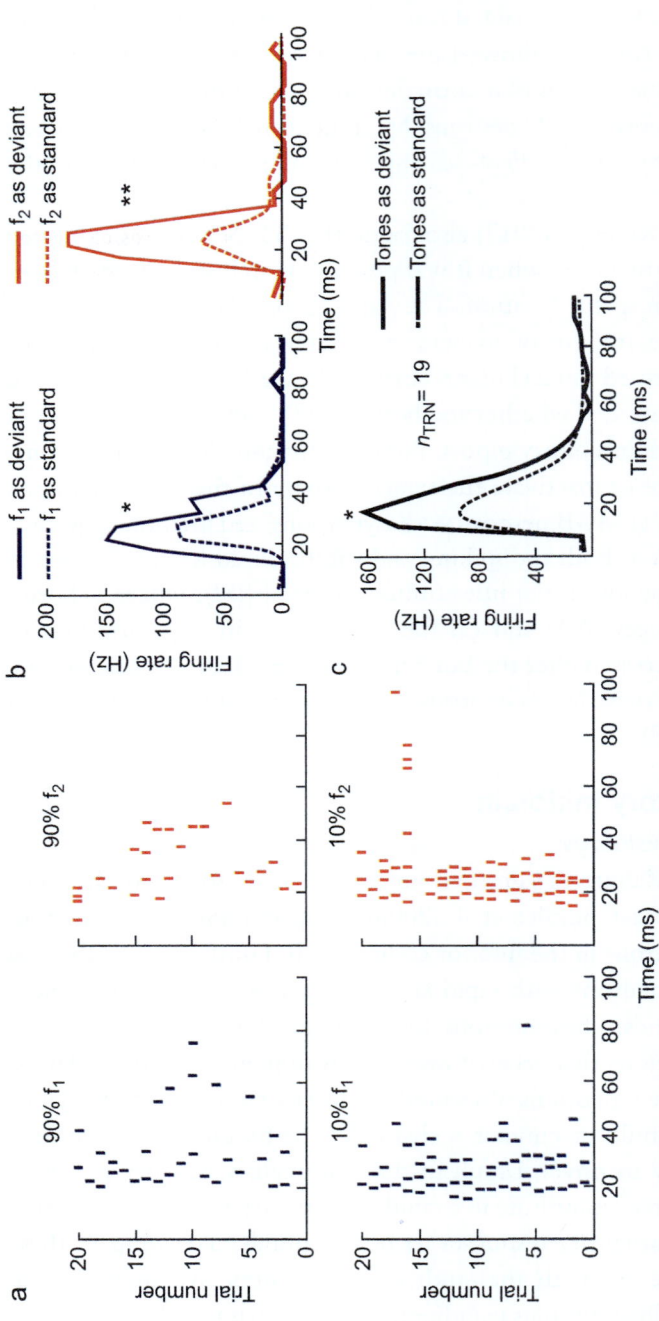

Fig. 4.8 Differential responses of TRN neurons to pure-tone stimuli of two frequencies presented in an oddball procedure. (A) Raster displays showing responses to tones of two frequencies (*f*1, 19.02 kHz; *f*2, 21.02 kHz) when presented as *f*1 standard (90% appearing probability) and *f*2 deviant (10% appearing probability) or *f*1 deviant and *f*2 standard. The ISI was 1 s. We sampled responses to the standard frequency from trial numbers 81 to 100 (20 trials), whereas we sampled all of the responses to the deviant frequency (20 trials, numbers 1–20). (B) Peristimulus time histograms (PSTHs) of responses shown in left. (C) PSTHs showing the mean responses of 19 TRN neurons. * *P* < .05 and ** *P* < .01 (ANOVA). The ISI was 1 s. *(From Yu, X.-J., Xu, X.-X., He, S., He, J., 2009. Change detection by thalamic reticular neurons. Nat. Neurosci. 12 (9), 1165–1170. Permission granted By Springer Nature.)*

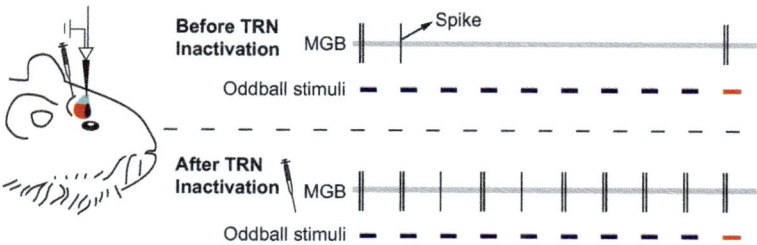

Fig. 4.9 Lidocaine injection in the thalamic reticular nucleus (TRN) reduces/eliminates the stimulus-specific adaptation in the medial geniculate body (MGB). *(From Jia, G., Li, X., Liu, C., He, J., Gao, L., 2021. Stimulus-specific adaptation in auditory thalamus is modulated by the thalamic reticular nucleus. ACS Chem. Nerosci. 12, 1688–1697.)*

SSA and were selective for novel stimuli were all onset responders (Pérez-González et al., 2005). Malmierca et al. (2009) showed a distinct relationship between the SI and the frequency tuning of neurons, i.e., neurons with the highest SIs were the most broadly tuned. SSA was greatest at the largest frequency contrast, the lowest probability of the oddball stimulus, and a repetition rate of 4/s. The existence of SSA at both the midbrain and cortex strongly suggested to Malmierca et al. (2009) two separate SSA mechanisms, one that is preattentive and operates subcortically, affecting primarily the onset portion of the response, and another that operates at the cortical level where it mainly affects the sustained portion of the response (Ulanovsky et al., 2003). To delineate the modulating role of auditory cortex, Anderson and Malmierca (2013) recorded the response from single IC neurons to stimuli presented in an oddball paradigm before, during, and after reversibly deactivating the ipsilateral ACx by cooling with a cryoloop (Lomber et al., 1999). While changes in the basic response properties of the IC neurons were widespread (in 89%), changes in SSA sensitivity were less common; approximately half of the neurons recorded showed a significant change in SSA, while the other half remained unchanged. Changes in SSA could be in either direction: 18% enhanced their SSA sensitivity, while 34% showed reduced SSA sensitivity. For the majority of this latter group, cortical deactivation reduced, but did not eliminate, significant SSA levels. Only eight neurons seemed to inherit SSA from the ACx, as their preexisting significant level of SSA became nonsignificant during cortical deactivation. Thus the presence of SSA in the IC is generally not dependent upon the cortico-collicular projection, suggesting the ACx is not essential for the generation of midbrain SSA.

Ayala et al. (2015) tested whether neurons exhibiting SSA and those which do not are part of the same networks in the IC. They recorded the responses to frequent and rare sounds and then marked the sites of these neurons with a retrograde tracer to correlate the source of projections with the physiological response. SSA neurons were confined to the nonlemniscal subdivisions of the IC and exhibited broad receptive fields, while the non-SSA were confined to the central nucleus (ICC) and displayed narrow receptive fields. SSA neurons receive strong inputs from auditory cortical areas and very poor or even absent projections from the brainstem nuclei. In contrast, the major sources of inputs to the neurons that lacked SSA were from the brainstem nuclei. These findings demonstrate that auditory cortical inputs to the IC are biased in favor of synaptic domains that are populated by SSA neurons enabling them to compare top-down signals with incoming sensory information from lower areas. Malmierca et al. (2019) showed that some neurons in the rat IC are sensitive to the history of patterned stimulation and to violations of patterned regularity, demonstrating that there is a population of subcortical neurons, located as early as the level of the midbrain, that can detect more complex stimulus regularities than previously supposed and that are as sensitive to complex statistics as some neurons in primary auditory cortex. Thus perceptual organization of sound already occurs at the IC level.

4.2.4.2 *The role of neurotransmitters and neuromodulators*

Pérez-González et al. (2012) found that the $GABA_A$ receptor antagonist gabazine slowed down the process of adaptation to high-probability stimuli but did not abolish it, with response magnitude and latency still depending on the probability of the stimulus. Blocking $GABA_A$ receptors increased the firing rate to high- and low-probability stimuli, but did not completely equalize the responses. Together, these findings suggest that $GABA_A$-mediated inhibition acts as a gain control mechanism that enhances SSA by modifying the responsiveness of the neuron. The same group (Ayala et al., 2016) found that the blockade of $GABA_A$ and glutamate receptors mediates an overall increase or decrease of the neural response, respectively, while acetylcholine affects only the response to the repetitive sounds (Fig. 4.10 top). These results demonstrate that GABAergic, glutamatergic, and cholinergic receptors play different and complementary roles on shaping SSA.

Using micro-iontophoresis, Ayala and Malmierca (2018) determined the role of $GABA_A$, $GABA_B$, and glycinergic receptors in stimulus-specific

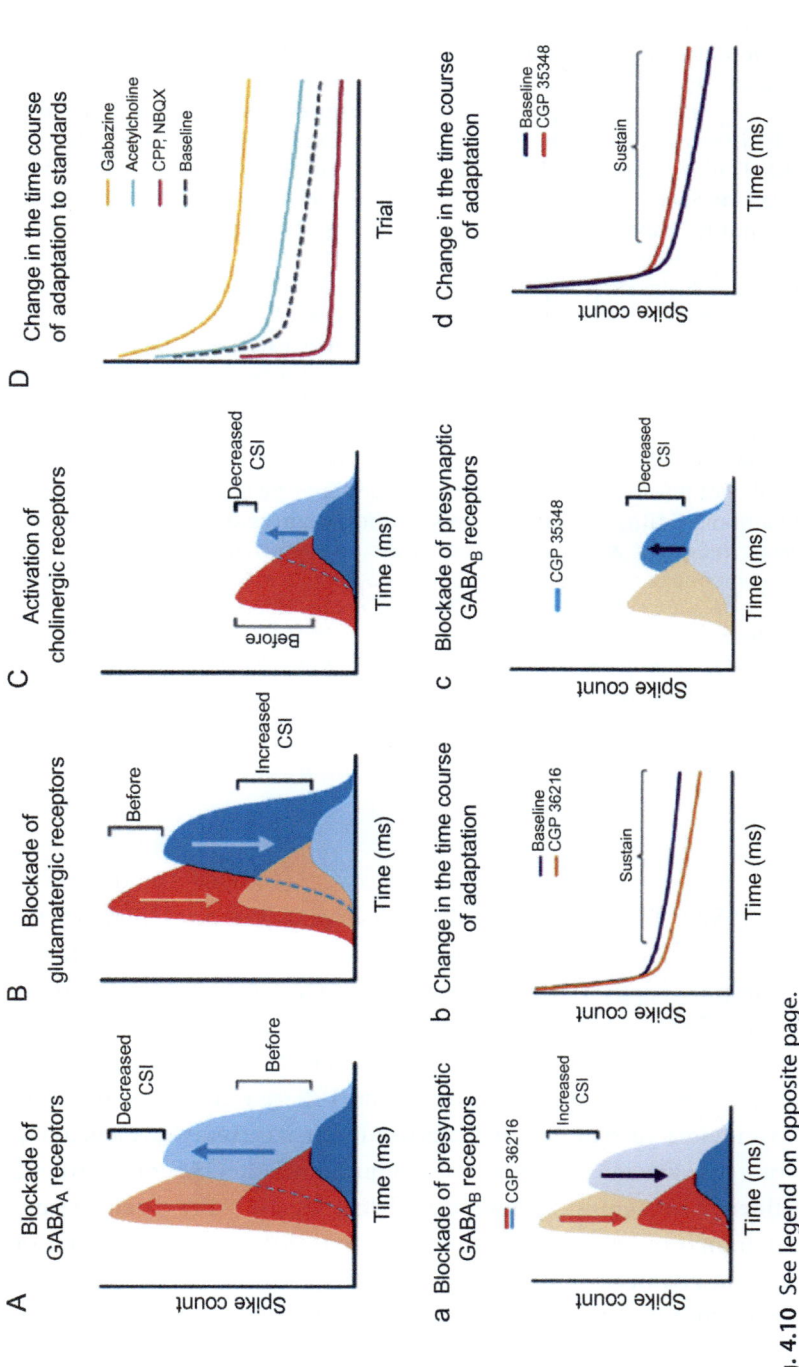

Fig. 4.10 See legend on opposite page.

Fig. 4.10, Cont'd Top. Schema of the effect of inhibition and acetylcholine on SSA. (A) The blockade of the GABA$_A$ receptor by the iontophoretic ejection of gabazine increased the overall response (shown here as PSTH) of IC neurons to both deviant *(red)* and standard *(blue)* stimuli, decreasing the CSI. (B) On the other hand, the application of glutamatergic receptor antagonists reduced the responses to deviants and standards and increased the CSI. (C) The activation of the cholinergic receptors by acetylcholine increases only the response to the standard tone of IC neurons with partial levels of SSA. (D) Gabazine exerts a drastic change in the time course of adaptation in the response to the standard tone by increasing the fast and slow components of adaptation as well as the steady state of adaptation. Acetylcholine affects only the sustained component of the time course of adaptation. CPP and NBQX make the time constants faster and reduce the steady state. Bottom. (a and b) Augmented inhibition decreases the overall neuronal excitability decreasing the response to deviant *(red)* and standard tone *(blue)* and accentuating the sustained component of adaptation. (c) Blockade of the postsynaptic GABA$_B$ receptors by CGP 35348 would only decrease the inhibition elicited by repetitively activated GABAergic inputs *(blue)*, thereby increasing the neuronal response to standard sounds while not affecting the response to deviant ones *(red)*. (d) Slow dynamics of the metabotropic GABA$_B$ receptors as well as the dynamics of GABA to reach the extrasynaptic receptors would contribute to the late effect of CGP 35348 on the sustained component of adaptation. *(Top Panel: Reprinted from Ayala, Y.A., Pérez-González, D., Malmierca, M.S., 2016. Stimulus-specific adaptation in the inferior colliculus: the role of excitatory, inhibitory and modulatory inputs. Biol. Psychol. 116, 10–22. © 2015, with permission from Elsevier; Bottom Panel: From Ayala, Y.A., Malmierca, M.S., 2018. The effect of inhibition on stimulus-specific adaptation in the inferior colliculus. Brain Struct. Funct. 223, 1391–1407, with permission from Springer Nature.)*

---

adaptation (SSA). They found that blockade of postsynaptic GABA$_B$ receptors selectively modulated response adaptation to repetitive sounds, whereas blockade of presynaptic GABA$_B$ receptors exerted a gain control effect on neuron excitability. Adaptation decreased when postsynaptic GABA$_B$ receptors were blocked, but increased if the blockade affected the presynaptic GABA$_B$ receptors (Fig. 4.10 bottom). A dual effect was elicited by blockade of glycinergic receptors, which showed either increase or decrease in adaptation. Moreover, simultaneous coapplication of GABA$_A$, GABA$_B$, and glycinergic antagonists demonstrated that local GABA- and glycine-mediated inhibition contributes to only about 50% of SSA. Therefore inhibition via chemical synapses dynamically modulates the strength and dynamics of stimulus–specific adaptation, but does not generate it (Ayala and Malmierca, 2018).

Dopaminergic neurons are traditionally thought to signal positive or negative prediction errors (PEs) when reward expectations are, respectively, exceeded or not matched. Valdés-Baizabal et al. (2020) tested how dopamine modulates midbrain processing of unexpected tones, by recording from the rat IC to oddball and cascade sequences, before, during, and after the microiontophoretic application of dopamine or eticlopride (a D2-like receptor antagonist). They found that dopamine reduces the net neuronal responsiveness exclusively to unexpected sensory input without significantly altering the processing of expected input. Valdés-Baizabal et al. (2020) conclude that dopaminergic projections from the thalamic subparafascicular nucleus to the inferior colliculus (Fig. 4.11) could encode the expected precision of unsigned PEs, attenuating via D2-like receptors the postsynaptic gain of sensory inputs forwarded by the auditory midbrain neurons. Valdés-Baizabal et al. (2020) propose that dopamine release in the nonlemniscal IC encodes uncertainty by reducing the postsynaptic gain of PE signals, thereby dampening their drive over higher processing stages. The dopaminergic projections from the thalamic subparafascicular nucleus to the IC could be the biological substrate of this early precision-weight mechanism.

As a summary, Liu et al. (2022) showed the inputs and outputs of the two subregions of the nonlemniscal IC, namely, the dorsal and external cortex of the IC. The nonlemniscal IC plays a crucial role in multisensory integration, animal behavior, and SSA, and the dorsal and external regions are distinct in many aspects, including molecular expression and neural circuits (Fig. 4.12).

4.2.5 Auditory brainstem

Ayala et al. (2013) explored SSA in the cochlear nucleus (CN; both dorsal and ventral parts) of rats and also recorded single-unit responses from the IC where SSA is known to occur. They reproduced the finding that many neurons in the IC exhibited SSA, but did not observe significant SSA in the CN sample. This suggests that SSA is not widespread along the entire auditory pathway. The same group (Duque et al., 2018) evaluated auditory brainstem responses (ABRs) to pure tones in anesthetized mice using an oddball paradigm. After applying a wide band-pass filter—allowing the visualization of a late slow wave (P0) in the ABR—they found a reduction of the P0 amplitude of the response to repetitive sounds, compared to rare ones. Because the P0 is temporally correlated with the sustained responses of IC, this suggests that the IC is the first to show SSA in the auditory pathway. Because the

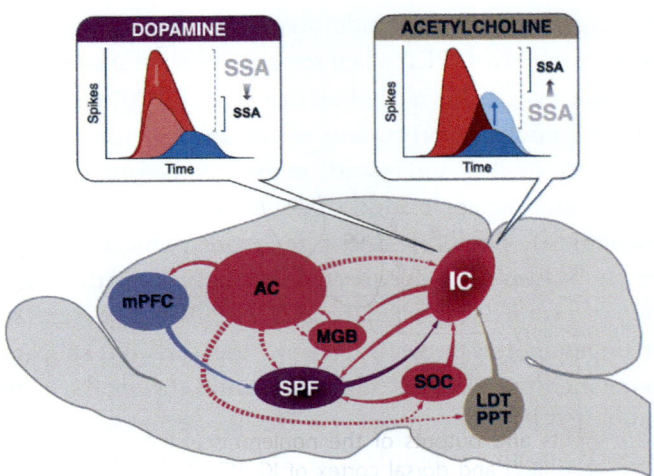

Fig. 4.11 Subparafascicular (SPF) auditory afferences. Schematic diagram of the main connections involved in SSA modulation and PE precision weighting in the IC. The auditory pathway *(in pink)* is composed of ascending *(solid arrows)* and descending connections *(dashed arrows)* between cortical and subcortical auditory structures. The hierarchical exchange of bottom-up PEs and top-down predictions postulated by the predictive coding framework could be extended to subcortical auditory neurons by means of these feedback loops. The SPF *(purple)* receives input from main auditory nuclei *(pink)* and from the mPFC *(violet)*, integrating information from manifold hierarchical processing levels. Such connectivity could allow SPF dopaminergic neurons to estimate the volatility of the probabilistic structure of the auditory context. In turn, the SPF sends dopaminergic projections back to the IC to signal the expected precision of PE signaling at low processing levels. Dopamine release in the nonlemniscal IC reduces the postsynaptic responses to surprising stimuli *(red)* but has no effect on repetitive ones *(blue)*, consequently reducing SSA indices. By contrast, cholinergic modulation from the LDT and PPT nuclei *(brown)* of the brainstem increases the responses to repetitive stimuli in the IC. Note that the net effect of both dopamine and acetylcholine is the reduction of SSA indices, decreasing the relative saliency of surprising input through complementary means. *AC*, auditory cortex; *CAS*, cascade sequence; *IC*, inferior colliculus; *LDT*, laterodorsal tegmental; *MGB*, medial geniculate body; *mPFC*, medial prefrontal cortex; *PE*, prediction error; *PPT*, pedunculopontine tegmental; *SOC*, superior olivary complex; *SPF*, subparafascicular nucleus of the thalamus; *SSA*, stimulus-specific adaptation. *(From Valdés-Baizabal, C., Carbajal, G.V., Pérez-González, D., Malmierca, M.S., 2020. Dopamine modulates subcortical responses to surprising sounds. PLoS Biol. 18 (6), e3000744. https://doi.org/10.1371/journal.pbio.3000744. Open Access.)*

amplitude of P0 to the deviant sound is not larger than that to the same sound presented infrequently in a deviant alone condition, the current data suggests that the phenomena observed in this chapter is more related to SSA than to genuine deviance detection (Duque et al., 2018).

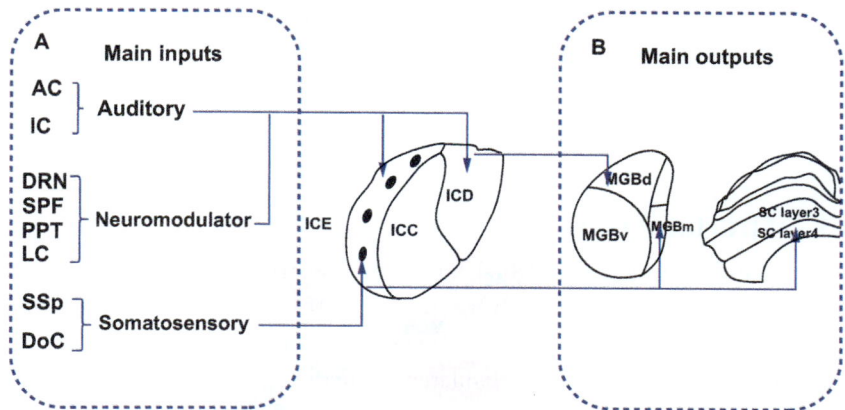

Fig. 4.12 Main inputs and outputs of the nonlemniscal IC. (A) Main inputs of the external cortex of IC (ICE) and dorsal cortex of IC (ICD). Auditory inputs contain the AC and ICC projects to the extramodule region of the ICE and ICD. Somatosensory inputs contain the primary somatosensory cortex (SSp) and column nuclei (DoC) mainly project to the module area of the ICE. Neuromodulator inputs contain the dorsal raphe nucleus (DRN), subparafascicular (SPF), pedunculopontine tegmental nucleus (PPT) and locus coeruleus (LC) mainly project to the ICE and ICD. (The *black shaded area* represents the module area in the ICE which highly expressed GAD-67, parvalbumin (PV), acetylcholinesterase (AChE), cytochrome oxidase (CO), and nicotinamide adenine dinucleotide phosphate diaphorase (NADPH-d). *Dashed lines* represent inputs of the nonlemniscal IC.) (B) Main outputs of the ICE and ICD. ICE mainly projects to the medial part of the medial geniculate body (MGBm) and deep layer of the superior colliculus (SC). ICD mainly project to the dorsal part of the MGB (MGBd). *(Reprinted from Liu, M., Dai, J., Zhou, M., Liu, J., Ge, X., Wang, N., Zhang, J., 2022, Mini-review: The neural circuits of the non-lemniscal inferior colliculus. Neurosci. Lett. 776, 136567. © 2022, with permission from Elsevier.)*

4.3 The role of auditory cortex in mismatch detection and predictive coding

4.3.1 Predictive coding

We follow here the review of Carbajal and Malmierca (2018) who described predictive coding (Chapter 1) as capable of integrating MMN and SSA data in a common theoretical framework underlying deviance detection at every level of measurement. Interaction of inputs and predictions forms a multilevel representation of sensory information. When top-down predictions do not fit the actual sensory input, the first-level neuronal populations—putatively subcortical—convey a prediction error signal to the higher levels to favor the processing of unpredicted features. This prediction error is functionally analogous to the genuine deviance detection. "Hence, lower and higher processing stages keep communicating iteratively in reciprocal pathways until the suppression of the error signal is accomplished, which indicates that perceptual encoding is optimized" (Carbajal and Malmierca, 2018).

4.3.2 Structural correlates

Carbajal and Malmierca (2018) also recalled the structural correlates of mismatch and novelty detection. Originating at the auditory midbrain level, two parallel pathways can be distinguished marking each station they cross with structural and functional characteristic features. These are based on the multiple subdivisions present in the inferior colliculus, the medial geniculate body of the thalamus, and the auditory cortex (Fig. 4.13).

Carbajal and Malmierca (2018) noted that the secondary ACx is at the top of the auditory nonlemniscal pathway (Fig. 4.13) and is where deviance detection and prediction error exhibit their uppermost expression. "The belt fields of the ACx show the greatest population indices of deviance detection, hosting a rich variety of context-driven responses, as it is characteristic of nonlemniscal neurons. SSA in the belt fields is not only the highest but also the swiftest in the ACx, insofar as repetition suppression obliterates the standard-evoked response almost completely from its onset (Nieto–Diego and Malmierca, 2016). Nonlemniscal ACx is also the only part of the auditory system where prediction error equalizes or supersedes repetition suppression as the main component of deviance detection, with population ratios of 50% in anesthetized preparations that swell up to 80% when stimulation intensity is near threshold or when animals are awake (Parras et al., 2017). This suggests that the non–lemniscal ACx hosts the highest order populations of neurons within the auditory system."

As we have seen, LFP and ERP data from animal models have shown the implication of prefrontal cortices in this auditory deviance-detection mechanism (Camalier et al., 2019), corroborating the putative prefrontal, modulating, sources of the human MMN, and the P3 (Dürschmid et al., 2016). A potential substrate for top–down predictive cues is the massive set of descending projections from the ACx to subcortical structures (cf. Fig. 1.1), although the role of this system in predictive processing has never been directly assessed. Lesicko et al. (2022) tested the effect of optogenetic inactivation of the auditory cortico-collicular feedback in awake mice on responses of IC neurons to stimuli designed to test prediction error and repetition suppression. Inactivation of the cortico-collicular pathway led to a

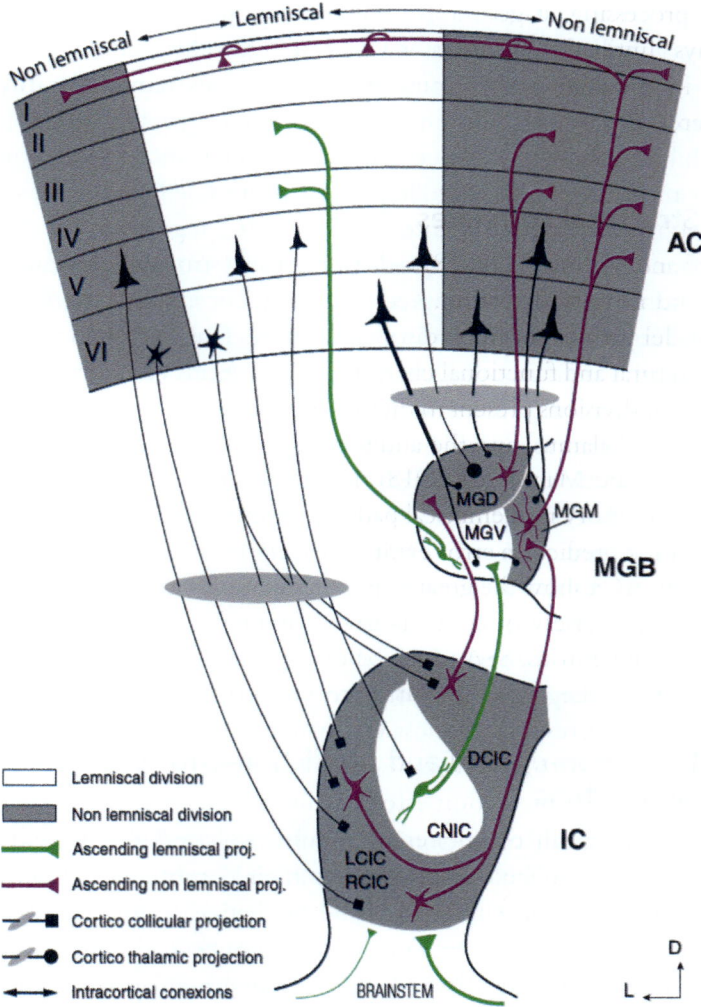

Fig. 4.13 Schematic diagram of the auditory pathway, showing the major stations and projections that constitute the lemniscal and nonlemniscal pathways. As a rule of thumb, lemniscal tonotopic laminae tend to project to their analogous lamina in the next lemniscal division and receive few cortical projections, shaping a sort of straightforward pathway to the cortex. Conversely, nonlemniscal divisions tend to project mostly to other nonlemniscal divisions and receive dense cortical projections, shaping a loop-like connectivity network ideal for hosting a generative mechanism of hierarchical inference. *AC*, auditory cortex; *CNIC*, central nucleus of the inferior colliculus; *DCIC*, dorsal cortex of the inferior colliculus; *IC*, inferior colliculus; *LCIC*, lateral cortex of the inferior colliculus; *MDM*, medial division of the MGB; *MGB*, medial geniculate body of the thalamus; *MGD*, dorsal division of the MGB; *MGV*, ventral division of the MGB; *RCIC*, rostral cortex of the inferior colliculus. *(From Carbajal, G.V., Malmierca, M.S., 2018. The neuronal basis of predictive coding along the auditory pathway: from the subcortical roots to cortical deviance detection. Trends Hear. 22, 1–33. doi:https://doi.org/10.1177/2331216518784822. Open Access.)*

decrease in prediction error in IC. Repetition suppression was unaffected by cortico-collicular inactivation, suggesting that this metric may reflect fatigue—adaptation—of bottom-up sensory inputs rather than predictive processing. Lesicko and Geffen (2022) discovered populations of IC units that exhibit repetition enhancement, a sequential increase in firing with stimulus repetition. Cortico-collicular inactivation led to a decrease in repetition enhancement in the central nucleus of IC, suggesting that it is a top-down phenomenon. Negative prediction error, a stronger response to a tone in a predictable rather than unpredictable sequence, was suppressed in shell IC units during cortico-collicular inactivation. These changes in predictive coding metrics arose from bidirectional modulations in the response to the standard and deviant contexts, such that the units in IC responded more similarly to each context in the absence of cortical input (Fig. 4.14).

Lesicko and Geffen (2022) also found that metrics of predictive coding and deviance detection differ depending on the anesthetic state of the animal, with negative prediction error emerging in the central IC and repetition enhancement and prediction error being more prevalent in the absence of anesthesia. Overall, these results demonstrate that the ACx provides cues about the statistical context of sound to subcortical brain regions via direct feedback, regulating processing of both prediction and repetition.

4.3.3 Prediction-update and prediction-error separation via brain rhythms

Predictions and prediction errors are simultaneous and interdependent processes, making it difficult to disentangle their constituent neural network organization. Chao et al. (2018) tested predictive coding by using high-density electrocorticography (ECoG) in monkeys during an auditory "local-global" paradigm in which the temporal regularities of the stimuli were controlled at two hierarchical levels. They decomposed the broadband data and identified lower- and higher-level prediction-error signals in early auditory cortex and anterior temporal cortex, respectively, and a prediction-update signal sent from prefrontal cortex back to temporal cortex. The prediction-error and prediction-update signals were transmitted via gamma ($>40\,Hz$) and alpha/beta ($<30\,Hz$) oscillations, respectively (Fig. 4.15).

Chao et al. (2018) found the higher-order error and update signals primarily in the frontopolar PFC (area 10) and DLPFC. Among prefrontal areas in macaque monkeys, the frontopolar area 10 has the densest interconnection with auditory association areas: it receives information from nearly all levels of auditory processing in the superior temporal gyrus, from the early

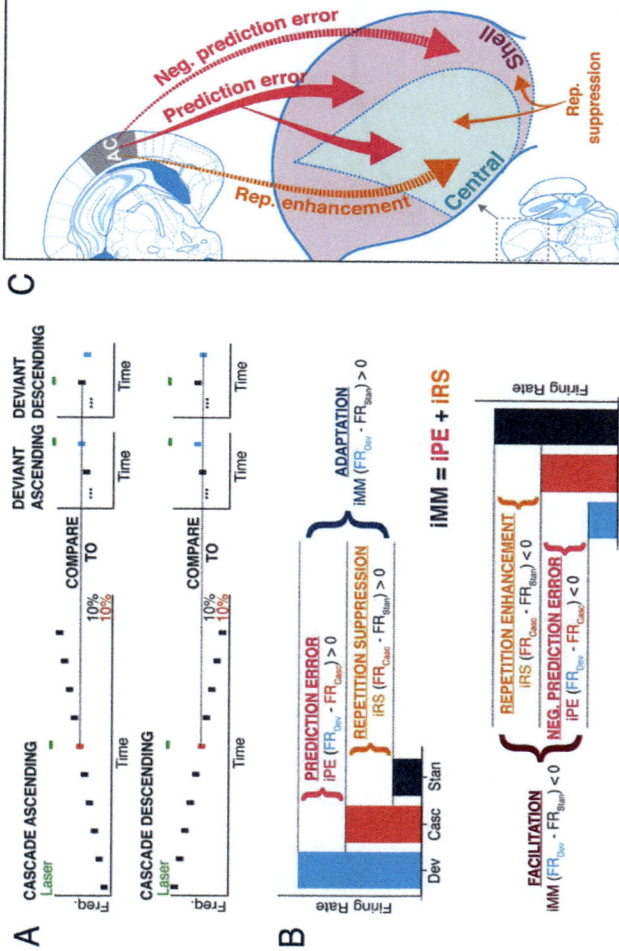

Fig. 4.14 Cortico-collicular (CC) input shapes predictive coding in IC. (A) All tones in the "cascade" sequence are presented with the same likelihood and no tone is repeated on adjacent tone presentations. Therefore the neuronal response to a tone in this sequence (*orange*) can be compared to the response to the same tone when it is presented as a standard (*gray*) to compute repetition effects. Similarly, prediction effects can be computed by comparing the response to a tone in the cascade sequence to the response to the same tone when it is presented as an unpredictable deviant (*blue*). (B) Prediction error is measured as a higher response to a deviant context than the cascade context, while the reverse relationship represents negative prediction error. A higher response to the cascade context than the standard context indicates repetition suppression, while a higher response to the standard context than the cascade context signifies repetition enhancement. (C) The AC sends information about prediction error, negative prediction error, and repetition enhancement to the IC via the CC pathway. Repetition suppression was unaffected by CC suppression, suggesting that it may be a bottom-up process. *(Reprinted from Lesicko, A.M.H., Geffen, M.N., 2022. Diverse functions of the auditory cortico-collicular pathway. Hear. Res. 108488. © 2022, with permission from Elsevier.)*

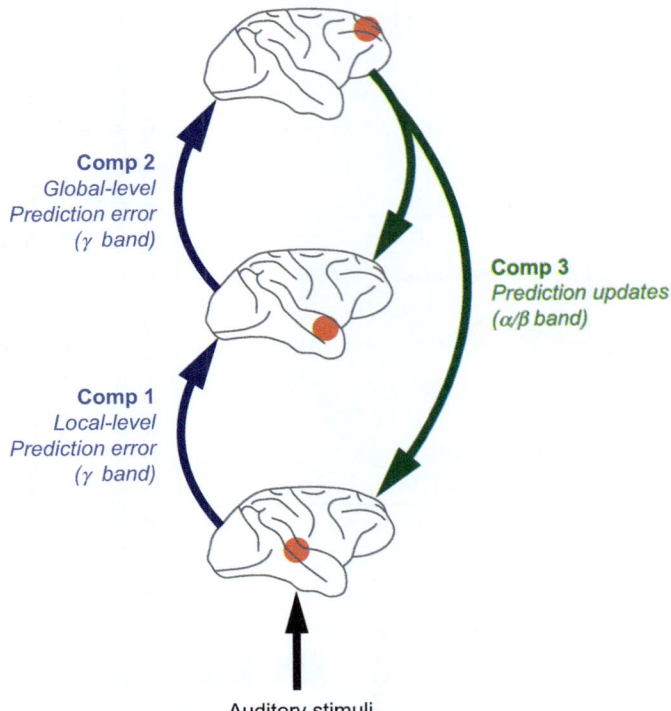

Fig. 4.15 Evaluation of functional correlations between activity components within and across trials. Schematics of the proposed bottom-up hierarchy of cortical signals coding for component 1, component 2, and prediction updates via component 3, and their corresponding cortical areas and frequency channels. *(Reprinted from Chao, Z.C., Takaura, K., Wang, L., Fujii, N., Dehaene, S., 2018. Large-scale cortical networks for hierarchical prediction and prediction error in the primate brain. Neuron 100, 1252–1266, © 2018, with permission from Elsevier.)*

sensory processing in the belt and parabelt areas to the higher-order processing of conspecific communication in the temporal polar areas and is also the main source of connections back to auditory cortices (Medalla and Barbas, 2014; Chapter 1). Chao et al. (2018) also noted that the ventrolateral PFC (VLPFC) is another key area for processing auditory sequences, particularly those with higher complexity (Wilson et al., 2017; Romanski and Averbeck, 2009). Studies with the local-global paradigm have shown that global novelty responses can be found in VLPFC in both monkeys and humans (Uhrig et al., 2014; Wang et al., 2015). Chao et al.'s (2018) analysis identified several electrodes in VLPFC in monkey 1 that showed late gamma-band power increases and beta-band power decreases in the global novelty response, although the responses were smaller than those in the frontopolar area 10 and DLPFC. In contrast to the frontopolar cortex and DLPFC, which showed signals associated with both the global prediction and its updates, VLPFC was only found involved in the update process. This suggested a distinctive modulatory role of VLPFC in auditory sequence encoding and storage.

Predictive coding suggests differences in message passing up versus down the cortical hierarchy. These differences result from the linear feedforward of prediction errors and the nonlinear feedback of predictions. This implies that cross-frequency interactions should predominate top-down. Márton et al. (2019) "revealed distinct patterns for bottom–up and top–down information processing among cross-frequency interactions. Both top-down and bottom-up interactions made prominent use of low frequencies: low-to-low-frequency (theta, alpha, beta) and low-frequency-to-high-gamma couplings were predominant top-down, while low-frequency-to-low-gamma couplings were predominant bottom-up." Márton et al. (2019) suggested that "the modulatory effect of low frequencies on gamma-rhythms in distant regions is important for bidirectional information transfer. The finding of low-frequency-to-low-gamma interactions in the bottom-up direction suggest that nonlinearities may also play a role in feedforward message passing."

Jiang et al. (2022) combined fMRI and high-density whole-brain ECoG in marmoset monkeys during an auditory local-global paradigm in which the temporal regularities of the stimuli were designed at two hierarchical levels. The prediction errors and prediction updates were examined as neural responses to auditory mismatches and omissions. Using fMRI, they identified a hierarchical gradient along the auditory pathway: midbrain and sensory regions represented local, shorter-timescale predictive processing followed by associative auditory regions, whereas anterior temporal and prefrontal areas represented global, longer-timescale sequence processing. The complementary ECoG recordings confirmed the activations at cortical surface areas and further differentiated the signals of prediction error and prediction update, which were transmitted via putative bottom-up gamma and top-down beta oscillations, respectively.

4.4 Summary

Several early studies in nonhuman primates reported MMN-like responses in auditory cortex, primarily originating from superficial layers and being abolished by NMDA antagonists. However, confusion of MMN with SSA required subsequent in-depth studies. Nonlemniscal cortical areas appear the dominant generators of the animal MMN, which were also distinct from areas generating CAEPs. In mouse cortex, SSA was similarly strong for excitatory and inhibitory neurons. SSA was also limited to the nonlemniscal auditory thalamus of guinea pigs and the nonlemniscal areas of the inferior colliculus. Inactivation of the thalamic reticular nucleus diminished the SSA in the MGB. SSA in IC was not dependent on cortico-collicular activity albeit that this modulated the strength of the SSA. GABAergic, glutamatergic, and cholinergic receptors play different and complementary roles on shaping SSA in IC. Simultaneous coapplication of $GABA_A$, $GABA_B$, and glycinergic antagonists demonstrated that local GABA- and glycine-mediated inhibition contributes to only about half of SSA. Therefore inhibition via chemical synapses dynamically modulates the strength and dynamics of stimulus-specific adaptation, but does not generate it.

Predictive coding allows integrating MMN and SSA data in a common theoretical framework underlying deviance detection at every level of measurement. Interaction of sensory input signals and top-down predictions forms a multilevel representation of sensory information. When top-down predictions do not fit the actual sensory input, the first-level neuronal populations—putatively subcortical—convey a prediction error signal to the higher levels to favor the processing of unpredicted features. This prediction error is functionally analogous to the genuine deviance detection. Predictive coding and deviance detection differ depending on the anesthetic state of the animal, whereby repetition enhancement and prediction error being more prevalent in the absence of anesthesia.

Higher-order error and update signals are found primarily in the frontopolar PFC (area 10) and DLPFC. Prediction errors and prediction signals in hierarchical coding can be separated in the frequency of the underlying brain rhythms. Top-down processes are represented by prediction updates via low-frequency ($<30\,Hz$) brain rhythms whereas the bottom-up prediction errors are conveyed by high-gamma ($>40\,Hz$) activity. The low frequencies have a modulatory effect on gamma rhythms in distant regions which is important for bidirectional information transfer that converges in minimizing prediction errors.

References

Anderson, L.A., Malmierca, M.S., 2013. The effect of auditory cortex deactivation on stimulus-specific adaptation in the inferior colliculus of the rat. Eur. J. Neurosci. 37, 52–62.

Anderson, L.A., Christianson, G.B., Linden, J.F., 2009. Stimulus-specific adaptation occurs in the auditory thalamus. J. Neurosci. 29 (22), 7359–7363.

Antunes, F.M., Malmierca, M.S., 2011. Effect of auditory cortex deactivation on stimulus-specific adaptation in the medial geniculate body. J. Neurosci. 31 (47), 17306–17316.

Antunes, F.M., Nelken, I., Covey, E., Malmierca, M.S., 2010. Stimulus-specific adaptation in the auditory thalamus of the anesthetized rat. PLoS One 5 (11), e14071. https://doi.org/10.1371/journal.pone.0014071.

Atencio, C.A., Schreiner, C.E., 2008. Spectrotemporal processing differences between auditory cortical fast-spiking and regular-spiking neurons. J. Neurosci. 28, 3897–3910. https://doi.org/10.1523/JNEUROSCI.5366-07.2008.

Ayala, Y.A., Malmierca, M.S., 2018. The effect of inhibition on stimulus-specific adaptation in the inferior colliculus. Brain Struct. Funct. 223, 1391–1407.

Ayala, Y.A., Pérez-González, D., Duque, D., Nelken, I., Malmierca, M.S., 2013. Frequency discrimination and stimulus deviance in the inferior colliculus and cochlear nucleus. Front. Neural Circuits 6, 119. https://doi.org/10.3389/fncir.2012.00119.

Ayala, Y.A., Udeh, A., Dutta, K., Bishop, D., Malmierca, M.S., Oliver, D.L., 2015. Differences in the strength of cortical and brainstem inputs to SSA and non-SSA neurons in the inferior colliculus. Sci. Rep. 5, 10383. https://doi.org/10.1038/srep10383.

Ayala, Y.A., Pérez-González, D., Malmierca, M.S., 2016. Stimulus-specific adaptation in the inferior colliculus: the role of excitatory, inhibitory and modulatory inputs. Biol. Psychol. 116, 10–22.

Bäuerle, P., von der Behrens, W., Kössl, M., Gaese, B.H., 2011. Stimulus-specific adaptation in the gerbil primary auditory thalamus is the result of a fast frequency-specific habituation and is regulated by the corticofugal system. J. Neurosci. 31 (26), 9708–9722.

Camalier, C.R., Scarim, K.C., Mishkin, M., Averbeck, B.A., 2019. A comparison of auditory oddball responses in dorsolateral prefrontal cortex, basolateral amygdala and auditory cortex of macaque. J. Cogn. Neurosci. 31 (7), 1054–1064.

Carbajal, G.V., Malmierca, M.S., 2018. The neuronal basis of predictive coding along the auditory pathway: from the subcortical roots to cortical deviance detection. Trends Hear. 22, 1–33. https://doi.org/10.1177/2331216518784822.

Casado-Román, L., Carbajal, G.V., Pérez-González, D., Malmierca, M.S., 2020. Prediction error signaling explains neuronal mismatch responses in the medial prefrontal cortex. PLoS Biol. 18 (12), e3001019. https://doi.org/10.1371/journal.pbio.3001019.

Chao, Z.C., Takaura, K., Wang, L., Fujii, N., Dehaene, S., 2018. Large-scale cortical networks for hierarchical prediction and prediction error in the primate brain. Neuron 100, 1252–1266.

Chen, I.W., Helmchen, F., Lütcke, H., 2015. Specific early and late oddball-evoked responses in excitatory and inhibitory neurons of mouse auditory cortex. J. Neurosci. 35 (36), 12560–12573.

Duque, D., Malmierca, M., Caspari, D., 2014. Modulation of stimulus-specific adaptation by $GABA_A$ receptor activation or blockade in the medial geniculate body of the anaesthetized rat. J. Physiol. 592, 729–743. https://doi.org/10.1113/jphysiol.2013.261941. In this issue.

Duque, D., Pais, R., Malmierca, M.S., 2018. Stimulus-specific adaptation in the anesthetized mouse revealed by brainstem auditory evoked potentials. Hear. Res. 370, 294–301.

Dürschmid, S., Zaehle, T., Hinrichs, H., Heinze, H.-J., Voges, J., Garrido, M.I., et al., 2016. Sensory deviancy detection measured directly within the human nucleus accumbens. Cereb. Cortex 26, 1168–1175.

Farley, B.J., Quirk, M.C., Doherty, J.J., Christian, E.P., 2010. Stimulus-specific adaptation in auditory cortex is an NMDA-independent process distinct from the sensory novelty encoded by the mismatch negativity. J. Neurosci. 30 (49), 16475–16484.

Fishman, Y.I., Steinschneider, M., 2012. Searching for the mismatch negativity in primary auditory cortex of the awake monkey: deviance detection or stimulus specific adaptation? J. Neurosci. 32 (45), 15747–15758.

Friston, K., 2005. A theory of cortical responses. Philos. Trans. R. Soc. B 360, 815–836. https://doi.org/10.1098/rstb.2005.1622.

Javitt, D.C., Schroeder, C.E., Steinschneider, M., Arezzo, J.C., Vaughan Jr., H.G., 1992. Demonstration of mismatch negativity in the monkey. Electroencephalogr. Clin. Neurophysiol. 83, 87–90.

Javitt, D.C., Steinschneider, M., Schroeder, C.E., Vaughan Jr., H.G., Arezzo, J.C., 1994. Detection of stimulus deviance within primate primary auditory cortex: intracortical mechanisms of mismatch negativity (MMN) generation. Brain Res. 667, 192–200.

Javitt, D.C., Steinschneider, M., Schroeder, C.E., Arezzo, J.C., 1996. Role of cortical N-methyl-D-aspartate receptors in auditory sensory memory and mismatch negativity generation: implications for schizophrenia. Proc. Natl. Acad. Sci. U. S. A. 93 (21), 11962–11967.

Jia, G., Li, X., Liu, C., He, J., Gao, L., 2021. Stimulus-specific adaptation in auditory thalamus is modulated by the thalamic reticular nucleus. ACS Chem. Nerosci. 12, 1688–1697.

Jiang, Y., Komatsu, M., Chen, Y., Xie, R., Zhang, K., Xia, Y., et al., 2022. Constructing the hierarchy of predictive auditory sequences in the marmoset brain. eLife 11, e74653. https://doi.org/10.7554/eLife.74653.

Kraus, N., McGee, T., Littman, T., Nicol, T., King, C., 1994. Nonprimary auditory thalamic representation of acoustic change. J. Neurophysiol. 72 (3), 1270–1277.

Lakatos, P., O'Connell, M.N., Barczak, A., McGinnis, T., Neymotin, S., Schroeder, C.E., et al., 2020. The thalamocortical circuit of auditory mismatch negativity. Biol. Psychiatry 87, 770–780.

Lesicko, A.M.H., Geffen, M.N., 2022. Diverse functions of the auditory cortico-collicular pathway. Hear. Res., 108488.

Lesicko, A.M.H., Angeloni, C.F., Blackwell, J.M., De Biasi, M., Geffen, M.N., 2022. Corticofugal regulation of predictive coding. eLife 11, e73289. https://doi.org/10.7554/eLife.73289.

Liu, M., Dai, J., Zhou, M., Liu, J., Ge, X., Wang, N., Zhang, J., 2022. Mini-review: the neural circuits of the non-lemniscal inferior colliculus. Neurosci. Lett. 776, 136567.

Lomber, S.G., Payne, B.R., Horel, J.A., 1999. The cryoloop: an adaptable reversible cooling deactivation method for behavioural or electrophysiological assessment of neural function. J. Neurosci. Methods 86, 179–194.

Malmierca, M.S., Cristaudo, S., Pérez-González, D., Covey, E., 2009. Stimulus-specific adaptation in the inferior colliculus of the anesthetized rat. J. Neurosci. 29 (17), 5483–5493.

Malmierca, M.S., Niño-Aguillón, B.E., Nieto-Diego, J., Porteros, A., Pérez-González, D., Escera, C., 2019. Pattern-sensitive neurons reveal encoding of complex auditory regularities in the rat inferior colliculus. Neuroimage 184, 889–900.

Márton, C.D., Fukushima, M., Camalier, C.R., Schultz, S.R., Averbeck, B.B., 2019. Signature patterns for top-down and bottom-up information processing via cross-frequency coupling in macaque auditory cortex. eNeuro 6 (2). e0467-18.2019.

Medalla, M., Barbas, H., 2014. Specialized prefrontal "auditory fields": organization of primate prefrontal-temporal pathways. Front. Neurosci. 8, 77. https://doi.org/10.3389/fnins.2014.00077.

Nieto-Diego, J., Malmierca, M.S., 2016. Topographic distribution of stimulus-specific adaptation across auditory cortical fields in the anesthetized rat. PLoS Biol. 14 (3), e1002397. https://doi.org/10.1371/journal.pbio.1002397.

O'Reilly, J.A., 2021. Roving oddball paradigm elicits sensory gating, frequency sensitivity, and long-latency response in common marmosets. IBRO Neurosci. Rep. 11, 128–136.

O'Reilly, J.A., Conway, B.A., 2021. Classical and controlled auditory mismatch responses to multiple physical deviances in anaesthetised and conscious mice. Eur. J. Neurosci. 53, 1839–1854. https://doi.org/10.1111/ejn.15072.

Parras, G.G., Nieto-Diego, J., Carbajal, G.V., Valdés-Baizabal, C., Escera, C., Malmierca, M.S., 2017. Neurons along the auditory pathway exhibit a hierarchical organization of prediction error. Nat. Commun. 8, 2148. https://doi.org/10.1038/s41467-017-02038-6.

Parras, G.G., Casado-Román, L., Schröger, E., Malmierca, M.S., 2021. The posterior auditory field is the chief generator of prediction error signals in the auditory cortex. Neuroimage 242, 118446.

Pérez-González, D., Malmierca, M.S., Covey, E., 2005. Novelty detector neurons in the mammalian auditory midbrain. Eur. J. Neurosci. 22, 2879–2885.

Pérez-González, D., Hernández, O., Covey, E., Malmierca, M.S., 2012. GABA$_A$-mediated inhibition modulates stimulus-specific adaptation in the inferior colliculus. PLoS One 7 (3), e34297. https://doi.org/10.1371/journal.pone.0034297.

Pérez-González, D., Parras, G.G., Morado-Díaz, C.J., Aedo-Sánchez, C., Carbajal, G.V., Malmierca, M.S., 2021. Deviance detection in physiologically identified cell types in the rat auditory cortex. Hear. Res. 399, 107997.

Pincze, Z., Lakatos, P., Rajkai, C., Ulbert, I., Karmos, G., 2001. Separation of mismatch negativity and the N1 wave in the auditory cortex of the cat: a topographic study. Clin. Neurophysiol. 112, 778–784.

Polterovich, A., Jankowski, M.M., Nelken, I., 2018. Deviance sensitivity in the auditory cortex of freely moving rats. PLoS One 13 (6), e0197678. https://doi.org/10.1371/journal.pone.0197678.

Romanski, L.M., Averbeck, B.B., 2009. The primate cortical auditory system and neural representation of conspecific vocalizations. Annu. Rev. Neurosci. 32, 315–346.

Rui, Y.-Y., He, J., Zhai, Y.-Y., Sun, Z.-H., Yu, X.-J., 2018. Frequency-dependent stimulus-specific adaptation and regularity sensitivity in the rat auditory thalamus. Neuroscience 392, 13–24.

Srivastava, H.K., Bandyopadhyay, S., 2020. Parallel lemniscal and non-lemniscal sources control auditory responses in the orbitofrontal cortex (OFC). eNeuro 7 (5), 1–19. ENEURO.0121-20.2020.

Szymanski, F.D., Garcia-Lazaro, J.A., Schnupp, J.W.H., 2009. Current source density profiles of stimulus-specific adaptation in rat auditory cortex. J. Neurophysiol. 102, 1483–1490.

Taaseh, N., Yaron, A., Nelken, I., 2011. Stimulus-specific adaptation and deviance detection in the rat auditory cortex. PLoS One 6 (8), e23369. https://doi.org/10.1371/journal.pone.0023369.

Uhrig, L., Dehaene, S., Jarraya, B., 2014. A hierarchy of responses to auditory regularities in the macaque brain. J. Neurosci. 34 (4), 1127–1132.

Ulanovsky, N., Las, L., Nelken, I., 2003. Processing of low-probability sounds by cortical neurons. Nat. Neurosci. 6 (4), 391–398.

Ulanovsky, N., Las, L., Farkas, D., Nelken, I., 2004. Multiple time scales of adaptation in auditory cortex neurons. J. Neurosci. 24, 10440–10453. https://doi.org/10.1523/JNEUROSCI.1905-04.2004.

Valdés-Baizabal, C., Carbajal, G.V., Pérez-González, D., Malmierca, M.S., 2020. Dopamine modulates subcortical responses to surprising sounds. PLoS Biol. 18 (6), e3000744. https://doi.org/10.1371/journal.pbio.3000744.

Wang, L., Uhrig, L., Jarraya, B., Dehaene, S., 2015. Representation of numerical and sequential patterns in macaque and human brains. Curr. Biol. 25, 1966–1974.

Wilson, B., Marslen-Wilson, W.D., Petkov, C.I., 2017. Conserved sequence processing in primate frontal cortex. Trends Neurosci. 40, 72–82.

Wolff, W., Berberian, N., Golesorkhi, M., Gomez-Pilar, J., Zilio, F., Northoff, G., 2022. Intrinsic neural timescales: temporal integration and segregation. Trends Cogn. Sci. 26 (2), 159–173. https://doi.org/10.1016/j.tics.2021.11.007.

Xu, X.-X., Zhai, Y.-Y., Kou, X.-K., Yu, X., 2017. Adaptation facilitates spatial discrimination for deviant locations in the thalamic reticular nucleus of the rat. Neuroscience 365, 1–11.

Yu, X.-J., Xu, X.-X., He, S., He, J., 2009. Change detection by thalamic reticular neurons. Nat. Neurosci. 12 (9), 1165–1170.

CHAPTER 5

Auditory cortical event-related potentials in the human brain

5.1 Introduction

5.1.1 Obligatory and endogenous event-related auditory potentials

Event-related brain potentials (ERPs) are evoked exogenously by environmental events (such as sensory stimuli) or endogenously by processes such as decision making. We will call the cortical obligatory responses to external stimuli "cortical auditory evoked potentials" (CAEPs), and leave the connotation of ERPs to endogenous evoked activity, e.g., from attention, stimulus mismatch, tasks, or novelty. Peaks within a CAEP or ERP have been classified according to their timing relative to stimulus onset and anatomical site of generation, or perceptual functions. The CAEPs represent peaks—transient activity—reflective of initial processing in auditory cortex, such as P50 (P1) followed by peaks such as N1, P2, N2, P3a, P3b, and N400. The latter four fall under the rubric of ERPs and reflect more integrative cognitive processing (Crowley and Colrain, 2004). The P2 is an outlier compared to its surrounding N1 and N2 peaks. In later chapters, we will also consider among others the error-related negativity (ERN), the reorienting negativity (RON), the contingent negative variation (CNV), and the readiness potential (RP), of which the properties are detailed in the Appendix.

5.1.2 Setting apart P2

Unlike the N1 or the other late CAEPs (cf. Fig. 2.5) and despite its longer latency, the P2 stands apart. P2 has multiple generators located in multiple auditory areas, including primary and secondary cortex, and is driven by the caudal mesencephalic reticular activating system (mRAS; Velasco et al., 1989; Crowley and Colrain, 2004). Our studies (Eggermont, 1988; Ponton et al., 2000a) show a shorter maturational time course of the P2 peak as compared to the N1 and provide strong support for different pathways and neural generators for these peaks. I noted that the short maturation time constant for P2 latency is similar to that of the auditory brain stem response I-V

Brain Responses to Auditory Mismatch and Novelty Detection
https://doi.org/10.1016/B978-0-443-15548-2.00005-3

interpeak interval, which may suggest that the maturation of the P2 latency is only limited by maturational processes affecting the brainstem (Eggermont, 1988). Thus Ponton et al. (2000a) have argued that—assuming that the proposed cortical generators of the P2 peak are correct—the early maturation of P2 latency may reflect the rapid maturation of pathways connecting the auditory brain stem, e.g., via the RAS to auditory cortex. In contrast, the N1 generated by the thalamocortical pathway matures more slowly and does not reach adult latency values until adolescence. For a stimulation rate of ∼1.5/s, Ponton et al. (2000a,b) showed an absence of a contralateral N1 in normal-hearing 9-year-olds (Chapter 6). Recently, Regev et al. (2021) noted that in a mismatch experiment N1 was highly affected by long-term context whereas P2 was not. Additionally, both were affected by short-term context but P2 more robustly so. Furthermore, the spectral context (frequency range in the sequence) had a distinct effect on both the N1 and P2 amplitudes. A smaller stimulus-sequence range reduced N1 more than P2 amplitudes, suggesting that adaptation affects the N1 more than the P2. These results imply that neural activity at the latencies of the N1 and P2 has distinct timescales of contextual influences, not surprising as the afferents to their cortical generation sites are completely different (Ponton et al., 2000a; Crowley and Colrain, 2004).

5.2 Effects of adaptation, attention, and memory

Hillyard et al. (1973) reported that the N1 was selectively enhanced in response to attended stimuli. Näätänen and Michie (1979) suggested a reinterpretation of this N1 effect, namely that it is not a "true" N1 component which is enhanced but the effect is produced by a summation of a negative shift with the N1 wave form. To explain the origin of this negative shift, Näätänen et al. (1982) noted that in detection conditions, deviant stimuli elicit two overlapping sequences of brain events: exogenous and endogenous. The exogenous ones mainly include the processes producing the N1 and the N2 (neuronal mismatch) components and represent an automatic, inflexible set of brain processes which appears as if providing a central-level stimulus to the endogenous sequence. The endogenous contribution includes a triphasic frontocentral complex overlapping the mismatch N2 and precedes the parietal late positive component and frontal late negative component. Thus Näätänen et al. (1982) described the N1, N2 complex as being composed of two sets of brain potentials. (1) The first is an exogenous series of waves, including the N1 and P2 components, which may be evoked by any stimulus. (2) The second set of brain waves is of endogenous nature and seems to reflect detection of a deviant stimulus in oddball paradigms. The earliest elements in the endogenous sequence are (i) the frontocentral waves, P2, N2b, and P3a, which appear to be superimposed on the slower mismatch-negativity (MMN, sometimes called N2a), and whose amplitude mainly depends on the salience and probability of the stimulus; and (ii) a slow parietal positivity, P3b (Fig. 5.1; Näätänen et al., 2014).

The first time that the mismatch-negative N2a (Näätänen et al., 1982) obtained the acronym of MNN is in Mäntysalo and Näätänen (1987) who stated: "Event-related brain potential studies have revealed a component called the mismatch negativity (MMN) in response to occasional deviant stimuli presented in a sequence of homogeneous repetitive, 'standard,' stimuli. This negativity probably reflects an automatic neuronal-mismatch

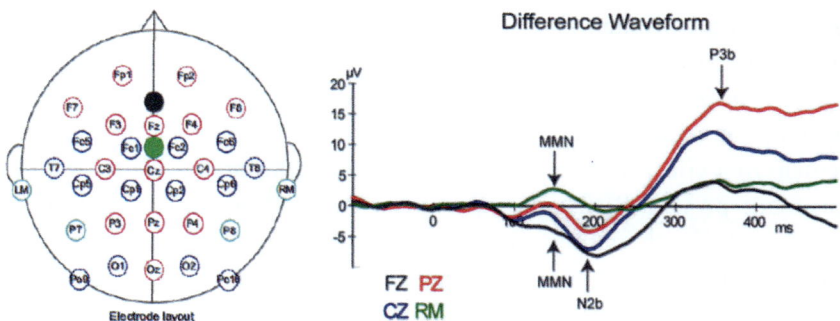

Fig. 5.1 Mismatch negativity (MMN) evoked during active detection of a deviant stimulus is elicited along with attention-related ERP components (N2b and P3b). Left. Standard 10-20 EEG-electrode placements. Right. Grand-mean difference waveform (obtained by subtracting the ERP elicited by the deviant stimulus from the ERP elicited by the standard stimulus) is displayed. The electrode positions used are all central line positions plus a mastoid electrode (RM). The MMN peak latency is denoted at the mastoid *(green trace)* because of overlap with N2b. The MMN is more frontally distributed and the peak of the MMN component is not well delineated at Fz *(black trace)* due to overlap with the N2b. The N2b component is seen clearest at the Cz electrode *(blue trace)*. The P3b *(red trace)* is largest at the midline parietal electrode (Pz). *(Left panel: From Virtala, P., Putkinen, V., Kailaheimo-Lönnqvist, L., Thiede, A., Partanen, E., Kujala, T., 2022. Infancy and early childhood maturation of neural auditory change detection and its associations to familial dyslexia risk. Clin. Neurophysiol. 137, 159–176, open access; Right panel: From Näätänen, R., Sussman, E.S., Salisbury, D., Shafer, V.L., 2014. Mismatch negativity (MMN) as an index of cognitive dysfunction brain. Brain Topogr. 27, 451–466. © Springer Science+Business Media New York 2014, with permission from Springer Nature.)*

process with a neuronal representation of the standard stimulus." Combined electromagnetic, hemodynamic, and psychophysical data (Jääskeläinen et al., 2004) indicated that "the MMN is generated in auditory cortex [potentially] as a result of differential adaptation of anterior and posterior N1 sources by preceding auditory stimulation. Early (~85 ms) neural activity within posterior auditory cortex adapts as sound novelty decreases. This alters the center of gravity of electromagnetic N1 source activity, creating an illusory difference between N1 and MMN source loci when estimated by using equivalent current dipole fits." A more precise localization of MMN source would allow a better comparison to known parts of the superior temporal (STG) and inferior frontal gyri (IFG) as used in dynamic causal modeling studies (Chapter 1).

From EEG recordings, Schönwiesner et al. (2007) concluded that pre-attentive change processing occurs in at least three stages: (1) initial detection in the primary auditory cortex (A1), (2) detailed analysis in the posterior STG and planum temporale, and (3) judgment of sufficient novelty for the allocation of attentional resources in the mid-ventrolateral prefrontal cortex (VLPFC). Hsu et al. (2021) also used EEG to investigate whether the attention-independent and attention-dependent components of stimulus-repetition effects previously described in the auditory modality remain in a context of low periodicity where temporary disruption might be absent/present. Participants were presented with repetition trains of various lengths, with/without temporary disruptions. They found attention-independent (P2) and attention-dependent repetition effects (P3a).

5.3 Deviance detection

5.3.1 Local vs global mismatch

Bekinschtein et al. (2009) designed an auditory paradigm for evaluating cortical responses to violations of temporal regularities that are either local in time or global across several seconds. The Local-Global experimental design is a variation of the auditory oddball task. It consists in presenting series of five-sound sequences which are composed of five identical sounds (local standard) or four identical sounds followed by a deviant one (local deviant). The global regularity is established across trials by making 80% of the trials identical (global standard). The design thus dissociates the violation of local predictions (change of sound in a given trial, local time) and global predictions (change of sequence across trials; cf. Fig. 2.11). "Local violations led to an early response in auditory cortex, independent of attention or the presence of a concurrent visual task, whereas global violations led to a late and spatially distributed response that was only present when subjects were attentive and aware of the violations. The global effect was detected in individual subjects using functional MRI and both scalp and intracerebral event-related potentials." In an EEG study with healthy young adults, Kompus et al. (2020) used such a hierarchical oddball paradigm with local (sequence-level) and global (block-level) violations in attended and unattended conditions. Amplitude of N2 and P3b was analyzed in a $2 \times 2 \times 2$ factorial model (local status, global status, attention condition). They found a significant interaction between the local and global status on the N2 amplitude, while there was no significant three-way interaction with attention, demonstrating that lower-level prediction error is modulated by detection of higher-order regularity but expressed independently of attention. By contrast, higher-level prediction error, indexed by P3b, was sensitive to global regularity violations if the auditory stream was attended.

5.3.2 Localization of cortical structures involved in deviance detection

Cacciaglia et al. (2019) implemented a modified version of the auditory frequency-oddball paradigm that enabled modeling the responses to both repeated standard and deviant stimuli. Fifteen subjects underwent functional magnetic resonance imaging (fMRI) while their attention was diverted from auditory stimulation. They found that deviants following a larger number of standard repetitions yielded a more robust and widespread activation bilateral in auditory cortex. Repetition of standard tones yielded a pattern of response comprising both suppression and enhancement effects depending on the predictability of upcoming stimuli. Furthermore, regularity encoding and deviance detection mapped onto spatially segregated cortical subfields (Fig. 5.2). Cacciaglia et al. (2019) found that regardless of the number of preceding stimuli, the comparison between all deviant (DEV) against all standard (STD) trials yielded significant activations in the superior frontal gyrus (STG) and Heschl's gyrus (HG), bilaterally. To their surprise, however, they did not find significant activations in the IFG, a result that was observed in previous classic oddball paradigms as well as in roving standard protocols and forms part of MMN-based prediction models (Chapter 1). Overall, Cacciaglia et al. (2019) suggested that compared to STG activations, IFG activation is less consistently detected in fMRI studies, which has led to the suggestion that the IFG response to auditory change might involve an increase in synchronization of neurons—detectable in EEG—rather than

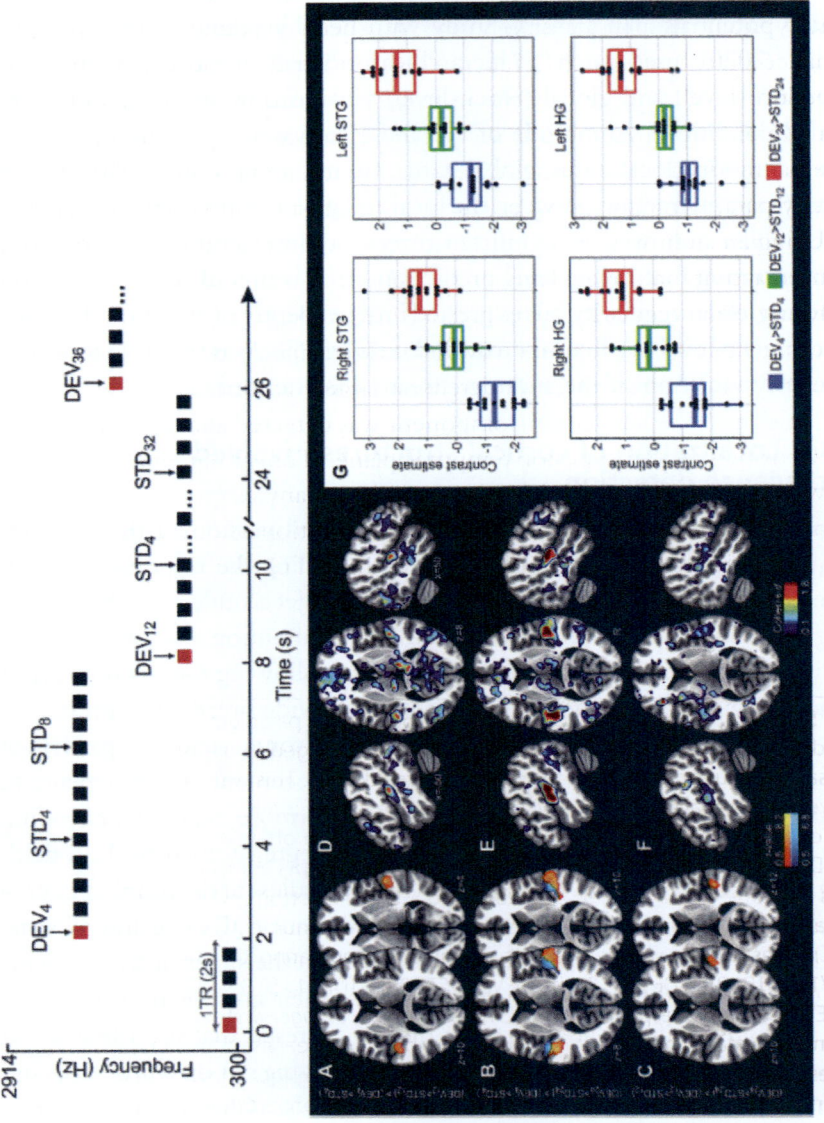

Fig. 5.2 See legend on opposite page.

an increase in the number activated or their firing rates (Deouell, 2007). Furthermore, in addition to auditory areas, a whole-brain analysis performed for all effects with DEV > STD trials showed a significant activation in the middle cingulate gyrus extending to the SFG. The cingulate cortex, although most typically its anterior subdivision, has been included in the auditory deviance-detection network and its function is related to automatic error detection as well as conflict monitoring (Kiehl et al., 2005). The middle portion of the cingulate cortex is reciprocally connected with the insula and other areas within the salience network (Menon and Uddin, 2010) and its activity paradigm may reflect a general mechanism of novelty detection.

Using an auditory oddball task paradigm, Ishishita et al. (2019) obtained induced activity in the high-gamma band as mismatch response (MMR) by analyzing electrocorticographic (ECoG) data from patients with refractory epilepsy. They found that the MMR localized mainly in the bilateral posterior STGs, and that deviance detection largely accounted for MMR. At the same time, the adaptation component was detected at a limited number of electrodes on the superior temporal plane. Tone-frequency difference showed no contribution to MMR generation at any electrode in the lateral cortices. These findings reveal a mixed contribution of deviance detection and adaptation depending on location in the STG, and demonstrate that

---

Fig. 5.2 Top. Schematic illustration of the experimental paradigm. A frequency roving standard paradigm was adopted with pure tones arranged in trains of stimuli having the same frequency but different number of repetitions. This way, the first stimulus of a given train always acts as a deviant tone. One event of interest was represented by a trial of four consecutive pure tones spanning a duration of 2000 ms. The first stimulus of a DEV trial is highlighted in *red color*. Numeric subscripts associated to STD and DEV trials indicate the number of preceding stimuli. Bottom. Comparison between deviance-detection responses associated to different number of preceding stimuli. (A–C) Statistical parametric maps as revealed by comparing the differential contrasts [DEV12>STD12] > [DEV4>STD4], [DEV24>STD24] > [DEV4>STD4], and [DEV24>STD24] >[DEV12>STD12], respectively. The STG is shown in *orange*, while the HG in *blue*. (D–F) Parametric maps of the effect sizes computed voxel-wise on the whole brain, corresponding to (A), (B), and (C), respectively. (G) Boxplots showing the change in magnitude of the hemodynamic response to distinct contrasts capturing the stimulus history. The lower and upper hinges correspond to the first and third quartile, while dots indicate individual subject data. *HG*, Heschl's gyrus; *STG*, superior temporal gyrus. *(From Cacciaglia, R., Costa-Faidella, J., Zarnowiec, K., Grimm, S., Escera, C., 2019. Auditory predictions shape the neural responses to stimulus repetition and sensory change. Neuroimage 186, 200–210. © 2018, with permission from Elsevier.)*

"deviance detection," which is higher-level neural processing than "adaptation," plays an important role in auditory contextual processing at the level of STG. The same group (Takasago et al., 2020) again used ECoG to spatiotemporally differentiate MMN from N1 adaptation. ERPs under the classical oddball task as well as the many-standards task were recorded in three patients with epilepsy whose lateral cortices were widely covered with high-density electrodes. They identified an electrode at which N1 adaptation was temporally separated from MMN, whereas N1 adaptation was partially incorporated into MMN at other electrodes. Using the many-standards task, instead of N1 adaptation resulting in the oddball task, they found N1 in a limited area around the Sylvian fissure adjacent to A1, whereas MMN was noted in wider areas, including the temporal, frontal, and parietal lobes (Fig. 5.3). This again suggests that MMN is not merely a product of the neural adaptation of N1 and instead represents higher-order processes in auditory deviance detection.

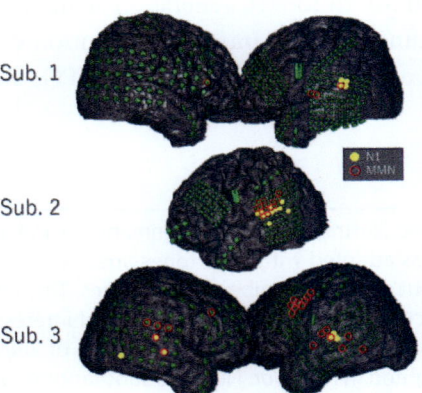

Fig. 5.3 Distribution of N1 and MMN electrodes in each subject. N1 and MMN electrodes are statistically selected and indicated on the three-dimensional brain surface as *yellow filled circles* and *red open circles*, respectively. N1 electrodes except for one electrode in the inferior temporal gyrus of Subject 3 were located in the superior temporal gyrus near the superior temporal plane, whereas MMN electrodes were widely distributed in the superior and middle temporal gyrus, parietal lobe, and frontal lobe. The distribution pattern of MMN electrodes differed markedly among subjects. *(From Takasago, M., Kunii, N., Komatsu, M., Tada, M., Kirihara, K., Uka, T., et al., 2020. Spatiotemporal differentiation of MMN from N1 adaptation: a human ECoG study. Front. Psychiatry 11, 586. doi:https://doi.org/10.3389/fpsyt.2020. 00586. Open Access.)*

5.3.3 Involvement of subcortical structures in human deviance detection

We have seen the dominance of nonlemniscal parts of the inferior colliculus (IC) and medial geniculate body (MGB) in the stimulus-specific adaptation in animals (Chapter 4) and will now show comparable subcortical effects on the MMN and/or P3. We will follow the hierarchy along the auditory pathway.

5.3.3.1 *The auditory brainstem*

Slabu et al. (2012) recorded the auditory brainstem frequency-following response (FFR) to consonant-vowel stimuli (/ba/, /wa/) in young adults (Fig. 5.4), with stimuli arranged in oddball and reversed oddball blocks (deviant probability, $P = .2$), allowing for the comparison of FFRs to the same physical stimuli presented in different contextual roles. Whereas no effect was observed for the /wa/ syllable, the /ba/ syllable showed a reduction in the brainstem FFR to deviant stimuli compared with standard ones and to similar stimuli arranged in a control block, with five equiprobable, rarely occurring sounds. These findings suggest that the human auditory brainstem is able to encode regularities in the recent auditory past to detect novel events and confirm the multiple anatomical and temporal scales of human deviance detection.

5.3.3.2 *The auditory midbrain and thalamus*

Using event-related fMRI imaging during a frequency oddball paradigm, Cacciaglia et al. (2015) reported that auditory deviance detection occurs in the inferior colliculus (IC) and medial geniculate body (MGB) of healthy human participants. By implementing a random condition controlling for neural refractoriness effects (Fig. 5.5 top), they showed that auditory change detection in these subcortical stations involves the encoding of statistical regularities from the acoustic input. Specifically, Cacciaglia et al. (2015) found significant activations in the left IC and bilateral in the MGB in response to deviant–containing trains of stimuli versus trains containing standard stimuli only (Fig. 5.5 bottom). In addition, they observed significant activations bilateral in the STG and left HG, matching previous imaging studies implementing frequency oddball tasks, where activation of these cortical regions has been consistently reported.

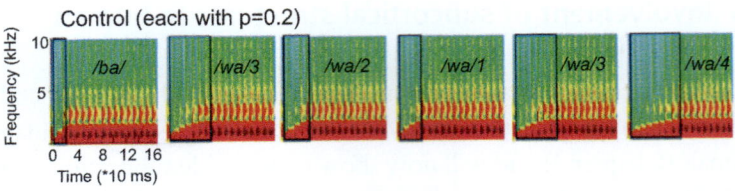

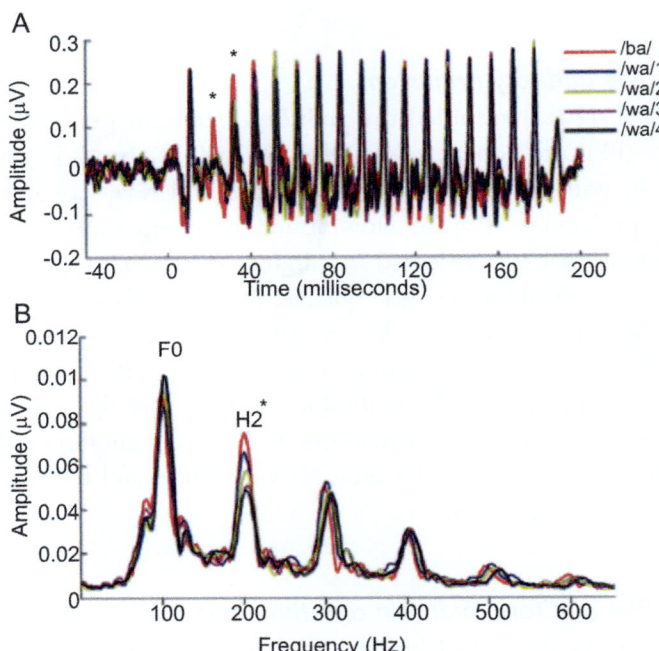

Fig. 5.4 Top. Stimuli and experimental design. In the control condition, five different syllables (/ba/, /wa/1, /wa/2, /wa/3, and /wa/4) were presented randomly, each with a probability of $P = .2$. Each *square* represents the spectrogram of the 170-ms stimulus. The *black rectangles* frame the transition duration of formants F1 and F2 that differentiated the five different syllables. Bottom. Grand-average FFRs (A) and FFR spectra (B) for the /ba/, /wa/1, /wa/2, /wa/3, and /wa/4 syllables in the control condition. (A) A significant enhancement of the response amplitude in two latency windows (18–22 ms and 27–31 ms) over the temporal transition of F1 and F2 was observed for the /ba/ syllable compared with the /wa/1 syllable (*$P < .05$). (B) The amplitude of the H2 (*$P < .05$) harmonic followed the increase of the formant transition duration in the control condition. *(From Slabu, L., Grimm, S., Escera, C., 2012. Novelty detection in the human auditory brainstem. J. Neurosci. 32 (4), 1447–1452.)*

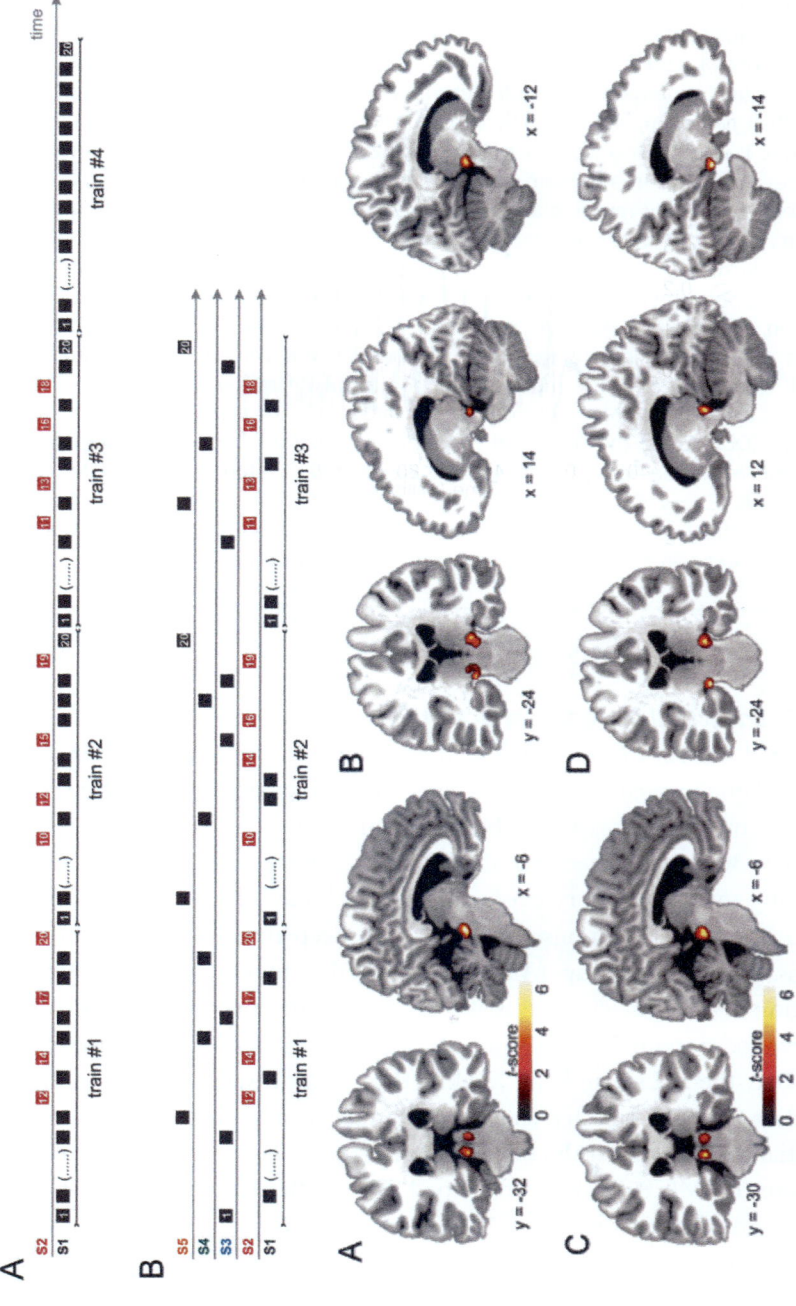

Fig. 5.5 See legend on next page.

Fig. 5.5 Top. Experimental paradigm. (A) The oddball paradigm consisted of 125 trains containing 20 identical stimuli acting as standard (i.e., train #4), randomly interleaved with 125 trains containing the standard plus the deviant stimulus (i.e., trains #1, #2, #3) (B) The control condition consisted of 125 trains each containing 5 randomly distributed and equiprobable stimuli presented with the same timing parameters as in the oddball paradigm, with the only constrain that S2 occurred in the same physical position as it was in the oddball paradigm, where it acted as deviant. Bottom. Auditory deviance detection in midbrain and thalamic structures. (A) For the DEV>STD condition, ROI analysis revealed a significant activation in the left inferior colliculus (IC). Contrasts of interest were corrected for multiple testing using a family-wise error (FWE) rate approach at $P < .05$. Color bar indicates t-values. (B) In the same condition (DEV>STD), both the right and left medial geniculate body (MGB) displayed significant activation, which survived FWE correction. (C) For the DEV>CON condition, ROI analysis yielded significant activation in the left inferior colliculus (IC). Contrasts of interest were corrected for multiple testing using a family-wise error (FWE) rate approach at $P < .05$. Color bar indicates t-values. (D) In the same experimental condition (DEV>CON) the right and left medial geniculate body (MGB) displayed a significant activation. CON, control; DEV, deviant; STD, standard. (Reprinted from Cacciaglia, R., Escera, C., Slabu, L., Grimm, S., Sanjuán, A., Ventura-Campos, N., Ávila, C., 2015. Involvement of the human midbrain and thalamus in auditory deviance detection. Neuropsychologia 68, 51–58. © 2015, with permission from Elsevier.)

5.3.3.3 Early thalamocortical activity

Slabu et al. (2010) measured the middle-latency response (MLR; reflecting early thalamocortical activity) to novel-frequency stimuli embedded in an oddball sequence. Occasional changes in auditory frequency were detected as early as 30 ms (Pa waveform of the MLR) after stimulus onset. Note that, according to the "auditory evoked potentials bible" (Picton, 2011, p. 117) the Pa is a cortical response originating in deep regions of core and belt regions. The same group (Althen et al., 2011) investigated whether the detection of intensity deviants is also reflected at these short latencies. Auditory evoked potentials in response to click sounds were analyzed regarding the ABR, the MLR, and the MMN. Rare stimuli with a lower intensity level than standard stimuli elicited (in addition to an MMN) a more negative potential in the MLR at the transition from the Na—afferent inflow to cortex from the MGB (Picton, 2011, p. 117)—to the Pa component at circa 24 ms from stimulus onset (Fig. 5.6).

Again, the same group (Grimm et al., 2012) measured the MLR and MMN to changes in sound location. Clicks were presented in either the left

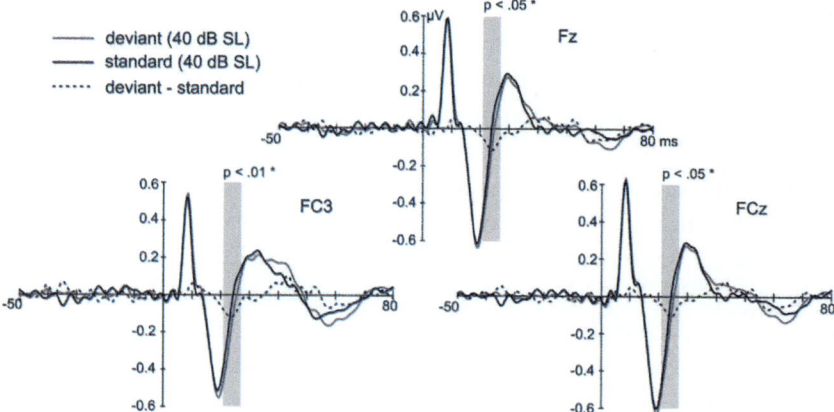

Fig. 5.6 Deviance-related changes in the MLR. Grand-average response ($N=23$) at FCz, FC3, and Fz elicited by deviants and standards. The *gray shaded fields* mark the time window of the mean amplitudes used for statistics. The difference waveforms reveal a negative displacement of the response to deviants compared to the one to standards. *MLR*, middle-latency response. *(From Althen, H., Grimm, S., Escera, C., 2011. Fast detection of unexpected sound intensity decrements as revealed by human evoked potentials. PLoS One 6 (12), e28522. doi:https://doi.org/10.1371/journal.pone.0028522. Open Access.)*

or right hemifields during oddball (rare 30-degree shifts in location), reversed oddball, and control (sounds occurring equiprobably from five locations) conditions. Clicks at deviant locations elicited an MMN and an enhanced Na component of the MLR peaking at 20 ms compared to clicks at standard or control locations. Whereas MMN was not significantly lateralized, the Na effect showed a contralateral dominance. Thus, also for sound location changes, early detection processes exist upstream of MMN. Once more, the same group (López-Caballero et al., 2016), also using an oddball location paradigm with three different deviant probabilities (5%, 10%, and 20%) and a reversed standard (91.5%), recorded MLRs in 24 healthy participants and analyzed differences in the MLRs elicited to each of the deviant stimuli and the reversed standard, as well as within deviant stimuli. They confirmed deviance detection at the level of both MLRs and MMN, but significant differences for deviant probabilities were found only for the MMN. These results suggest a functional dissociation between regularity encoding, already present at early stages of auditory processing, and the encoding of the probability with which this regularity is disrupted, which is only processed at higher stages of the auditory hierarchy.

5.3.3.4 Limbic structures

Dürschmid et al. (2016) reported event-related potentials obtained from the nucleus accumbens (NAc) using direct intracranial recordings in 5 human participants while they listened to trains of auditory stimuli differing in their degree of deviation from repetitive background stimuli. NAc recordings revealed an early mismatch signal (50–220 ms) in response to all deviants. NAc activity in this time window was also sensitive to the statistics of stimulus deviancy, with larger amplitudes as a function of the level of deviancy. Importantly, this NAc mismatch signal also predicted generation of longer-latency scalp potentials (300–400 ms). This suggests that the NAc is a key component of a network engaged in encoding statistics of the sensory environment.

5.3.4 Effect of stimulus regularities and probabilities

Does behavioral and cortical auditory deviance detection (the latter indexed by the MMN) relies on probabilities of sound patterns or on transitional probabilities, i.e., changes in regularity? To investigate this, Mittag et al. (2016) presented healthy adult volunteers with three types of rare tone triplets among frequent standard triplets of high-low-high (H-L-H) or L-H-L frequency structure. These types were (1) proximity deviant (H-H-H/L-L-L), (2) reversal deviant (L-H-L/H-L-H), and (3) first-tone deviant (L-L-H/H-H-L). They argued that if deviance detection is based on pattern probability, reversal and first-tone deviants should be detected with similar latency because both differ from the standard at the first pattern position. If deviance detection is based on transitional probabilities, then reversal deviants should be the most difficult to detect because, unlike the other two deviants, they contain no low-probability pitch transitions. Mittag et al. (2016) reported that "both behavioral and cortical auditory deviance detection uses transitional probabilities. Thus, the memory traces underlying cortical deviance detection may provide a link between stimulus probability-based change/novelty detectors operating at lower levels of the auditory system and higher auditory cognitive functions that involve predictive processing." Schröger and Roeber (2021) studied a simple stochastic regularity—two tone pitches (standards, each occurring on 45% of trials)—that was occasionally violated by another tone pitch (deviant, occurring on 10% of trials). MMN was obtained when the deviant's pitch was outside those of the standards, but not when it was between them. Alternating the occurrence of the same two standards—making them deterministic—caused the deviant to elicit an MMN, even when its pitch was between those of the standards.

Fig. 5.7 shows in the top panel a typical trial run and in the middle panel the ERPs from frontal-frontocentral and right-mastoid electrodes to standard and deviant trials and their difference waves in all conditions. Voltage maps for both stimulus types in each condition averaged across three time windows of interest (N1, MMN, N2b) are illustrated in the bottom panel in Fig. 5.7. "The ERPs to all stimuli show an early positivity (P1) at about 60 ms at the frontal-frontocentral electrodes that is accompanied by a slight negativity at the mastoid electrode. P1 amplitudes are about the same for standards and deviants in each condition. ERPs to standard and deviant stimuli do not differ much in the tightly enclosing oddball condition in which P1 is followed by a negativity at about 90 ms (N1) and another positivity (P2) at about 150 ms. N1's and P2's are also observed in the other condition. In the widely enclosing oddball condition, deviants elicited larger N1 amplitudes than standards but there is no apparent P2 amplitude difference. In the alternating enclosing oddball condition there is no apparent N1 difference but deviants have smaller P2 amplitudes than standards. In the active-listening tightly enclosing oddball condition, ERPs to deviants and standards did not differ from each other for the first 200 ms." Schröger and Roeber (2021) concluded that the MMN system—unlike other regularity encoding systems indicated by N1 and N2b—does not apply basic probability as in the existing stochastic MMN studies (e.g., Garrido et al., 2013) but rather the learning of a category.

5.4 Novelty processing

5.4.1 The P3 complex

ERP studies have shown that target and novel auditory stimuli elicit a sequence of positive components, the most prominent of which is a large complex peaking approximately 300-ms poststimulus (Fig. 5.1). These positive components for novel and target stimuli are commonly referred to as the P3a and P3b, respectively. The P3a component reflects an automatic orienting response (Halgren et al., 1998), whereas the P3b reflects directed, effortful processing (Knight, 1996), often the result of a required button press on detection. Thus the scalp P3b is linked to memory storage and detecting behavioral-relevant targets (Polich and Criado, 2006; Polich, 2007). This processing may be associated with stimulus evaluation/categorization, specifically with cognitive processes necessary to update mental

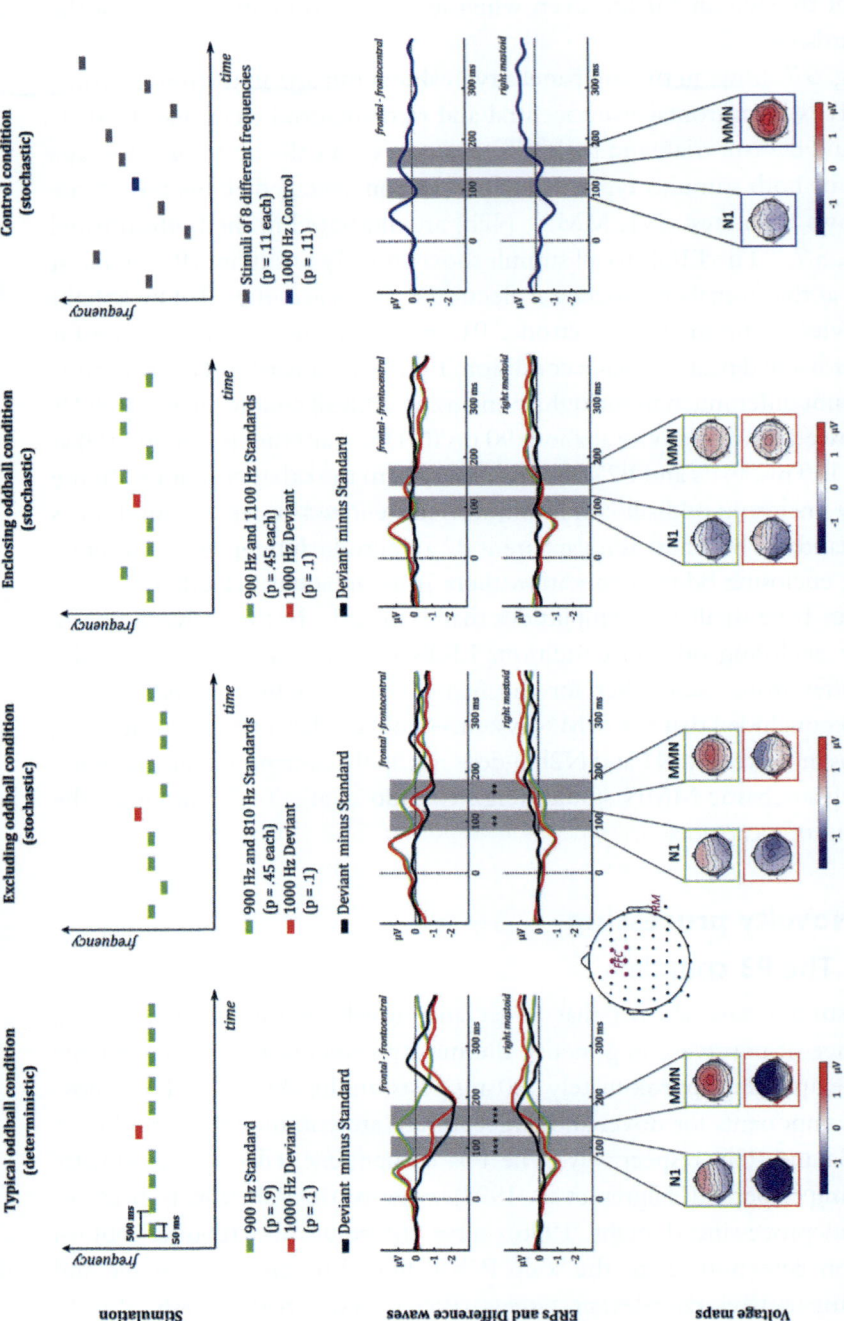

Fig. 5.7 Top panel: Typical trial sequence for each condition in Experiment 2. Middle panel: Event-related potentials (ERPs) and difference waves, from a cluster of six frontal and frontocentral electrodes and from the electrode at the right mastoid (see schematic head between the middle and the bottom panel between the first two conditions). Bottom panel: Voltage maps for standard and deviant stimuli in three 40-ms time windows of interest (N1, MMN, and N2b). *(Reprinted from Schröger, E., Roeber, U., 2021. Encoding of deterministic and stochastic auditory rules in the human brain: the mismatch negativity mechanism does not reflect basic probability. Hear. Res. 399, 107907. © 2010, with permission from Elsevier.)*

models of context within working memory (Donchin and Coles, 1988). Frontal gamma-band generators may be associated with attentional control for the binding of consecutive cognitive stages corresponding to earlier P3a and later P3b components, which have distinct but somewhat overlapping source distributions (Lee et al., 2007).

In general, the ERPs to auditory novelty comprise besides a P3—representing working memory updating—an early MMN reflecting a prediction error. These two responses predict fundamentally different dynamics: prediction errors are thought to propagate serially through several distinct brain areas, while working memory activity is sustained over time within a stable set of brain areas. Using the local/global paradigm described in Chapter 2 (Fig. 2.11), King et al. (2014) showed that the mismatch evoked by the *global novelty*—an unexpected sequence of five sounds—elicits a sustained state of brain activity that lasts for several hundreds of milliseconds (Fig. 5.8, bottom middle). Whereas global auditory violations (global standard–global deviant) led to a sustained activity from ~300 ms after the onset of the fifth sound, in contrast, "the average activity elicited by local auditory violations (local standard–local deviant) led, on average, to the traditional mismatch field, peaking at around 120 ms after the onset of the fifth sound (Fig. 5.8, top middle).

Crucially, generalization-across-time demonstrated remarkably different dynamics for the local and global effects (Fig. 5.8, right top versus bottom). In the local contrast (decoding of local standards versus local deviants, corresponding to the classical mismatch response), none of the classifiers generalized over the full-time window. Although the "diagonal" classifiers decoded information about local auditory novelty within a long-time interval of approximately 400 ms, each classifier significantly generalized for ~100 ms on average ($p_{FDR} < 0.05$) and did not significantly differ from the corresponding diagonal classifiers (Fig. 5.8, top right) over a time window of only ~50 ms ($p_{FDR} < 0.05$). Six classifiers, trained between 100 ms and 600 ms, are presented in Fig. 5.8 (top middle) and correspond to six horizontal lines—training times—of the temporal generalization matrix (Fig. 5.8, top right). King et al. (2014) noted that "The results showed a clear diagonal pattern of temporal generalization and thus indicated that each classifier only generalized for a limited amount of time: each time sample was thus associated with a slightly different pattern of MEG activity. This result suggests that different brain regions are serially recruited, each for a short-lived time period, in response to a local auditory violation. [...] Applying these analyses to the global contrast (global standard – global deviant;

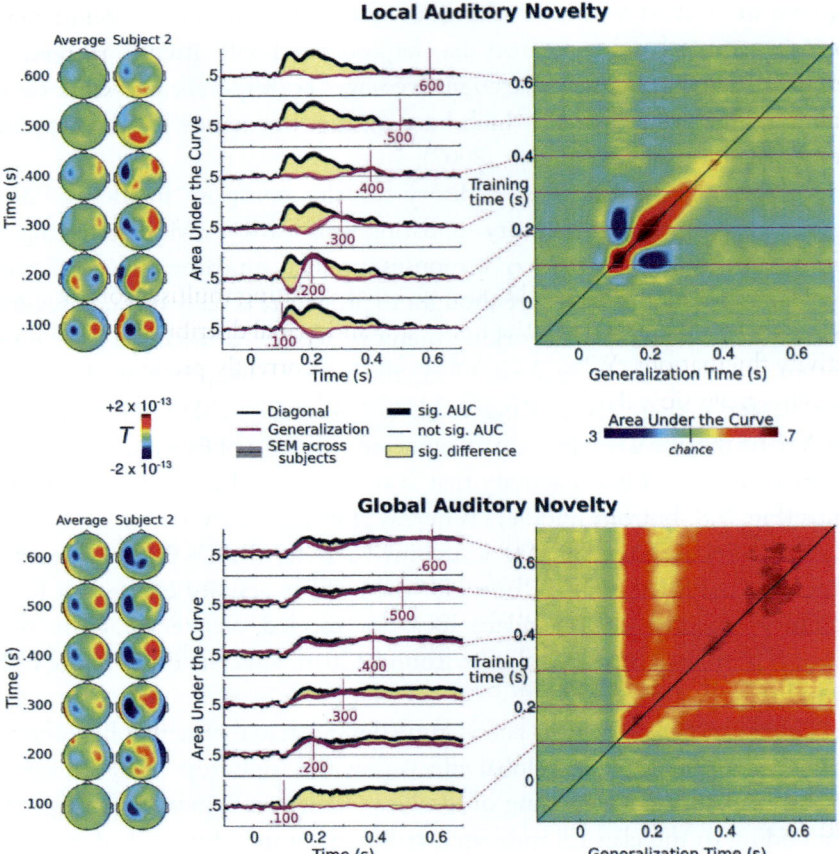

Fig. 5.8 Generalization across time of the local and global responses to auditory novelty. At each time point, a classifier was trained to extract the pattern of MEG activity that distinguishes local-standard from local-deviant trials (mismatch effect, top) or to contrast global-standard from global-deviant trials (bottom). Each classifier was subsequently tested on its ability to generalize this discrimination to all other time samples. (left) Differential patterns (standard − deviant) of brain activity across subjects as well as in a single representative subject using classic univariate analyses. For simplicity purposes, only the magnetometers are plotted ($n = 102/306$ channels). Note that, unlike subject-specific decoding, classic event-related fields (ERF) analyses are tested across subjects and are thus insensitive to interindividual variability of subjects' topographies. (middle) Generalization of six different classifiers trained at regularly spaced times between 100 ms and 600 ms *(purple)*, compared to the traditional "diagonal" decoding method where a different classifier is trained and tested at the same time point *(black)*. The thick lines indicate significant decoding scores. The *yellow areas* indicate when the diagonal performance was significantly different from the generalization across time. *Error bars* indicate the standard error of the mean (SEM) across subjects. (right) Generalization matrices.

(*Continued*)

Fig. 5.8, bottom right) led to a strikingly different pattern of decoding performance. Within a broad temporal window, a nearly 'square' pattern of temporal generalization indicated that most classifiers, regardless of their training time, produced very similar decoding performance across all testing times" (King et al., 2014).

5.4.2 Brain rhythms

Using MEG, Arnal et al. (2011) showed that violating multisensory predictions causes a change in both the frequency and spatial distribution of cortical activity fluctuations. When visual speech input correctly predicted auditory speech signals, slow delta activity (3–4 Hz) developed in higher-order speech areas. In contrast, when auditory signals invalidated predictions inferred from vision, low-beta (14–15 Hz)/high-gamma (60–80 Hz) coupling appeared locally in a multisensory area (area STS). These findings are consistent with the predictive-coding (Chapter 1) notion that bottom-up prediction errors are communicated in predominantly high (gamma) frequency ranges, whereas top-down predictions are mediated by slower (beta) frequencies. I have elaborated on the role of various brain rhythms in cognition in a previous book (Eggermont, 2021).

Nourski et al. (2018) investigated cortical responses to auditory novelty using the local–global deviant paradigm, which engages the hierarchical network underlying auditory predictive coding over short ("local deviance") and long ("global deviance") timescales. Electrocorticographic responses to auditory stimuli were obtained in neurosurgical patients from regions of interest (ROIs), including auditory, auditory related, and prefrontal cortex. Local and global deviance effects were assayed in averaged cortical

Fig. 5.8, Cont'd Decoding performance is plotted as a function of training time (vertical axis) and testing time (horizontal axis) for all classifiers. Decoding of the local-violation effect leads to a diagonal-shaped decoding performance from 82 ms to 508 ms (AUC over 50% in *red*), demonstrating that each classifier was only able to predict trials' classes for a short amount of time. Decoding of the global-violation effect leads to a square generalization matrix, suggesting that the underlying brain activity is essentially stable during this time period. Early classifiers of the global violation (<350 ms) are slightly lower than the traditional "diagonal" decoding performance, thus suggesting only a small change in the underlying pattern of activity. AUC, area under curve. *(From King, J.-R., Gramfort, A., Schurger, A., Naccache, L., Dehaene, S., 2014. Two distinct dynamic modes subtend the detection of unexpected sounds. PLoS One 9 (1), e85791. doi:https://doi.org/10.1371/journal.pone. 0085791. Open Access.)*

evoked potential and high-gamma (70–150 Hz) signals, the former likely dominated by local synaptic currents and the latter largely reflecting local spiking activity. "CAEP local deviance effects were distributed across all ROIs, with greatest percentage of significant sites in core and non-core auditory cortex. High gamma local deviance effects were localized primarily to auditory cortex in the superior temporal plane and on the lateral surface of the superior temporal gyrus" (STG; Fig. 5.9).

Nourski et al. (2018) concluded: "Local deviance effects exhibited progressively longer latencies in core, non-core, auditory-related and prefrontal cortices, consistent with feedforward signaling. The spatial distribution of CAEP global deviance effects overlapped that of local deviance effects, but high gamma global deviance effects were more restricted to noncore areas. High gamma global deviance effects had shortest latencies in STG and preceded CAEP global deviance effects in most ROIs. This latency profile, along with the paucity of high gamma global deviance effects in the superior temporal plane, suggest that the STG plays a prominent role in initiating novelty detection signals over long time scales. Thus, the data demonstrate distinct patterns of information flow in human cortex associated with auditory novelty detection over multiple time scales."

Nourski et al. (2021) then compared cortical responses to auditory novelty during passive versus active listening conditions in awake listeners. Local/global deviant stimuli were sequences of four identical vowels followed by a fifth identical or different vowel. In the passive condition, the stimuli were presented to subjects as they watched a silent TV program and were instructed to attend to its content. In the active condition, stimuli were presented in the absence of a TV program, and subjects were instructed to press a button in response to global deviance target stimuli. Intracranial recordings were made from multiple brain regions, including core and non-core auditory, auditory related, prefrontal, and sensorimotor cortex. Metrics of task performance included hit rate, sensitivity index, and reaction times. Cortical activity was measured as averaged CAEPs and event-related band power in high gamma (70–150 Hz) and alpha (8–14 Hz) frequency bands. "The vowel stimuli and local deviance elicited robust CAEPs in all studied brain areas in both passive and active conditions. High gamma responses to stimulus onset and local deviance were localized predominantly to the auditory cortex in the superior temporal plane and had a comparable prevalence and spatial extent between the two conditions. In contrast, global deviance effects (CAEPs, high gamma and alpha suppression) were greatly enhanced during the active condition in all studied brain areas. The prevalence of

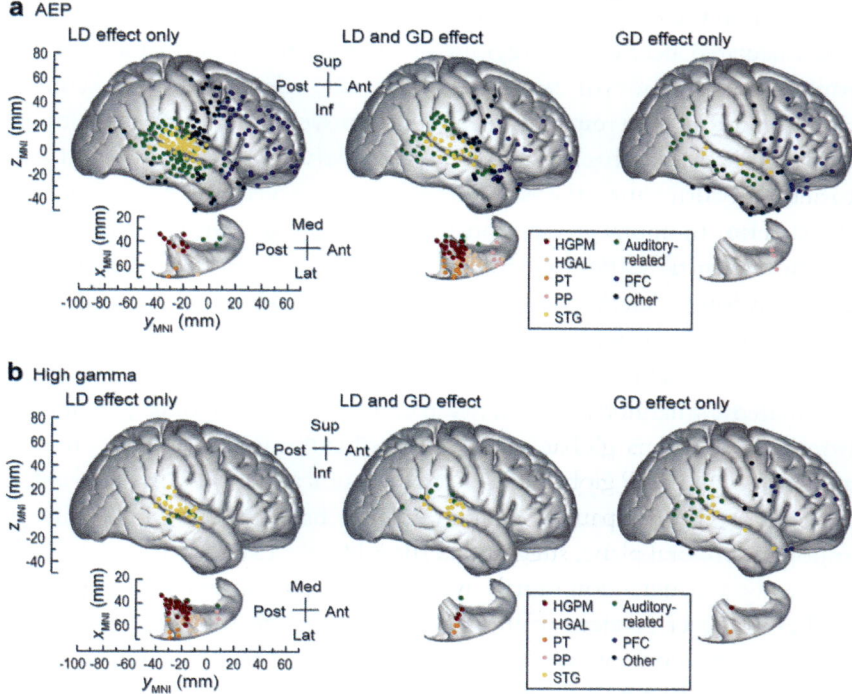

Fig. 5.9 Spatial distribution of sites exhibiting significant local deviance (LD) and global deviance (GD) effects as measured by the AEP and high-gamma activity (panels A and B, respectively). Summary of data from six subjects, plotted in MNI coordinate space and projected onto FreeSurfer average template brain. Top-down views of the right superior temporal plane are plotted underneath side views of the right lateral hemispheric convexity, aligned with respect to the y_{MNI} coordinate. Sites exhibiting significant LD effects only, both LD and GD effects, and GD effects only are depicted in the left, middle, and right column, respectively. Sites are *color coded* based on their ROI assignment in each individual subject. *AEP*, averaged evoked potential; *GD*, global deviance; *HGAL*, anterolateral Heschl's gyrus; *HGPM*, posteromedial Heschl's gyrus; *LD*, local deviance; *MNI*, Montreal Neurological Institute; *PFC*, prefrontal cortex; *PP*, planum polare; *PT*, planum temporale; *STG*, superior temporal gyrus. *(From Nourski, K.V., Steinschneider, M., Rhone, A.E., Kawasaki, H., Howard III, M.A., Banks, M.I., 2018. Processing of auditory novelty across the cortical hierarchy: an intracranial electrophysiology study. Neuroimage 183, 412–424. ©2018, with permission from Elsevier.)*

high-gamma global deviance effects was positively correlated with individual subjects' task performance" (Nourski et al., 2021).

Mamashli et al. (2020) instructed participants to detect a predictable harmonic target sound embedded among standard tones in one ear and to ignore the standard tones and occasional unpredictable novel sounds presented in the opposite ear. Phase coherence of estimated source activity was calculated between subregions of superior temporal, frontal, inferior parietal, and superior parietal cortex ROIs. The inferred functional connectivity was stronger in response to target than novel stimuli between left superior temporal and left parietal ROIs and between left frontal and right parietal ROIs, with the largest effects observed in the beta band (15–35 Hz). In contrast, functional connectivity was stronger in response to novel than target stimuli in interhemispheric connections between left and right frontal ROIs, observed in early time windows in the alpha band (8–12 Hz). This suggests that auditory processing of expected target vs unexpected novel sounds involves different spatially, temporally, and spectrally distributed oscillatory connectivity patterns across temporal, parietal, and frontal areas (Fig. 5.10). Mamashli et al. (2020) noted that these "findings are in line with the hypothesis that communication between the cortical areas involves functionally distinct frequency bands during (predominantly endogenous) target detection vs (predominantly exogenous) orienting to physically varying, unexpected novel sounds. Consistent with previous studies in the visual domain, the larger phase synchronization during trials requiring greater degrees of endogenous or top-down processing was most prominent in the beta band, whereas for trials involving suppression of irrelevant information, the largest effects were observed in the alpha band."

5.5 Summary

Event-related potentials are, depending on latency, reflections of prediction errors. Consequently, they are dependent on corticofugal activity from higher-order cortex to lower levels. This potentially excludes short-latency ERPs such as the middle-latency responses—latencies < 50 ms—and the P2 which is the product of the reticular activating system and purely a bottom-up response, albeit that these are affected by stimulus-specific adaptation as well. All long-latency ERPs are affected by adaptation, attention, and memory. Some of those such as the MMN and P3 complex have been set apart as the canonical prediction error signals. The early MMN appears for local (in time) deviances, whereas the P3b reflects global (in time) oddballs and novelty signals. ERPs underlying the local deviances originate from a network composed of A1, bilateral STG, and right IFG. However, for fMRI-based analysis the IFG falls by the wayside as its signal is too small to be reliably detected. Combining ERP and fMRI evidence suggests that the IFG changes are likely based on neural synchrony—which affects the ERP

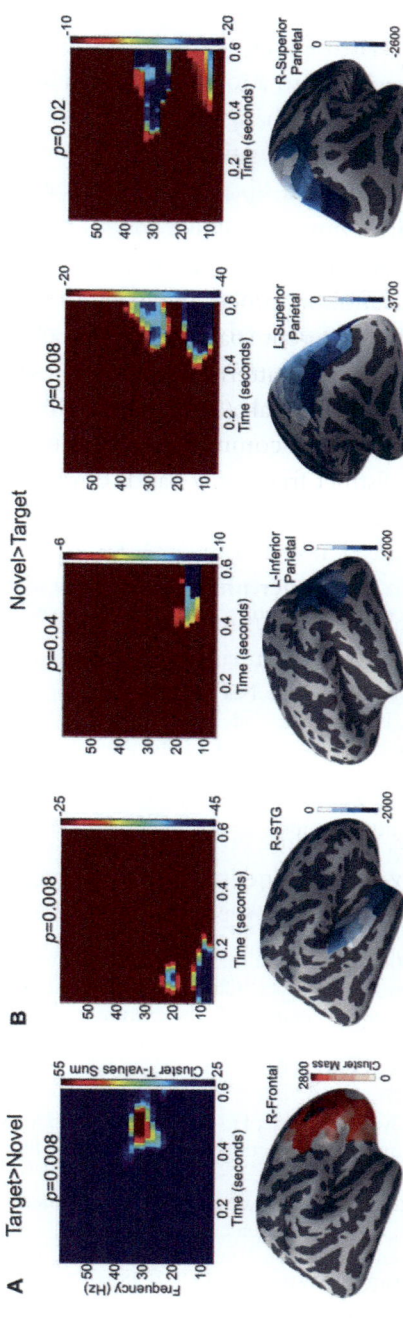

Fig. 5.10 ROIs with significant differences in the power of regional oscillatory activity between the conditions: (A) Target > Novel and (B) Novel > Target. The upper row shows time-frequency maps obtained by summing the weighted cluster masks for significant clusters for all sub-ROIs. In the lower row, color coding indicates the cluster mass for each sub-ROI. (*From Mamashli, F., Huang, S., Khan, S., Hämäläinen, M.S., Ahlfors, S.P., Ahveninen, J., 2020. Distinct regional oscillatory connectivity patterns during auditory target and novelty processing. Brain Topogr. 33, 477–488. © Springer Science+Business Media, LLC, part of Springer Nature 2020, with permission.*)

and is not easy detectable by fMRI—rather than increase in the number of active neurons and/or their firing rates. Comparison of the MMN, P3a, and P3b suggests that different brain regions are serially recruited, each for a short-lived time period, in response to a local or global auditory violation. Violating sensory predictions causes a change in both the frequency and spatial distribution of cortical activity fluctuations. This is consistent with the predictive-coding notion that bottom–up prediction errors are communicated in predominantly high (gamma) frequency ranges, whereas top–down predictions are mediated by slower (beta) frequencies. High-gamma responses to stimulus onset and local deviance are localized predominantly to the auditory cortex and had a comparable prevalence and spatial extent between the two conditions. In contrast, global deviance effects (CAEPs, high gamma and alpha suppression) were also enhanced during the active condition in higher-order brain areas.

References

Althen, H., Grimm, S., Escera, C., 2011. Fast detection of unexpected sound intensity decrements as revealed by human evoked potentials. PLoS One 6 (12), e28522. https://doi.org/10.1371/journal.pone.0028522.

Arnal, L.H., Wyart, V., Giraud, A.-L., 2011. Transitions in neural oscillations reflect prediction errors generated in audiovisual speech. Nat. Neurosci. 14 (6), 797–801.

Bekinschtein, T.A., Dehaene, S., Rohaut, B., Tadela, F., Cohen, L., Naccache, L., 2009. Neural signature of the conscious processing of auditory regularities. PNAS 106 (5), 1672–1677.

Cacciaglia, R., Escera, C., Slabu, L., Grimm, S., Sanjuán, A., Ventura-Campos, N., Ávila, C., 2015. Involvement of the human midbrain and thalamus in auditory deviance detection. Neuropsychologia 68, 51–58.

Cacciaglia, R., Costa-Faidella, J., Zarnowiec, K., Grimm, S., Escera, C., 2019. Auditory predictions shape the neural responses to stimulus repetition and sensory change. Neuroimage 186, 200–210.

Crowley, K.E., Colrain, I.M., 2004. A review of the evidence for P2 being an independent component process: age, sleep and modality. Clin. Neurophysiol. 115, 732–744.

Deouell, L.Y., 2007. The frontal generator of the mismatch negativity revisited. J. Psychophysiol. 21 (3–4), 188–203. https://doi.org/10.1027/0269-8803.21.34.188.

Donchin, E., Coles, M.G.H., 1988. Is the P300 component a manifestation of context updating? Behav. Brain Sci. 11, 357–374.

Dürschmid, S., Zaehle, T., Hinrichs, H., Heinze, H.-J., Voges, J., Garrido, M.I., et al., 2016. Sensory deviancy detection measured directly within the human nucleus accumbens. Cereb. Cortex 26, 1168–1175.

Eggermont, J.J., 1988. On the rate of maturation of sensory evoked potentials. Electroencephalogr. Clin. Neurophysiol. 70 (4), 293–305.

Eggermont, J.J., 2021. Brain Oscillations, Synchrony and Plasticity. Basic Principles and Application to Auditory-Related Disorders. Academic Press, London, ISBN: 978-0-12-819818-6, pp. 1–250.

Garrido, M.I., Sahani, M., Dolan, R.J., 2013. Outlier responses reflect sensitivity to statistical structure in the human brain. PLoS Comput. Biol. 9 (3), e1002999. https://doi.org/10.1371/journal.pcbi.1002999.

Grimm, S., Recasens, M., Althen, H., Escera, C., 2012. Ultrafast tracking of sound location changes as revealed by human auditory evoked potentials. Biol. Psychol. 89, 232–239.

Halgren, E., Marinkovic, K., Chauvel, P., 1998. Generators of the late cognitive potentials in auditory and visual oddball tasks. Electroencephalogr. Clin. Neurophysiol. 106 (2), 156–164.

Hillyard, S.A., Hink, R.F., Schwent, V.L., Picton, T.W., 1973. Electrical signs of selective attention in the human brain. Science 182, 177–180.

Hsu, Y.-F., Darriba, Á., Waszak, F., 2021. Attention modulates repetition effects in a context of low periodicity. Brain Res. 1767, 147559.

Ishishita, Y., Kunii, N., Shimada, S., Ibayashi, K., Tada, M., Kirihara, K., et al., 2019. Deviance detection is the dominant component of auditory contextual processing in the lateral superior temporal gyrus: a human ECoG study. Hum. Brain Mapp. 40, 1184–1194.

Jääskeläinen, I.P., Ahveninen, J., Bonmassar, G., Dale, A.M., Ilmoniemi, R.J., Levänen, S., et al., 2004. Human posterior auditory cortex gates novel sounds to consciousness. PNAS 101 (17), 6809–6814.

Kiehl, K.A., Stevens, M.C., Laurens, K., Pearlson, G., Calhoun, V.D., Liddle, P.F., 2005. An adaptive reflexive processing model of neurocognitive function: supporting evidence from a large scale (n = 100) fMRI study of an auditory oddball task. Neuroimage 25, 899–915.

King, J.-R., Gramfort, A., Schurger, A., Naccache, L., Dehaene, S., 2014. Two distinct dynamic modes subtend the detection of unexpected sounds. PLoS One 9 (1), e85791. https://doi.org/10.1371/journal.pone.0085791.

Knight, R.T., 1996. Contribution of human hippocampal region to novelty detection. Nature 383 (6597), 256–259.

Kompus, K., Volehaugen, V., Todd, J., Westerhausen, R., 2020. Hierarchical modulation of auditory prediction error signaling is independent of attention. Cogn. Neurosci. 11 (3), 132–142. https://doi.org/10.1080/17588928.2019.1648404.

Lee, B., Park, K.S., Kang, D.-H., Kang, K.W., Kim, Y.Y., Kwon, J.S., 2007. Generators of the gamma-band activities in response to rare and novel stimuli during the auditory oddball paradigm. Neurosci. Lett. 413, 210–215.

López-Caballero, F., Zarnowiec, K., Escera, C., 2016. Differential deviant probability effects on two hierarchical levels of the auditory novelty system. Biol. Psychol. 120, 1–9.

Mamashli, F., Huang, S., Khan, S., Hämäläinen, M.S., Ahlfors, S.P., Ahveninen, J., 2020. Distinct regional oscillatory connectivity patterns during auditory target and novelty processing. Brain Topogr. 33, 477–488.

Mäntysalo, S., Näätänen, R., 1987. The duration of a neuronal trace of an auditory stimulus as indicated by event-related potentials. Biol. Psychol. 24, 183–195.

Menon, V., Uddin, L.Q., 2010. Saliency, switching, attention and control: a network model of insula function. Brain Struct. Funct. 214, 655–667. https://doi.org/10.1007/s00429-010-0262-0.

Mittag, M., Takegata, R., Winkler, I., 2016. Transitional probabilities are prioritized over stimulus/pattern probabilities in auditory deviance detection: memory basis for predictive sound processing. J. Neurosci. 36 (37), 9572–9579.

Näätänen, R., Michie, P.T., 1979. Early selective-attention effects on the evoked potential: a critical review and reinterpretation. Biol. Psychol. 8, 81–136.

Näätänen, R., Simpson, M., Loveless, N.E., 1982. Stimulus deviance and evoked potentials. Biol. Psychol. 14, 53–98.

Näätänen, R., Sussman, E.S., Salisbury, D., Shafer, V.L., 2014. Mismatch negativity (MMN) as an index of cognitive dysfunction brain. Brain Topogr. 27, 451–466. https://doi.org/10.1007/s10548-014-0374-6.

Nourski, K.V., Steinschneider, M., Rhone, A.E., Kawasaki, H., Howard III, M.A., Banks, M.I., 2018. Processing of auditory novelty across the cortical hierarchy: an intracranial electrophysiology study. Neuroimage 183, 412–424.

Nourski, K.V., Steinschneider, M., Rhone, A.E., Krause, B.M., Kawasaki, H., Banks, M.I., 2021. Cortical responses to auditory novelty across task conditions: an intracranial electrophysiology study. Hear. Res. 399, 107911.

Picton, T.W., 2011. Human Auditory Evoked Potentials. Plural Publishing, San Diego, CA.

Polich, J., 2007. Updating P300: an integrative theory of P3a and P3b. Clin. Neurophysiol. 118, 2128–2148.

Polich, J., Criado, J.R., 2006. Neuropsychology and neuropharmacology of P3a and P3b. Int. J. Psychophysiol. 60, 172–185.

Ponton, C.W., Eggermont, J.J., Kwong, B., Don, M., 2000a. Maturation of human central auditory system activity: evidence from multi-channel evoked potentials. Clin. Neurophysiol. 111 (2), 220–236.

Ponton, C.W., Eggermont, J.J., Don, M., Waring, M.D., Kwong, B., Cunningham, J., Trautwein, P., 2000b. Maturation of mismatch negativity: effects of profound deafness and cochlear implant use. Audiol. Neurootol. 5, 167–185.

Regev, T.I., Markusfeld, G., Deouell, L.Y., Nelken, I., 2021. Context sensitivity across multiple time scales with a flexible frequency bandwidth. Cereb. Cortex, 1–18.

Schönwiesner, M., Novitski, N., Pakarinen, S., Carlson, S., Tervaniemi, M., Näätänen, R., 2007. Heschl's gyrus, posterior superior temporal gyrus, and mid-ventrolateral prefrontal cortex have different roles in the detection of acoustic changes. J. Neurophysiol. 97, 2075–2082.

Schröger, E., Roeber, U., 2021. Encoding of deterministic and stochastic auditory rules in the human brain: the mismatch negativity mechanism does not reflect basic probability. Hear. Res. 399, 107907.

Slabu, L., Escera, C., Grimm, S., Costa-Faidella, J., 2010. Early change detection in humans as revealed by auditory brainstem and middle-latency evoked potentials. Eur. J. Neurosci. 32, 859–865.

Slabu, L., Grimm, S., Escera, C., 2012. Novelty detection in the human auditory brainstem. J. Neurosci. 32 (4), 1447–1452.

Takasago, M., Kunii, N., Komatsu, M., Tada, M., Kirihara, K., Uka, T., et al., 2020. Spatiotemporal differentiation of MMN from N1 adaptation: a human ECoG study. Front. Psychiatry 11, 586. https://doi.org/10.3389/fpsyt.2020.00586.

Velasco, M., Velasco, F., Velasco, A.L., 1989. Intracranial studies on potential generators of some vertex auditory evoked potentials in man. Stereotact. Funct. Neurosurg. 53 (1), 49–73. https://doi.org/10.1159/000099517.

CHAPTER 6

Development and maturation aspects of predictive coding

This chapter's coverage is partly based on an extensive review of histological, electrophysiological, and behavioral maturation of the auditory system by Eggermont and Moore (2012).

6.1 Introduction

In terms of auditory capability, infants in the months before and after birth respond to environmental sounds and accurately distinguish different sounds (Table 6.1). Electrophysiological testing reveals that infants exhibit auditory evoked potentials (AEPs) arising both in the brainstem and in the cortex. These behavioral and physiological findings are in basic agreement with anatomical and imaging studies showing early and rapid maturation of auditory structures in the brain stem and in some elements of cortex. In this chapter, we review the maturation of the auditory brain stem response (ABR), the middle-latency response (MLR), the obligatory—or exogenous—cortical auditory evoked potential and magnetic field components (CAEPs/ CAMFs), the mismatch response (MMR), and the P300 complex. This background information is valuable in using the potentials in assessment of developmental disorders such as dyslexia, autism, and attention-deficit/ hyperactivity disorder, which will be presented in Chapter 8.

6.2 Maturation of subcortical activity

6.2.1 The auditory brainstem response

A typical mature ABR consists of a sequence of up to seven vertex-positive waves (recorded from the top of the head relative to the mastoids), separated by negative valleys (Eggermont, 2019; Fig. 6.1). The first five peaks are typically labeled with roman numerals, while the valleys are generally not labeled.

Brain Responses to Auditory Mismatch and Novelty Detection
https://doi.org/10.1016/B978-0-443-15548-2.00006-5

Table 6.1 Structural, electrophysiological, and behavioral correlates in human maturation.

Age	Structural	Electrophysiological	Behavioral
<6 months	• Layer I cortical axons mature	ABR wave I mature, positive MMRs	Discrimination of speech sounds
6 months– 5 years	• Brainstem is mature • Acoustic radiation matures • Cortex layers IV–VI mature • Cortical synaptic density peaks	ABR, MLR, P2, N2, MMN, and T-complex mature N1 emerges for low-stimulus rates	Onset and development of perceptual language
5–12 years	• Cortical layers II and III mature • Decrease in cortical synaptic density	P1 matures, N1 emerges for fast stimulus rates	Processing of masked and degraded speech improves
>12 years	• Cortex matures	Asymptotic N1 maturation	Speech in reverberation and noise matures

Our interpretation of the ABR morphology has been described in Eggermont and Moore (2012). Briefly, wave I is the compound action potential (CAP) generated by the initial part of the auditory nerve before it enters the petrous bone canal. Wave II originates from the auditory nerve at the point where it leaves the petrous bone to enter the intracranial space. Peaks with numbers from III to V likely are generated sequentially in the auditory brain stem, but waves IV and V may also reflect parallel activation of ipsilateral and contralateral pathways. Wave III is likely generated by nerve branching within the cochlear nucleus. Wave IV is putatively generated by axons passing directly from the dorsal cochlear nucleus (DCN) and ventral cochlear nucleus (VCN) to the contralateral lateral lemniscus, as the track bends in passing around the superior olivary complex. Wave V probably arises from fibers synapsing in the medial superior olive and running into the lateral lemnisci on both sides of the brain toward the inferior colliculus (IC).

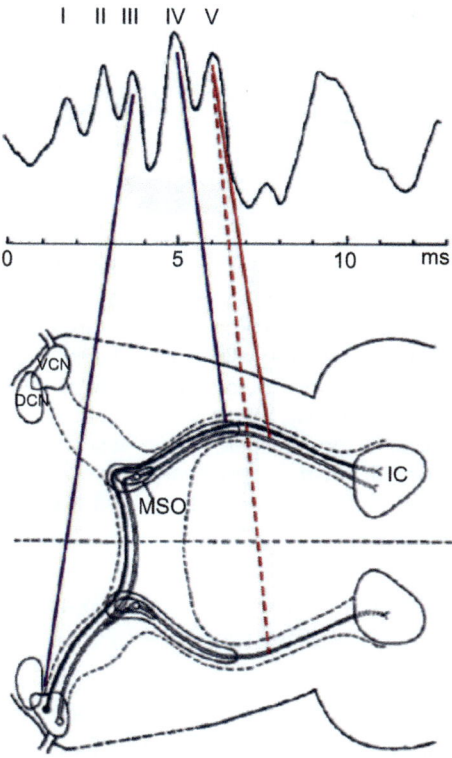

Fig. 6.1 ABR sources. Wave III cochlear nucleus (blue line); wave IV dominantly from the purely contralateral axonal pathway at the bend near, or ending in, the lateral lemniscus (blue line). Wave V from the pathway synapsing in the MSO and continuing to the lateral lemniscus, dominantly contralateral to the CN (full red line) but also a minor contribution from the ipsilateral lemniscus (dashed red line). *(Reprinted from Eggermont, J.J. Auditory Brainstem Response, Chapter 30 in Audition, vestibular, and language testing. In: Levin, K.H., Chauvel, P. (Eds.), Handbook of Clinical Neurology, vol. 160 (3rd series) Clinical Neurophysiology: Basis and Technical Aspects, vol. 160, 451–464, ©2019, with permission from Elsevier.)*

The ABR can reliably be recorded in premature infants from the 28th to 29th weeks conceptual age (CA; Starr et al., 1977; Pasman et al., 1991; Ponton et al., 1992; Fig. 6.2). At ages 35 weeks CA and older, waves I, III, and V at both sides are clearly present (Pasman et al., 1991; Ponton et al., 1992). For all practical purposes, the maturational process underlying the I–V interval reaches adult values about 1–1.5 years postterm. For the apical part of the cochlea, represented by the octave band centered at 0.7 kHz, and for the basal part of the cochlea, that is, the frequency band centered

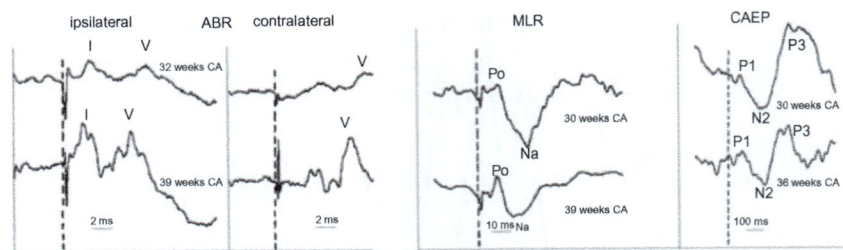

Fig. 6.2 Prenatal changes in ABR (ipsi- and contralateral), MLR, and late cortical potentials at 30–32 weeks (top row in each panel) and around term. Some of the peaks have been identified. Note the dramatic changes in the ABR waveforms where contralateral activity lags behind the ipsilateral one and the much more modest ones for the OEPs. ABR, auditory brainstem response; MLR, middle-latency response; OEP, obligatory evoked potential. *(From Eggermont, J.J., Moore, J.K., 2012. Morphological and functional development of the auditory nervous system. In: L.A. Werner et al. (Eds.), Human Auditory Development, Springer Handbook of Auditory Research, vol. 42, pp. 61–105, ©2012 Springer Science+Business Media, with permission from Springer Nature.)*

around 11.3 kHz, maturity is reached at approximately 2–2.5 years and 1.5–2 years postterm, respectively (Ponton et al., 1992).

Recent longitudinal studies (Krizman et al., 2015; Thompson et al., 2021) suggest that the auditory brainstem has not even fully matured by the end of adolescence and that subcortical auditory development continues throughout childhood, with different facets of auditory processing following distinct developmental trajectories.

6.2.2 The middle-latency response

The ABR's maturational time course, reflecting activity in brain stem structures up to the level of the superior olivary complex, is closely followed by the middle-latency response (MLR). The MLR components that are most clearly detectable in infants are the P0-Na complex (Fig. 6.2). Because recordings from the surface of the human brain stem match the P0 peak to a postsynaptic potential at the inferior colliculus (Hashimoto et al., 1981), it seems likely that the P0-Na waves reflect transmission in the pathway from the inferior colliculus to the thalamus (Picton, 2011; p.117). The P0-Na waves are barely detectable at the 25–27th fetal weeks, but are fairly well defined by the 33rd fetal week, and more pronounced by the time of term birth (Rotteveel et al., 1985, 1987; Pasman et al., 1991). The Na peak latency decreases steadily from about 28 ms at the 30th fetal week to approximately 20 ms at term (Rotteveel et al., 1985, 1987; Pasman et al., 1991).

By the third postnatal month, the Na peak achieves a latency of approximately 18 ms, a value that remains unchanged throughout childhood, teen years, and adulthood (Kraus et al., 1985).

6.3 Cortical auditory evoked potentials

6.3.1 Early cortical activity maturation

The summary in Table 6.1 indicates that electrical or magnetic evoked activity in auditory cortex is related to the behavioral change detection capabilities of infants. Early studies (Barnet et al., 1975; Ohlrich et al., 1978; Rotteveel et al., 1987; Novak et al., 1989; Pasman et al., 1991) showed that in preterm babies at 24 weeks the CAEP is dominated by a large negative wave with a peak latency of 200 ms, followed by a positive peak at 600 ms that by 30 weeks is reduced to a latency approaching 300 ms and increases in amplitude (Fig. 6.2). By the time of term birth, this positive peak has a latency of about 250 ms and dominates the response because the earlier negative wave (labeled N2) has nearly completely disappeared.

This dominance of positive components in the CAEP continues until about 5 months of age, when a new negative peak with latency of approximately 400 ms begins to increase in amplitude and reduce in latency. Even by age 5 years, this sequence of positive (100 ms)–negative (200 ms)–positive (~350 ms) peaks is still very much the standard morphology for CAEPs recorded at the vertex (Pang and Taylor, 2000; Ponton et al., 2000a; Kushnerenko et al., 2002b). Eggermont (1988) showed that an exponential decrease with age (time constant of ~40 weeks) comprehensively described the changes in P2 and N2 latency and was interpretable in biological terms (Ponton and Eggemont, 2007; Eggermont and Moore, 2012). This time constant is only slightly longer than that of the maturation of the wave I–V latency difference in the ABR, suggesting that the maturation of this early cortical activity is either limited by that of the auditory brain stem or "cotuned" to it. The lower brain stem is always the first step in processing, giving rise to both the Reticular Activating System (RAS) pathway involved in the P2 (Chapter 5) and the late maturing thalamocortical pathways underlying P1, N1, etc.

6.3.2 Electrophysiological measures of cortical maturation

6.3.2.1 Amplitude and latency changes

Across early childhood, there is increasing prominence of short-latency over longer-latency positive cortical CAEP components, which may be related to the newly maturing system of thalamocortical connections. The MLR components typically mature earlier than the long-latency CAEPs, consisting of the P1–N1–P2–N2 complex with mature latencies of about 50, 100, 150, and 200 ms. The adult waveform P1–N1–P2–N2 complex (Fig. 6.3; see also

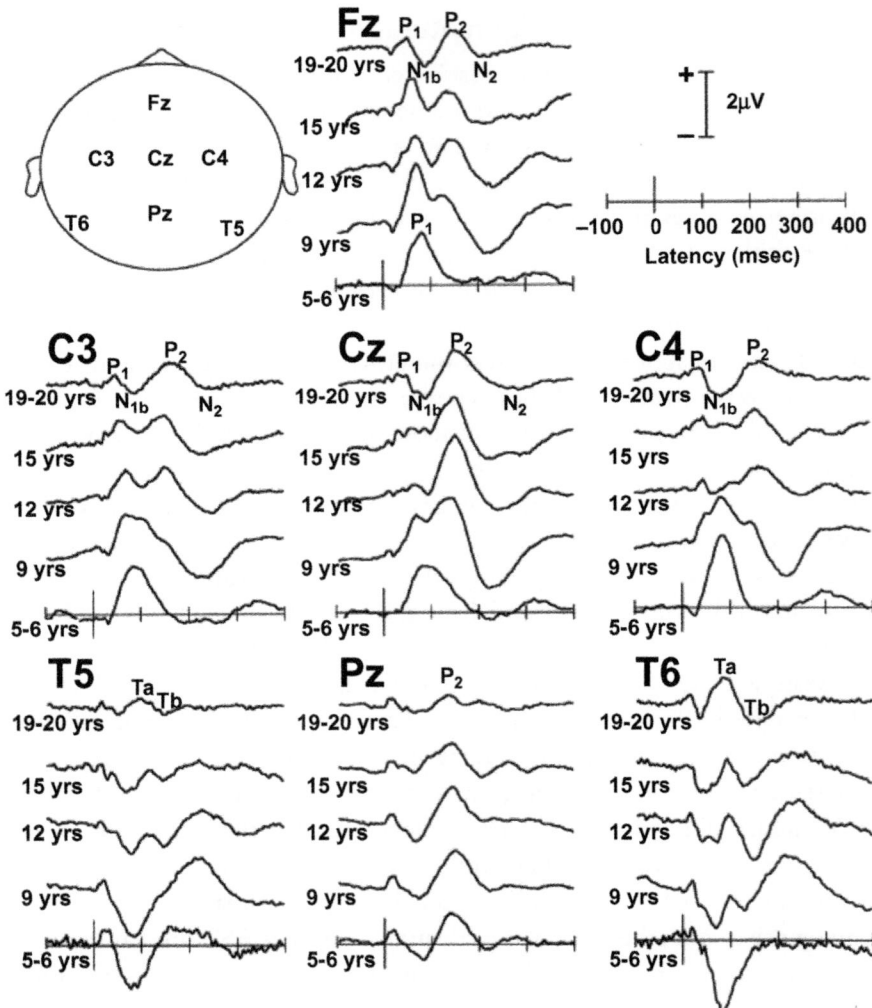

Fig. 6.3 Grand-mean waveforms for selected electrode locations and selected age groups. All waveforms are on the same scale. For the midline and central electrodes peaks are indicated as P1 to N2. For the temporal electrodes the peak indications are those for the T-complex. *(Reprinted from Ponton, C.W., Eggermont, J.J., Kwong, B., Don, M., 2000a. Maturation of human central auditory system activity: evidence from multi-channel evoked potentials. Clin. Neurophysiol. 111, 220–236. © 2000, with permission from Elsevier.)*

Fig. 2.5) is achieved between 14 and 16 years of age (Pasman et al., 1999; Ponton et al., 2000a). P1 can be distinguished at 30 weeks conceptional age as a broad wave with a latency of 80–100 ms (Pasman et al., 1991). Over time, its peak latency gradually shortens to reach an adult level of 50 ms. The late appearance of N1 is partly due to its greater sensitivity to interstimulus interval (ISI) in children than in adults. Therefore most studies using ISIs shorter than 1 s do not observe N1 reliably before age 9 (Pang and Taylor, 2000; Ponton et al., 2000a), but with a longer ISI, N1 becomes visible from age 6 on (Paetau et al., 1995; Gomes et al., 2001; Ceponiené et al., 2002; Gilley et al., 2005) or even earlier with ISIs in the 3–6 s range (Wunderlich et al., 2006). Note that during early maturation it is difficult to identify a component as N2 when the N1 is not yet visible. However, their maturational trajectories are different; the N2 is mature much earlier (time constant 2 years) than the N1 (time constant >4 years), as shown in Ponton et al. (2002). Further, the scalp distributions of both the N1 and N2 were found to change with maturation, which might indicate different component structure (and thus function) of children's and adults' N1 and N2 (Ceponiené et al., 2002). Although the latency changes for P1 and N1 peaks are similar, the maturational changes in magnitude are opposite; P1 magnitude decreases while N1 increases with increasing age. Because N1 emerges in the CAEP at about 9–10 years of age—for fast stimulation rates—when the neural generators producing the P1 peak are essentially adult-like, and given the partial temporal overlap and common tangential dipole orientation of these two components, it is possible that the magnitude and latency changes of the maturing N1 peak are superimposed on those of P1 thereby decreasing P1 amplitude (Ponton and Eggermont, 2001).

Wunderlich et al. (2006) recorded CAEP components P1, N1, P2, and N2 as a function of participant age, stimulus type, and electrode montage. CAEP component latencies were relatively stable from birth to 6 years, but adults demonstrated significantly shorter latencies compared to infants and children. Components P1 and N2 decreased in amplitude, while components N1 and P2 increased in amplitude from birth to adulthood. Words evoked significantly larger CAEPs in newborns compared to responses evoked by tones, but in other age groups the effects of stimulus type on component amplitudes and latencies were less consistent. They used very long interstimulus intervals, random between 3.1 and 6 s, and were able to see N1 even in newborns, although in my opinion the latency of this putative N1 appears too short, compared to ages 1–3 and 4–6 years (Wunderlich and Cone-Wesson, 2006; Fig. 6.4).

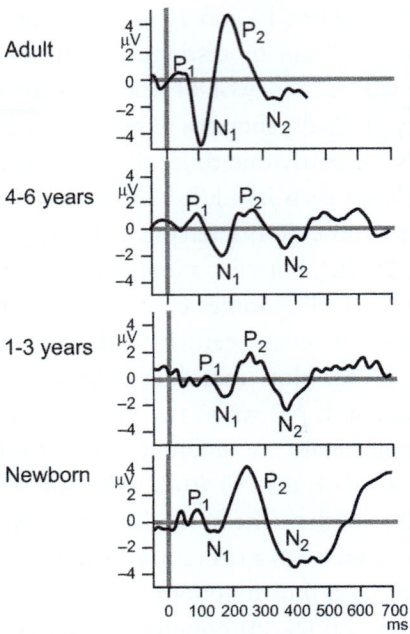

Fig. 6.4 Grand mean-average CAEP responses to the word "bad" presented with long interstimulus interval (ISI 3100–6050 ms, median 4600 ms) showing age-related changes in waveform morphology, peak amplitude, and latency of the components. When tested with tone bursts the CAEP recorded in newborns typically lacked P1 and N1. Adults (n=9), 4–6 year olds (n=20), 1–3 year olds (n=19), newborns (n=8). All responses were recorded at Cz referred to the right mastoid with left mastoid as ground. The 200-ms words were presented to the right ear via an insert earphone at 60 dB above threshold. *(Reprinted from Wunderlich, J.L., Cone-Wesson, B.K., 2006. Maturation of CAEP in infants and children: a review. Hear. Res. 212, 212–223, © 2005, with permission from Elsevier.)*

Bishop et al. (2007) distinguished three developmental periods: 5–12 years, 13–16 years, and adulthood (see Table 6.1). They showed a fairly sharp change in the CAEP waveform at Fz around 12 years of age, when N1–P2 was fully mature. The results also demonstrated the independence of the T-complex components, represented in the radial dipoles, from the P1, N1, and P2 components, contained in the tangentially oriented dipole sources. Sussman et al. (2008) found that both age and stimulus rate produced profound changes in CAEP morphology. Between the ages of 8 and 11 years, the P1 and N2 components dominated the ERP waveform at all stimulus rates. N1, the dominant CAEP component in adults, appeared as a bifurcation in a broad positive peak at earlier ages and did not emerge as a separate component until adolescence. While the P1–N1–P2 components are more "adult-like" than "child-like" in the adolescent subjects, the N2 component, a hallmark of the child obligatory response, was still present. Ruhnau et al. (2011) recorded MEG and EEG in adults and 9- to 10-year-old children, while presenting pure tones either repetitively or randomly among tones of different pitch, at an SOA of 0.5 s. In adults, P1(m) and N1(m) components were localized in auditory cortex (ACx) regions, with the N1(m) largely attenuated for repetitive tones. The P1(m) and N2(m) components observed in children were also localized in ACx regions. Most importantly, ERP modulations in the N1 time window (i.e., larger responses for random than repetitive tones) were remarkably similar for adults and children, both in amplitude and latency as was demonstrated in the difference between the random and repetitive stimulation responses.

6.3.2.2 The concept of an equivalent dipole

A dipole is a configuration with a positive and (numerically equal) negative voltage separated in space. Pyramidal cell neurons in auditory cortex have large apical (top) dendrites that are oriented perpendicular to the surface of the brain, so for those parts of the cortex that are not folded inward large numbers of them will have a near-parallel orientation. The elongated dendrites form the anatomical substrate of a pyramidal cell's dipole, with for instance its negative polarity near the cortical surface and positive polarity near the cell body. The spatial alignment of the organization of positive and negative voltages at individual pyramidal cells makes it possible to record a compound postsynaptic potential (PSP), i.e., an LFP, at a distance. An example of such a compound PSP is the N1 component of the CAEP. For a perfect spherical cortex with a smooth surface, all dipoles would be perpendicular to the surface. These directions are called radial and cannot be recorded using MEG (Eggermont, 2007). The human cortex is characterized by extensive folding. This folding of the cortical surface results in dipole orientations that are tangential, i.e., parallel, to the scalp. An important surface structure of human auditory cortex is called Heschl's gyrus; it is flanked by sulci so that its surface is curved. Heschl's gyrus is so oriented that its pyramidal cells produce a large, tangentially oriented, dipole. However, neurons on both sides of a fold or on the edges of a gyrus will have their dendrites oriented at an angle to each other. If these two sides are activated simultaneously the currents of the two sides will flow in opposite direction and partial or complete cancellation of the currents can occur. This occurs to some extent also in Heschl's gyrus and is likely the reason why the MLRs, which are generated in Heschl's gyrus, are relatively small at the scalp. Thus MLRs and CAEPs are based on the PSPs. These PSPs reflect typically relatively slow monophasic changes lasting 10–15 ms and are either reflecting depolarization (EPSP, the cell interior becomes less negative than its resting value) or hyperpolarization (IPSP, the cell interior becomes more negative than its resting value). These changes result in a current flow across the membrane, so if the interior of the cell becomes more positive (depolarization), then the cell exterior at that location has to become more negative to counterbalance the charge across the membrane. Depolarization of the neuron produces an inflow of Na^+ ions into the dendrite through specialized ion channels in the postsynaptic membrane and the location where this happens is called a "sink." As a result of the local current flow into the cell, the other end of the dendrite will become relatively negative and this site is called a (passive) "source." The sink-source configuration forms a dipole. One pole (sink) is positive, the other (source) is negatively charged. If all the cells show the same pattern of activation along their spatially aligned dendrites the total effect can be described as resulting from an "equivalent dipole": one dipole to describe the compound effect of many similarly oriented individual cell dipoles (Eggermont, 2007).

6.3.2.3 Dipole sources of MLRs and CAEPs

Results of dipole source modeling (Ponton et al., 2002) demonstrated that the three orthogonal dipole components of the sources in each hemisphere isolate three distinct sets of CAEP components. The MLR peaks Pa and Pb are best represented by the sagittal (anteroposterior) dipole sources; the "classic" P1–N1–P2–N2 sequence is isolated to the tangential (to the scalp) sources that are perpendicular to the superior surface of the temporal lobe; and the T-complex peaks Ta and Tb, together with the TP200, are represented in the radial (perpendicular to the scalp) dipole sources (Fig. 6.5). Ponton et al. (2000a, 2002) distinguished three maturation groups of potentials: one group reaching maturity at age 5–6 and consist of the MLR components Pa and Pb, the CAEP component P2, and the T-complex, recorded from temporal electrodes. The temporally recorded Ta/Tb complex, which is distinct from N1, is already mature at age 5–6 (Pang and Taylor, 2000; Gomes et al., 2001; Tonnquist-Uhlen et al., 2003). Because of its radial-oriented dipole it is likely generated in secondary cortical area BA22. A second group that was relatively fast to mature (time constant 2 years) was represented by N2 only. A third group was characterized by a slower

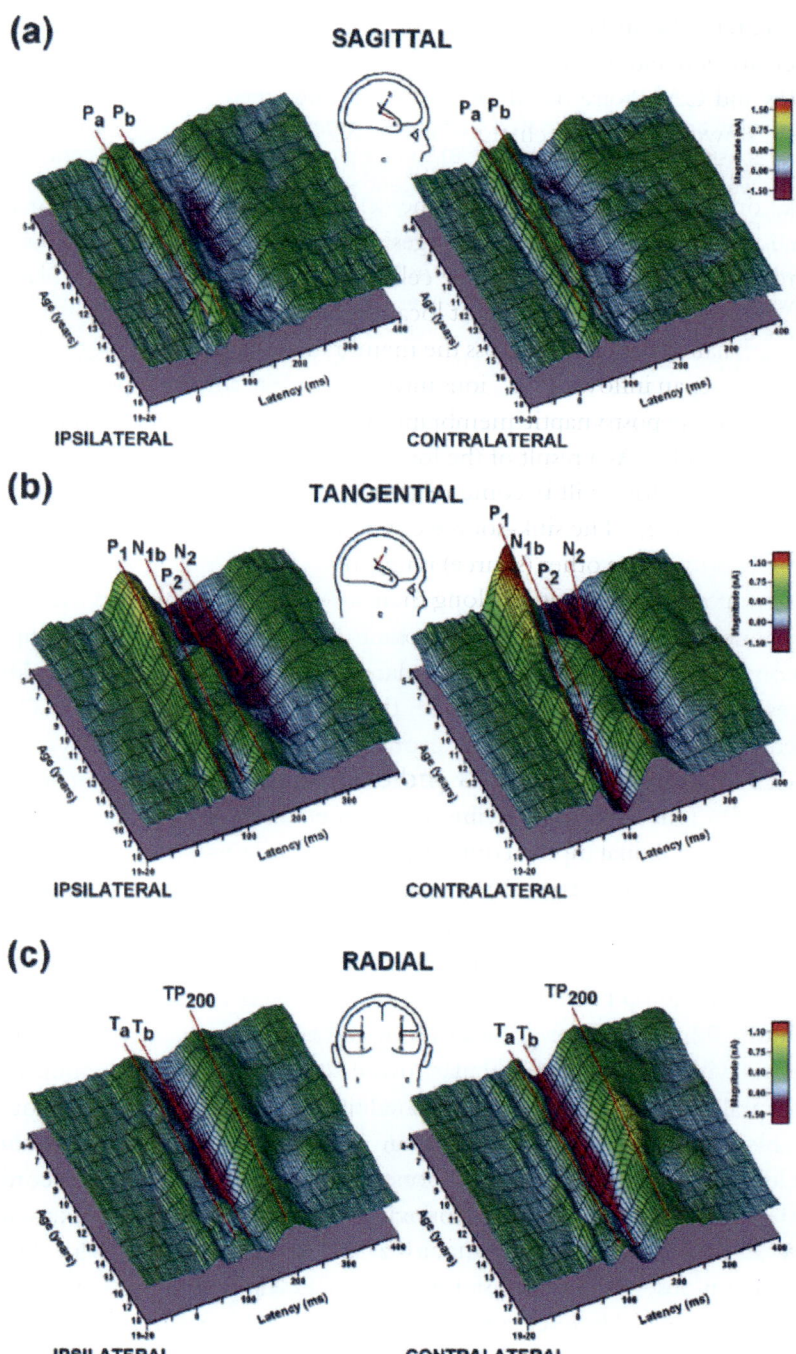

Fig. 6.5 See figure legend on next page.

Fig. 6.5 (A) Surface plots of the ipsilateral (left) and contralateral (right) source waveforms for the sagittally oriented dipoles. These source waveforms contain activity that corresponds in latency to the scalp-recorded Pa and Pb of the MLR. While there are some age-related changes in magnitude, latencies for both peaks are nearly constant as a function of age. (B) Surface plots of the ipsilateral and contralateral source waveforms are shown for the tangential dipoles. In younger children, a large positive peak labeled P1 with a latency similar to that of the adult N1b peak dominates response. As P1 magnitude decreases, the N1b peak begins to emerge between 9 and 11 years of age. (C) Ipsilateral and contralateral surface plots are shown for radially oriented dipole sources. The two T-complex components, Ta and Tb, are clearly represented in these source waveforms. A third peak, labeled TP200, is also apparent both ipsilateral and contralateral to the stimulated ear. *(Reprinted from Ponton, C., Eggermont, J.J., Khosla, D., Kwong, B., Don, M., 2002. Maturation of human central auditory system activity: separating auditory evoked potentials by dipole source modeling. Clin. Neurophysiol. 113, 407–420. © 2002 Elsevier Science Ireland Ltd, with permission from Elsevier.)*

pattern of maturation (time constants of 4–9 years) and included the CAEP components P1, N1, and TP200, the long-latency component following the T-complex. The observed latency differences combined with the differences in maturation rate indicate that P2 is not identical to TP200 (Picton and Taylor, 2007). The grouping of CAEP components isolated in each orthogonal dipole remained the same across a 5- to 20-year age span (Fig. 6.5). This suggests that the orientations of the CAEP generators are essentially adult-like by 5 years of age. Based on a global measure of similarity between individuals' CAEP waveforms, Adibpour et al. (2020) measured CAEPs to syllables in 1- to 6-month-old infants and related them to the maturational properties of underlying neural substrates measured with diffusion tensor imaging (DTI). Firstly, they observed a decrease in the latency of the auditory P2, and in the diffusivities in the auditory tracts and perisylvian regions with age. Secondly, they showed some of the early functional and structural substrates of lateralization. Contralateral responses to monaural syllables were stronger and faster than ipsilateral responses, particularly in the left hemisphere. Besides, the acoustic radiations, arcuate fasciculus, middle temporal, and angular gyri showed DTI asymmetries with a more complex and advanced microstructure in the left hemisphere, whereas the reverse was observed for the inferior frontal (IFG) and superior temporal gyri (STG; Fig. 6.6). After accounting for the age-related variance, they concluded that P2 responses might depend both on callosal connectivity and on the maturation of IFG.

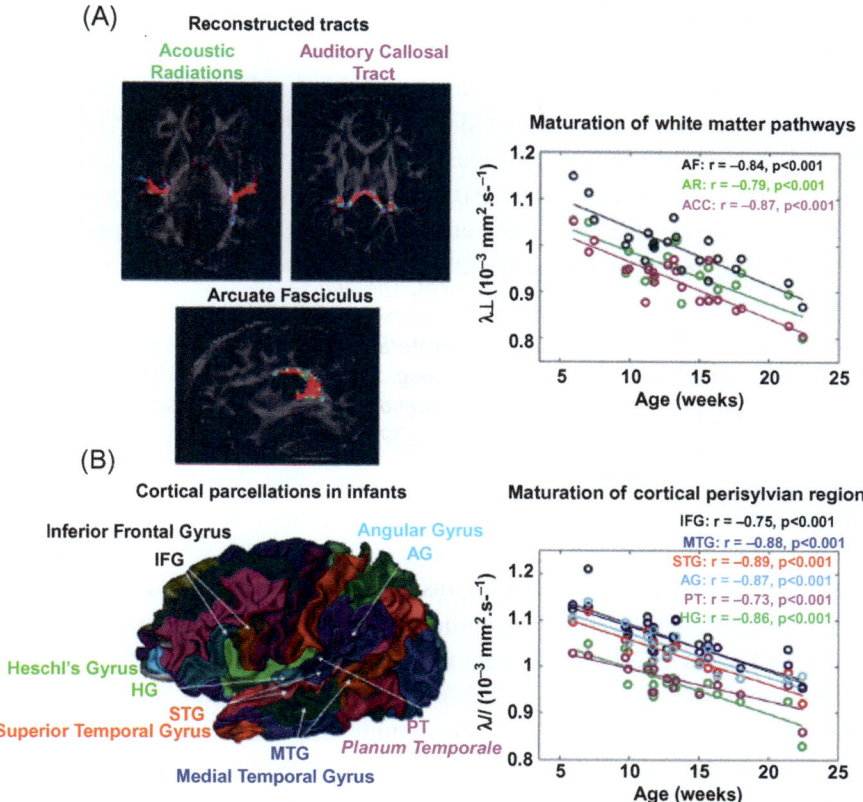

Fig. 6.6 Maturation of the linguistic pathways and perisylvian regions quantified with diffusion MRI. (A): Left panel: Example of tract reconstructions in a 12-week-old infant superposed on the map of DTI anisotropy: acoustic radiations (AR), auditory fibers of the corpus callosum (ACC), and arcuate fasciculus (AF). Right panel: Transverse diffusivity ($\lambda_\perp$) plotted as a function of infants' age in each tract (left and right values are averaged). Major age-related decreases are observed ($n = 22$). (B): Left panel: 3D parcellation atlas of a 10-week-old infant showing the studied regions: Heschl's gyrus (HG), planum temporale (PT), superior and middle temporal gyri (STG, MTG), angular and inferior frontal gyri (AG, IFG). Right panel: Longitudinal diffusivity ($\lambda_{||}$) plotted as a function of infants' age in each region (left and right values are averaged). Major age-related decreases are observed ($n = 21$). *(From Adibpour, P., Lebenberg, J., Kabdebon, C., Dehaene-Lambertz, G., Dubois, J., 2020. Anatomo-functional correlates of auditory development in infancy. Dev. Cogn. Neurosci. 42, 100752. Open Access.)*

6.3.2.4 Maturation of brain rhythms

By means of MEG, Fujioka and Ross (2008) investigated event-related synchronization and desynchronization (ERS/ERD) in auditory cortex activity, represented as amplitude increase and decrease, respectively, and recorded from twelve children aged 4–6 years, while they passively listened to a violin tone and a noise-burst stimulus. Time-frequency analysis using wavelet transforms was applied to single trials of source waveforms observed from left and right auditory cortices. Stimulus-induced changes in non-phase-locked activities were evident. ERS in the beta range (13–30 Hz) lasted only for 100 ms after stimulus onset. This was followed by prominent alpha ERD, which showed a clear dissociation between the upper (12 Hz) and lower (8 Hz) alpha range in both left and right auditory cortices for both stimuli. The time courses of the alpha ERD (onset around 300 ms, peak at 500 ms, offset after 1500 ms) were similar to those previously found for older children and adults with auditory memory related tasks. For the violin tone only, the ERD lasted longer in the upper than the lower alpha band. The findings suggest that induced alpha ERD indexes auditory stimulus processing in children without specific cognitive task requirement. The left auditory cortex showed a larger and longer-lasting upper alpha ERD than did the right auditory cortex, likely reflecting hemispheric differences in maturational stages of neural oscillatory mechanisms. Shahin et al. (2010) examined the development of phase locking—again reflected in increase and decrease in amplitude—of oscillatory responses to music sounds and to pure tones matched to the fundamental frequency of the music sounds. Phase locking for theta (4–8 Hz), alpha (8–14 Hz), lower-to-mid beta (14–25 Hz), and upper-beta and gamma (25–70 Hz) bands strengthened with age. Phase locking in the upper-beta and gamma range matured later than in lower frequencies and was stronger for music sounds than for pure tones, likely reflecting the maturation of neural networks that code spectral complexity. Phase locking for theta, alpha, and lower-to-mid beta was sensitive to temporal onset (rise time) sound characteristics. Tang et al. (2016) recorded stimulus envelope following responses to white noise which was amplitude modulated at rates of 1–80 Hz in healthy 3–5 year old children and adults using a pediatric MEG system and a conventional MEG system, respectively. For children, there were envelope following responses to slow modulations but no significant responses to rates higher than about 25 Hz, whereas adults showed significant envelope following responses to almost the entire range of stimulus rates. These neurophysiological results are consistent with previous psychophysical evidence for a protracted maturational time course for auditory temporal processing (Eggemont, 2015).

6.4 Maturation of the mismatch response

Presentation of a novel (oddball or deviant) stimulus produces a larger event-related response (ERP) compared to that for a frequent stimulus just preceding it. The difference between the ERP to oddball and frequent stim-uli in adults is called the mismatch negativity (MMN). The presence of an MMN has been seen as an indicator of preattentive detection of stimulus change, be it acoustical, phonetic, or contextual (Näätänen, 2001). Recall that the MMN is never recorded as such from the scalp; it is a construct designed by the investigator and is obtained by subtracting two nonsimul-taneously recorded ERPs. It also assumes that the N1 to the frequent and the oddball are the same, which is unlikely (see also Chapter 2.3; May and Tii-tinen, 2010). Thus, in interpreting the MMN one should always inspect the individually recorded CAEPs, or their magnetic field equivalent, that are at the basis of this construct. The MNN is often considered an ERP related to the adaptation of activity of the frequent stimulus as a result of its (quasi-) periodic presentation (May, 2021). For other interpretations, see Chapters 2 and 4.

6.4.1 Early age maturation studies: From positive to negative mismatch responses

6.4.1.1 Newborns

Early ERP studies in newborns (Leppänen et al., 1997) and infants (Alho and Cheour, 1997; Cheour et al., 1998) indicate passive detection of even a small pitch change based either on refractoriness to repetition or dishabituation to change, or both. Some evidence was also found for a mismatch negativity-like response overlapping with the positive response and appearing as a reduction of this positive deflection at a latency of a typical mismatch neg-ativity. Leppänen et al. (2004) then measured ERPs of 21 newborns during quiet sleep to rarely occurring deviant tones of 1100 Hz (probability 12%) embedded among repeated standard tones of 1000 Hz in an oddball sequence. They found that during quiet sleep, maturational factors explain a significant portion of the ERP difference wave amplitude in terms of its polarity, indicating that the more mature the ERPs are, the more positive the amplitude.

Suppanen et al. (2019) compared newborn infants' learning from a song, a nursery rhyme, and normal speech in the same study. Infants' electrophys-iological brain responses revealed that the nursery rhyme condition facili-tated learning from auditory input, and thus led to successful detection of deviations (Fig. 6.7). These findings suggest that coincidence of prosodic cue patterns and to-be-learned items is more important than the format of the input. This supports the view that rhythm is likely to create a template for future events, which allows auditory system to predict prospective input and thus facilitates language development.

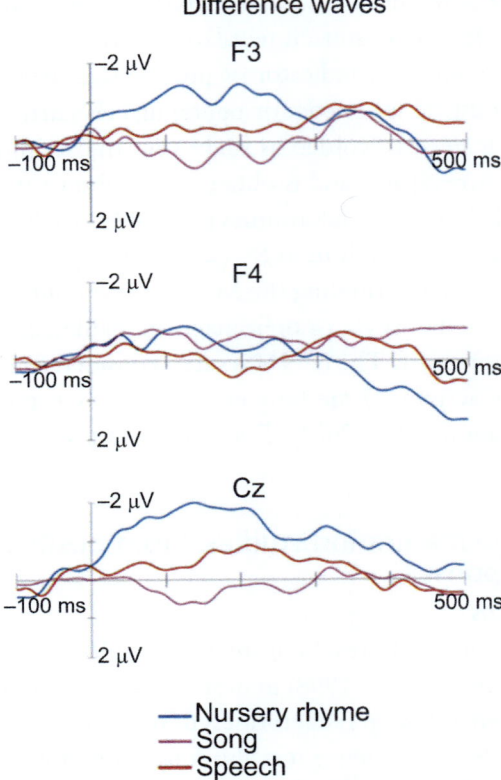

Fig. 6.7 The brain responses of newborn infants to word changes in stressed syllables from electrodes Cz, F3, and F4. *(From Suppanen, E., Huotilainen, M., Ylinen, S., 2019. Rhythmic structure facilitates learning from auditory input in newborn infants. Infant Behav. Dev. 57, 101346. Open Access.)*

6.4.1.2 Infants

In infants, the early difference response is of positive polarity, and to avoid confusion, it is often called the mismatch response (MMR). Dehaene-Lambertz and Dehaene (1994) described high-density recordings of ERPs in three-month-old infants listening to syllables in which the first consonants differed in place of articulation. Two processing stages were identified and localized to the temporal lobes. A late frontal response to novelty was also observed. Cheour et al. (1998) compared MMRs in preterm infants (conceptional age at the time of recording, 30–35 weeks), full-term newborns, and full-term 3-month-old infants. They found no significant differences in

MMR amplitudes between the three age groups either. The mean MMR latency, however, decreased significantly with age, although in 3-month-old infants it was not much longer than in a previous study conducted in adults with the same stimuli, suggesting very early maturation. The suppression of the frequent CAEP is clearly present in 3-month-old infants (Dehaene-Lambertz and Gliga, 2004) and can form the basis for finding an MMR in infants and even in fetuses as young as 28 weeks' gestational age (Draganova et al., 2005, 2007).

In two-month-olds, using gap duration as the contrast, the standard stimuli evoked only a positive slow wave, and its amplitude was increased in response to the deviant stimuli. By 6 months, the deviant stimuli evoked an increased negativity at approximately 200 ms, similar to the MMN response in adults (Trainor et al., 2003). However, eight-month-old infants still showed a slow positive MMR to /da/-/ta/phoneme contrasts (Pang et al., 1998). In term infants, the initial positivity recorded from midline electrodes and the negativities recorded from ipsi- and contralateral temporal electrodes did not correlate in their peak and offset latencies, suggesting independent generators for each of these components (Novak et al., 1989). With respect to the positive vs negative MMR, Friederici et al. (2002) recorded event-related potentials (ERPs) in 2-month-old infants in two different states of alertness: awake and asleep. Syllables varying in vowel duration (long vs short) were presented in an oddball paradigm, known to elicit a mismatch brain response. ERPs of both groups showed a mismatch response reflected in a positivity followed by a frontal negativity. While the positivity was present as a function of the stimulus type (present for long deviants only), the negativity varied as a function of the state of alertness (present for awake infants only). Thus there may be two overlapping mismatch responses (Kushnerenko et al., 2002a): a negative MMR that is later than the MMN of an older child or adult, and a positive MMR. They found evidence for this in a portion of 3- to 9-month-old infants, a large-amplitude positive component commenced at the latency of the MMN and thus might have masked it. The results of the additional measurement, employing distracting novel sounds in 2-year-old infants and newborns, suggested that the observed positive component could represent an infant analogue of the adult P3a response, indexing an involuntary orienting of attention.

Morr et al. (2002) determined whether an adult-like MMN can be reliably elicited in typically developing awake infants and preschool children (aged 2–47 months) and examined whether maturational changes exist in

MMN latency and amplitude. In Experiment 1 (small deviant in frequency) a negativity was not elicited in the majority of the infants and preschoolers tested. In Experiment 2 (large deviant in frequency) a negativity was reliably elicited in the infants and preschoolers across all ages. A significant negative correlation was observed between age and latency, but not for age and amplitude for this negativity. Thus, depending on the relative timing and sizes of these waves, a negative or positive MMR might result (Picton and Taylor, 2007). Again, one has to keep in mind that the MMR does not exist in the brain, but rather is constructed by subtraction. One can only argue that activity in one part of the brain is going more negative while another part is going more positive, and that the distribution of activity across cortical layers and regions determines the summed changes observed in scalp recordings.

He et al. (2007) investigated the mismatch polarity changes to piano tones by recording 2-, 3-, and 4-month-old infants' CAEP responses to frequent and infrequent pitch changes (Fig. 6.8). In all age groups, the infants' responses to deviant tones were significantly different from the responses to the frequent tones, suggesting that the two tones were processed as different.

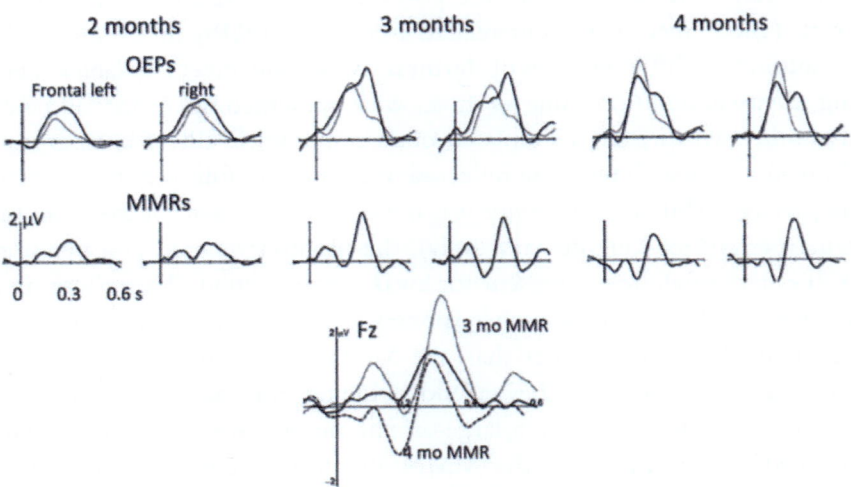

Fig. 6.8 CAEPs to frequent (dotted lines) and deviant (solid lines) stimuli are shown in the top layer panels for 2-, 3-, and 4-month-olds. The middle layers show the corresponding MMRs. In the bottom central panel, the MMRs for the Fz recording site are shown. FL and FR stand for left and right frontal region, which bracket the midline location Fz. *(From Eggermont, J.J., Moore, J.K., 2012. Morphological and functional development of the auditory nervous system. L.A. Werner et al. (Eds.), Human Auditory Development, Springer Handbook of Auditory Research, vol. 42, pp. 61–105, ©2012 Springer Science+Business Media, with permission from Springer Nature.)*

However, two types of MMR were observed simultaneously in the constructed difference waves. By low-pass filtering of the MMR, an increase in the left lateralized positive slow MMR emerged that was prominent in 2-month-olds, present in 3-month-olds, but insignificant in 4-month-olds. By high-pass filtering a faster, more adult-like MMR, lateralized to the right hemisphere, emerged at 2 months of age and became earlier and stronger with increasing age. The coexistence but spatial dissociation of the two types of MMR suggested different underlying neural mechanisms. The infant cortex may differentially process incoming information, as infants are prone to alternate sleep-wake cycles. During sleep, the positive MMR dominates, whereas the negative polarity was more obvious in waking infants (Friederici et al., 2002).

He et al. (2009) then measured "ERPs from adults, 2-month-olds, and 4-month-olds to a repeating piano tone (standard) that occasionally changed in pitch (deviant). The pitch changes between standards and deviants were either small (1/12 octave) or large (1/2 octave) in magnitude, and the stimulus presentation rate was either slow (800 ms SOA) or fast (400 ms SOA). As the presentation rate increased, both adults and 4-month-olds showed an MMN response that decreased in latency, but was unaffected in amplitude. As the magnitude of the pitch change increased, MMN increased in amplitude. On the other hand, only a broad positive mismatch response was seen in 2-month-olds. As the presentation rate increased, 2-month-olds' responses to standard tones decreased in amplitude while their responses to deviant tones were unaffected. The magnitude of the pitch change did not affect 2-month-olds' responses. These results suggest that pitch is processed differently in auditory cortex by 2-month-olds and 4-month-olds, and that a cortical change-detection mechanism for pitch discrimination similar to that of adults emerges between 2 and 4 months of age" (Eggermont and Moore, 2012).

Virtala et al. (2022) recorded MMRs to vowel, duration, and frequency deviants in pseudo-words at 0, 6, and 28 months and compared MMRs. A broad positivity (MMR, positive (P)-MMR) is the most prevalent MMR in infancy, its amplitude increased by 6 and decreased by 28 months; latency decreased with increasing age. The P-MMR was predominant in infancy; an early negativity, MMN, emerges preceding it, growing in amplitude and decreasing in latency with maturation. MMN was elicited by more deviants only at 28 months. A late negativity (LDN) emerges following the positivity and is a prominent MMR in early childhood. Neonatal MMN to the duration deviant became larger and earlier by 28 months (Fig. 6.9).

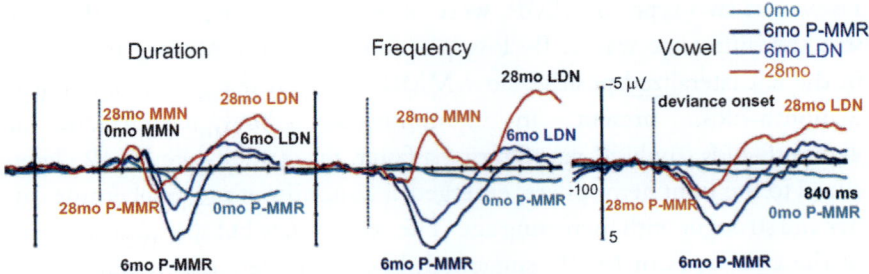

Fig. 6.9 Maturation of the mismatch responses (MMRs), mismatch negativity (MMN), positive mismatch response (P-MMR), and late discriminative negativity (LDN). Subtraction waveforms illustrating MMR maturation at 0 month (red line), 6 months (black thin line/blue thin line), and 28 months (black thick line) at the large ROIs, with baseline at deviance onset (marked with the vertical dashed line). *(From Virtala, P., Putkinen, V., Kailaheimo-Lönnqvist, L., Thiede, A., Partanen, E., Kujala, T., 2022. Infancy and early childhood maturation of neural auditory change detection and its associations to familial dyslexia risk. Clin. Neurophysiol. 137, 159–176. Open access.)*

To determine whether 5–7 month infants' expectations about future sensory input modulate their sensory cortices without the confounds of stimulus novelty or repetition suppression, Emberson et al. (2015) used a cross-modal (audiovisual) omission paradigm and functional near-infrared spectroscopy (fNIRS) to record hemodynamic responses in the infant cortex. Emberson et al. (2015) show that the occipital cortex of 6-month-old infants exhibits the signature of expectation-based feedback. Crucially, this region does not respond to auditory stimuli if they are not predictive of a visual event. Overall, these findings suggest that the young infant's brain is already capable of some rudimentary form of expectation-based feedback. Subsequently, Emberson et al. (2019) investigated whether repetition suppression in the infant brain is similarly sensitive to top-down mechanisms. They again used fNIRS in two experimental conditions, one in which variability in stimulus presentation is expected (occurs 75% of the time) and a control condition where variability and repetition are equally likely (50% of the time). They show that 6-month-old infants exhibit attenuated frontal lobe response to blocks of variable auditory stimuli during contexts when variability is expected as compared to the control condition. This suggests that young infants' neural responses are modulated by predictions gained from experience and not simply by bottom-up mechanisms (Emberson et al., 2019).

6.4.2 From childhood onward

Ceponiené et al. (1998) studied the component structure of auditory event-related potentials in children of 7–9 years old by presenting stimuli with different interstimulus intervals (ISI) of 350, 700, and 1400 ms. For the 350-ms ISI, the ERP waveform to the standard stimulus consisted of P100–N250 peaks. For longer ISIs, in addition, a frontocentral N160—likely a correlate of the N1—and N460 peaks were observed. In difference waves, obtained by subtracting ERP to standard stimuli from ERP to deviant stimuli, two negativities were found. The first was the MMN, which suggested that neural traces of auditory sensory memory lasted for at least 1400 ms, probably considerably longer, as no MMN attenuation was found across the ISIs used. The second, later negativity was similar to MMN in all aspects, except for the scalp distribution, which was posterior to that of the MMN. In children aged 5–7 years, Gomot et al. (2000) found that the amplitudes of the temporal components were significantly greater than in adults, whereas the frontal components were similar at all ages. This suggests that MMN is mediated by at least two separate neural systems and that the frontal system (driven by the reticular formation?) matures earlier than the sensory-specific (thalamo-cortical) system. In school-age children (4–10 years) Shafer et al. (2000) found that the MMN decreased in latency by 11 ms/year from 4 to 10 year of age. No developmental change in MMN amplitude was seen in that same group. However, the MMN amplitude was significantly smaller in adults than in children. The prominent negativity in children was significantly later than the adult N1 component and did not change in latency from 4 to 10 year of age. This finding adds to a body of evidence suggesting that this prominent negativity and the adult N1 are not the same component. This was highlighted by Ponton et al. (2000b) who studied the MMN in a group of late implanted children who have good spoken language perception through their cochlear implant, and where no N1 could be found, nevertheless, the MMN was robustly present.

In 2- to 6-year-olds, Glass et al. (2008) examined the maturation of auditory sensory memory in normally developing children between the ages of 2 and 6 years. To probe the lifetime of auditory sensory memory they used the MMN for tone stimuli of two different frequencies presented with various interstimulus intervals between 500 and 5000 ms. Findings suggested that memory traces for tone characteristics have a duration of 1–2 s in 2- and 3-year-old children, more than 2 s in 4-year-olds and 3–5 s in 6-year-olds (Fig. 6.10).

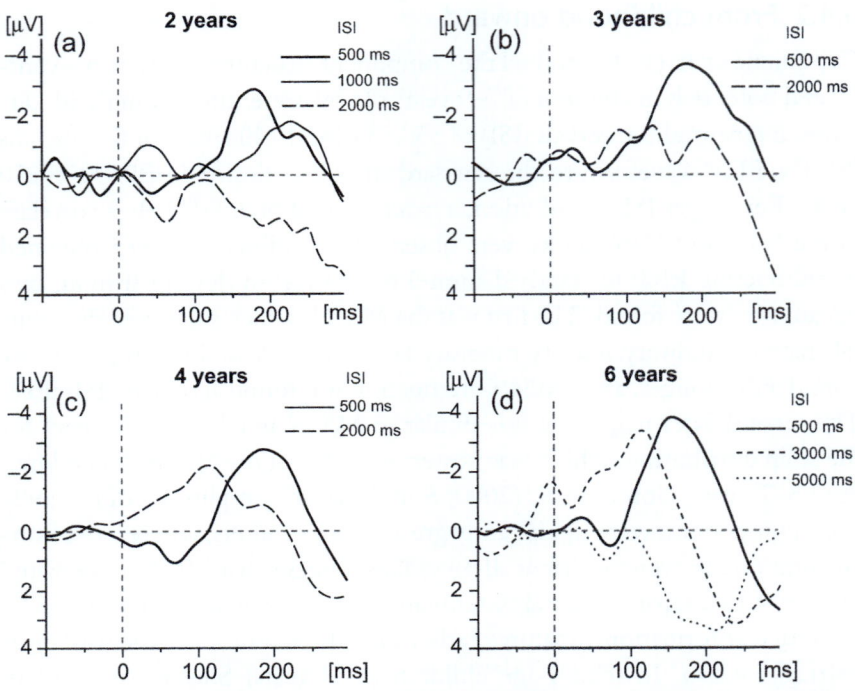

Fig. 6.10 MMN for different ISI conditions at electrode Fz by age group (reference: linked mastoids). *(From Glass, E., Sachse, S., von Suchodoletz, W., 2008. Development of auditory sensory memory from 2 to 6 years: an MMN study. J. Neural Transm. 115, 1221–1229, © Springer-Verlag 2008, with permission from Springer Nature.)*

In 4- to 7-year-old children, Shafer et al. (2010) recorded ERPs to a standard /ɛ/ and deviant /i/ vowel presented in trains of 10 stimuli at a rate of 1/650 ms and with an intertrain interval of 1.5 s. Each train contained two deviant vowels. They observed significantly greater negativity, consistent with the adult MMN, to the deviants between 300 and 400 ms for both younger (4- and 5-year-old) and older (6- and 7-year-old) children. This MMN-like negativity shifted to shorter latencies with increasing age by ~25 ms/year. Most of the children younger than 5.5 years and some of the older children also showed a positive MMR peaking between 100 and 300 ms. The positive MMR diminished in amplitude with increasing age. Bishop et al. (2011) measured auditory ERPs in children (7–12 years), teenagers (13–16 years), and adults (35–56 years) in an oddball paradigm with tone or syllable stimuli. For each stimulus type, a standard stimulus (1000-Hz tone or syllable [ba]) occurred on 70% of trials, and one of two deviants (1030- or 1200-Hz tone, or syllables [da] or [bi]) equiprobably on the remaining trials. For the traditional MMN interval of 100–250 ms postonset, size of mismatch responses increased with age, whereas the opposite trend was seen for an interval from 300 to 550 ms postonset, corresponding to the late discriminative negativity (LDN; see Appendix).

In 6–8 year old children and adults, Dercksen et al. (2022) used a motor-auditory omission paradigm. In an identity-specific condition, the sound coupled to the motor action was predictable, while in an identity unspecific condition the sound was unpredictable. Results of a temporal principal component analysis revealed that sound-related brain responses underlying the N1 complex differed considerably between age groups. Despite these developmental differences, omission responses (oN1) were similar between age groups. Two subcomponents of the oN1 were differently affected by specific and unspecific predictions. Results demonstrate that children, independent from the maturation of sound processing mechanisms, can implement specific and unspecific predictions as flexibly as adults (Fig. 6.11). This supports action and prediction error as important drivers of cognitive development.

6.4.3 An MMR polarity explanation based on changes in auditory cortex

Based on all of the above, a set of mechanisms (Section 6.3.2.2 for details about sinks and sources) that could accommodate the various MMR and obligatory CAEPs observed in the preterm and postnatal period is the following (from Eggermont and Moore, 2012):

1. During the perinatal period, there are functional glutaminergic Cajal-Retzius (C-R) cell synapses in layer I on dendrites of pyramidal cell in layers II, III, and V. Because these synapses release glutamate, they affect the slow NMDA receptors. Activation of these synapses will induce a layer I sink (depolarizing synapse; extracellular negative potential) that will be visible as a scalp negative deflection. This can explain the large, broad, and very long latency negative obligatory cortical evoked potentials (OCEPs) recorded in preterm infants.

2. After 4.5 months of age, glutamate C-R cells disappear while GABA cells remain and now only cause transient hyperpolarizations in layer I synapses. This results in layer I sources that are visible as scalp-positive deflections, which could explain the emerging positive obligatory CAEPs and consequently the resulting positive MMRs. It should be noted that a positive MMR generated by this mechanism would require increasing hyperpolarization for the deviant stimulus. Alternatively,

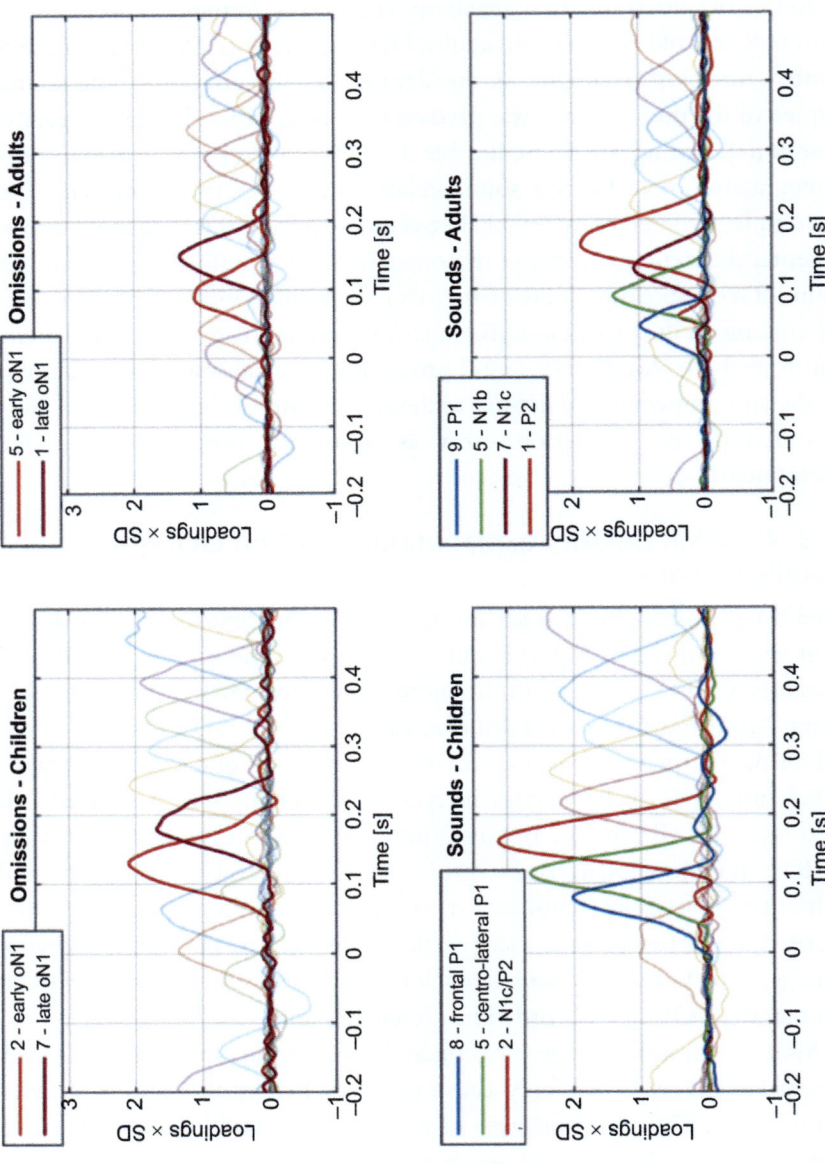

Fig. 6.11 Overview of component loadings for omissions and sounds in children and adults. Omission independent principal components identified as oN1 and sound-related components of interest are shown in opaque. *(From Dercksen, T.T., Widmann, A., Scharf, F., Wetzel, N., 2022. Sound omission related brain responses in children. Dev. Cogn. Neurosci. 53, 101045. Open Access.)*

positive MMRs can be derived from increased positive obligatory CAEPs resulting from a depolarization site (sink) in layer IV, the entrance of (unmyelinated) thalamocortical activation, which is likely the earliest specific or lemniscal auditory input to the cortex.

3. With increasing maturity, other largely excitatory inputs may arrive in layer I from the MGBm, which will again produce layer I sinks that would give rise to negative polarity MMNs. This, however, cannot explain the persistent positive polarity CAEPs.

4. Slow-conducting, thalamocortical inputs that start to myelinate at 3–4 months of age could produce poorly synchronized depolarization in layer IV with the required scalp-positive CAEP polarity (see under point 2). However, with the MMR resulting from a repetition-related decrease in the CAEP to the frequent stimulus, this would also produce a positive MMR, which is observed rarely after the first 6 months of age. Though this is an incomplete mechanism, an interaction with the mechanism mentioned under point 3 can account for both negative MMRs and positive OCEPs. This would be accomplished by partial cancellations at the scalp of the activity generated by these two mechanisms.

5. The dominance of a negative MMR after 6–8 months of age would require that, at this early age as well as in adults, it is generated in superficial layers I and/or II and results from excitatory modulations of activity in pyramidal cells with cell bodies in layers II, III, and V.

In light of recent evidence, the origin of this MMN activity might well be driven by prefrontal or orbitofrontal cortex, via the inferior frontal gyrus (Chapter 1 and next section) ultimately affecting the nonlemniscal auditory cortex with a polarity following from the previously mentioned description.

6.4.4 Predictive coding and development

"A predictive neural and psychological system is a system that changes over time as a function of the experience it accrues and integrates in persisting predictive models, which, in turn, shape the following experiences. This definition is strikingly similar to a general definition of development as a dynamical process of increasing neural and psychological complexity through interactions with the physical and social environment, and shaping future interactions based on the development of those neural and psychological structures" (Pereira et al., 2019).

Kouider et al. (2015) studied how the infant brain responds to prediction violations using a cross-modal cueing paradigm. They recorded EEG responses to expected and unexpected visual events preceded by auditory cues in 12-month-old infants and found an increased response for unexpected events. However, this effect of prediction error was only observed during late processing stages (N290 and P400) associated with conscious access mechanisms. In contrast, early perceptual response components (P1) reveal an amplification of neural responses for predicted relative to surprising events, suggesting that selective attention enhances perceptual processing for expected events. Kouider et al.'s (2015) results "fit within a framework where perception, at least in infants, involves two distinct mechanisms that rely upon prior information: first, an early mechanism of selective attention aimed at amplifying target features that match template information generated by the auditory cue, and then a second mechanisms of Bayesian integration aimed at generating an error signal when the unexpected target appears instead of the expected one." As an alternative possibility, both these effects of attention and expectation may be explained by predictive coding. Indeed, recent accounts of predictive coding have proposed that attention is actually subserved by the same mechanisms as those underlying expectation (Feldman and Friston, 2010). According to this approach, "attention reflects an increase in the precision of prior information, while surprise reflects an increase in the error signal resulting from the mismatch between prior and novel information" (Kouider et al., 2015).

Pendl et al. (2017) characterized the development of the brain's hierarchical organization using a modified stepwise functional connectivity approach based on resting-state fMRI in a fully longitudinal sample of infants ($N=28$, with scans after birth, and at 1 and 2 years) and adults. Results obtained from seeds in early sensory cortices revealed hierarchical patterns of adult brain organization ultimately converging in limbic, paralimbic, basal ganglia, and frontoparietal brain regions. These findings are consistent with predictive coding accounts of neural processing that place these regions at the top of predictive coding hierarchies. Infants gradually developed toward this architecture in a region- and step (subsequent links) -dependent manner and displayed many of the same regions as adults in top hierarchical positions, starting from 1 year of age. Across all three sensory systems, step 1 matured earlier than steps involving multiple links (shown for the auditory system in Fig. 6.12). Pendl et al. (2017) described this as: "Among multi-link steps, those associated with the visual system showed more deterministic and statistically significant growth toward their mature forms than sensorimotor and auditory systems. Inter-sensory convergence, particularly for Step 1, included sensory integration 'analogs,' and in several instances inter-sensory

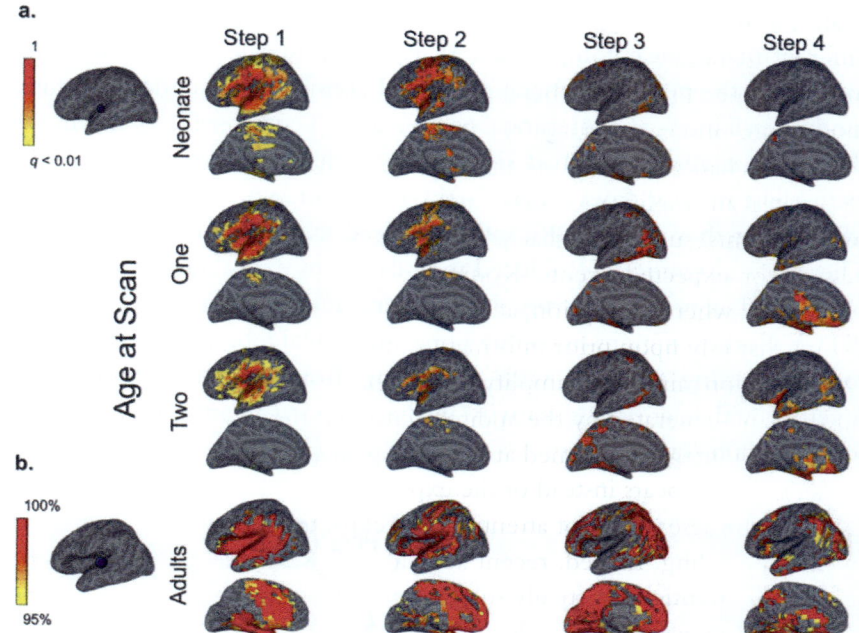

Fig. 6.12 Modified stepwise functional connectivity (mSFC) of the auditory (AUD) seed in infants and adults. (A) The AUD seed in infant template space is shown on the top left. Infant results (right) are thresholded at $P < 0.01$ (FDR corrected) and scaled to the maximum t-value. (B) The adult AUD seed is displayed to the left of adult maps, which are scaled to show mSFC that was significant ($P < 0.01$, FDR corrected) at the group level and for at least 95% of 1000 iterations, including 28 randomly selected adults for each iteration. *(From Pendl, S.L., Salzwedel, A.P., Goldman, B.D., Barrett, L.F., Lin, W., Gilmore, J.H., Gao, W., 2017. Emergence of a hierarchical brain during infancy reflected by stepwise functional connectivity. Hum. Brain Mapp. 38, 2666–2682. © 2017 Wiley Periodicals, Inc., with permission from John Wiley and Sons.)*

connection densities were related to cognitive performance, especially between perisylvian audio–motor regions and language-dependent scales. Finally, the emergence of consistent and adult-like Step 4 overlap in limbic, paralimbic and striatum areas starting from 1 year provides support for a unique developmental role of these dopamine-pathway-related brain areas."

6.4.5 An MMN-generation DCM model

Using dynamical causal modeling (DCM; Chapter 1), Cooray et al. (2016) investigated the neurophysiological mechanisms underpinning the generation of the MMN and its development from adolescence to early adulthood. MMN was elicited with an auditory oddball paradigm in two groups of healthy subjects with mean age 14 and 26 years. They tested models with different hierarchical complexities including up to five cortical nodes and found that the network generating MMN consisted of 5 nodes that could modulate all intra- and internodal connections. The inversion of this model showed that adolescents had significantly reduced backward connection from right inferior frontal gyrus (rIFG) to right superior temporal gyrus (rSTG) together with significantly increased excitatory activity in rSTG. There was a significantly reduced modulation of excitability in rSTG and of forward connectivity from left primary auditory cortex (lA1) to left (STG). Cortical regions in the temporal and frontal lobes, involved in auditory processing, mature with increasing frontotemporal connectivity together with increased sensitivity in the temporal regions for changes in sound stimuli (Fig. 6.13).

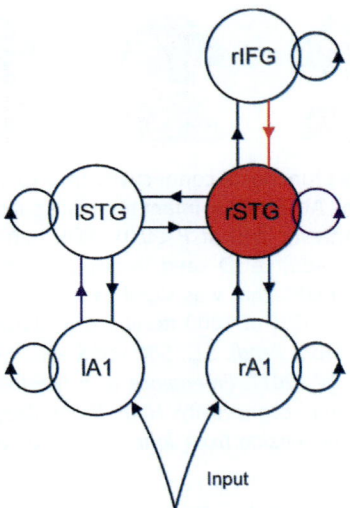

Fig. 6.13 The DCM of standards vs. deviants showed a significant reduction in modulation of intrinsic connectivity at rSTG and of forward connection between lA1 and lSTG for adolescents compared to adults (P < 0.02, corrected), arrows marked in blue. Furthermore, there was a significant reduction in backward connectivity between rIFG and rSTG for adolescent subjects (−0.10 ± 0.28) compared to adult subjects (0.091 ± 0.36) (P < 0.04) for the inversion of the standard-tone ERP, arrows marked in red. There was also increased activity of the excitatory interneurons at rSTG for the adolescent subjects compared to the adult subjects (adolescent group 4.1 nV and adult group 2.3 nV, P < 0.03), node filled in red. (Reprinted from Cooray, G.K., Garrido, M.I., Brismar, T., Hyllienmark, L., 2016. The maturation of mismatch negativity networks in normal adolescence. Clin. Neurophysiol. 127(1), 520–529. © 2015 International Federation of Clinical Neurophysiology, with permission from Elsevier.)

6.5 Novelty detection. Maturation of the P300 complex

In children, P3 latency decreases with increasing age. This decrease may be linked with the maturation of cognitive processes. Martin et al. (1993) recorded P3 in gifted children (IQs over 140; aged 4.4–7.7 years) and control children (IQ not mentioned; aged 3.9–12.2 years). They found that P3 component latency was significantly shorter in the gifted children at Cz. Interpeak interval N1-P3 was significantly shorter at all three recording sites (Fz, CZ, and Pz). These results suggested to them a relationship between the P3 component and cognitive ability in children. Gumenyuk et al. (2004) investigated distractibility in three age groups of children (8–9, 10–11, and 12–13 years) with ERPs and performance measures in a forced-choice visual task. The amplitude of the late portion of the P3a elicited by novel sounds was largest for the youngest group and showed a centrally dominant scalp distribution and smallest for the oldest group with a frontal scalp distribution. A frontally dominant late negativity (LN; see also Fig. 6.9) was largest in the youngest group and followed the P3a. Correlation between the RT increase caused by the distracting novel sounds and the amplitude of the LN elicited by these sounds suggested that the LN is associated with the degree of attention engaged by the distracting sounds (Fig. 6.14).

Wetzel et al. (2006) investigated auditory involuntary and voluntary attention in children aged 6–8, 10–12, and young adults. Pitch changed sounds caused prolonged reaction times and decreased hit rates in all age groups. Larger distractors (20%) caused stronger distraction in children, but not in adults. The amplitudes of MMN, P3a, and reorienting negativity (RON; Appendix) were modulated by age and by voluntary attention. P3a was additionally affected by distractor strength. Maturational changes were also observed in the amplitudes of P1 (decreasing with age) and N1 (increasing with age). P2 modulation by voluntary attention was opposite in young children and adults. They concluded that "the processing steps involved in distraction (pre-attentive change detection, attention switch, reorienting) are functional in children aged 6–8 but reveal characteristic differences to those of young adults. In general, distractibility as indicated by behavioral and ERP measures decreases from childhood to adulthood." In the same three age groups, Wetzel and Schröger (2007) then investigated ERPs to short (100 ms) and long (500 ms) novels and pitch deviant tones. Age-specific distributions of P3a demonstrate developmental differences in the processing of unexpected sounds. Moreover, long compared with short novel sounds (but not long compared with short pitch deviant tones) elicited

P3a and LN difference waveforms in children

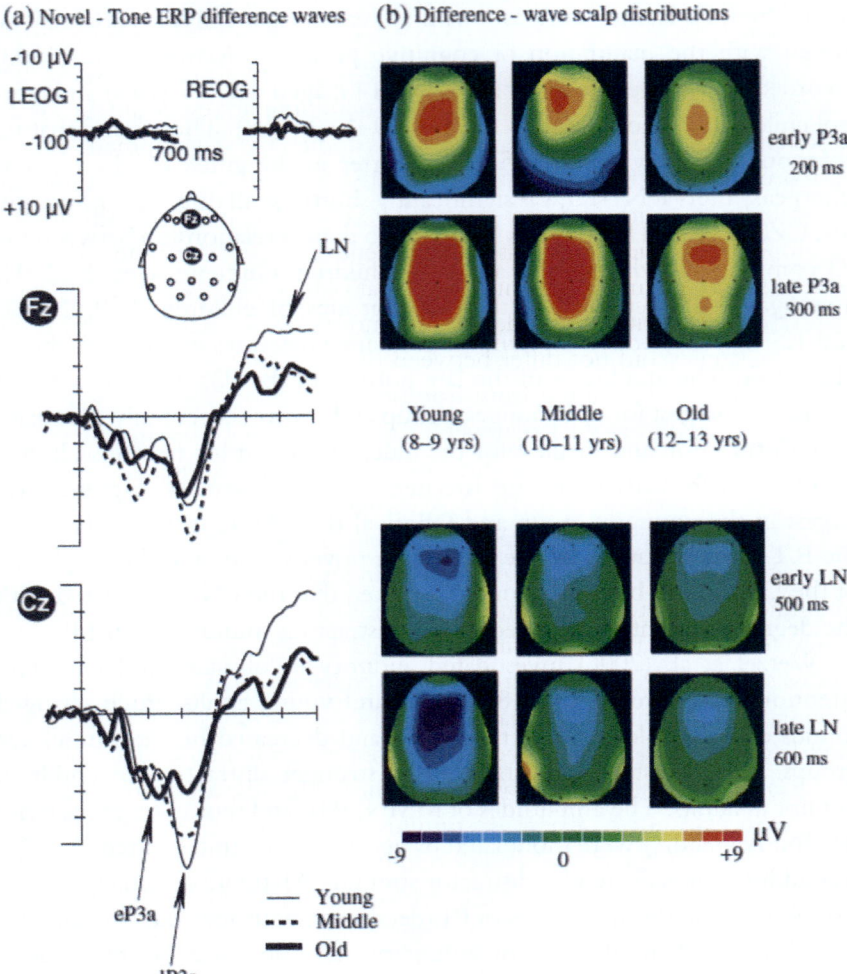

Fig. 6.14 Grand-average difference waves (novel minus tone) for the distracting novel sound. (A) Grand-average ERP difference waves recorded from young (solid line), middle (dashed line), and old (heavy line) groups of children at the frontal (Fz) and central (Cz) midline electrodes. The P3a elicited by novel sounds consists of two phases: early P3a (eP3a) and late P3a (IP3a), which was followed by the late negativity (LN). Horizontal EOG was recorded with electrodes attached to the left (LEOG) and right (REOG) canthi. (B) P3a and LN scalp-distribution maps for each group of children. The maps demonstrate average voltage at 200 ms (early P3a), 300 ms (late P3a), 500 ms (early phase of LN), and 600 ms (late phase of LN). Small open circles on the schematic scalp depict locations of electrodes; large filled circles depict the Fz and Cz electrode positions. *(From Gumenyuk, V., Korzyukov, O., Alho, K., Escera, C., Näätänen, R., 2004. Effects of auditory distraction on electrophysiological brain activity and performance in children aged 8–13 years. Psychophysiology. 41, 30–36. ©2003 Society for Psychophysiological Research, with permission from John Wiley and Sons.)*

enhanced positive brain waves in early (200–300 ms) and late (300–400 ms) P3a as well as in post-P3a (400–600 ms) windows. This finding suggests stronger attentional capture for unexpected sounds with higher information content. The fact that in the post-P3a window this duration effect was largest for the 6–8 years old indicates that young children are especially prone for distraction. Brinkman and Stauder (2008) studied the development of novelty processing in 5–7, 8–9, 10–12, and 18–29 year olds. Firstly, their data indicated two—early and late—novelty P3 components, each with a different topography and a different development. Secondly, both novelty components were still not mature in 10–12 year olds. The early novelty P3 had a central focus and its amplitude became more positive with increasing age. Also, its latency did not differ between the four age groups. The focus of the late novelty P3 shifted from frontocentral in 5–7 year olds to parietal in adults. In addition, the late novelty P3 amplitude at Pz became more positive with age, while the late novelty P3 latency was longer in 5–7 and 8–9 year olds compared to 10–12 year olds and adults.

Mahajan and McArthur (2015) tracked the maturation of the auditory mismatch negativity (MMN) and P3a brain responses—both posited as neural indices of auditory discrimination—in 90 adolescents aged 10–18 years. They found that P3a mean amplitude and latency decreased significantly across adolescence, but there was no reliable change in the MMN (Fig. 6.15). This suggested that the MMN is relatively stable in amplitude and latency across adolescence. In contrast, the amplitude and latency of the P3a reliably decreased across adolescence. Thus the ability to orient passively toward a novel or deviant stimulus continues to develop throughout adolescence, presumably as a result of the reduction in gray matter volume, synaptic pruning, and frontal cortex maturation that is known to occur during adolescence (Mahajan and McArthur, 2015).

6.6 Summary

Infants have some well-developed cognitive capabilities as reflected in mismatch responses. The polarity of these MMRs changes from positive to negative in the first 6 months of life, largely due to early development of cortical superficial layers, which receives early input of the reticular activating system. After 6 months of age the thalamocortical tract matures slowly, reflected in the maturation of the long-latency obligatory ERPs, particularly in their underlying tangential dipoles and white matter

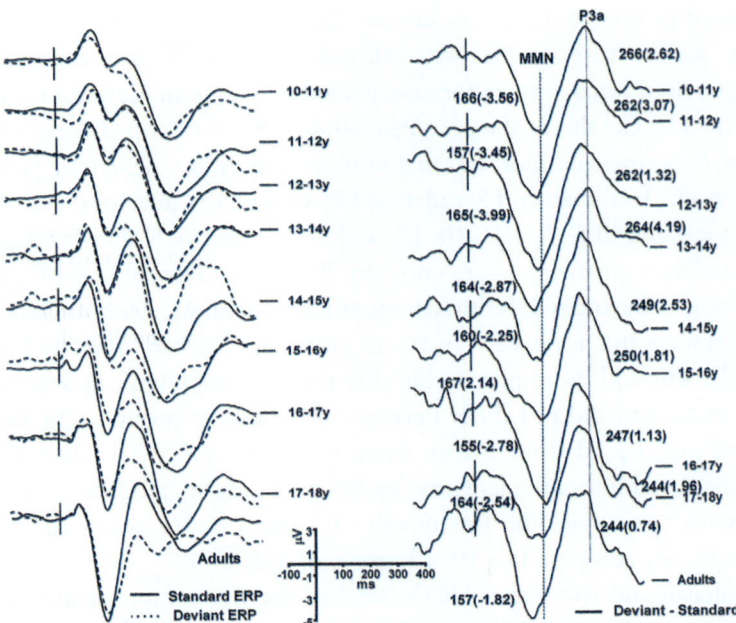

Fig. 6.15 Grand-mean standard and deviant waveforms (left panel) and difference waveforms (right panel). The mean MMN and P3a peak latency and mean amplitude values (in parenthesis) are also shown for each age group measured from Fz. (*Reprinted from Mahajan, Y., McArthur, G., 2015. Maturation of mismatch negativity and P3a response across adolescence. Neurosci. Lett. 587, 102–106. © 2014 Elsevier Ireland Ltd., with permission from Elsevier.*)

pathways. The maturation of the sagittal and radial dipoles appears to be much faster. Changes in oscillatory brain responses also strengthen with age but initially only frequencies <25 Hz are present. MMN can be used to measure the maturation of auditory sensory memory and suggests that memory traces for tone characteristics have a duration of 1–2 s in 2- and 3-year-old children, more than 2 s in 4-year-olds and 3–5 s in 6-year-olds. Predictive coding is already present in 1-year-old children, but the MMN network is still maturing in adolescence. This may be partially due to the late maturation of multilink steps in connectivity. Novelty processing maturation comprised early and late P3 components, each with a different topography and a different development, which were still not mature in 10–12 year olds.

References

Adibpour, P., Lebenberg, J., Kabdebon, C., Dehaene-Lambertz, G., Dubois, J., 2020. Anatomo-functional correlates of auditory development in infancy. Dev. Cogn. Neurosci. 42, 100752.

Alho, K., Cheour, M., 1997. Auditory discrimination in infants as revealed by the mismatch negativity of the event-related brain potential. Dev. Neuropsychol. 13 (2), 157–165. https://doi.org/10.1080/87565649709540675.

Barnet, A.B., Ohlrich, E.S., Weiss, I.P., Shanks, B., 1975. Auditory evoked potentials during sleep in normal children from ten days to three years of age. Electroencephalogr. Clin. Neurophysiol. 39, 29–41.

Bishop, D.V.M., Hardiman, M., Uwer, R., von Suchodoletz, W., 2007. Maturation of the long-latency auditory ERP: step function changes at start and end of adolescence. Dev. Sci. 10 (5), 565–575.

Bishop, D.V.M., Hardiman, M.J., Barry, J.G., 2011. Is auditory discrimination mature by middle childhood? A study using time-frequency analysis of mismatch responses from 7 years to adulthood. Dev. Sci. 14 (2), 402–416. https://doi.org/10.1111/j.1467-7687.2010.00990.x.

Brinkman, M.J.R., Stauder, J.E.A., 2008. The development of passive auditory novelty processing. Int. J. Psychophysiol. 70, 33–39.

Ceponiené, R., Cheour, M., Näätänen, R., 1998. Interstimulus interval and auditory event-related potentials in children: evidence for multiple generators. Electroencephalogr. Clin. Neurophysiol. 108, 345–354.

Ceponiené, R., Rinne, T., Näätänen, R., 2002. Maturation of cortical sound processing as indexed by event-related potentials. Clin. Neurophysiol. 113, 870–882.

Cheour, M., Alho, K., Ceponiene, R., Reinikainen, K., Sainio, K., Pohjavuori, M., et al., 1998. Maturation of mismatch negativity in infants. Int. J. Psychophysiol. 29, 217–226.

Cooray, G.K., Garrido, M.I., Brismar, T., Hyllienmark, L., 2016. The maturation of mismatch negativity networks in normal adolescence. Clin. Neurophysiol. 127 (1), 520–529. https://doi.org/10.1016/j.clinph.2015.06.026.

Dehaene-Lambertz, G., Dehaene, S., 1994. Speed and cerebral correlates of syllable discrimination in infants. Nature 370, 292–295.

Dehaene-Lambertz, G., Gliga, T., 2004. Common neural basis for phoneme processing in infants and adults. J. Cogn. Neurosci. 16, 1375–1387.

Dercksen, T.T., Widmann, A., Scharf, F., Wetzel, N., 2022. Sound omission related brain responses in children. Dev. Cogn. Neurosci. 53, 101045.

Draganova, R., Eswaran, H., Murphy, P., Huotilainen, M., Lowery, C., Preissl, H., 2005. Sound frequency change detection in fetuses and newborns, a magnetoencephalographic study. Neuroimage 28, 354–361.

Draganova, R., Eswaran, H., Murphy, P., Lowery, C., Preissl, H., 2007. Serial magnetoencephalographic study of fetal and newborn auditory discriminative evoked responses. Early Hum. Dev. 83, 199–207.

Eggermont, J.J., 1988. On the rate of maturation of sensory evoked potentials. Electroencephalogr. Clin. Neurophysiol. 70, 293–305.

Eggermont, J.J., 2007. Electric and magnetic fields of synchronous neural activity propagated to the surface of the head: peripheral and central origins of AEPs. Chapter 1. In: Burkard, R.R., Don, M., Eggermont, J.J. (Eds.), Auditory Evoked Potentials. Lippincott Williams & Wilkins, Baltimore, pp. 2–21.

Eggermont, J.J., 2015. Auditory Temporal Processing and Its Disorders. Oxford University Press, Oxford, UK.

Eggermont, J.J., 2019. Auditory brainstem response, Chapter 2 Audition, vestibular, and language testing. In: Levin, K.H., Chauvel, P. (Eds.), Handbook of Clinical Neurology. Clinical Neurophysiology: Basis and Technical Aspects. vol. 160 (3rd series), pp. 451–464, https://doi.org/10.1016/B978-0-444-64032-1.00030-8.

Eggermont, J.J., Moore, J.K., 2012. Morphological and functional development of the auditory nervous system. In: Werner, L.A., et al. (Eds.), Human Auditory Development. Springer Handbook 61 of Auditory Research 42, https://doi.org/10.1007/978-1-4614-1421-6_3,pp. 61–105.

Emberson, L.L., Boldin, A.M., Robertson, C.E., Cannon, G., Aslin, R.N., 2019. Expectation affects neural repetition suppression in infancy. Dev. Cogn. Neurosci. 37, 100597.

Emberson, L.L., Richards, J.E., Aslin, R.N., 2015. Top-down modulation in the infant brain: learning-induced expectations rapidly affect the sensory cortex at 6 months. Proc. Natl. Acad. Sci. U. S. A. 112 (31), 9585–9590.

Feldman, H., Friston, K.J., 2010. Attention, uncertainty, and free-energy. Front. Hum. Neurosci. 4, 215. https://doi.org/10.3389/fnhum.2010.00215.

Friederici, A.D., Friedrich, M., Weber, C., 2002. Neural manifestation of cognitive and precognitive mismatch detection in early infancy. Neuroreport 13, 1251–1254.

Fujioka, T., Ross, B., 2008. Auditory processing indexed by stimulus-induced alpha desynchronization in children. Int. J. Psychophysiol. 68, 130–140.

Gilley, P.M., Sharma, A., Dorman, M., Martin, K., 2005. Developmental changes in refractoriness of the cortical auditory evoked potential. Clin. Neurophysiol. 116, 648–657.

Glass, E., Sachse, S., von Suchodoletz, W., 2008. Development of auditory sensory memory from 2 to 6 years: an MMN study. J. Neural Transm. 115, 1221–1229. https://doi.org/10.1007/s00702-008-0088-6.

Gomes, H., Dunn, M., Ritter, W., Kurtzberg, D., Brattson, A., Kreuzer, J.A., Vaughan Jr., H.G., 2001. Spatiotemporal maturation of the central and lateral N1 components to tones. Devel. Brain Res. 129, 147–155.

Gomot, M., Giard, M.H., Roux, S., Barthélémy, C., Bruneau, N., 2000. Maturation of frontal and temporal components of mismatch negativity (MMN) in children. Neuroreport 11. 3109–3012.

Gumenyuk, V., Korzyukov, O., Alho, K., Escera, C., Näätänen, R., 2004. Effects of auditory distraction on electrophysiological brain activity and performance in children aged 8–13 years. Psychophysiology 41, 30–36.

Hashimoto, I., Ishiyama, Y., Yoshimoto, T., Nemoto, S., 1981. Brain-stem auditory-evoked potentials recorded directly from human brain-stem and thalamus. Brain 104, 841–859.

He, C., Hotson, L., Trainor, L.J., 2007. Mismatch responses to pitch changes in early infancy. J. Cogn. Neurosci. 19, 878–892.

He, C., Hotson, L., Trainor, L.J., 2009. Maturation of cortical mismatch responses to occasional pitch change in early infancy: effects of presentation rate and magnitude of change. Neuropsychologia 47, 218–229.

Kouider, S., Long, B., Le Stanc, L., Charron, S., Fievet, A.-C., Barbosa, L.S., Gelskov, S.V., 2015. Neural dynamics of prediction and surprise in infants. Nat. Commun. 6, 8537. https://doi.org/10.1038/ncomms9537.

Kraus, N., Smith, D.I., Reed, N.L., Stein, L.K., Cartee, C., 1985. Auditory middle latency responses in children: effects of age and diagnostic category. Electroencephalogr. Clin. Neurophysiol. 62, 343–351.

Krizman, J., Tierney, A., Fitzroy, A.B., Skoe, E., Amar, J., Kraus, N., 2015. Continued maturation of auditory brainstem function during adolescence: a longitudinal approach Clin. Neurophysiol. 126, 2348–2355.

Kushnerenko, E., Ceponiene, R., Balan, P., Fellman, V., Huotilaine, M., Näätänen, R., 2002a. Maturation of the auditory change detection response in infants: a longitudinal ERP study. Neuroreport 13, 1843–1848.

Kushnerenko, E., Ceponiene, R., Balan, P., Fellman, V., Huotilaine, M., Näätänen, R., 2002b. Maturation of the auditory event-related potentials during the first year of life. Neuroreport 13, 47–51.

Leppänen, P.H.T., Eklund, K.M., Lyytinen, H., 1997. Event-related brain potentials to change in rapidly presented acoustic stimuli in newborns. Dev. Neuropsychol. 13 (2), 175–204. https://doi.org/10.1080/87565649709540677.

Leppänen, P.H.T., Guttorm, T.K., Pihko, E., Takkinen, S., Eklund, K.M., Lyytinen, H., 2004. Maturational effects on newborn ERPs measured in the mismatch negativity paradigm. Exp. Neurol. 190 (Suppl 1), S91–S101.

Mahajan, Y., McArthur, G., 2015. Maturation of mismatch negativity and P3a response across adolescence. Neurosci. Lett. 587, 102–106.

Martin, F., Delpont, E., Suisse, G., Richelme, C., Dolisi, C., 1993. Long latency event-related potentials (P300) in gifted children. Brain Dev. 15, 173–177.

May, P.J.C., 2021. The adaptation model offers a challenge for the predictive coding account of mismatch negativity. Front. Hum. Neurosci. 15, 721574. https://doi.org/10.3389/fnhum.2021.721574.

May, P.J.C., Tiitinen, H., 2010. Mismatch negativity (MMN), the deviance-elicited auditory deflection, explained. Psychophysiology 47, 66–122.

Morr, M.L., Shafer, V.L., Kreuzer, J.A., Kurtzberg, D., 2002. Maturation of mismatch negativity in typically developing infants and preschool children. Ear Hear. 23, 118–136.

Näätänen, R., 2001. The perception of speech sounds by the human brain as reflected by the mismatch negativity (MMN) and its magnetic equivalent (MMNm). Psychophysiology 38, 1–21.

Novak, G.P., Kurtzberg, D., Kreuzer, J.A., Vaughan Jr., H.G., 1989. Cortical responses to speech sounds and their formants in normal infants: maturational sequence and spatiotemporal analysis. Electroencephalogr. Clin. Neurophysiol. 73, 295–305.

Ohlrich, E.S., Barnet, A.B., Weiss, I.P., Shanks, B.L., 1978. Auditory evoked potential development in early childhood: a longitudinal study. Electroencephalogr. Clin. Neurophysiol. 44, 411–423.

Paetau, R., Ahonen, A., Salonen, O., Sams, M., 1995. Auditory evoked magnetic fields to tones and pseudowords in healthy children and adults. J. Clin. Neurophysiol. 12, 177–185.

Pang, E.W., Edmonds, G.E., Desjardins, R., Khan, S.C., Trainor, L.J., Taylor, M.J., 1998. Mismatch negativity to speech stimuli in 8-month-old infants and adults. Int. J. Psychophysiol. 29, 227–236.

Pang, E.W., Taylor, M.J., 2000. Tracking the development of the N1 from age 3 to adulthood: an examination of speech and non-speech stimuli. Clin. Neurophysiol. 111, 388–397.

Pasman, J.W., Rotteveel, J.J., de Graaf, R., Maassen, B., Notermans, S.L.H., 1991. Detectability of auditory response components in preterm infants. Early Hum. Dev. 26, 129–141.

Pasman, J.W., Rotteveel, J.J., Maassen, B., Visco, Y.M., 1999. The maturation of auditory cortical evoked responses between (preterm) birth and 14 years of age. Eur. J. Paediatr. Neurol. 3, 79–82.

Pendl, S.L., Salzwedel, A.P., Goldman, B.D., Barrett, L.F., Lin, W., Gilmore, J.H., Gao, W., 2017. Emergence of a hierarchical brain during infancy reflected by stepwise functional connectivity. Hum. Brain Mapp. 38, 2666–2682.

Pereira, M.R., Barbosa, F., de Haan, M., Ferreira-Santos, F., 2019. Understanding the development of face and emotion processing under a predictive processing framework. Dev. Psychol. 55(9), 1868–1881. https://doi.org/https://doi.org/10.1037/dev0000706.

Picton, T.W., 2011. Human auditory evoked potentials. Plural Publishing, San Diego, CA.

Picton, T.W., Taylor, M.J., 2007. Electrophysiological evaluation of human brain development. Dev. Neuropsychol. 31 (3). 249–27.

Ponton, C.W., Eggermont, J.J., 2001. Of kittens and kids: altered cortical maturation following profound deafness and cochlear implant use. Audiol. Neurootol. 6, 363–380.

Ponton, C.W., Eggermont, J.J., 2007. Electrophysiological measures of human auditory system maturation: Relationship with neuroanatomy and behavior. In: Burkard, R.R., Don, M., Eggermont, J.J. (Eds.), Auditory evoked potentials. Lippincott Williams & Wilkins, Baltimore, pp. 385–402.

Ponton, C.W., Eggermont, J.J., Coupland, S.G., Winkelaar, R., 1992. Frequency-specific maturation of the eighth nerve and brain-stem auditory pathway: evidence from derived auditory brain-stem responses (ABRs). J. Acoust. Soc. Am. 91. 1576–158.

Ponton, C.W., Eggermont, J.J., Kwong, B., Don, M., 2000a. Maturation of human central auditory system activity: evidence from multi-channel evoked potentials. Clin. Neurophysiol. 111, 220–236.

Ponton, C.W., Don, M., Eggermont, J.J., Waring, M.D., Kwong, B., Cunningham, J., Trautwein, P., 2000b. Maturation of the mismatch negativity: effects of profound deafness and cochlear implant use. Audiol. Neurootol. 5, 167–185.

Ponton, C., Eggermont, J.J., Khosla, D., Kwong, B., Don, M., 2002. Maturation of human central auditory system activity: separating auditory evoked potentials by dipole source modeling. Clin. Neurophysiol. 113, 407–420.

Rotteveel, J.J., Colon, E.J., Notermans, L.H., Stoelinga, G.B.A., Visco, Y.M., 1985. The central auditory conduction at term date and three months after birth. I. Composite group averages of brainstem ABR, middle latency (MLR) and auditory cortical responses (ACR). Scand. Audiol. 14, 179–186.

Rotteveel, J.J., Stegeman, D.F., de Graaf, R., Colon, E.J., Visco, Y.M., 1987. The maturation of the central auditory conduction in preterm infants until three months post term. III. The middle latency auditory evoked response (MLR). Hear. Res. 27, 245–256.

Ruhnau, P., Herrmann, B., Maess, B., Schröger, E., 2011. Maturation of obligatory auditory responses and their neural sources: evidence from EEG and MEG. Neuroimage 58, 630–639.

Shafer, V.L., Morr, M.L., Kreuzer, J.A., Kurtzberg, D., 2000. Maturation of mismatch negativity in school-age children. Ear Hear. 21, 242–251.

Shafer, V.L., Yu, Y.H., Datta, H., 2010. Maturation of speech discrimination in 4- to 7-yr-old children as indexed by event-related potential mismatch responses. Ear Hear. 31, 735–745.

Shahin, A.J., Trainor, L.J., Roberts, L.E., Backer, K.C., Miller, L.M., 2010. Development of auditory phase-locked activity for music sounds. J. Neurophysiol. 103, 218–229.

Starr, A., Amlie, R.N., Martin, W.H., Sanders, S., 1977. Development of auditory function in newborn infants revealed by auditory brainstem potentials. Pediatrics 60, 831–839.

Suppanen, E., Huotilainen, M., Ylinen, S., 2019. Rhythmic structure facilitates learning from auditory input in newborn infants. Infant Behav. Dev. 57, 101346.

Sussman, E., Steinschneider, M., Gumenyuk, V., Grushko, J., Lawson, K., 2008. The maturation of human evoked brain potentials to sounds presented at different stimulus rates. Hear. Res. 236, 61–79.

Tang, H., Brock, J., Johnson, B.W., 2016. Sound envelope processing in the developing human brain: a MEG study. Clin. Neurophysiol. 127, 1206–1215.

Thompson, E.C., Estabrook, R., Krizman, J., Smith, S., Huang, S., White-Schwoch, T., et al., 2021. Auditory neurophysiological development in early childhood: a growth curve modeling approach. Clin. Neurophysiol. 132, 2110–2122.

Tonnquist-Uhlen, I., Ponton, C.W., Eggermont, J.J., Kwong, B., Don, M., 2003. Maturation of human central auditory system activity: the T-complex. Clin. Neurophysiol. 114, 685–701.

Trainor, L., McFadden, M., Hodgson, L., Darragh, L., Barlow, J., Matsos, L., et al., 2003. Changes in auditory cortex and the development of mismatch negativity between 2 and 6 months of age. Int. J. Psychophysiol. 51, 5–15.

Virtala, P., Putkinen, V., Kailaheimo-Lönnqvist, L., Thiede, A., Partanen, E., Kujala, T., 2022. Infancy and early childhood maturation of neural auditory change detection and its associations to familial dyslexia risk. Clin. Neurophysiol. 137, 159–176.

Wetzel, N., Widmann, A., Berti, S., Schröger, E., 2006. The development of involuntary and voluntary attention from childhood to adulthood: a combined behavioral and event-related potential study. Clin. Neurophysiol. 117, 2191–2203.

Wetzel, N., Schröger, E., 2007. Modulation of involuntary attention by the duration of novel and pitch deviant sounds in children and adolescents. Biol. Psychol. 75, 24–31.

Wunderlich, J.L., Cone-Wesson, B.K., 2006. Maturation of CAEP in infants and children: a review. Hear. Res. 212, 212–223.

Wunderlich, J.L., Cone-Wesson, B.K., Shepherd, R., 2006. Maturation of the cortical auditory evoked potential in infants and young children. Hear. Res. 212, 185–202.

CHAPTER 7

Role of event-related potentials and brain rhythms in predictive coding

7.1 Auditory stream segregation

7.1.1 The problem

Previously (Eggermont, 2015; see also Chapter 2) I wrote: "There are two conceptually distinct challenges for a listener in a 'cocktail party' situation (McDermott, 2009). The first is the problem of sound segregation. The auditory system must derive the properties of individual sounds from the mixture entering the ears, i.e., through a binaural filter to assess spatial proximity. The second challenge is that of directing attention to the sound source of interest while ignoring the others, i.e., by applying an attentional filter, and of switching attention between sources, as when intermittently following two conversations. Most of our cognitive processes can operate only on one thing at a time, so we typically select a particular sound source on which to focus. This process affects sound segregation, which is biased by what we attend to. Because speech is first and foremost a temporal stimulus, decoding the momentary spectral content ('fine structure') of speech benefits greatly from considering its position within the longer temporal context of an utterance (Shannon et al., 1995; Shamma et al., 2011; Giraud and Poeppel, 2012). In other words, the timing of events within a speech stream is critical for its decoding, attribution to the correct source, and ultimately to whether it will be included in the attentional 'spotlight' or forced to the background of perception. Cocktail parties and other comparably noisy environments thus present extreme examples of the challenges inherent to sound segregation. In many cases the bottom–up segregation cues may not be enough, and listeners must also rely on their knowledge of specific sounds or sound classes (McDermott, 2009). This is especially important for speech understanding in noise, which is substantially easier if the words form coherent sentences. Listeners are also better at comprehending speech in a cocktail party setting if

Brain Responses to Auditory Mismatch and Novelty Detection
https://doi.org/10.1016/B978-0-443-15548-2.00007-7

the attended speech is in the listener's native language, as personal experience testifies."

Fritz et al. (2007) identified two cortical mechanisms of streaming—an automatic "preattentive" segregation of sounds and a streaming mechanism that builds up over a period of up to several seconds that can either be preattentive or modulated by attention. This process of auditory scene analysis (ASA) sets the stage for further attentional selection. Automatic preattentive scene analysis can free up attentional resources to "fine-tune" segmentation of a complex acoustic scene or focus on individual streams and extract meaning from the attended stream. Micheyl and Oxenham (2010) reiterated that sounds with common properties are more likely to be integrated by the auditory system. When properties differ enough between sound elements, they probably arose from different sources and are more likely to be assigned to different auditory objects or streams. In humans, some of the acoustic properties of sound that play important roles in auditory scene analysis include fundamental frequency and harmonic relationships among spectral components, temporal onsets/offsets, timbre, and patterns of amplitude modulation. Timing synchrony of frequency partials in a sound allows fusion into a more complex sound, and if the frequency partials are harmonic the fusion is more likely. If the partials have identical amplitude and frequency modulations they are also more likely to fuse into one object (Cooke, 1993).

Consequently, timing asynchrony is a major element to distinguish one stream from two streams. Zion Golumbic et al. (2013) investigated the manner in which speech streams are represented in brain activity and the way that selective attention governs the brain's representation of speech using a "Cocktail Party" paradigm, coupled with direct recordings from the cortical surface in surgical epilepsy patients. This brain activity dynamically tracks speech streams using both low-frequency phase and high-frequency amplitude fluctuations and that optimal encoding likely combines the two. In and near low-level auditory cortices, attention "modulates" the representation by enhancing cortical tracking of attended speech streams, but ignored speech remains represented (Fig. 7.1, see also Fig. 2.6). In higher-order cortical regions, the representation appears to become more "selective," in that there is no detectable tracking of ignored speech.

Nie et al. (2014) measured event-related potentials (ERPs) to investigate auditory stream segregation of noise stimuli with or without clear spectral contrast. Sequences of alternating A and B noise bursts were presented to elicit stream segregation in normal-hearing listeners (Fig. 7.2 top). The successive B bursts in each sequence maintained an equal amount of temporal

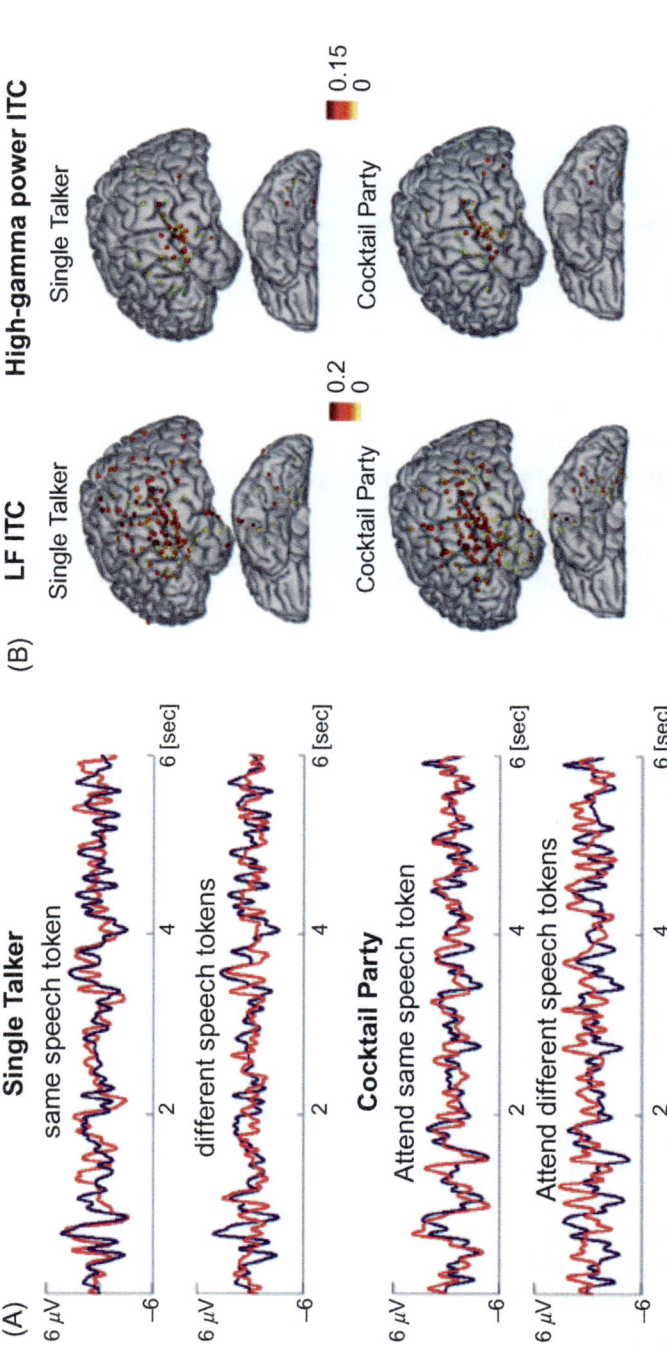

Fig. 7.1 Intertrial correlation analysis (ITC). (A) Traces of single trials from one sample electrode, filtered between 1 and 10 Hz. The top panel shows the similarity in the time course of the neural response in two single trials (*blue and red traces*) where the same stimulus was presented versus two trials in which different stimuli were presented in the Single Talker condition. The bottom—cocktail party—panel demonstrates that a similar effect is achieved by shifting attention in the cocktail party condition. Two trials in which attention was focused on the same talker elicit similar neural responses, whereas trials in which different talkers were attended generate different temporal patterns in the neural response, despite the identical acoustic input. (B) Location of sites with significant low-frequency (LF) phase-ITC (left) and high-gamma power-ITC (right) in both conditions. The colors of the dots represent the ITC value (see color bars) at each site. (*Reprinted from Zion Golumbic, E.M., Ding, N., Bickel, S. Lakatos, P., Schevon, C.A., McKhann, G.M., et al., 2013. Mechanisms underlying selective neuronal tracking of attended speech at a "cocktail party". Neuron 77, 980—991. ©2013, with permission from Elsevier.*)

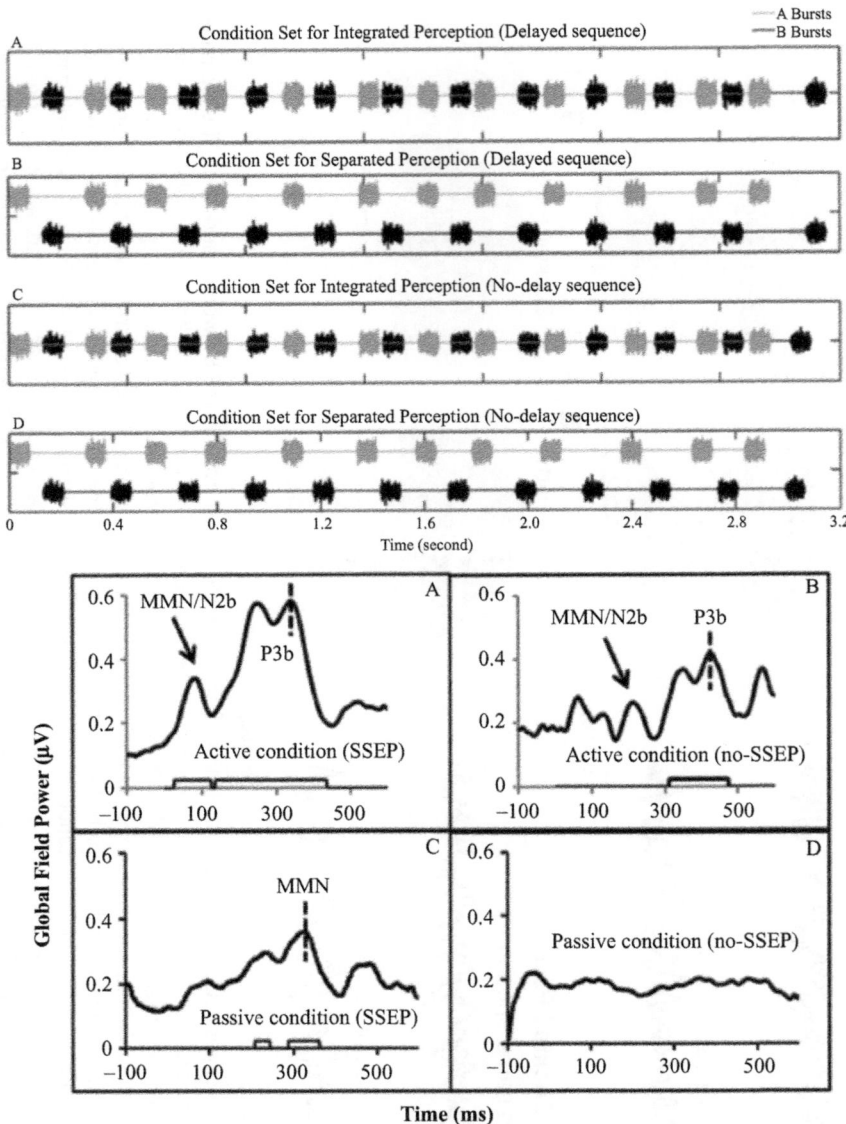

Fig. 7.2 Top. Schematic illustration of the stimulus paradigm. Waveforms of noise bursts are plotted. Panels (A and B) show the delayed sequences with the *black solid line pieces* indicating the duration of the delay for the 12th B burst. Panels (C and D) show the no-delay sequences. Panels (A and C) depict the condition when individual sequence elements result in an integrated perception, and panels (B and D) depict the condition when individual spectrally separated elements result in a segregated perception. Bottom. Global field power (GFP) data obtained from the grand-mean ERP deviant-to-standard difference waveforms. Panels (A and B), respectively,

(Continued)

separation with manipulations introduced on the last stimulus. The last B burst was either delayed for 50% of the sequences or not delayed for the other 50%. The A bursts were jittered in between every two adjacent B bursts. In the passive listening condition, a trend for a possible late mismatch negativity (MMN) or late discriminative negativity (LDN, see Appendix) response was observed only when the A and B bursts were spectrally separate, suggesting that spectral separation in the A and B burst sequences could be conducive to stream segregation at the preattentive level. In the attentive condition, a P3b response was consistently elicited regardless of whether there was spectral separation between the A and B bursts, indicating the facilitative role of voluntary attention in stream segregation (Fig. 7.2 bottom). The results suggest that reliable ERP measures can be used as indirect indicators for auditory stream segregation in conditions of weak spectral contrast.

Hsu et al. (2019) conducted an MEG experiment which independently manipulated prime probe relation (for contextual precision) and stimulus repetition (for perceptual learning which decreases prediction error). Participants were presented with cycles of tone quartets (Fig. 7.3) which consisted of three prime tones (black) and one probe tone X (gray) of randomly selected frequencies (Hsu et al., 2019). The three prime tones remained identical within each cycle while the probe tones changed from high frequency (X) to low frequency (Y) once at some point (e.g., from repetition of 123X to repetition of 123Y). Thus the repetition of probe tones can reveal the development of perceptual inferences in low and high precision contexts depending on their position within the cycle. Hsu et al. (2019) found that "the two conditions resemble each other in terms of N1m modulation (as both were associated with N1m suppression) but differ in terms of N2m modulation. While repeated probe tones in low precision context did

---

Fig. 7.9, Cont'd show GFPs for the spectral separation evoked potential (SSEP) and no-SSEP stimuli during attentive listening. Panels (C and D) show GFPs for the SSEP and no-SSEP stimuli during passive listening. The bars along the x-axes in (A–C) show significant activities as determined from z-scores relative to the 100-ms prestimulus baseline. Panel (D) shows an absence of significant GFP activities. The *dashed vertical lines* in (A and B) mark the GFP peaks of P3b. The *dashed vertical line* in (C) indicates the GFP peak of MMN for the SSEP stimuli. The GFP peaks falling in the MMN/N2b time window are indicated for the SSEP (A) and no-SSEP stimuli (B) by the *arrows*. *(From Nie, Y., Zhang, Y., Nelson, P.B., 2014. Auditory stream segregation using bandpass noises: evidence from event-related potentials. Front. Neurosci. 8, 277. https://doi.org/10.3389/fnins.2014.00277. Open Access.)*

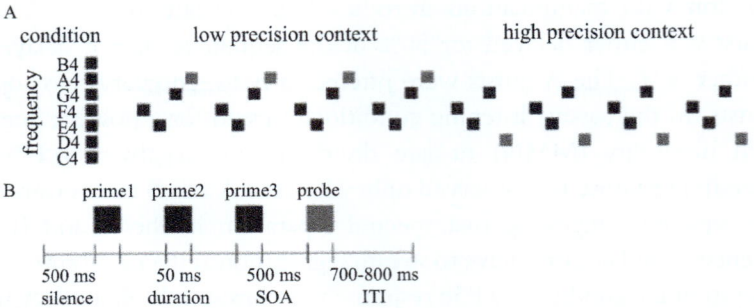

Fig. 7.3 (A) Schematic representation of a cycle. In this example, the first tone quartet (F4-E4-G4-A4) was repeated four times before the second tone quartet (F4-E4-G4-D4) was repeated four times. (B) Schematic representation of a tone quartet. *(From Hsu, Y.-F., Waszak, F., Hämäläinen, J.A., 2019. Prior precision modulates the minimization of auditory prediction error. Front. Hum. Neurosci. 13, 30. https://doi.org/10.3389/fnhum. 2019.00030. Open Access.)*

not exhibit any modulatory effect, repeated probe tones in high precision context elicited a suppression and rebound of the N2m source power. The differentiation suggested that the minimization of prediction error in low and high precision contexts likely involves distinct mechanisms" (Hsu et al., 2019).

During EEG recording Brace and Sussman (2021) presented participants with a tone sequence that had two different tone feature patterns, one based on the sequential rhythmic variation in tone duration and the other on sequential rhythmic variation in tone intensity. They found that "the neural entrainment to the rhythm of the standard attended patterns had similar power to the standard of the unattended feature patterns. In addition, the infrequent pattern deviants elicited an MMN. The MMN elicited by task-based feature pattern deviants had a similar amplitude to MMNs elicited by pattern deviants that were unattended because they were not the target pattern or because the participant ignored the sounds and watched a movie."

7.1.2 Can predictive coding explain stream segregation?

Predictive coding can be indexed with the MMN and P3a that are elicited by infrequent deviant sounds (e.g., differing in pitch, duration, and loudness) in a stream of frequent sounds. If these components reflect prediction error, they should also be elicited by omitting an expected sound (Prete et al., 2022).

7.1.2.1 EEG/MEG data

Winkler et al. (2012) considered the nature of sound representations and how they compete with each other. They noted that predictive processing (Friston, 2005) helps to maintain perceptual stability by signaling the continuation of previously established patterns as well as the emergence of new sound sources. It also provides a measure of how well each of the competing representations describes the current acoustic scene. Bendixen (2014) argued that predictability affects auditory scene analysis both when it is present in the sound source of interest and when it is present in other sound sources that the listener wishes to ignore. In this spirit, Bendixen (2014) linked predictive processing with an important principle of ASA the "old-plus-new" heuristic being a cue for simultaneous sound grouping (Bregman, 1990) as follows: "Predictive coding modeling of perception postulates that the primary objective of the brain is to infer the causes of sensory inputs by reducing prediction errors (i.e., the discrepancy between expected and actual information). Moreover, prediction errors are weighted by their precision (i.e., inverse variance), which quantifies the degree of certainty about the variables. The reduction of precision-weighted prediction errors can be affected by contextual regularity (as an external factor) and selective attention (as an internal factor)" (Bendixen, 2014).

Sohoglu and Chait (2016) presented listeners with cluttered, artificial auditory scenes comprised of several sound sources. Listeners were more accurate and quicker to detect source appearance in scenes comprised of temporally regular (REG), rather than random (RAND), sources. MEG in passive listeners and those actively detecting appearance events revealed increased sustained activity in auditory and parietal cortex for REG relative to RAND scenes, emerging ~400 ms of scene onset. "This finding is opposite to what would be expected based on adaptation i.e. decreased neural responses for temporally regular events, which has previously been observed for isolated tone sequences. It is however consistent with a mechanism that infers the precision (predictability) of sensory input and uses this information to up-regulate neural processing towards more reliable sensory signals" (Sohoglu and Chait, 2016; see also Barascud et al., 2016). Overall, the results demonstrate that the enhanced detection performance observed in behavior is not solely the result of changes in neural responses occurring prior to source appearance but also due to enhanced neural responses to novel events themselves. From a literature review Denham and Winkler (2020) noted that "Overall, the picture emerging about the role of predictable patterns in sound processing is that (1) the human auditory system is sensitive to patterns and statistical regularities in sequences of sounds, (2) detecting regularities does not require attention to be focused on the sounds (Sussman, 2007) and listeners are not necessarily aware of the detected regularities, and (3) the utilization of regularities varies with context and experimental details in ways that have yet to be fully understood. Taken together, what these studies show is that the influence of predictability on perceptual organization is not mandatory." Hsu and Hämäläinen (2021) found that contextual regularity and selective attention independently modulated the N1/MMN where the repetition effect was absent. On the P2, the two factors, respectively, interacted with the repetition effect without interacting with each other. The results showed that contextual regularity and selective attention likely affect the reduction of precision-weighted prediction errors in distinct manners.

Prete et al. (2022) compared ERPs elicited by infrequent randomly occurring unexpected silences in tone sequences presented at two tones per second to ERPs elicited by frequent, regularly occurring omissions (expected silences) within a sequence of tones presented at one tone per second. They found that unexpected silences elicited significant MMN and P3a, although the magnitude of these components was quite small and variable. These results provide evidence for hierarchical predictive coding, indicating that the brain predicts both silences and sounds.

7.1.2.2 fMRI data

Using fMRI, Hesselmann et al. (2010) found that neuronal activity in sensory areas (extra-striate visual and early auditory cortex) biases perceptual decisions toward correct inference and not toward a specific percept. In accord with predictive coding models, this suggested to them that cortical activity in early sensory brain areas reflects the precision of prediction errors and not just the sensory evidence or prediction errors per se. This was based on two experiments; the first experiment involved detecting motion coherence in random dot kinematograms with coherent motion at threshold (periliminal) in most trials and above or below threshold (supra and subliminal) in a smaller number of trials. They measured cortical activity, prior to evoked responses (gray ellipse in Fig. 7.4 left), in the human visual motion areas V5/hMT+. According to accumulation models, higher prestimulus activity levels should be observed more for hits and false alarms compared to misses and correct rejections. However, according to predictive coding correct rejections and hits, prestimulus activity levels should be larger than for misses and false alarms. Empirical observations confirmed the latter.

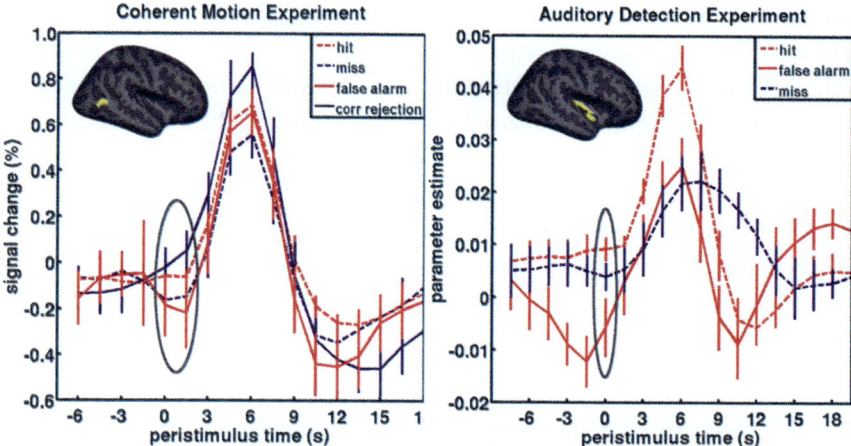

Fig. 7.4 Left. Peristimulus fMRI signal time courses from the visual motion experiment. Data were normalized to grand mean and averaged across nine subjects (bars represent standard error of the mean) performing a motion coherence judgment task. The insert specifies the conditions as a function of stimulus and percept. The inflated right hemisphere rendering of the group result shows the right hMT+ region of interest, which was identified subject by subject in a localizer procedure employing coherent motion stimuli vs static displays. The *gray ellipse* covers the prestimulus period submitted to statistical testing. Right. Peristimulus fMRI signal time courses from the auditory experiment. Data were estimated under a finite response model and averaged across nine subjects (bars represent standard error of the mean) performing an auditory stimulus detection task. Data are plotted for conditions specified by an insert. The inflated right hemisphere rendering of the group result shows the location of the region of interest, which includes early auditory cortex with parts of Heschl's gyrus (identified bilaterally subject by subject). The *gray ellipse* covers the prestimulus period submitted to statistical testing. *(From Hesselmann, G., Sadaghiani, S., Friston, K.J., Kleinschmidt, A., 2010. Predictive coding or evidence accumulation? False inference and neuronal fluctuations. PLoS One 5 (3), e9926. https:// doi.org/10.1371/journal.pone.0009926. Open access.)*

In the second experiment, Hesselmann et al. (2010) studied detection of auditory signals presented at threshold against ongoing scanner noise. This detection paradigm can be reconciled with the form of the previous experiment by regarding it as a continuous discrimination, with two alternatives of stimulus "present" or "absent." However, this free-response paradigm does not furnish correct rejection trials (i.e., subjects are not required to indicate the stimulus is absent). They expected the difference between hits and false alarms to be even more pronounced than in the first experiment. This is because in the auditory fMRI experiment ongoing sensory noise levels were higher due to scanner noise than in the visual experiment, where interstimulus intervals contained a stationary dot pattern. Under predictive coding, this higher sensory noise should suppress the gain of error units and reduce activity levels, accentuating the effect of endogenous fluctuations. As before, the predictions of the two theoretical accounts differ: Evidence accumulation would expect hits and false alarms (i.e., an auditory percept) to follow higher baseline levels, relative to misses (no percept). Conversely, the predictive coding account suggests that (incorrect) misses and false alarms are foreshadowed by significantly lower prestimulus activity than (correct) hits. Hesselmann et al.'s (2010) findings in this experiment supported the latter prediction (Fig. 7.4 right).

7.2 Networks underlying error responses in predictive coding

7.2.1 The MMN network

Opitz et al. (2002) identified sources for the MMN, with fMRI and EEG measures, in both left and right superior temporal gyrus (STG), and right inferior frontal gyrus (IFG). For each individual subject, Garrido et al. (2007) used these source locations as priors to estimate the posterior locations and moments of the equivalent current dipoles (cf. Fig. 6.5). Fig. 7.5 shows, "for the best model FB [with Forward and Backward connections], the predicted responses at each node of the network for each trial type (i.e., standard or deviant) for a single subject. The coupling gains and the conditional probability of the gains—being greater or smaller than one—are shown for each connection in the network. The values on each connection represent a scaling effect, for example, a coupling change of 2.04 from left primary auditory cortex (A1) to left STG in the model means that the effective connectivity increased 104% for rare events relative to frequent events" (Garrido et al., 2007). In a review of studies that focus on neuronal mechanisms underlying the MMN generation, Garrido et al. (2009) discuss the "adaptation hypothesis" (Chapter 2) and the "model adjustment hypothesis," and proposed predictive coding as a general framework to unify both hypotheses for which current inputs are predicted from past inputs. In the case of a prediction error, i.e., when there is a mismatch between the predicted—expected—and the actual sensory input, the neural system implementing the model must be adjusted (for example, by short-term synaptic plasticity). During the repetition of subsequent events, that adjustment is reflected in the suppression of prediction error and the disappearance, but mostly just reduction, of the MMN. The underlying network, as found by

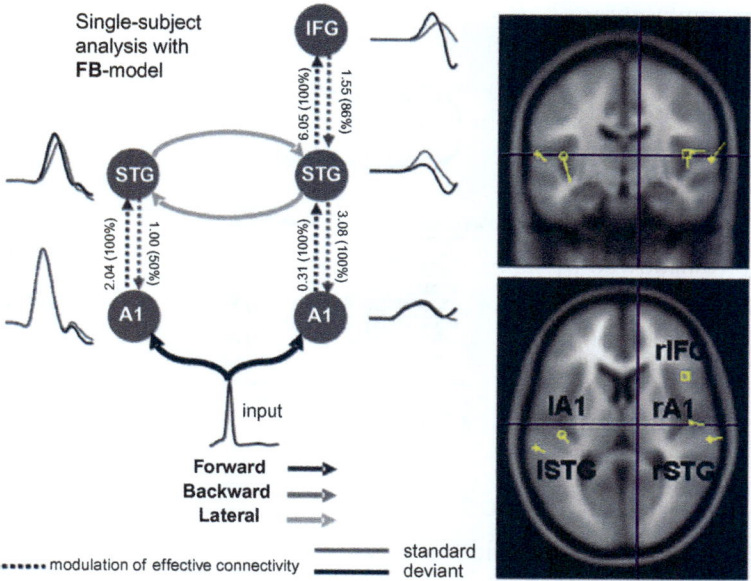

Fig. 7.5 Left. DCM results for a single subject (FB model). Reconstructed responses for each source and changes in coupling adjacent to the connections during oddball processing relative to standards. The mismatch response is expressed in nearly every source. Right. Sources of activity, modeled as dipoles (estimated posterior moments and locations), are superimposed in an MRI of a standard brain in MNI space. *(From Garrido, M.I., Kilner, J.M., Kiebel, S.J., Stephan, K.E., Friston, K.J., 2007. Dynamic causal modelling of evoked potentials: a reproducibility study. Neuroimage 36, 571–580. Open Access.)*

dynamic causal modeling (DCM), includes left and right A1, left and right STG, and right IFG (Fig. 7.5). This mechanistic model attempts to explain the generation of each individual response (i.e., responses to standards and responses to deviants). Therefore both left and right A1 were chosen as cortical input stations for processing the auditory information.

Based on MEG recordings, Phillips et al. (2015) found evidence for including A1, STG, and IFG in the MMN network, but supplemented it with modulatory prefrontal cortex (PFC) input (Fig. 7.6). The model comprised reciprocal feedforward and feedback connections between A1 and STG, connections between STG and IFG bilaterally, as well as interhemispheric interactions. Internally generated expectations drive inputs to PFC at the uppermost level of the model's hierarchy to represent temporal structure violations. Notably, the prefrontal temporal expectancy input was important for explaining the response to duration and gap deviants, which, unlike the other deviants, are defined by a violation in the temporal structure of stimuli.

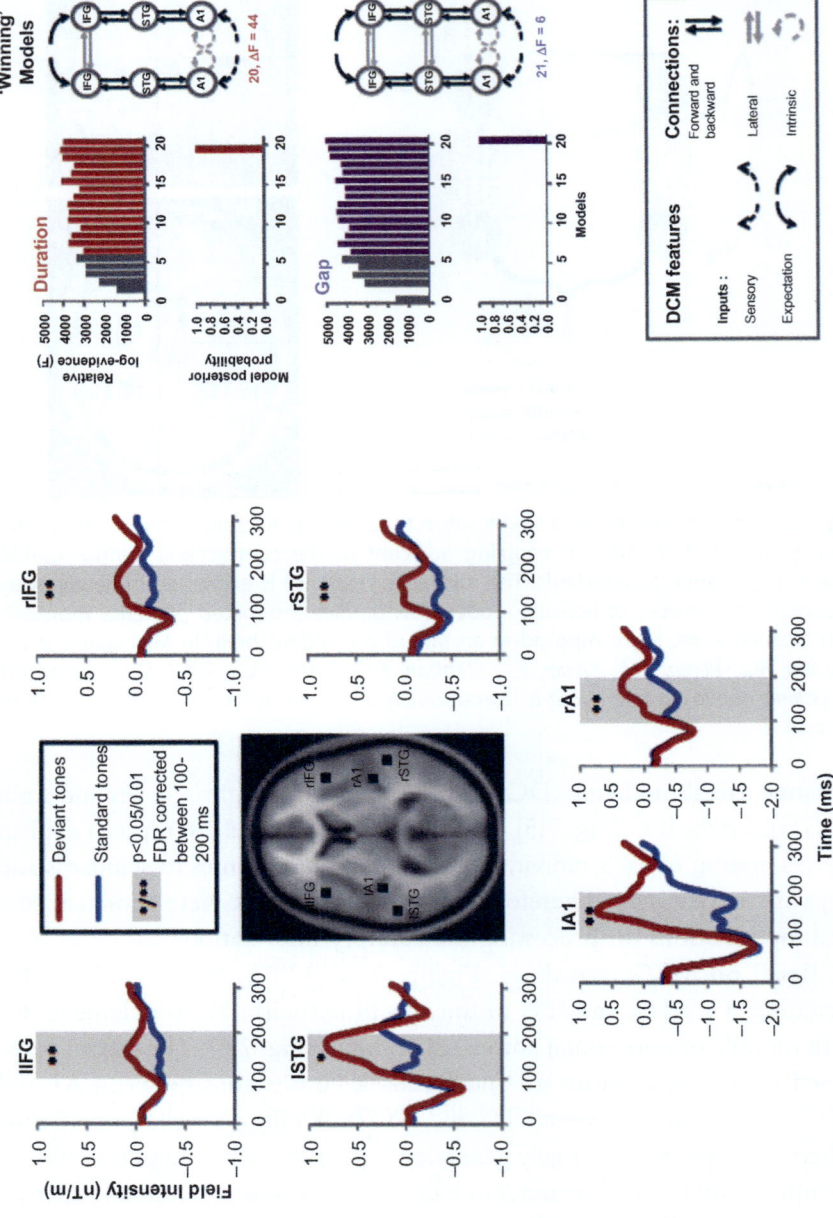

Fig. 7.6 See legend on opposite page.

Using auditory data, Lecaignard et al. (2021) evaluated multimodal (EEG/MEG) fusion and showed (1) the superiority of a combined EEG/MEG inference, (2) the greater spatial sensitivity of MEG compared to EEG, (3) the ability of EEG data alone to source reconstruct temporal lobe activity, and (4) the usefulness of EEG to improve MEG-based source reconstruction. Importantly, Lecaignard et al. (2021) largely reproduced MMN findings over two different—frequency (FRQ) and intensity (INT)—experimental mismatch conditions. Fig. 7.7 shows the results of the statistical analysis projected on the inflated cortical surface in each modality (EEG, MEG, and EEG-MEG, Fusion) and each mismatch condition (FRQ, INT). In both mismatch conditions, EEG and MEG inversions led to different (but not inconsistent) reconstructed activity, and more focal clusters were found with fused inversion. Precisely, in condition FRQ, EEG inversion revealed bilateral activity in the anterior part of the supratemporal plane and in the lower bank of the posterior STG. No frontal area was found significant. MEG inversion indicated a large cluster in the supratemporal plane expanding from the lateral part of HG through the planum polare (PP) in both hemispheres. A bilateral frontal area was located in the posterior IFG. The fused distribution comprised smaller supratemporal clusters (right: a single cluster including the lateral part of HG and PP; left: separate clusters for HG and PP) and bilateral clusters similar to MEG ones in the frontal lobe.

Fig. 7.6 Cont'd Left. Mean source waveforms to standard *(blue)* and all deviant *(red)* events in bilateral A1, STG, and IFG MMN sources, with anatomical locations illustrated in the center. The *shaded areas* indicate the 100–200 ms period of the characteristic MMN-evoked response. Significant differences between standard and deviant tones using paired *t*-tests were evident at each location as indicated (*$P < .05$ and **$P < .01$), with FDR correction for multiple comparisons. *lA1*, left A1; *lIFG*, left IFG; *lSTG*, left STG; *rA1*, right A1; *rIFG*, right IFG; *rSTG*, right STG. Right. Bayesian model selection for each deviant type with full model space. For each deviant type, we include the relative log evidence and model posterior probabilities for every model, a diagram of the winning model, and the difference between winning and second place relative log evidence (ΔF). In each plot of log evidence, the first six bars in *gray* indicate the models used by Garrido et al. (2008). Both duration and gap deviants (left) reveal bilateral prefrontal expectancy inputs into the IFG sources, with lateral connections between IFG sources. The gap deviant also has lateral connections between STG sources. For all deviant types, the winning models have a $\Delta F > 5$ and, therefore, are considered to have very strong evidence in favor of those models. *(From Phillips, H.N., Blenkmann, A., Hughes, L.E., Bekinschtein, T.A., Rowe, J.B., 2015. Hierarchical organization of frontotemporal networks for the prediction of stimuli across multiple dimensions. J. Neurosci. 35 (25), 9255–9264.)*

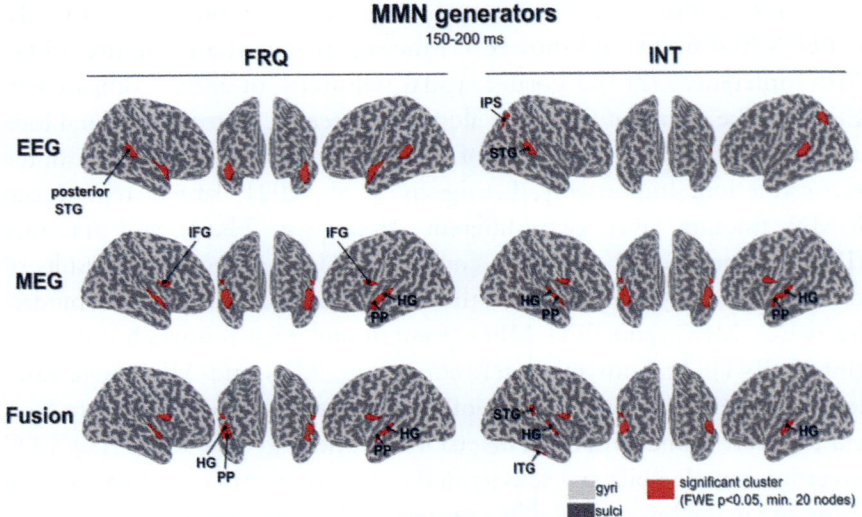

Fig. 7.7 Mean source reconstructions of the MMN obtained in unimodal and multimodal inversions. Left/Right panel: frequency/intensity MMN ([150, 200] ms, condition FRQ/INT). *Red clusters* indicate the significant source activity over the group (*N* = 20) projected on the inflated cortical surface (*HG*, Heschl's gyrus; *IFG*, inferior frontal gyrus; *IPS*, inferior parietal sulcus; *ITG*, inferior temporal gyrus; *PP*, planum polare; *STG*, superior temporal gyrus). (*Reprinted from Lecaignard, F., Bertrand, O., Caclin, A., Mattout, J., 2021. Empirical Bayes evaluation of fused EEG-MEG source reconstruction: application to auditory mismatch evoked responses. Neuroimage 226, 117468. © 2020 The authors, with permission from Elsevier.*)

In the INT mismatch (Fig. 7.7 right), the EEG solution indicated bilateral activity in the posterior STG and the intraparietal sulcus (IPS). There was a similar distribution to condition FRQ with MEG. Fused inversion gave largest contributions in the lateral part of HG in both hemispheres, but also right clusters located in posterior IFG, posterior STG, and in the inferior temporal gyrus (ITG). Thus this validates the accuracy of the inversion scheme in each modality, and suggests that EEG and MEG contributed equally to the fusion inversion, and finally reveals expected differences in EEG and MEG mean reconstructions that qualitatively motivate the fusion of these modalities to take benefit of their respective sensitivity (Lecaignard et al., 2021).

7.2.2 The role of the IFG

We will now look at the specific role of the IFG in speech perception and in the MMN network. Clos et al. (2014) investigated the effect of prior information on fMRI-based brain activity during the decoding of degraded speech stimuli. The participants performed a delayed-matching-to-sample task in which a target sentence had to be compared with a preceding reference sentence. The stimuli comprised 25 sentences, each in a nondegraded and a degraded version. To identify regions implicated in the processing of all six sentence-type events, the conjunction (∩) "nondegraded reference sentence ∩ degraded reference sentence ∩ structural match ∩ structural mismatch ∩ propositional match ∩ propositional mismatch" was used (Fig. 7.8A left). This analysis should thus reflect regions commonly activated by the sound stimuli or recruited by the general task demands (e.g., working memory, decision making). The contrast "degraded reference sentence > nondegraded reference sentence" (Fig. 7.8B left) was employed to isolate regions that are more activated by the unintelligible sounds as compared to meaningful verbal information. The inverse contrast "nondegraded reference sentence > degraded reference sentence" (Fig. 7.8C left) was analyzed to discern regions more tuned to intelligible speech than to (unintelligible) dynamic intonation contour. The latter contrast should thus identify regions that are selectively involved in processing the lexical-semantic aspects of speech. When prior information enabled the comprehension of the degraded sentences (Fig. 7.8 right), all activated areas were left localized: the left middle temporal gyrus (MTG) and the left angular gyrus, highlighting a role of these areas in extraction of meaning. In contrast, the activation of the left IFG appeared to reflect the search for meaningful information in degraded speech material that could not be decoded because of mismatches with the prior information (Clos et al., 2014).

Lui et al. (2021) applied Transcranial Magnetic Stimulation (TMS) at the IFG to disrupt the processing of the initial 2 or 5 standards of a 3-, 6-, or 9-standard train, while the MMN responses to pitch deviant presented after the standard trains were recorded and compared. An abolishment of MMN was only observed when TMS was delivered to the IFG at the initial 2 standards of the 3-standard train, but not at the initial 5 standards, or when TMS at the vertex or a TMS-sound recording was applied. The MMNs were also preserved when IFG TMS, vertex TMS, or TMS sound recording was applied at the initial 2 or 5 standards of longer trains. Thus the IFG plays a critical role in processing the initial standards of a short standard train for subsequent deviant detection. This result is consistent with the prediction violation account that the IFG is important for establishing the predictive coding network.

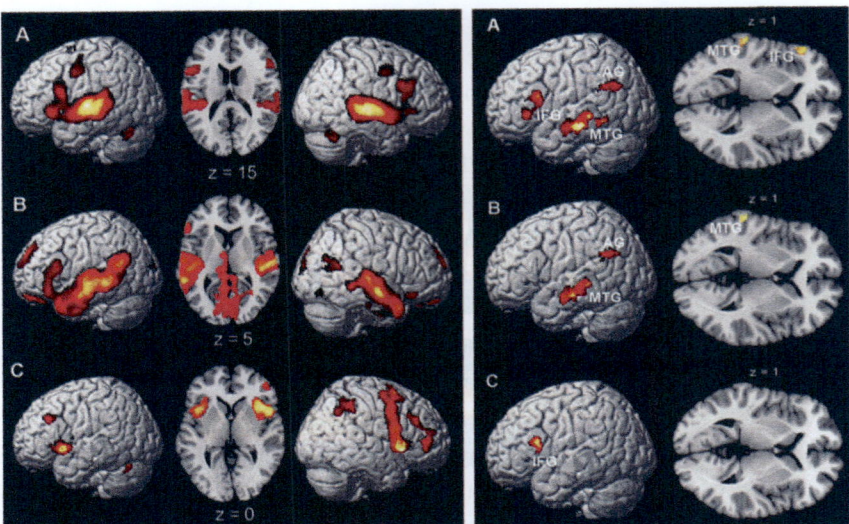

Fig. 7.8 Left. Overview of the general fMRI findings. (A) The conjunction nondegraded reference sentence ∩ degraded reference sentence ∩ structural match ∩ structural mismatch ∩ propositional match ∩ propositional mismatch revealed bilateral activations reflecting auditory processing common to all six sentence-type events. (B) Regions representing the lexical-semantic rather than the prosodic aspects of speech (nondegraded reference sentence > degraded reference sentence) and (C) the inverse contrast (degraded reference sentence > nondegraded reference sentence). Right. Effects of propositional prior were all left lateralized. (A) Regions within the lexical-semantic network responding more to degraded targets that were preceded by propositional compared to structural priors. Dissociation of this network into (B) left MTG and AG for propositional matches and (C) Broca's area for propositional mismatches. All images are thresholded at *P* < .05 (FWE corrected at cluster level; cluster-forming threshold at voxel level: *P* < .001). *(From Clos, M., Langner, R., Meyer, M., Oechslin, M.S., Zilles, K., Eickhoff, S.B., 2014. Effects of prior information on decoding degraded speech: an fMRI study. Hum. Brain Mapp. 35, 61–74. Open Access.)*

7.2.3 Complex mismatch conditions

Barascud et al. (2016) used behavior, MEG, and fMRI to investigate how human listeners discover temporal patterns and statistical regularities in complex sound sequences. Sensitivity to patterns is fundamental to auditory processing because most auditory signals only have meaning as successions over time. "Subjects listened to random–regular (RAND-REG) stimuli, and two types of controls (REG-RAND and STEP; Fig. 7.9 Top) presented in random order. They were instructed to detect the changes within the stimuli (transitions from RAND to REG or vice versa) and respond as quickly as

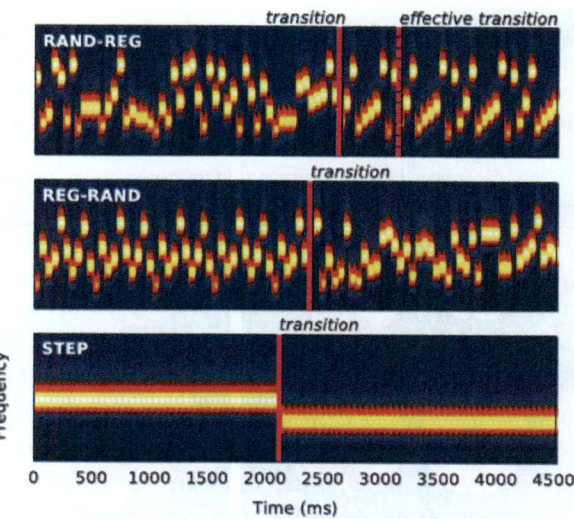

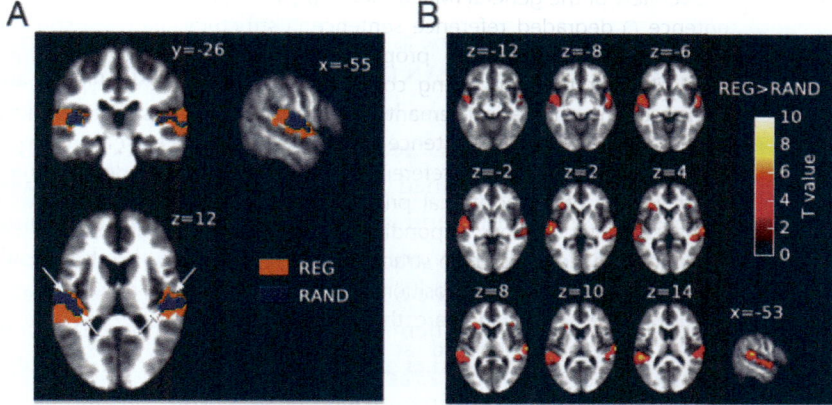

Fig. 7.9 Top. Spectrograms of the stimuli used in the experiment. Top: Change from random to regular (RAND-REG); Middle: regular to random (REG-RAND); Bottom: tone-frequency step (STEP). Transition times are marked with a *solid line* in each exemplar. The transition in RAND-REG is not detectable until one regularity cycle has elapsed (i.e., until the pattern starts repeating). This time point is labeled "effective transition" in the figure. REG (and RAND) patterns were generated anew for each trial. Bottom. fMRI results. (A) fMRI group activation for REG and RAND conditions, superimposed onto coronal, sagittal, and axial sections of the average structural image. The height threshold for activation was $P < .001$ (uncorrected) at the peak level, $P < .05$ at the cluster level. *Blue*, activation for RAND; *orange*, activation for REG. The *white arrowheads* on the axial section indicate the midline of Heschl's gyrus in each hemisphere. (B) Regions showing increased hemodynamic responses to REG sequences compared with RAND, thresholded at $P < .001$ (uncorrected). Results are shown for different axial sections on the subjects' average structural anatomy.

(Continued)

Fig. 7.9, Cont'd The *color bar* indicates peak level significance. The contrast revealed increased hemodynamic responses to REG bilaterally in PT, STG, and PP. Additional activation in the left inferior frontal gyrus (IFG; right IFG was also present but does not survive correction for multiple comparisons). *PP*, planum polare; *PT*, planum temporale; *STG*, superior temporal gyrus. *(From Barascud, N., Pearce, M.T., Griffiths, T.D., Friston, K.J., Chait, M., 2016. Brain responses in humans reveal ideal observer-like sensitivity to complex acoustic patterns. Proc. Natl. Acad. Sci. 113 (5), E616–E625.)*

possible by pressing a keyboard button. From source analysis they found an interaction between A1, hippocampus, and IFG in the process of discovering the regularity within the ongoing sound sequence." In detail, Fig. 7.9 Bottom A presents the activation elicited by the regular (REG) and random (RAND) conditions, overlaid on the subjects' average anatomy. Both signals produced very similar patterns of activation along Heschl's gyrus (HG) and planum temporale (PT) bilaterally (the approximate position of HG is indicated by two arrowheads on the axial section in Fig. 7.9 Bottom A). For REG, the activation cluster was larger, extending into more posterior regions of the superior temporal plane as well as the superior temporal gyrus. In contrast, group activation for RAND was mostly centered on HG and extended less into PT, this fits with the MEG results showing increased activation for REG relative to RAND sequences. In Fig. 7.9 Bottom B, a REG/RAND contrast was examined to isolate regions associated with the processing of regularity. This contrast revealed increased hemodynamic responses to REG in the left hemisphere in the lateral part of HG, bilaterally in PT, all along the upper bank of the STG, extending to PP. Additionally, IFG activation was found in the left hemisphere (right-hemisphere IFG activation was not significant). Barascud et al. (2016) found this consistent with precision-based PC accounts of perceptual inference, and the underlying network (e.g., Fig. 7.5) and to provide evidence of the brain's capacity to encode high-order temporal structure in sensory signals.

Hofmann–Shen et al. (2020) employed an auditory sequence oddball paradigm to investigate different levels of cortical auditory processing and the contribution of neuronal habituation and prediction error mechanism to N1 and MMN. Their findings suggest that N1 reflects a lower cortical process primarily involved in the encoding of simple physical features and is thus mainly modulated by neuronal attenuation and not complex top-down mechanisms. By analyzing within-sequence signal differences, they divided the MMN into distinct subcomponents reflecting different hierarchical levels of auditory processing. Hofmann–Shen et al. (2020) determined a "first-order" MMN (Fig. 7.10 top A) that reflects the processing of simple

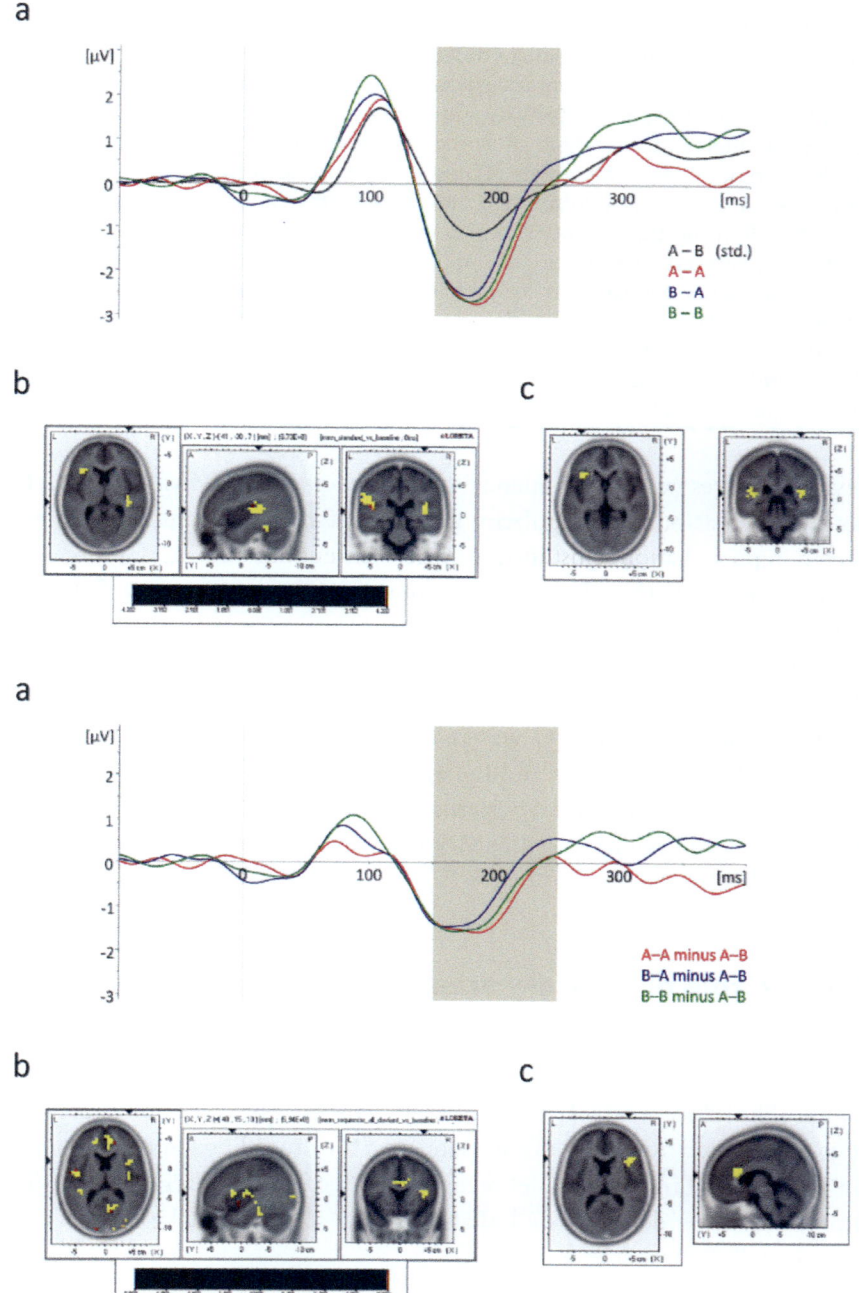

Fig. 7.10 Top. (A) First-order MMN components, where 0 ms corresponds to onset of second stimuli of the standard A-B *(black)* and the deviant A-A *(red)*, B-A *(blue)*, and B-B *(green)* sequences. The standard sequence A-B served as subtrahend

(Continued)

Fig. 7.10, Cont'd for subsequent isolations of higher-order MMN components and was identified at 150–250 ms after stimulus onset. (B) Cortical source analysis of the standard sequence A-B revealed significant activations of primary auditory cortex bilaterally and left inferior frontal gyrus. (C) Left IFG and bilateral auditory cortex were defined as regions of interest for analysis of cortical sources of first-order MMN. Bottom. (A) Higher-order MMN components elicited in response to second stimuli in deviant sequences A-A *(red)*, B-A *(blue)*, and B-B *(green)*; the standard sequence AB is subtracted from all signals. (B) Cortical source analysis averaged across all deviant sequences showed activations of anterior cingulate cortex and right IFG. (C) Corresponding cortical ROIs in right IFG and in ACC for source estimation of higher-order MMN. *ACC, anterior cingulate cortex; IFG, inferior frontal gyrus. (From Hofmann-Shen, C., Vogel, B.O., Kaffes, M., Rudolph, A., Brown, E.C., Tas, C., et al., 2020. Mapping adaptation, deviance detection, and prediction error in auditory processing. Neuroimage 207, 116432. Open Access.)*

deviant features (such as frequency) and "higher-order" MMNs (Fig. 7.10 bottom A) that occur at regularity violation of complex patterns or unexpected inputs that do not allow further predictions. Cortical sources of the first-order MMN following the standard A–B sequence (Fig. 7.10 top B and C) were A1 bilaterally as well as left IFG. In source-localization analysis of higher-order MMN (Fig. 7.10 bottom B and C), both the A1 and left IFG were primarily involved in the detection of simple, physically deviant features, while the right IFG was associated with the processing of novel, unexpected auditory inputs and the anterior cingulate cortex (ACC) with regularity violation of known patterns. It is noted that the right IFG is generally part of the DCM-based MMN network (Fig. 7.5), whereas the left IFG (Fig. 7.6) rarely is.

7.3 The role of auditory areas in predictive coding

"One example, in which prediction-related concepts have been implicated early on in human electrophysiology studies is the auditory oddball paradigm. The discovery of the auditory MMN, an electrophysiological response elicited by stimuli that deviate from the preceding context, exposed the existence of internal models keeping track of the regularities in sensory input and have demonstrated that deviance detection takes place preattentively" (Schröger et al., 2015).

7.3.1 Auditory cortex

In the framework of predictive coding, Wacongne et al. (2012) proposed a detailed neuronal model of auditory cortex—interacting with brain areas subserving memory—that accounts for the critical features of the MMN. The model is entirely composed of spiking excitatory and inhibitory neurons interconnected in a layered cortical architecture with distinct input, predictive, and prediction error units. Fig. 7.11 left "shows an implementation of the model for an input composed of two pure tones, A and B. Each column of the network represents a cortical column with its thalamic input responding maximally to one of the two frequencies of the input. The two frequencies A and B are supposed to be different enough to activate only one of the two columns. In each column, three populations of neurons are simulated. The essential component of the model is the population of neurons involved in prediction, which are proposed as part of the supragranular layers of the cortex. This population constantly tries to anticipate the upcoming auditory inputs. A prediction of sound A consists in an increase in the population firing rate coding for this stimulus. At every moment, the continuously variable predictions arising from the predictive populations of neurons are compared with the incoming inputs. This comparison is achieved at the level of a population of neurons called the 'prediction error' population, which receives two sets of inputs: excitatory inputs coming from the thalamus and conveying the current sensory stimulus, and inhibitory inputs that reflect the activity of the predictive population. Through this scheme, whenever the thalamic input is not canceled by predictive signals, the prediction error population fires." The activity of the prediction error population is transmitted to the predictive population as a feedback and this error signal is used to adapt the internal model of this population, and this error signal may account for the MMN effect (Fig. 7.11 right). The predictive population needs to build an internal model of the regularities of the incoming stimulus to form relevant predictions.

Wacongne et al. (2012) "propose that this model is based on learning the statistical temporal dependencies linking the stimuli within the past few hundred milliseconds. A memory of the recent past is needed to achieve such a goal. This memory has to keep the trace of two properties: the identity of the past inputs and the time elapsed since they occurred." They choose to model this function in the simplest manner possible, using a delay line for each frequency, where activation propagates linearly from one neuron to the next as a function of time. ... Memory neurons—putatively also located in auditory cortex—are connected to both predictive subpopulations so that predictions of one frequency (A) can be based on the recent occurrence of a sound of the other frequency (B). The internal model of the predictive population is built by adapting the synaptic weights linking the memory neurons and the predictive populations. A spike-timing dependent (STDP) learning rule,

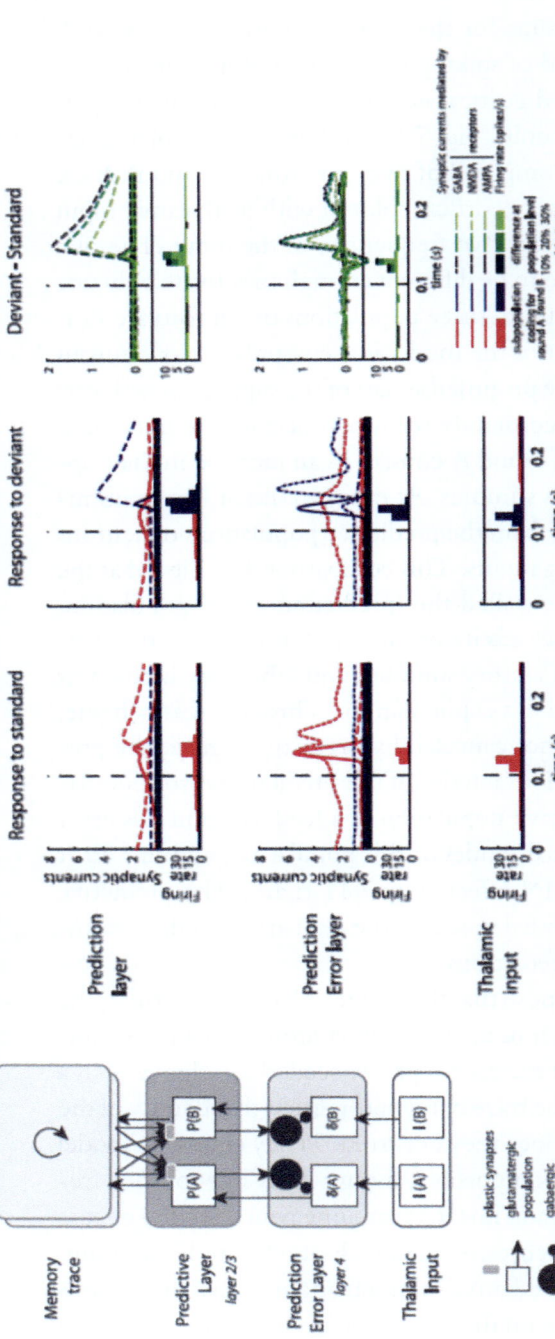

Fig. 7.11 Left. Scheme of the predictive coding model for two sounds. For each layer, two subpopulations are modeled that respond, respectively, to the frequencies of sounds A and B. Prediction error activity in layer 4 is the result of the difference between thalamic inputs and predictive activity arising from the supragranular layer, whose sign is inverted through inhibitory interneurons *(black circles)*. Prediction error is then fed back to adjust the activity of predictive populations. Dynamic predictions are made possible in the model because predictive units send and receive projections with a recurrent network serving as a short-term memory. NMDA-dependent plasticity adjusts the synaptic weights onto predictive units until their dynamics matches that of the inputs and therefore minimizes the prediction error. Right. Simulating the MMN in an oddball paradigm: mean synaptic currents and firing rates. The figure shows the mean simulated response to a standard tone (first column), a deviant tone (second column), and their difference (third column) after 200 learning trials in an oddball paradigm. Each line shows the response of a different layer of units in the model. For each layer, the top part of the plot represents the synaptic currents received by the subpopulation, separately for the different types of postsynaptic receptors that mediate these currents: AMPA *(continuous line)*, NMDA *(dashed line)*, or GABA *(dotted line)*. The bottom part of each plot displays the mean firing rate of each subpopulation. In the first and second columns, subpopulations responding to the frequent A sound (90% of trials) are represented in *red* and those responding to the rare B sound (10%) in *blue*. The third column shows the results of simulations in which the percentage of deviants was varied (10%, 20%, or 30%). *(From Wacongne, C., Changeux, J.-P., Dehaene, S., 2012. A neuronal model of predictive coding accounting for the mismatch negativity. J. Neurosci. 32 (11), 3665–3678. https://doi.org/10.1523/JNEUROSCI.5003-11.2012.)*

relying upon NMDA receptor synaptic transmission, allows the network to adjust its internal predictions and use a memory of the recent past inputs to anticipate on future stimuli based on transition statistics. Wacongne et al. (2012) showed how the subtraction of observed versus predicted signals can be implemented through a specific architecture of inhibitory interneurons. They also showed that a NMDA-dependent STDP plasticity rule is well adapted for learning of stimulus associations, leading to the prediction of a precise and essential contribution of NDMA receptors to PC. In principle the memory trace unit could cover both STG and IFG activity, optionally modulated by the PFC.

Stein et al. (2022) used fMRI to systematically investigate prediction error to subjective expectations—following an earlier paper of the same group (Tabas et al., 2020) and shown in Fig. 7.12—in human auditory cortical (ACx) fields Te1.0, Te1.1, Te1.2, and Te3, and two types of stimuli: pure tones and frequency modulated (FM) sweeps. They showed that prediction error is elicited with respect to the participants' expectations independently of stimulus repetition and similarly expressed across auditory fields. In addition, despite the radically different strategies underlying the decoding of pure tones and FM sweeps, both stimulus modalities were encoded as prediction error in most fields of ACx. Stein et al. (2022) concluded that predictive coding is the general encoding mechanism in ACx.

7.3.2 Auditory subcortical areas

Supporting Cacciaglia et al.'s (2015) findings (Chapter 5), Tabas et al. (2020) used abstract rules to manipulate expectations independently of local stimulus statistics. Their 7-Tesla fMRI data show that abstract expectations can drive the response amplitude to tones in the human auditory pathway. They tested two opposing views on the mechanism of sensory processing in the auditory midbrain (IC) and auditory thalamus (MGB). In one view, sensory processing can be explained by habituation to local stimulus statistics (Fig. 7.12C, h1), in the other by predictive coding (Fig. 7.12C, h2). Tabas et al. (2020) found that: "First, mean BOLD responses in IC and MGB correlated with the subjects' expectations of the probability of the stimulus occurrence but not with the local stimulus statistics. Second, events deviating from local stimulus statistics did not lead to increased responses in IC and

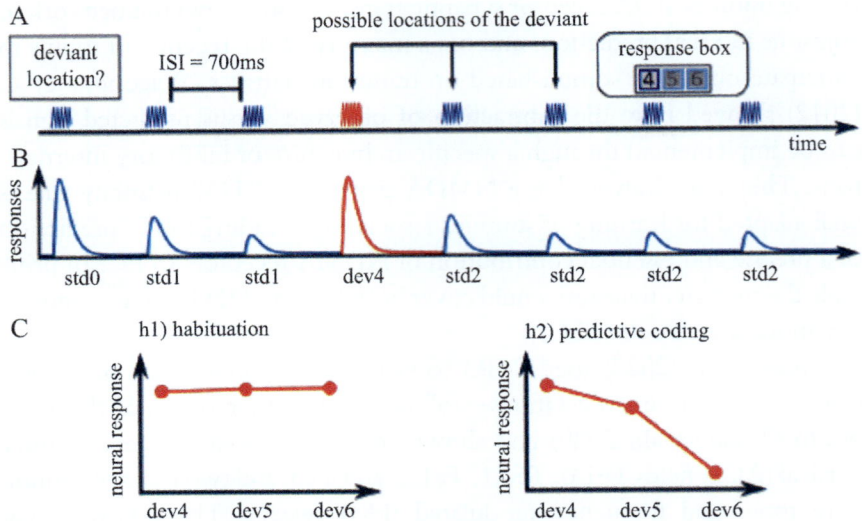

Fig. 7.12 Experimental design and hypotheses. (A) Example of a trial, consisting of a sequence of seven pure tones of a standard frequency *(blue waveform)* and one pure tone of a deviant frequency (fourth tone in the example; *red waveform*), that could be located in positions 4, 5, or 6. Subjects had to report, in each trial, the position of the deviant. Each subject completed 240 trials in total, 80 per deviant position. All tones had a duration of 50 ms and were separated by 700-ms interstimulus intervals (ISIs). (B) Schematic view of the expected underlying responses in the auditory pathway for the sequence shown in (A), together with the definition of the experimental variables *(dev x:* deviant in position x; *std0,* first standard; *std1,* repeated standards preceding the deviant; *std2,* standards following the deviant). (C) Expected responses in the auditory pathway nuclei corresponding to the habituation (h1) and predictive coding (h2) hypotheses. Since the posterior probability of finding a deviant at locations 4, 5, or 6 after hearing 3, 4, or 5 standards is 1/3, 1/2, and 1, respectively, predictive coding predicts different BOLD responses to different deviant locations. *(From Tabas, A., Mihai, G., Kiebel, S., Trampel, R., von Kriegstein, K., 2020. Abstract rules drive adaptation in the subcortical sensory pathway. eLife 9, e64501. https://doi.org/10.7554/eLife.64501. Open Access.)*

MGB if subjects expected these events. Third, Bayesian model comparison showed that the responses of the majority of voxels in IC and MGB are best explained by a predictive coding model" (Fig. 7.13).

7.4 Are brain oscillations a substitute for ERPs in predictive coding?

According to the predictive coding theory, top-down predictions are conveyed by backward connections and prediction errors are propagated

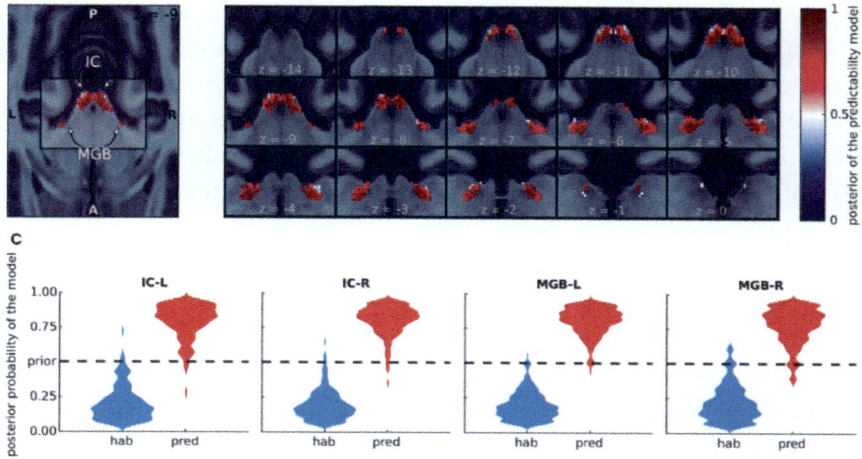

Fig. 7.13 Bayesian model comparison analysis of the BOLD responses. Top. Posterior probability map of the predictive coding model. Since only two models were used to compute the posteriors, $P < .5$ means that the habituation model *(blue)* is the most likely explanation of the data, and $P > .5$ means that the predictive coding model *(red)* is the most likely explanation of the data. Bottom. Histograms showing the prevalence of each of the two models in each of the SSA regions. *Hab*, habituation model; *pred*, predictive coding model. *(From Tabas, A., Mihai, G., Kiebel, S., Trampel, R., von Kriegstein, K., 2020. Abstract rules drive adaptation in the subcortical sensory pathway. eLife 9, e64501. https://doi.org/10.7554/eLife.64501. Open Access.)*

forward across the cortical hierarchy. Bastos et al. (2012) reported that feed-forward connections originating from lower-level areas mainly show neural activation in the gamma frequencies (≥ 40 Hz), whereas feedback connections originating from higher-order brain areas prefer mostly alpha and beta frequencies (about 8–25 Hz). They found a remarkable correspondence between the anatomy and physiology of the canonical microcircuit (cf. Fig. 1.11) and the formal constraints implied by generalized predictive coding. Bastos et al. (2012) stressed that there are many variations on the mapping between computational and neuronal architectures: even if predictive coding is an appropriate implementation of Bayesian filtering. For example, feedback connections could arise directly from cells encoding conditional expectations in supragranular layers. Indeed, there is emerging evidence that feedback connections between proximate hierarchical levels originate from both deep and superficial layers (cf. Wacongne et al., 2012).

Using direct (ECoG) recordings from human auditory cortex, Sedley et al. (2016) found that surprise due to prediction violations is encoded by oscillations in the gamma band (>30 Hz), changes to predictions are encoded in the beta band (12–30 Hz), and that the precision of predictions appears to quantitatively relate to alpha-band oscillations (8–12 Hz). These results confirm oscillatory codes for critical aspects of generative models of perception. Also using ECoG recordings, Dürschmid et al. (2016) found deviance-related responses in both frequency bands—high gamma (80–150 Hz) and low-frequency, LF-ERPs—over lateral temporal and inferior frontal cortex, with an earlier latency for high gamma than for LF-ERPs. Critically, frontal high-gamma activity but not LF-ERPs discriminated between fully predictable and unpredictable changes, with frontal cortex sensitive to unpredictable events. The results highlight the role of frontal cortex and high-gamma activity in deviance detection and prediction error generation. Dürschmid et al. (2019) provided further evidence, obtained from intracranial cortical recordings in human surgical patients, that the lateral PFC shows prediction signals while anticipating an event. Patients listened to task-irrelevant sequences of repetitive tones including infrequent predictable or unpredictable pitch deviants. The broadband high-frequency amplitude decreased prior to the onset of expected relative to unexpected deviants in the frontal cortex only, and its amplitude was sensitive to the increasing likelihood of deviants following longer trains of standards in the unpredictable condition. Because the train length of standards under the irregular condition varied pseudo-randomly, they could test whether prestimulus predictive activity varies gradually as a function of train length. Two effects could be operative; temporally local effects suggest that the probability of a standard tone increases the more standard tones which are played in a row. In contrast, using a more global strategy, the so-called hazard function suggests that, given that deviations will happen eventually, expectation of a deviant increases the longer it is since the last deviation. To test whether and where such effects prevail, Dürschmid et al. (2019) correlated predeviant high-frequency activity (HFA; 80–150 Hz) with train length of standards before deviants. Fig. 7.14 shows that the direction of correlation between HFA and standard train length was different between temporal electrodes, showing mostly positive correlations, and frontal electrodes, showing mostly negative correlations. Individually, only the negative correlations in frontal channels reached the permutation critical r-values of $r_{crit} = \pm 0.19$ (*white-in-blue dots*, in Fig. 7.14). "Considering that the analysis of the regular versus irregular condition indicated that a decrease in HFA indicates proactive prediction of a deviant, these results suggest that frontal electrodes 'apply' predictions even under the irregular condition based on the more global hazard function strategy." These results provide

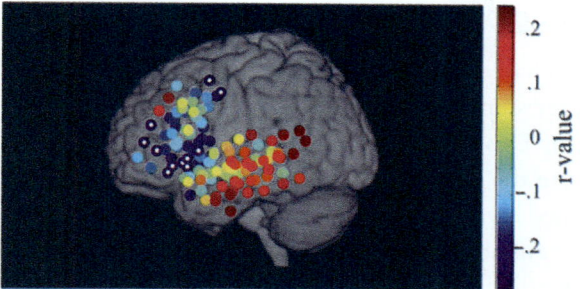

Fig. 7.14 Prefrontal electrodes reflect the hazard function in irregular sequences. Each *circle* depicts channel positions with the *color-coding* Pearson's correlation coefficient between train length and predeviant HFA. Channels with a *white dot* show a statistically significant correlation. HFA significantly decreased after longer trains of standards in frontal cortex, while HFA tended to increase with longer trains of standards in temporal cortex. HFA, high-frequency amplitude. *(From Dürschmid, S., Reichert, C., Hinrichs, H., Heinze, H.-J., Kirsch, H.E., Knight, R.T., Deouell, L.Y., 2019. Direct evidence for prediction signals in frontal cortex independent of prediction error. Cereb. Cortex 29 (11), 4530–4538. © The n(s) 2018, with permission from Oxford University Press.)*

evidence of high-level prediction, modifying the poststimulus comparison between the actual input and the ongoing prediction.

Dürschmid et al. (2019) therefore suggested that the flow of information up and down the hierarchy of the network is not as simple as gleaned from typical DCM diagrams (Fig. 7.5; Garrido et al., 2007). They speculated on the functional advantage of maintaining spatially segregated predictions. Specifically, maintaining predictions that account for global regularities allows the prefrontal cortex to efficiently direct attention only to unexpected events (e.g., Fig. 7.6; Phillips et al., 2015), whereas for the auditory cortex, detecting all local changes is advantageous for parsing the auditory input into meaningful chunks (e.g., in speech perception).

As described in Chapter 4, Chao et al. (2018) used high-density electro-corticography (ECoG) in monkeys during an auditory "local-global" paradigm in which the temporal regularities of the stimuli were controlled at two hierarchical levels. They decomposed the broadband data and identified lower- and higher-level prediction-error signals in early auditory cortex and anterior temporal cortex, respectively, and a prediction-update signal sent from prefrontal cortex back to temporal cortex. The prediction-error and prediction-update signals were transmitted via gamma (>40 Hz) and alpha/beta (<30 Hz) oscillations, respectively (cf. Fig. 4.15). "These findings provide strong support for hierarchical predictive coding and outline how it is dynamically implemented using distinct cortical areas and frequencies."

7.5 Local vs global deviance

Earlier in this chapter the local vs global deviance was labeled as resulting from "low precision context" and "high precision context" (Hsu et al., 2019; Fig. 7.3). This section presents an expansion on Chapter 5 Section 3.

To elucidate the neural mechanisms of predictions, Auksztulewicz et al. (2018) combined invasive ECoG with computational modeling while manipulating predictions about content ("what") and time ("when"). They found that "when" predictions increased evoked activity over motor and prefrontal regions both at early (~180 ms) and late (430–450 ms) latencies. "What" predictability, however, increased evoked activity only over prefrontal areas late in time (420–460 ms). Beyond these dissociable influences, they found that "what" and "when" predictability interactively modulated the amplitude of early (165 ms) evoked responses in the superior temporal gyrus (Fig. 7.15). Thus content and temporal predictions engage complementary neural mechanisms in different regions, suggesting domain-specific prediction signaling along the cortical hierarchy. Auksztulewicz et al. (2018) noted that "the effects of 'what' and 'when' predictability on connectivity between regions were also dissociable. Stimuli with unpredictable contents were linked to increased excitability in frontal region to ascending drive from the STG, whereas stimuli with unpredictable onsets were associated with increased excitability in motor region to ascending inputs from the STG. These results indicate that 'what' prediction errors, likely signaled in the unpredictable condition, propagate primarily to prefrontal regions, whereas 'when' prediction error signaling relies on sensorimotor processing, likely involving indirect subcortical or cortical connections linking auditory and motor cortices. Content predictability also increased the strength of connections from sensory to precentral regions, possibly facilitating transmission of information to downstream areas."

Tabas and von Kriegstein (2021) also described two possible models of the organizational topology of the predictive processing auditory network: (1) the global view, which assumes that predictions on the sensory input are generated at high-order levels of the cerebral cortex and transmitted in a cascade of generative models to the subcortical sensory pathways; and (2) the local view, which assumes that independent local models, computed using local information, are used to perform predictions at each processing stage. In the global view, information encoding is optimized globally but biases sensory representations along the entire brain according to the subjective views of the observer. The local view results in a diminished coding efficiency, but guarantees in return a robust encoding of the features of sensory

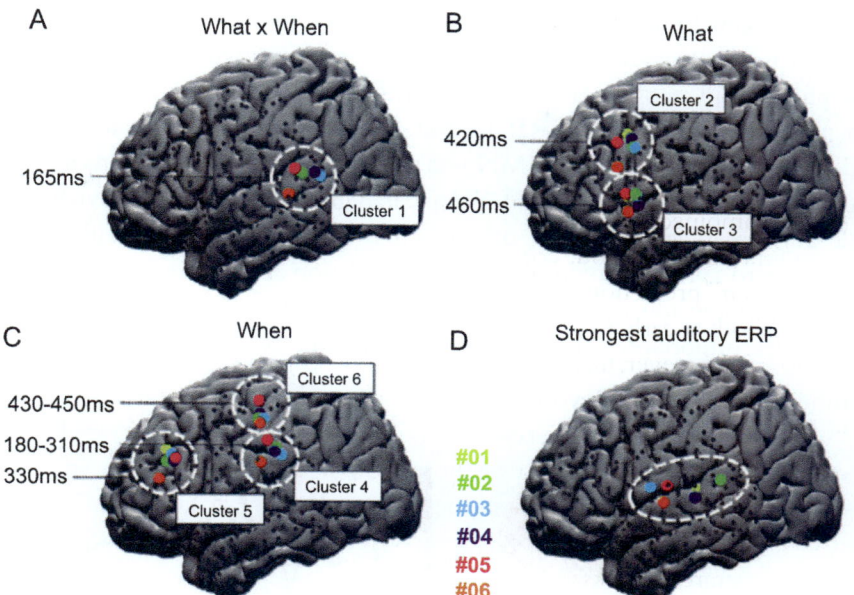

Fig. 7.15 Group-level main and interaction effects of "what" and "when" predictability. Plots represent single participants' *(color-coded)* electrodes *(gray dots* indicate all recording sites) adjacent to the peak group effects (used in subsequent analyses), significant at the family-wise error-corrected peak-level threshold $P < .05$. (A) The amplitude of the auditory evoked response in STG showed an interaction effect of "what" and "when" peaking at 165 ms after stimulus. *Colored dots* indicate individual subjects. (B) "What" predictability increased evoked amplitudes at frontal electrodes in two clusters peaking at 420 and 460 ms. (C) "When" predictability increased evoked amplitudes over the supramarginal gyrus and motor regions in two clusters peaking at 180–310 ms and 430–450 ms, and at frontal electrodes (over the middle frontal gyrus) peaking at 330 ms. (D) Location of electrodes with strongest auditory evoked responses. *(From Auksztulewicz, R., Schwiedrzik, C.M., Thesen, T., Doyle, W., Devinsky, O., Nobre, A.C., et al., 2018. Not all predictions are equal: "what" and "when" predictions modulate activity in auditory cortex through different mechanisms. J. Neurosci. 38 (40), 8680–8693.)*

input at each processing stage. Tabas and von Kriegstein (2021) note that "Although most experimental results to-date are ambiguous in this respect, recent evidence favors the global model." Liaukovich et al. (2022) applied the local–global auditory paradigm to obtain electrophysiological signatures of implicit detection of hardly distinguishable auditory stimuli. ERPs were recorded from 20 healthy volunteers during active discrimination of deviant sounds in the oddball sequence and passive listening of the same sounds in the sequence with local-global irregularity. The discrimination task consisted of two blocks with different deviant sounds targeted to respond. The sound discrimination accuracy was at an average of 40%, implying the difficulty of explicit sound recognition. Comparing ERPs to standard and deviant sounds, a posterior negativity in ERP around 450–600 ms in response to targeted deviant sounds was found. MMN was significant only in response to nontarget deviants. In the passive local–global paradigm (Fig. 7.16), Liaukovich et al. (2022) "observed an anterior positivity (284–412 ms), compatible with P3a, in response to a violation of local regularity. Violation of global regularity elicited an anterior negative response (228–586 ms), resembling the N400 component of ERPs. Importantly, the other indexes of auditory discrimination, such as

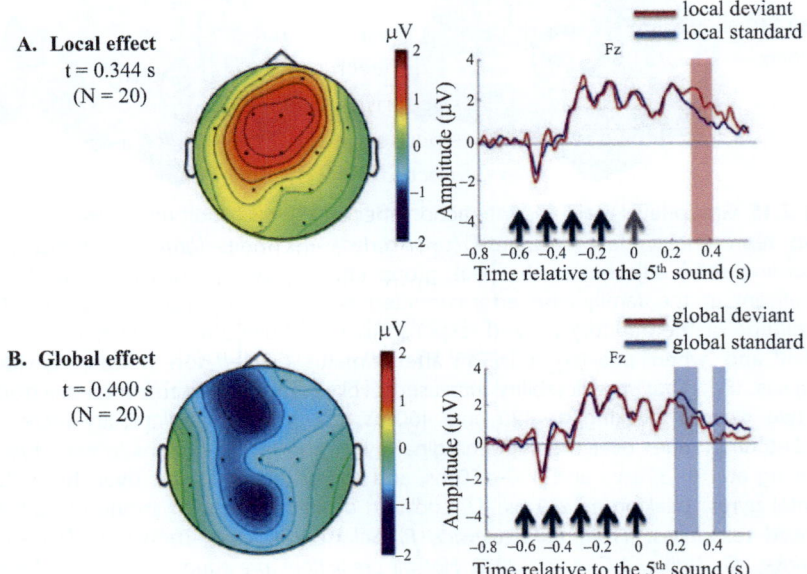

Fig. 7.16 Event-related responses to local and global irregularities in a passive listening condition. (A) Local mismatch response in EEG. The topography of the local effect (local deviants − local standards) is shown at $t = 344$ ms after the fifth sound onset. Cluster-based significant effects ($p_{uncorr} < 0.05$) in 296–412 ms time window are represented by *pink-shaded region*. *Black arrows* are standard sounds, a *gray arrow* is a deviant sound. (B) Global MMR in EEG. The topography of the global effect (global deviants − global standards) is shown at $t = 400$ ms after the fifth sound onset. Cluster-based significant effects ($p_{uncorr} < 0.05$) in 212–292 and 400–468 ms time windows are represented by *blue-shaded region*. *Black arrows* are standard sounds. X-axis, time in seconds; Y-axis, amplitude in microvolts. *(Reprinted from Liaukovich, K., Ukraintseva, Y., Martynova, O., 2022. Implicit auditory perception of local and global irregularities in passive listening condition. Neuropsychologia 165, 108129. © 2021, with permission from Elsevier.)*

MMN and P3b, were insignificant in ERPs to both regularity violations. The observed P3a and N400 components of ERPs may reflect prediction error signals in the implicit perception of sound patterns even if behavioral recognition was poor."

7.6 Summary

Some of the more salient findings are recalled here. Timing asynchrony is a major element to distinguish one auditory stream from two streams. Intracortical brain activity dynamically tracks speech streams using both low-frequency phase and high-frequency amplitude fluctuations and that optimal encoding likely combines the two. In and near low-level auditory cortices, attention "modulates" the representation by enhancing cortical tracking of attended speech streams, but ignored speech remains represented. ERPs elicited by infrequent randomly occurring unexpected silences in tone sequences presented at two tones per second to ERPs elicited by frequent, regularly occurring omissions (expected silences) within a sequence of tones presented at one tone per second. These unexpected silences elicit significant MMN and P3a. These results provide evidence for hierarchical predictive coding, indicating that the brain predicts both silences and sounds.

During the repetition of subsequent events, reduction of the MMN occurs. The underlying network, as found by dynamic causal modeling, includes left and right A1, left and right STG, and right IFG. One may define a "first-order" MMN that reflects the processing of simple deviant features (such as frequency) and a "higher-order" MMN that occurs at regularity violation of complex patterns or unexpected inputs that do not allow further predictions. First-order and higher-order MMN may be related to local and global deviance, respectively. Cortical sources of the first-order MMN are A1 bilaterally as well as left IFG. In source-localization analysis of higher-order MMN both the A1 and left IFG were primarily involved in the detection of simple, physically deviant features, while the right IFG was associated with the processing of novel, unexpected auditory inputs and the anterior cingulate cortex with regularity violation of known patterns. It is noted that the right IFG is generally part of the DCM-based MMN network whereas the left IFG rarely is.

Prediction-error and prediction-update signals are transmitted via gamma ($>40\,Hz$) and alpha/beta ($<30\,Hz$) oscillations, respectively. These findings provide strong support for hierarchical predictive coding and outline how it is dynamically implemented using distinct cortical areas and frequencies.

References

Auksztulewicz, R., Schwiedrzik, C.M., Thesen, T., Doyle, W., Devinsky, O., Nobre, A.C., et al., 2018. Not all predictions are equal: "what" and "when" predictions modulate activity in auditory cortex through different mechanisms. J. Neurosci. 38 (40), 8680–8693.

Barascud, N., Pearce, M.T., Griffiths, T.D., Friston, K.J., Chait, M., 2016. Brain responses in humans reveal ideal observer-like sensitivity to complex acoustic patterns. Proc. Natl. Acad. Sci. 113 (5), E616–E625.

Bastos, A.M., Usrey, W.M., Adams, R.A., Mangun, G.R., Fries, P., Friston, K.J., 2012. Canonical microcircuits for predictive coding. Neuron 76 (4), 695–711.

Bendixen, A., 2014. Predictability effects in auditory scene analysis: a review. Front. Neurosci. 8, 60. https://doi.org/10.3389/fnins.2014.00060.

Brace, K.M., Sussman, E.S., 2021. The brain tracks multiple predictions about the auditory scene. Front. Hum. Neurosci. 15, 747769. https://doi.org/10.3389/fnhum.2021.747769.

Bregman, A.S., 1990. Auditory Scene Analysis: The Perceptual Organization of Sound. Bradford Books, MIT Press, Cambridge, MA.

Cacciaglia, R., Escera, C., Slabu, L., Grimm, S., Sanjuán, A., Ventura-Campos, N., Ávila, C., 2015. Involvement of the human midbrain and thalamus in auditory deviance detection. Neuropsychologia 68, 51–58.

Chao, Z.C., Takaura, K., Wang, L., Fujii, N., Dehaene, S., 2018. Large-scale cortical networks for hierarchical prediction and prediction error in the primate brain. Neuron 100, 1252–1266.

Clos, M., Langner, R., Meyer, M., Oechslin, M.S., Zilles, K., Eickhoff, S.B., 2014. Effects of prior information on decoding degraded speech: an fMRI study. Hum. Brain Mapp. 35, 61–74.

Cooke, M., 1993. Modelling Auditory Processing and Organization. Cambridge University Press, Cambridge.

Denham, S.L., Winkler, I., 2020. Predictive coding in auditory perception: challenges and unresolved questions. Eur. J. Neurosci. 51, 1151–1160.

Dürschmid, S., Edwards, E., Reichert, C., Dewar, C., Hinrichs, H., Heinze, H.J., et al., 2016. Hierarchy of prediction errors for auditory events in human temporal and frontal cortex. Proc. Natl. Acad. Sci. 113 (24), 6755–6760. https://doi.org/10.1073/pnas.1525030113.

Dürschmid, S., Reichert, C., Hinrichs, H., Heinze, H.-J., Kirsch, H.E., Knight, R.T., Deouell, L.Y., 2019. Direct evidence for prediction signals in frontal cortex independent of prediction error. Cereb. Cortex 29 (11), 4530–4538.

Eggermont, J.J., 2015. Auditory Temporal Processing and Its Disorders. Oxford University Press, Oxford.

Friston, K.J., 2005. A theory of cortical responses. Philos. Trans. R. Soc. B 360, 815–836. https://doi.org/10.1098/rstb.2005.1622.

Fritz, J.B., Elhilali, M., David, S.V., Shamma, S.A., 2007. Auditory attention—focusing the searchlight on sound. Curr. Opin. Neurobiol. 17 (4), 437–455. https://doi.org/10.1016/j.conb.2007.07.011.

Garrido, M.I., Kilner, J.M., Kiebel, S.J., Stephan, K.E., Friston, K.J., 2007. Dynamic causal modelling of evoked potentials: a reproducibility study. Neuroimage 36, 571–580.

Garrido, M.I., Friston, K.J., Kiebel, S.J., Stephan, K.E., Baldeweg, T., Kilner, J.M., 2008. The functional anatomy of the MMN: a DCM study of the roving paradigm. Neuroimage 42, 936–944.

Garrido, M.I., Kilner, J.M., Stephan, K.E., Friston, K.J., 2009. The mismatch negativity: a review of underlying mechanisms. Clin. Neurophysiol. 120 (3), 453–463. https://doi.org/10.1016/j.clinph.2008.11.029.

Giraud, A.-L., Poeppel, D., 2012. Cortical oscillations and speech processing: emerging computational principles and operations. Nat. Neurosci. 15, 511–517.

Hesselmann, G., Sadaghiani, S., Friston, K.J., Kleinschmidt, A., 2010. Predictive coding or evidence accumulation? False inference and neuronal fluctuations. PLoS One 5 (3), e9926. https://doi.org/10.1371/journal.pone.0009926.

Hofmann-Shen, C., Vogel, B.O., Kaffes, M., Rudolph, A., Brown, E.C., Tas, C., et al., 2020. Mapping adaptation, deviance detection, and prediction error in auditory processing. Neuroimage 207, 116432.

Hsu, Y.-F., Hämäläinen, J.A., 2021. Both contextual regularity and selective attention affect the reduction of precision-weighted prediction errors but in distinct manners. Psychophysiology 58, e13753. https://doi.org/10.1111/psyp.13753.

Hsu, Y.-F., Waszak, F., Hämäläinen, J.A., 2019. Prior precision modulates the minimization of auditory prediction error. Front. Hum. Neurosci. 13, 30. https://doi.org/10.3389/fnhum.2019.00030.

Lecaignard, F., Bertrand, O., Caclin, A., Mattout, J., 2021. Empirical Bayes evaluation of fused EEG-MEG source reconstruction: application to auditory mismatch evoked responses. Neuroimage 226, 117468.

Liaukovich, K., Ukraintseva, Y., Martynova, O., 2022. Implicit auditory perception of local and global irregularities in passive listening condition. Neuropsychologia 165, 108129.

Lui, T.K.-Y., Shum, Y.-H., Xiao, X.-Z., Wang, Y., Cheung, A.T.-C., Chan, S.S.-M., 2021. The critical role of the inferior frontal cortex in establishing a prediction model for generating subsequent mismatch negativity (MMN): a TMS-EEG study. Brain Stimul. 14, 161–169.

McDermott, J.H., 2009. The cocktail party problem. Curr. Biol. 19, R1024–R1027.

Micheyl, C., Oxenham, A.J., 2010. Objective and subjective psychophysical measures of auditory stream integration and segregation. J. Assoc. Res. Otolaryngol. 11, 709–724.

Nie, Y., Zhang, Y., Nelson, P.B., 2014. Auditory stream segregation using bandpass noises: evidence from event-related potentials. Front. Neurosci. 8, 277. https://doi.org/10.3389/fnins.2014.00277.

Opitz, B., Rinne, T., Mecklinger, A., von Cramon, D.Y., Schröger, E., 2002. Differential contribution of frontal and temporal cortices to auditory change detection: fMRI and ERP results. Neuroimage 15, 167–174.

Phillips, H.N., Blenkmann, A., Hughes, L.E., Bekinschtein, T.A., Rowe, J.B., 2015. Hierarchical organization of frontotemporal networks for the prediction of stimuli across multiple dimensions. J. Neurosci. 35 (25), 9255–9264.

Prete, D.A., Heikoop, D., McGillivray, J.E., Reilly, J.P., Trainor, L.J., 2022. The sound of silence: predictive error responses to unexpected sound omission in adults. Eur. J. Neurosci. 55, 1972–1985.

Schröger, E., Kotz, S.A., San Miguel, I., 2015. Bridging prediction and attention in current research on perception and action. Brain Res. 1626, 1–13.

Sedley, W., Gander, P.E., Kumar, S., Kovach, C.K., Oya, H., Kawasaki, H., et al., 2016. Neural signatures of perceptual inference. eLife 5, e11476. https://doi.org/10.7554/eLife.11476.

Shamma, S.A., Elhilali, M., Micheyl, C., 2011. Temporal coherence and attention in auditory scene analysis. Trends Neurosci. 34, 114–123.

Shannon, R.V., Zeng, F.-G., Kamath, V., Wygonski, J., Ekelid, M., 1995. Speech recognition with primarily temporal cues. Science 270, 303–304.

Sohoglu, E., Chait, M., 2016. Detecting and representing predictable structure during auditory scene analysis. eLife 5, e19113. https://doi.org/10.7554/eLife.19113.

Stein, J., von Kriegstein, K., Tabas, A., 2022. Predictive encoding of pure tone and FM sweeps in the human auditory cortex. Cereb. Cortex Commun. 3, 1–14. https://doi.org/10.1093/texcom/tgac047.

Sussman, E.S., 2007. A new view on the MMN and attention debate: the role of context in processing auditory events. J. Psychophysiol. 21, 164–175.

Tabas, A., von Kriegstein, K., 2021. Adjudicating between local and global architectures of predictive processing in the subcortical auditory pathway. Front. Neural Circuits 15, 644743. https://doi.org/10.3389/fncir.2021.644743.

Tabas, A., Mihai, G., Kiebel, S., Trampel, R., von Kriegstein, K., 2020. Abstract rules drive adaptation in the subcortical sensory pathway. eLife 9, e64501. https://doi.org/10.7554/eLife.64501.

Wacongne, C., Changeux, J.-P., Dehaene, S., 2012. A neuronal model of predictive coding accounting for the mismatch negativity. J. Neurosci. 32 (11), 3665–3678. https://doi.org/10.1523/JNEUROSCI.5003-11.2012.

Winkler, I., Denham, S., Mill, R., Böhm, T.M., Bendixen, A., 2012. Multistability in auditory stream segregation: a predictive coding view. Philos. Trans. R. Soc. B 367, 1001–1012. https://doi.org/10.1098/rstb.2011.0359.

Zion Golumbic, E.M., Ding, N., Bickel, S., Lakatos, P., Schevon, C.A., McKhann, G.M., et al., 2013. Mechanisms underlying selective neuronal tracking of attended speech at a "cocktail party". Neuron 77, 980–991.

CHAPTER 8

Predictive coding in autism spectrum disorder, attention-deficit/hyperactivity disorder, and dyslexia

8.1 Introduction

In this chapter we will compare brain structures and their activity involved in predictive coding of the following developmental disorders: autism spectrum disorder (ASD), attention-deficit/hyperactivity disorder (ADHD), and developmental dyslexia (DD).

First of all, I note that ASD, ADHD, and DD all show disorder-specific changes in cerebellar structure as outlined by Stoodley (2014), who conducted a metaanalysis on voxel-based morphometry (VBM) studies. ASD studies revealed reduced gray matter (GM) in the inferior cerebellar vermis (lobule IX), left lobule VIIIB, and right Crus I. In ADHD, significantly decreased GM was found bilaterally in lobule IX, whereas participants with developmental dyslexia showed GM decreases in left lobule VI. The cerebellar regions identified in ASD showed functional connectivity with frontoparietal, default mode, somatomotor, and limbic networks; in ADHD, the clusters were part of dorsal and ventral attention networks; and in dyslexia, the clusters involved ventral attention, frontoparietal, and default mode networks. The results suggest that different cerebellar regions are affected in ASD, ADHD, and dyslexia, and these cerebellar regions participate in functional networks that are consistent with the characteristic symptoms of each disorder. Because predictive coding deals mainly with neocortex the remainder of this chapter does not dwell on cerebellar involvement, although it is important to draw attention to it.

8.2 Characterizing auditory aspects of ADHD and ASD

Attention-deficit/hyperactivity disorder is a behavioral and neurodevelopmental disorder characterized by inattention, hyperactivity, and impulsivity,

Brain Responses to Auditory Mismatch and Novelty Detection
https://doi.org/10.1016/B978-0-443-15548-2.00008-9

which are pervasive, impairing, and typically age inappropriate (Sergeant, 2000). Other neurodevelopmental conditions such as autism are common comorbidities. Ralli et al. (2020) evaluated the presence of hyperacusis in a small sample of children affected by ADHD compared to a control group of healthy children. Hearing was assessed using pure-tone audiometry and measuring tympanic membrane (eardrum) mobility. Hyperacusis was diagnosed in ~37% of children in the study group and in ~13% of children in the control group. In general, ADHD children and adolescents show severe impairment for duration-difference detection in the milliseconds range. In the seconds range, duration-difference thresholds are indistinguishable between ADHD and typical controls, and not correlated with milliseconds-range thresholds. Because time perception of milliseconds-duration stimuli might have a smaller effect on cognitive control and working memory, compared to longer durations, this is consistent with a pure timing deficit in individuals with ADHD (Anobile et al., 2022) specifically detrimental in multisensory processing (Panagiotidi et al., 2017). A recent literature review (Hours et al., 2022) noted that 50%–70% of individuals with ASD also have ADHD, which affects social skills, ability to communicate, behavior, and interests.

Autism spectrum disorder involves impairments in communication and social interaction, as well as high levels of repetitive, stereotypic, and ritualistic behaviors, and extreme resistance to change (Gomot et al., 2006). Regarding temporal processing deficits in ASD, I wrote (Eggermont, 2015): "Current theories and experimental data suggest that dysfunctional integrative mechanisms in autism may be the result of reduced neural synchronization. It has also been proposed that cortical networks in autism may be characterized by an imbalance between excitation and inhibition, which leads to hyperexcitability and unstable cortical networks (Rubenstein and Merzenich, 2003). Autistic subjects have aberrant perception, especially in the auditory domain, with both hypo- and hypersensitive features (Kujala et al., 2013). Speech reception thresholds (SRTs) in background noise for high-functioning autists (HFA) and Asperger syndrome (AS) patients were generally worse than for controls. The SRT is the average speech level at which 50% of the sentences are repeated correctly by the listener. A statistically significant difference in SRTs between the subject groups was found only for those background sounds that contained temporal or spectro-temporal dips. SRTs for the HFA/AS individuals were 2–3.5 dB higher than for the controls, which is equivalent to a substantial decrease (~10%–20%) in speech recognition. Thus, the HFA/AS individuals required a higher signal-to-noise ratio, whenever there were temporal dips in the background sound, to perform at the same level as the controls (Alcántara et al., 2004)." Recently (Eggermont, 2022), I noted that: "Hyperacusis is one of the most commonly identified auditory responses in children with ASD (Kujala et al., 2013; Dabbous, 2012). Unusual responses to sensory stimuli are experienced by up to 90% of individuals with ASD (Ben-Sasson et al., 2009; Williams et al., 2021). Abnormal behavioral responses to sensory stimuli, including sound, have become so prevalent, that the most recent criteria in the Diagnostic and Statistical Manual of Mental Disorders 5th edition (DSM-V) for ASD added a diagnostic component of hyper- and hypo-reactivity to sensory stimuli (Pfeiffer et al., 2019)."

8.3 Changes in auditory brainstem responses and cortical auditory evoked potentials

8.3.1 A general overview

Before discussing event-related potential (ERP) findings in ASD and ADHD, I will present data from ~100 papers in Table 8.1; however, I only describe some of those in upcoming sections. The ERPs include both obligatory (exogenous) potentials from auditory brainstem response (ABR), to P1, N1 to P2, and endogenous ones; N2, MMN, and the P3 complex. In Table 8.1, I use bold font in findings for which there is an overwhelming consensus, but showing that P3a amplitude changes in ASD have wide but contradictory support. Albeit that there are much more results on ASD compared to ADHD, the data suggest that using ERPs to diagnose between these two, often comorbid, disorders does not seem possible. We will however present a few studies where recordings have been made in ASD and ADHD patients under similar conditions that at least suggest some diagnostic potential based on ERPs. I also recall here that endogenous ERPs reflect the cognitive aspects of these disorders.

8.3.2 Auditory brainstem responses

The scalp-recorded response to stimulation of the auditory nerve and brainstem is called the auditory brainstem response (ABR). A typical ABR consists of a sequence of up to seven vertex positive waves with negative valleys in between. The subsequent peaks are typically labeled with Roman numerals. Wave I equates the compound action potential (CAP) of the auditory nerve generated peripherally from the petrous bone, whereas wave II is the CAP at the site where the auditory nerve leaves the petrous bone. Peaks

Table 8.1 Amplitude and latency changes in exogenous and endogenous ERPs.

	Autism			ADHD		
	Children	Adolescents	Adults	Children	Adolescents	Adults
ABR (I–V)	↑ 53,54,55,66,67,69 ↑ 70,72,75,84,86,89,92 ≈ 63,79		≈ 59	→ 31,44		
P1 (M50)	↑ 73,74,85,90 ↓ 80 → 91		↓ 62			↓ 39
N1 (M100)	↓ 3,58 ↑ 5,78 ↑ 65 ↑ 14,71,74,77,85		→ 22	↓ 32		
P2	→ 83	↓ 52				
N2	↓ 5,80,87 ↑ 90			↑ 30,36,40 ↓ 32 → 88 ↓ 27,35,40 → 41	↓ 35	↓ 39 ↓ 37 → 38
MNN	↓ 2,9,82 ↑ 7,68 ↓ 4,6,15,16,17,19,24,82 ↓ 10,14,20,23,24	↓ 11,25 ↑ 18 → 25	↑ 21 → 22	≈ 30 ↑ 42,47,48,49 ↓ 47,48		↓ 50,51
P3a	↓ 1,4,58,76,78,87 ≈ 17 ↑ 3,8,9,15,19,82 → 58	↓ 34 ↑ 81	↓ 1,2	↓ 31,32,33,34,40,45,49 ≈ 43 → 31,41	≈ 35	≈ 36,38 → 38
P3b	↓ 1,64 ↑ 8,20	↓ 52,56,57	≈ 21,61,60	↓ 26,27,28,33,34,40,46	↓ 34	↓ 37

↓↑ ≈ amplitude changes; ← ~ → latency changes.

References for ASD: [1]Kenner et al. (1995), [2]Gomot et al. (2002), [3]Ferri et al. (2003), [4]Lepistö et al. (2006), [5]Orekhova et al. (2009), [6]Dunn et al. (2008), [7]Lepistö et al. (2008), [8]Whitehouse and Bishop (2008), [9]Gomot et al. (2011), [10]Roberts et al. (2011), [11]Ludlow (2014), [12]Ford et al. (2017), [13]Galilee et al. (2017), [14]Matsuzaki et al. (2017), [15]Vlaskamp (2017), [16]Yoshimura et al. (2017), [17]Huang et al. (2018), [18]Leno et al. (2018), [19]Lindström et al. (2018), [20]Riva et al. (2018), [21]Grisoni et al. (2019), [22]Matsuzaki et al. (2019a), [23]Matsuzaki et al. (2019b), [24]Chen et al. (2020), [25]Di Lorenzo et al. (2020), [52]Novick et al. (1980), [53]Skoff et al. (1980), [54]Tanguay et al. (1982), [55]Taylor et al. (1982), [56]Niwa et al. (1983), [57]Dawson et al. (1988), [58]Oades et al. (1988), [59]Grillon et al. (1989), [60]Ciesielski et al. (1990), [61]Erwin et al. (1991), [62]Buchwald et al. (1992), [63]McClelland et al. (1992), [64]Lincoln et al. (1993), [65]Bruneau et al. (1999), [66]Maziade et al. (2000), [67]Rosenhall et al. (2003), [68]Lepistö et al. (2005), [69]Kwon et al. (2007), [70]Tas et al. (2007), [71]Gandal et al. (2010), [72]Roth et al. (2011), [73]Roberts et al. (2013), [74]Matsuzaki et al. (2012), [75]Ververi et al. (2015), [76]Donkers et al. (2015), [77]Port et al. (2016), [78]Irwin et al. (2017), [79]Santos et al. (2017), [80]Yu et al. (2018), [81]Hudac et al. (2018), [82]Charpentier et al. (2018), [83]Foss-Feig et al. (2018), [84]Ramezani et al. (2019), [85]Roberts et al. (2019), [86]Kamita et al. (2019), [87]Donkers et al. (2020), [88]Aykan et al. (2020), [89]Li et al. (2020), [90]Dwyer et al. (2021), [91]Yoshimura et al. (2021), [92]Miron et al. (2016).

References for ADHD: [26]Holcomb et al. (1986), [27]Johnstone and Barry (1996), [28]Kemner et al. (1996), [29]Jonkman et al. (1997), [30]Oades et al. (1996), [31]Puente et al. (2002), [33]Brown et al. (2005), [34]Gumenyuk et al. (2005), [35]Alexander et al. (2008), [36]Barry et al. (2009), [37]Itagaki et al. (2011), [38]Fisher et al. (2011), [39]Retz et al. (2012), [40]Senderecka et al. (2012), [41]Tsai et al. (2012), [42]Gomes et al. (2012), [43]Gomes et al. (2013), [44]Jafari et al. (2015), [45]Yang et al. (2015), [46]Janssen et al. (2016), [47]Yamamuro et al. (2016), [48]Lee et al. (2020), [49]Zhang et al. (2020), [50]Hsieh et al. (2021), [51]Kim et al. (2021).

with numbers from III up to and including wave V are generated in the auditory brainstem (cf. Fig. 6.1; Eggermont, 2019).

8.3.2.1 ABR changes in ASD

In a review on the ABRs in autism, Pillion et al. (2018) noted two main trends: (1) abnormalities occur mainly at higher levels of the auditory brainstem, according to structural imaging, and in the I-V interval, specifically caused by changes in the III-V interval (Table 8.1; Skoff et al., 1980) and (2) brainstem abnormalities appear to be more common in younger than older children with ASD. This suggests delayed maturation of neural transmission pathways between lower and higher levels of the brainstem, consistent with atypical sound sensitivity, poor sound localization, and difficulty listening in background noise in ASD.

Compared to typical developing (TD) children, ASD children have a significant higher incidence of hearing loss as reflected in a prolonged wave I latency (Fig. 8.1; Taylor et al., 1982). Routine ABR testing is therefore indicated, especially because increased I-V intervals were found in 27% of

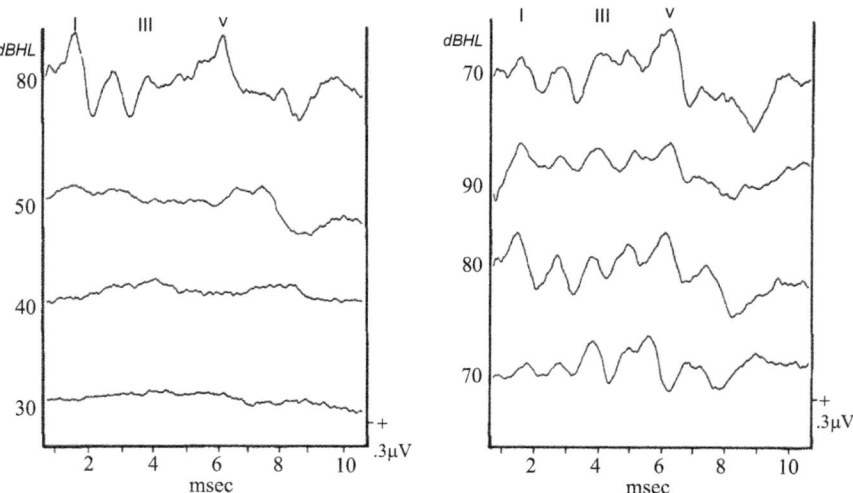

Fig. 8.1 Left. Auditory threshold determination in an autistic child. Wave V is clearly seen at 50- and 40-dB HL, but absent at 30-dB HL, indicating a raised threshold at 40-dB HL. The I-V interval is clearly enlarged, norm is ~4ms. Right. ABRs from three autistic children (three top traces) showing the increased I-V interwave latencies compared to a normal control child's response (bottom trace). *(From Taylor, M.J., Rosenblatt, B., Linschoten, L., 1982. Auditory brainstem response abnormalities in autistic children. Can. J. Neurol. Sci. 9 (4), 429–433, reproduced with permission from Cambridge University Press.)*

49 children who were difficult to test or who had hearing loss (Rosenhall et al., 2003). From a metaanalysis, Talge et al. (2018) found that "ASD was associated with longer ABR latencies for waves III, I–III and I–V. All components showed significant heterogeneity. Associations with ASD symptoms were strongest among participants ≤8 years of age and those without middle ear abnormalities or elevated auditory thresholds." Note that hearing loss on its own typically does not result in increased I–V intervals (Eggermont, 2017).

8.3.2.2 ABR changes in ADHD

Puente et al. (2002) compared ABRs in school children with and without ADHD. They found that the latencies of waves III and V were significantly prolonged in children with ADHD. Furthermore, a significant difference was found between the mean I–III and I–V intervals of children with ADHD as compared with controls. From a metaanalysis, Talge et al. (2022) concluded that ADHD was associated with longer latencies for waves III and V and waves I–III and I–V. Effect sizes from the ASD (Talge et al., 2018) and ADHD metaanalyses did not differ from each other.

8.3.3 Changes in cortical auditory evoked potentials

The cortical auditory evoked potentials (CAEPs) considered here are the P1 (P50), N1 (N100), P2, and N2. Data from the P1 in the cat, and complementary data from humans, link the substrate of the P1 to cholinergic components of the ascending reticular activating system (RAS) and their thalamic target cells (Buchwald et al., 1992). The N1 is dominantly generated in association auditory cortex (component N1a, which if there is no confusion I will indicate as N1) but also has a frontal component N1b (Picton et al., 1974). P2 is also generated by activity from the RAS (Ponton et al., 2000; Chapter 4) and has a source in auditory association cortex. The N2 consists of several subcomponents; the one involved in oddball tasks (Chapter 7) is related to stimulus discrimination and elicited by unexpected events (Näätänen and Picton, 1986). The N2 in response to attended or unattended deviant auditory stimuli is also referred to as the early mismatch negativity (MMN), so some confusion may exist in the entries of Table 8.1. More generally, the N2 component has been described in tasks that reflect stimulus identification, attentional shifts, and response selection timing (details in Appendix). Park et al. (2022) noted that loudness dependence—which plays a role in hyperacusis—of the CAEPs is associated with central serotonergic neurotransmission (O'Neill et al., 2008). Park et al. (2022) found that in boys with ADHD, the CAEP is associated with symptom severity, as they showed a significant association with impulsivity and inattention. Importantly, CAEPs and symptoms decreased after treatment with a dopamine uptake inhibitor. Studies in mice also have proposed that the CAEP is influenced by dopaminergic activity (Mereu et al., 2017).

8.3.3.1 Findings in ASD

I only discuss here some of the many results presented in Table 8.1. For P1 and N1, in general, the latencies are prolonged in ASD compared to TD. In an early study, Buchwald et al. (1992) found that the P1 component was significantly smaller in the autistic subjects at slow rates of stimulation, and that the autistic P1 did not further change as rates of click stimulation increased from 0.5 to 10/s, in contrast to the normally produced P1 decrement (Fig. 8.2). The abnormal P1 responses strongly suggest that the RAS and/or its postsynaptic thalamic targets may be dysfunctional in ASD. Note that the Pa component belonging to the middle–latency response (MLR)

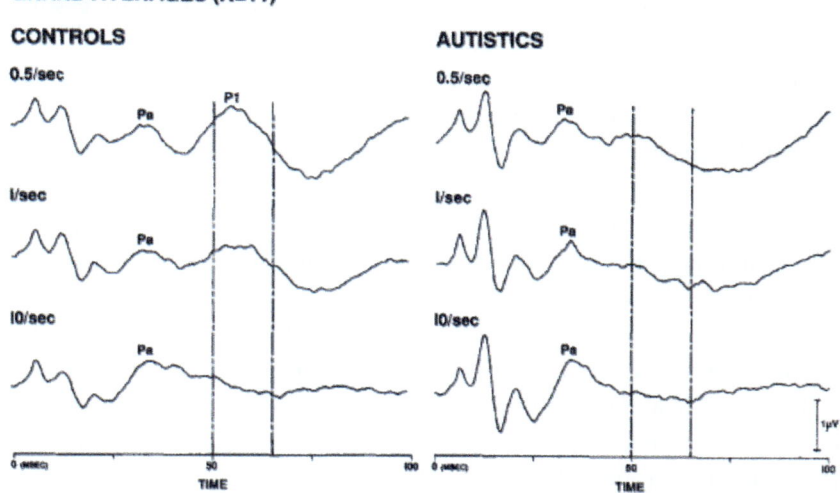

Fig. 8.2 MLR grand averages for control and autistic groups to three click rates (0.5, 1, 10/s). The *vertical lines* indicate the 50–65 ms time period within which PI peak amplitudes occurred in all control subjects. Each trace represents the grand average of 500 trials from each of 11 subjects. (*Reprinted from Buchwald, J.S., Erwin, R., Van Lancker, D., Guthrie, D., Schwafel, J., Tanguay, P., 1992. Midlatency auditory evoked responses: P1 abnormalities in adult autistic subjects. Electroencephal. Clin. Neurophysiol. 84, 164–171. © 1992 Elsevier Scientific Publishers Ireland, Ltd., with permission from Elsevier.*)

and likely originating in the auditory radiation (Picton, 2011) is not affected. Bruneau et al. (1999, 2003) investigated auditory processing at the cortical level in 4–8-year-old autistic children with mental retardation. They found that autistic children with very disturbed verbal communication had greater Na amplitude in right auditory cortex. The scalp potential field configuration observed for Na (Deiber et al., 1988) suggested a deep generator, so the thalamocortical radiation is a more likely source (Appendix). This suggested to Bruneau et al. (1999, 2003) that the auditory associative cortex in the lateral part of the superior temporal gyrus shows more specific left side defects when auditory stimulus has to be processed.

Roberts et al. (2013) compared the magnetic equivalent of P1 (M50) with magnetic resonance diffusion tensor imaging (DTI). They characterized white matter diffusion anisotropy in the acoustic radiations as a function of development in autistic and typically developing children. It appeared that only in TD controls, white matter fractional anisotropy—indicating nerve-tract patency—increased with age and was associated with earlier M50 responses. Whereas the patency of this nerve tract did not differ between groups, individuals with autism displayed a significant M50 latency increase. In both groups, M50 latency decreased with age, again suggesting a maturation delay in the autism group.

In 6–17-year-old individuals with ASD, Brandwein et al. (2015) examined the relationship between auditory-ERP measures and autism severity. They performed a median split of the ERP data as a function of the participant's autism severity score to illustrate the effects significantly related to autism symptom severity. These are shown in the N1a (temporal) and N1b (frontal) components and the parietally focused audiovisual integration peak (between 100 and 130 ms). Waveforms for these three groups—including typical development—show that the peak amplitude of the N1a in the "ASD-moderate" group is midway between the peak amplitude of the N1a in the "ASD-severe" and TD group (Fig. 8.3A). The N1b component (Fig. 8.3B) in the "ASD-moderate" group is very similar to, and in fact overlapping with, that of the TD group. The N1b is strikingly smaller in the "ASD-severe" group. The parietally focused audiovisual integration peak between 100 and 130 ms is largest in the TD group and smallest in the "ASD-severe" group, with the peak amplitude of the "ASD-moderate" group falling midway between the severely autistic and the TD children (Fig. 8.3C). Brandwein et al. (2015) suggested that deficits in audiovisual integration "may lead to experiences of a disorganized sensory environment and 'sensory overload', which in turn may lead to withdrawal and defensive sensory behaviors."

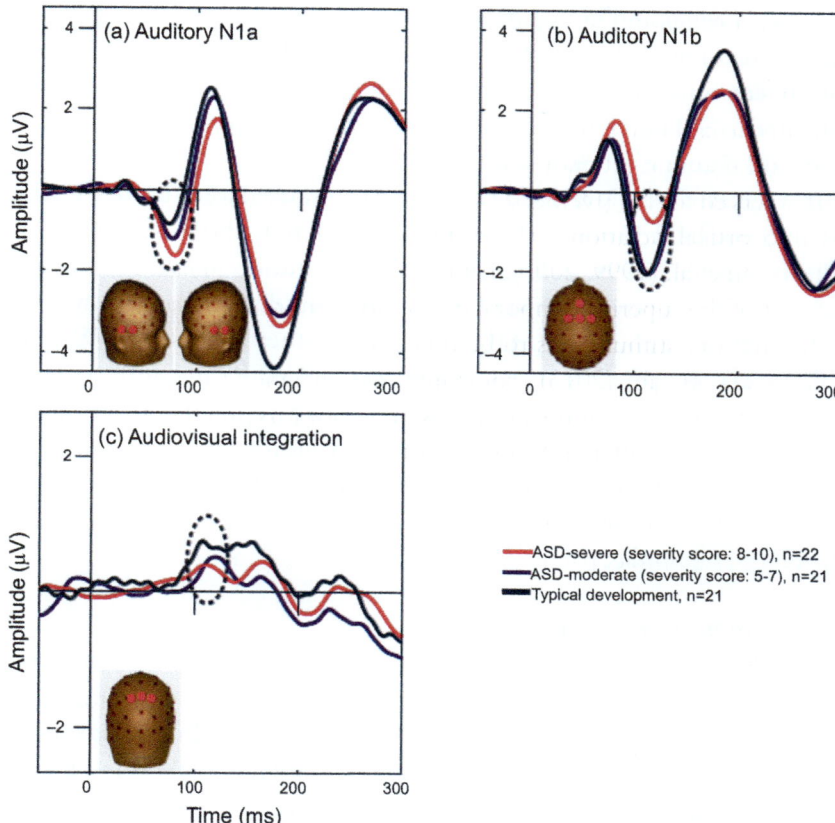

Fig. 8.3 Mean ERPs for the ASD-severe, ASD-moderate, and typically developing groups. (A and B) The three groups' responses to the auditory-alone condition, with *dashed ellipses* indicating the component of interest (the auditory N1a and N1b). (C) A measure of audiovisual integration, represented by a difference wave over the parietal scalp, with a *dashed ellipse* to indicate the response window of interest. Traces represent the composite signal from adjacent electrodes, the locations of which are indicated on the head models. *(From Brandwein, A.B., Foxe, J.J., Butler, J.S., Frey, H.-P., Bates, J.C., Shulman, L.H., et al., 2015. Neurophysiological indices of atypical auditory processing and multisensory integration are associated with symptom severity in autism. J. Autism Dev. Disord. 45, 230–244. https://doi.org/10.1007/s10803-014-2212-9. © Springer Science+Business Media New York 2014, with permission from Springer Nature.)*

Because autistic people often have atypical experiences of sound intensity (e.g., hyperacusis; Eggermont, 2022), Dwyer et al. (2021) examined ERPs evoked by tones of differing intensities (50-, 60-, 70-, and 80-dB SPL) in a large sample of young children (2–5 years) with and without an autism diagnosis. They found no significant latency-age effects in TD. Autistic participants had longer-latency P1 responses to 80-dB sounds over the right hemisphere. Smaller N2 responses were also observed in autistic participants compared to TD children, which exhibited larger negative N2 amplitudes in the 60-dB through 80-dB conditions.

In a metaanalysis, Jorgensen et al. (2021) compared long-latency CAEPs and magnetic fields produced by autistic and TD individuals to nonlinguistic auditory stimuli. Their findings indicated that autistic individuals demonstrate bilaterally delayed P1(M50) peaks and lateralized delays in the right but not left hemisphere N1(M100) peak. In another metaanalysis Williams et al. (2021) compared latencies/amplitudes of CAEPs and corresponding magnetic field components. They found that in response to pure tones, autistic individuals exhibited prolonged latencies for P1(M50) and M100 and reduced amplitudes for N1a and N2. There were no significant group differences in P2(M200) and N2 latencies, and in P1(M50), N1b, M100, or P2(M200) amplitudes.

8.3.3.2 Findings in ADHD

The predominantly inattentive type of ADHD shows either cortical hypoarousal, or maturational lag (with EEGs resembling those of younger TD children). Brown et al. (2005) investigated whether CAEPs from an audiovisual oddball task could differentiate inattentive-type ADHD in children (8–12 yrs of age) from controls. Stimuli (20% targets) were presented at a fixed interstimulus interval of 1.03 s and the children were required to count all targets. The task successfully differentiated the inattentive-type ADHD children from controls, with inattentive-type ADHD children having smaller N1, and P2 amplitudes to both the auditory targets and the visual nontargets. Senderecka et al. (2012), however, reported that ADHDs showed enhanced P2 and reduced N2 component to auditory two-tone oddball stimuli, compared to controls. Gomes et al. (2012) found a reduced amplitude of Ta in children with ADHD. This Ta component of the auditory T-complex matures relatively early (Tonnquist-Uhlen et al., 2003) and is best represented by radial dipoles located in parabelt area of secondary auditory cortex (cf. Fig. 6.5; Ponton et al., 2002).

The inability in ADHD to filter out extraneous stimuli and to attend to salient features of the environment implicates a deficit in gating—just as in ASD, which is characterized by a general reduction of the ability to gate intrusive sensory, motor, and/or cognitive information. Holstein et al. (2013) compared two distinct operational measures designed to assess gating or central inhibition in ADHD patients and healthy controls. These are pre-pulse inhibition (PPI) of the acoustic startle response, considered to be a form of sensorimotor gating, and P50 suppression. Compared to healthy controls, ADHD patients exhibited impaired P50 suppression, performed worse in cognitive tasks, and reported more psychopathological symptoms, but were normal in the test of PPI. These findings indicate the differences between P50 gating and PPI as measures of the gating construct (Holstein et al., 2013).

8.4 Changes in MMN and the P3 complex

Reflecting endogenous responses, deviant sounds elicit an MMN as an index of automatic sound-change discrimination, the P3a as an index of attentional orienting, and the P3b in task orienting (Chapter 5).

8.4.1 Findings in ASD

The MMN and P3 complex has been extensively studied in ASD as shown in the numerous references in Table 8.1. Here we highlight only some typical findings, without pretending to be complete. We will start with a comparison of findings in low to moderate- and high-level functioning, i.e., Asperger. Then we follow up with the effects on brain rhythms and end with recent studies focusing on speech and language discrimination.

8.4.1.1 Standard ERPs in low-to-moderate functioning ASD

Courchesne et al. (1984) showed that compared to the control group, low-to-moderate functioning autistics (13–21-year-olds) had smaller N1 and P2 amplitudes to novel sounds and smaller P3b amplitudes to target sounds. Dawson et al. (1988) found that compared to normal subjects, 8–19-year-old autistic subjects showed a significantly smaller P3b amplitude to phonetic stimuli for Cz and left hemisphere recording sites. However, no group difference in P3b amplitude to the phonetic stimulus was found for the right hemisphere. In contrast, Gomot et al. (2011) examined auditory change detection in children with ASD with MMN and P3a. ASD children had significantly shorter MMN latency and larger P3a than controls, indicating a greater tendency to switch attention to deviant events. Gomot and Wicker (2012) then proposed that dysfunction in the ability to build flexible predictions in ASD may originate from impaired top-down influence over a variety of sensory and higher-level information processing.

More recently, Galilee et al. (2017) investigated the detection and discrimination of speech and nonspeech sounds in children (4–6-year-olds) with ASD, compared with gender and verbal-age matched controls. Control participants exhibited late match/mismatch (N330) responses measured from temporal electrodes, reflecting speech versus nonspeech detection, bilaterally, whereas children with ASD exhibited this effect only over temporal electrodes in the left hemisphere (*blue curves*, in Fig. 8.4). These findings suggest that children with ASD fail to activate right hemisphere mechanisms, likely associated with social or emotional aspects of speech detection, when distinguishing nonspeech from speech stimuli.

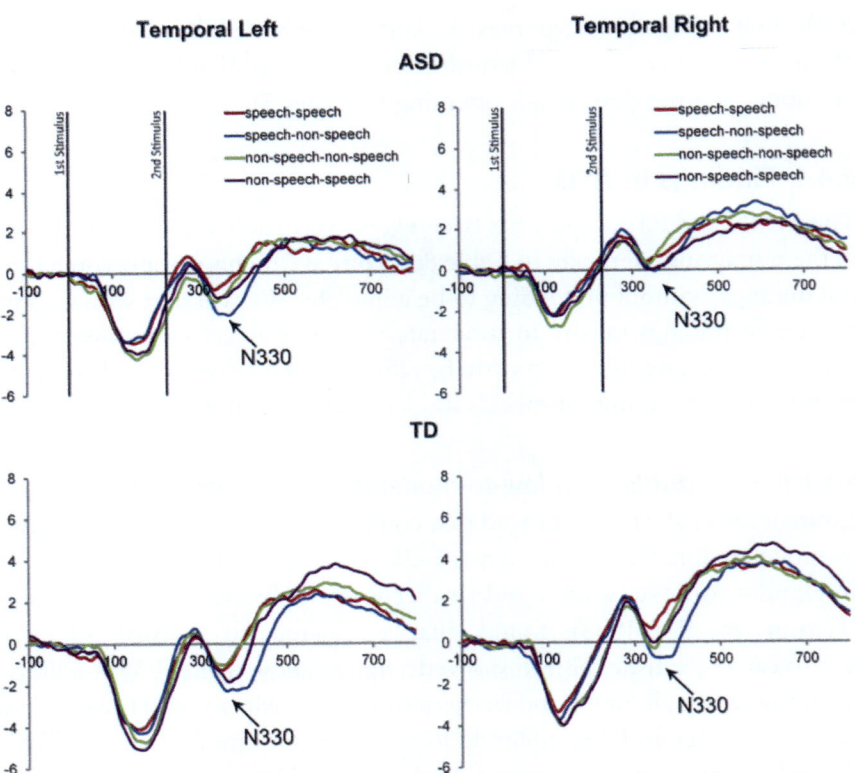

Fig. 8.4 ERP waveforms in the temporal area. The figure represents the ERP waveforms recorded from the temporal left (left side) and temporal right (right side) electrodes in the ASD and the TD control groups. *(From Galilee, A., Stefanidou, C., McCleery, J.P., 2017. Atypical speech versus non-speech detection and discrimination in 4- to 6-yr old children with autism spectrum disorder: an ERP study. PLoS One 12 (7), e0181354. https://doi.org/10.1371/journal.pone.018135 Open access.)*

Using MEG, Mamashli et al. (2017) focused on both the evoked mismatch field (MMF) response in temporal and frontal cortical locations, and functional connectivity (FC) with spectral specificity between those locations. In the quiet condition (Fig. 8.5), they found common MMF sources (white outline) in the right superior temporal gyrus (STG, *red outline*) and inferior frontal gyrus (IFG, *yellow outline*) response in ASD and TD. In the noise condition, the MMF response in the right IFG was preserved in the TD group, but strongly reduced relative to the quiet condition in ASD group. The MMF response in the right IFG also correlated with severity of ASD. Considering the role of the IFG in predictive coding (Chapters 1 and 7) and in fine tuning temporal cortex, Mamashli et al.'s (2017) findings in ASD indicate that reduced feedback connections and consequent weak top-down processing may contribute to auditory and language impairments in background noise.

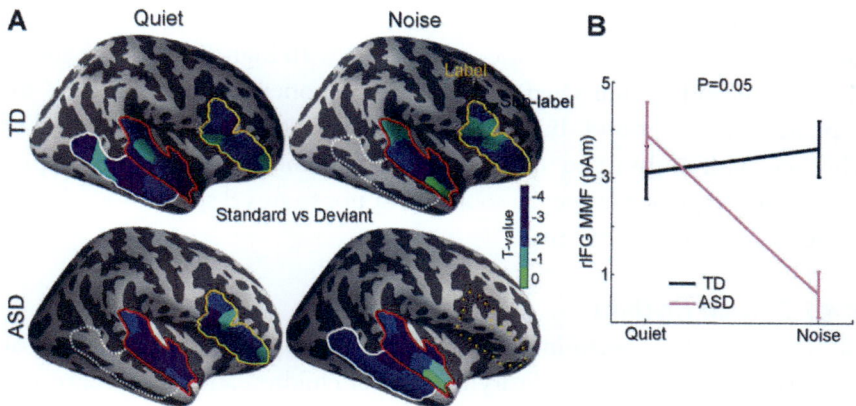

Fig. 8.5 (A) MMF response, obtained by contrasting the responses to deviant versus standard tones, in each condition (quiet and noise), and within each group. MMF was observed in the rSTG *(red outline)*, rMTG *(white outline)*, and rIFG *(yellow outline)*. *Solid outlines: P<.05 corrected. Dashed outlines: P<.1 (trend). Dotted outlines:* No significant MMF. *T*-values of the sublabels within each significant label are color coded as indicated in the *color bar.* (B) Mean and standard error of the MMF in rIFG for the TD and ASD group in the two conditions (quiet, noise). *(From Mamashli, F., Khan, S., Bharadwaj, H., Michmizos, K., Ganesan, S., Garel, K.-L.A., et al., 2017. Auditory processing in noise is associated with complex patterns of disrupted functional connectivity in autism spectrum disorder. Autism Res. 10. 631–647. © 2016 International Society for Autism Research, Wiley Periodicals, Inc., with permission from John Wiley and Sons.)*

8.4.1.2 Asperger as a special case?

Using an ERP paradigm for pitch, duration, and phonetic changes in vowels and to corresponding changes in nonspeech sounds, Lepistö et al. (2006) investigated auditory discrimination and orienting in children with Asperger syndrome (AS). In contrast to the TD controls, the MMN (N2) in the AS group showed right-hemisphere dominance. Furthermore, the children with AS had diminished MMN amplitudes and decreased hit rates for stimulus-duration changes. In contrast, their MMN to speech-pitch changes was parietally enhanced. The P3a, reflecting preattentive orienting to changes, was diminished in the children with AS for speech-pitch and pho-neme changes, but not for the corresponding nonspeech changes. In response to vowels and complex tones, Whitehouse and Bishop (2008) recorded ERPs of children with high-functioning autism (mean nonverbal IQ = 109.87, aged 7:6–14:3 yrs) and TD children (mean nonverbal IQ = 115.73, aged 7:0–14:3 years). In each condition, repetitive "standard" sounds were replaced by a within stimulus-type "deviant" sound and a between stimulus-type "novel" sound. Participants' level of attention was also varied between conditions. Children with high-functioning autism had significantly diminished obligatory components in response to the repetitive speech sound, but not to the repetitive nonspeech sound. The children with autism also showed reduced orienting to novel tones pre-sented in a sequence of speech sounds, but not to novel speech sounds pre-sented in a sequence of tones.

8.4.1.3 Brain rhythms

Wilson et al. (2007) examined the integrity of local circuitry by focusing on gamma-band activity in auditory cortices of children and adolescents with autism. Children and adolescents with autism showed significantly reduced left hemispheric 40-Hz power from 200- to 500-ms poststimulus onset. In contrast, no significant between-group differences were observed for right hemispheric cortices. In two separate studies, Orekhova et al. (2007) found that young boys with autism demonstrated a pathological increase of beta (13.2–24 Hz) in one study and gamma (24.4–44.0 Hz) in both studies. The amount of gamma activity correlated positively with degree of devel-opmental delay. Using child-size MEG equipment, Kikuchi et al. (2013) obtained measurements of the brain activity of young children (3–7 years old) with ASD and age-matched TD children. Physiological connectivity and the laterality of physiological connectivity were assessed using intrahe-mispheric coherence for nine frequency bands. In contrast to Stroganova et al. (2007), who studied a similar age group, Kikuchi et al. (2013) found significant rightward connectivity between the parieto-temporal areas, via increased gamma-band oscillations, in the ASD group.

Clarke et al. (2016) recorded EEG in 20 boys with Asperger syndrome, aged 7–12 years, and an age- and sex-matched control group in the eyes closed condition. They found that "the Asperger's group had a global increase in absolute delta and an anterior increase in relative delta. Both absolute and relative theta was globally increased and relative alpha was globally decreased. Subjects with Asperger's Syndrome exhibited reduced anterior inter-hemispheric coherence in the alpha and beta bands." These results suggested to Clarke et al. (2016) the existence of frontal lobe abnormalities in children with Asperger syndrome and possible abnormalities in normal brain maturational processes. Results also suggest that these children may have a group of brain abnormalities different from children with other forms of ASD. Edgar et al. (2016) binaurally presented 40-Hz amplitude-modulated tones of 1-s duration while MEG data were obtained from ASD and TD children. Using single dipoles anatomically constrained to each participant's left and right Heschl's gyrus, left and right 40-Hz ASSR total power and intertrial coherence measures were obtained. They found that TD and ASD did not differ in 40-Hz ASSR total power or intertrial coherence.

8.4.1.4 *Effects of language impairment*

Lepistö et al. (2008) found that children with autism had enhanced MMNs for pitch changes, as well as for phoneme-category changes in the constant-feature condition. However, children with autism lose their advantage in the phoneme discrimination when the context of the stimuli is speech-like and requires abstracting invariant speech features from varying input. This suggests that enhanced low-level perceptual skills present in children with autism might affect their perceptual processing of speech sounds at a higher level. Using MEG, Roberts et al. (2011) assessed associations between language impairment and the amplitude and latency of the superior temporal gyrus MMN in response to changes in an auditory stream of tones or vowels in children (aged 6–15 years) with ASD and controls. MMN latency was significantly prolonged in children with ASD, compared with neurotypical control subjects. Furthermore, this delay was most pronounced ($\sim$50 ms) in children with concomitant language impairment (LI). Lindström et al. (2018) explored the processing of emotional speech in school-aged children with ASD but without marked language impairments (ASD/−LI). Overall, children with ASD/−LI were slower in behaviorally discriminating prosodic features of speech stimuli than TD children. Further, smaller standard-stimulus ERPs and MMNs were found in children with ASD/−LI than in controls. In addition, the amplitude of the P3a was diminished and differentially distributed on the scalp in children with ASD/−LI than in control children. Green et al. (2020) investigated the relationship between MMN latency in children ages 5–10 with ASD and language impairment (ASD/+LI), ASD/−LI, and TD children during an auditory oddball experiment presenting speech and pure-tone sounds. They observed that children with ASD/+LI had decreased MMN latency in the left hemisphere in response to novel vowel sounds compared to children with ASD/−LI and TD controls. Parent responses to the Sensory Experiences Questionnaire revealed that all participating individuals with ASD were hypersensitive to sounds (hyperacusis; Eggermont, 2022). This may underlie the increased connectivity in primary sensory cortices at the expense of connectivity to association areas of the brain in some children with ASD and language impairment. Matsuzaki et al. (2019b) examined MMF responses bilaterally during an auditory oddball paradigm with vowel stimuli in extremely language impaired (minimally verbal/nonverbal) children who have ASD (ASD/MVNV). This was compared with a verbal ASD cohort without cognitive impairment that was split into those demonstrating considerable language impairment versus those with less or no language impairment. Delayed MMF latencies were found bilaterally in ASD/MVNV compared to verbal ASD—both ASD/−LI and ASD/+LI—and TD children. Delayed MMF responses were associated with diminished language and communication skills. Furthermore, whereas the TD children showed leftward lateralization of MMF amplitude, ASD/MVNV and verbal ASD—ASD/−LI and ASD/+LI—showed abnormal rightward lateralization.

8.4.2 Findings in ADHD

8.4.2.1 Metaanalyses

We start with two metaanalyses. The first (Hart et al., 2013) focused on fMRI during inhibition and attention tasks. They found that "patients with ADHD relative to controls showed reduced activation for inhibition in the right IFG, supplementary motor area, and anterior cingulate cortex (ACC), as well as striato-thalamic areas, and showed reduced activation for attention in the right dorsolateral prefrontal cortex (DLPFC), posterior basal ganglia, and thalamic and parietal regions." They also reported that "the supplementary motor area and basal ganglia were under activated solely in children with ADHD relative to controls, while the IFG and thalamus were under activated solely in adults with ADHD relative to controls." Hart et al. (2013) concluded that "patients with ADHD have consistent functional abnormalities in 2 distinct domain-dissociated right hemispheric fronto-basal ganglia networks, including the IFG, supplementary motor area, and ACC for inhibition and DLPFC, parietal, and cerebellar areas for attention."

A more recent metaanalysis (Kaiser et al., 2020) summarized the relevant literature on earlier (P100, N100, P200, N200, error-related negativity, ERN) versus later (P300, Error-Processing, and contingent negative variation, CNV) cognitive ERP differences (see Appendix for details) between children, adolescents, and adults with ADHD and without ADHD (non-ADHD). Via database search 52 relevant articles were identified including $N = 1576$ ADHD and $N = 1794$ non-ADHD individuals. Various task conditions included Cue, Go, and NoGo. [Cue: Stimulus presented to signal upcoming task, typically before target or NoGo stimulus. Go: Target stimulus presented that requires a response (e.g., motor). NoGo: Stimulus presented that requires to inhibit prepared or prepotent response (e.g., motor)]. For earlier components, individuals with ADHD showed shorter Go-P100 latencies than non-ADHD. For later ERPs, individuals with ADHD showed smaller Cue-P300 amplitudes, longer Go-P300 latencies, smaller NoGo-P300 amplitudes, longer NoGo-P300 latencies, smaller CNV amplitudes, and smaller Error-Processing amplitudes. Findings showed substantial heterogeneity and moderate effect sizes ($d < 0.6$) that limit the use for clinical application.

8.4.2.2 Auditory ERP data papers

Holcomb et al. (1986) investigated reading disabled children, ADHD children with hyperactivity, ADHD children without hyperactivity, and normal controls. They were instructed to press a button to a low probability tone (target, $P = 16.8\%$) and to ignore all other events which included a high-probability tone (nontarget, $P = 66.4\%$) and unexpected novel sounds ($P = 16.8\%$). They found that the amplitude of the P3b component was significantly smaller in all the clinical groups compared to controls. P3 latency did not differ from controls. Johnstone and Barry (1996) found that children with ADHD relative to control children had a P3b component to target stimuli that was smaller in the posterior region and larger in the frontal region for the ADHD group compared with the control group. Kemner et al. (1996) showed that ADDH children had smaller P3 amplitudes and (marginally) smaller MMN to auditory deviant stimuli, irrespective of task relevance, so smaller P3's in ADDH children are due to stimulus deviancy per se. Huttunen-Scott et al. (2008) found no MMN difference, but larger P3a, in an ADHD group than in controls. They argued that P3a, as an indication of attention switch, would be expected in children with ADHD, because of their known distractibility.

Gomes et al. (2012) examined electrophysiological and behavioral measures during an auditory selective attention task in ADHD children, TD children, and adults. Negative difference wave waveforms (Nd; see Appendix) were elicited from adults and children with TD, but strongly reduced from children with ADHD (Fig. 8.6). Further, those with ADHD exhibited significantly smaller auditory Ta responses. P3b's were elicited in all three groups by targets but not by unattended deviants. Children with ADHD showed poorer attention allocation, as measured by Nd and hits, but were not more distracted by unattended deviants, as measured by P3b and false alarms, than children with TD. Findings for Nd, P3b, and Ta considered together suggest that deficits in auditory selective attention in children with ADHD may be attributable to reduced information early in the processing stream.

Yang et al. (2015) noted that there are at least three processing steps underlying attentional control for auditory change detection, namely preattentive change detection (MMN), involuntary attention orienting (P3a), and attention reorienting (late discriminative negativity, LDN, Appendix) for further evaluation. Two types of stimuli—pure tones and Mandarin lexical tones—were used to examine if the deficits were general across linguistic and nonlinguistic domains. The results showed no MMN difference, but did show attenuated P3a and enhanced LDN to the large deviants for both pure and lexical tone changes in the ADHD group. Correlation analysis showed that children with higher ADHD tendency, as indexed by parents' and teachers' ratings on ADHD symptoms, showed less positive P3a amplitudes when responding to large lexical tone deviants. In an auditory oddball task, Janssen et al. (2016) showed reduced P3b (310–410 ms) amplitudes in response to targets in the ADHD group compared to the TD group. Reductions in P3b amplitudes were related to more inattention and hyperactivity/impulsivity problems in the ADHD group. Differences were located primarily in frontopolar (cingulo-opercular network, BA10) and temporo-parietal regions (ventral attention network, BA39 and 19) in the left hemisphere. Note that the extended cingulo-opercular network is composed of anterior insula/operculum, dorsal ACC, and thalamus. Its function has been particularly difficult to characterize due to the network's pervasive activity and frequent coactivation with other control-related networks (Sadaghiani and D'Esposito, 2015). Previous studies (e.g., Lee et al., 2019; Testo

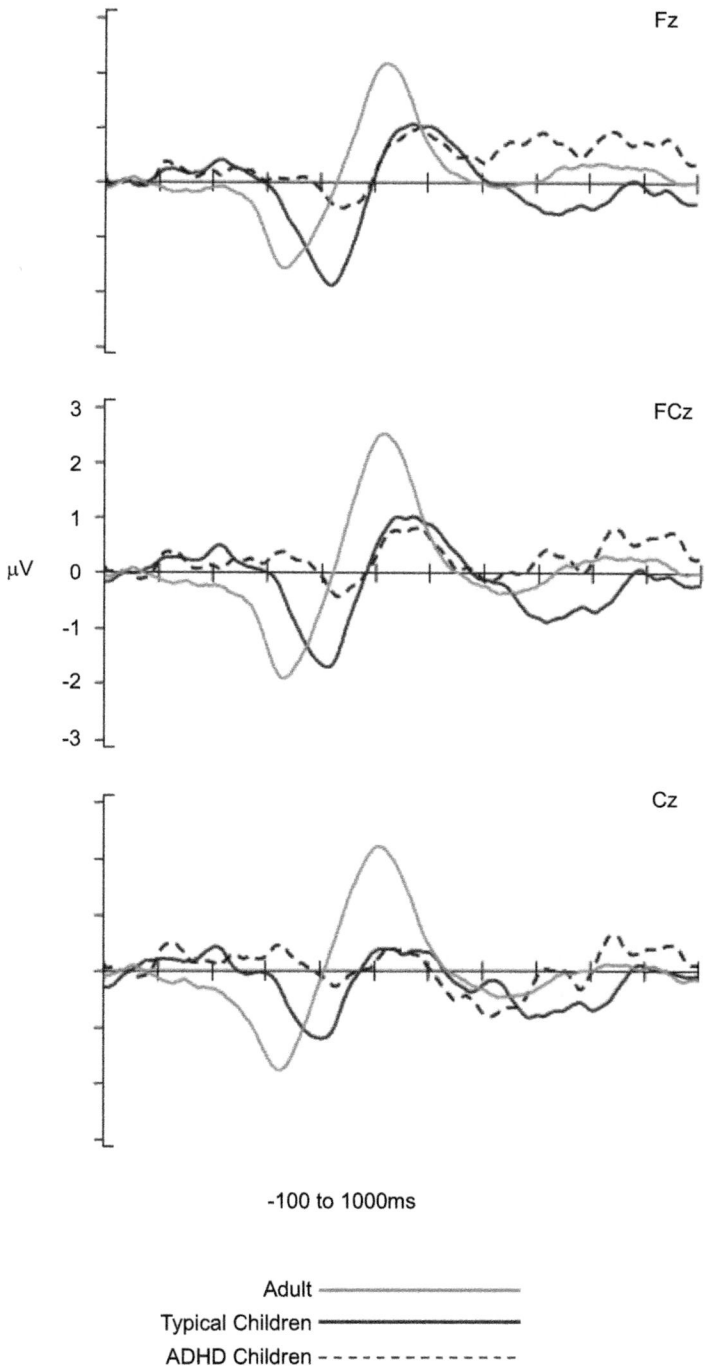

Fig. 8.6 See legend on next page.

et al., 2020) showed reduced activity of the ACC in children with attention-deficit/hyperactivity disorder (ADHD). Janssen et al. (2016) concluded with evidence for alterations in both top-down (frontopolar cortex) and bottom-up (inferior parietal/superior temporal cortex) attention-related brain areas, which may underlie P3b amplitude reductions in children with ADHD. These brain areas seem to play central roles in a cingulo-opercular and ventral attention network, respectively.

Yamamuro et al. (2016) correlated the amplitude and latency of MMN with the clinical severity of ADHD, as measured by the ADHD Rating Scale IV—Japanese version. Participants were treatment-naïve children and adolescents with ADHD and age- and sex-matched TD children. In the ADHD group, MMN amplitudes were reduced at the central electrode (Cz) and MMN latencies prolonged at the parietal electrode (Pz) relative to those in the control group. Furthermore, MMN amplitudes at Pz were negatively correlated with ADHD full-scale and hyperactivity–impulsivity and inattention subscale scores, and MMN latency at Pz was positively correlated with ADHD hyperactivity–impulsivity subscale scores. Similar results were obtained by Lee et al. (2020), Zhang et al. (2020), Hsieh et al. (2021), and Kim et al. (2021).

8.4.2.3 Visual ERP data papers

Liu et al. (2020) recruited children and adolescents with ADHD and age- and gender-matched healthy controls, ages 8–18 years, who performed an arrow flanker task, in which arrows appeared on a personal computer display with congruent (e.g., →→→→→) and incongruent (e.g., →→←→→) conditions during EEG recording. Compared to healthy controls, participants with ADHD responded more slowly and showed larger reaction time variability and reduced posterror slowing; they also exhibited reduced error-related negativity (ERN, Appendix) and error positivity effects and reduced N2 and P3 congruency effects (Fig. 8.7). With increasing age, participants

---

Fig. 8.6 Grand-mean difference waveforms (Nd) elicited from participants in all three groups obtained by subtracting the ERPs elicited by the standard tones when they were unattended from the ERPs elicited by the standard tones when they were attended at selected electrode sites (Fz, FCz, Cz). The gray, black, and dotted lines are the waveforms elicited from the adult, children with TD, and children with ADHD, respectively. (Reprinted from Gomes, H., Duff, M., Ramos, M., Molholm, S., Foxe, J.J., Halperin, J., 2012. Auditory selective attention and processing in children with attention-deficit/hyperactivity disorder. Clin. Neurophysiol. 123, 293–302. © 2011 International Federation of Clinical Neurophysiology, with permission from Elsevier.)

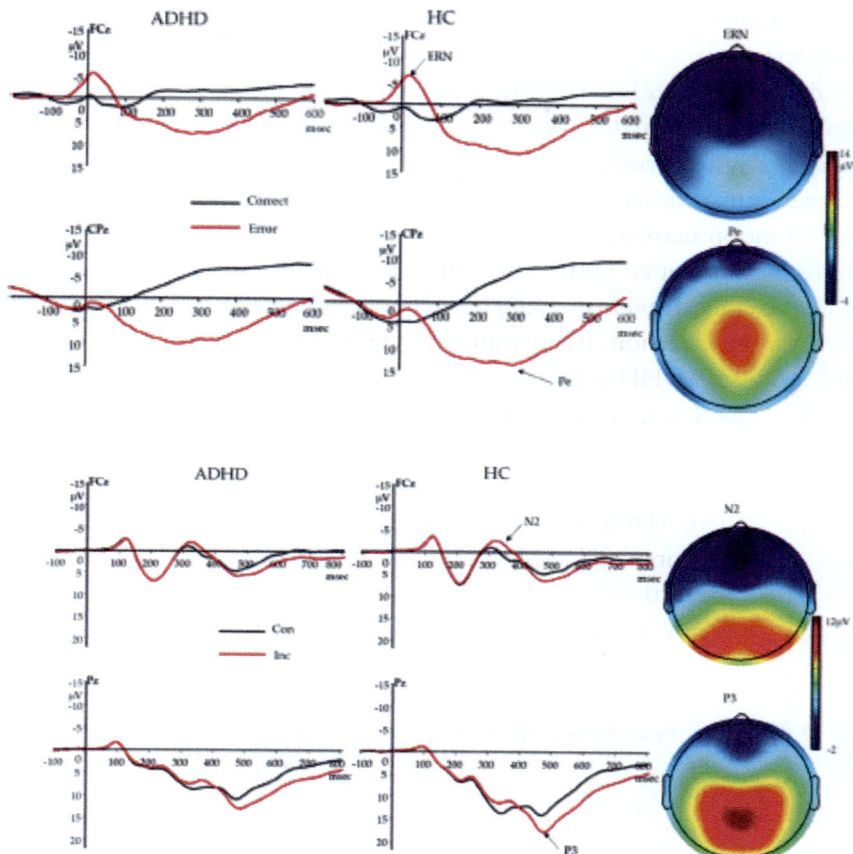

Fig. 8.7 Top. ERN and Pe waveforms for participants with ADHD and HC, and topography for error response (ERN: 0–80 ms; Pe: 200–400 ms; Baseline: −200 to −50 ms) in all participants. Responses occurred at 0 ms. Bottom. N2 and P3 waveforms for participants with ADHD and HC, and topography for incongruent correct trials (N2: 300–400 ms mean amplitude; P3: 400–600 ms mean amplitude; Baseline: −100 to 0 ms) in all participants. Stimuli onset occurred at 0 ms. *ADHD,* attention deficits/ hyperactive disorder; *Con,* congruent correct trials; *ERN,* error-related negativity; *HC,* healthy controls; *Inc,* incongruent correct trials; *Pe,* error positivity. *(From Liu, Y., Hanna, G.L., Hanna, B.S., Rough, H.E., Arnold, P.D., Gehring, W.J., 2020. Behavioral and electrophysiological correlates of performance monitoring and development in children and adolescents with Attention-Deficit/Hyperactivity Disorder. Brain Sci. 10, 79. https:// doi.org/10.3390/brainsci10020079. Open Access.)*

responded faster, with less variability, and with increased posterror slowing. Older participants also exhibited increased ERN effects and increased N2 and P3 congruency effects. Increased reaction time variability and reduced

P3 amplitude in incongruent trials were associated with increased ADHD Problems Scale scores on the Child Behavior Checklist across groups.

Zarka et al. (2021) investigated generators underlying alterations of ERP components triggered by visual Go/NoGo tasks in children with ADHD compared with typically developing (TD) children. Low–resolution electromagnetic tomography (LORETA) source analysis showed that lower Go-P3 component in children with ADHD was explained not only by a reduced contribution of the frontal areas but also by a stronger contribution of the anterior part of the caudate nucleus in these children compared with TD children. While the reduction of the NoGo-P3 component in children with ADHD was essentially explained by a reduced contribution of the dorsal ACC, the higher NoGo-P2 amplitude (Fig. 8.8) in these children was

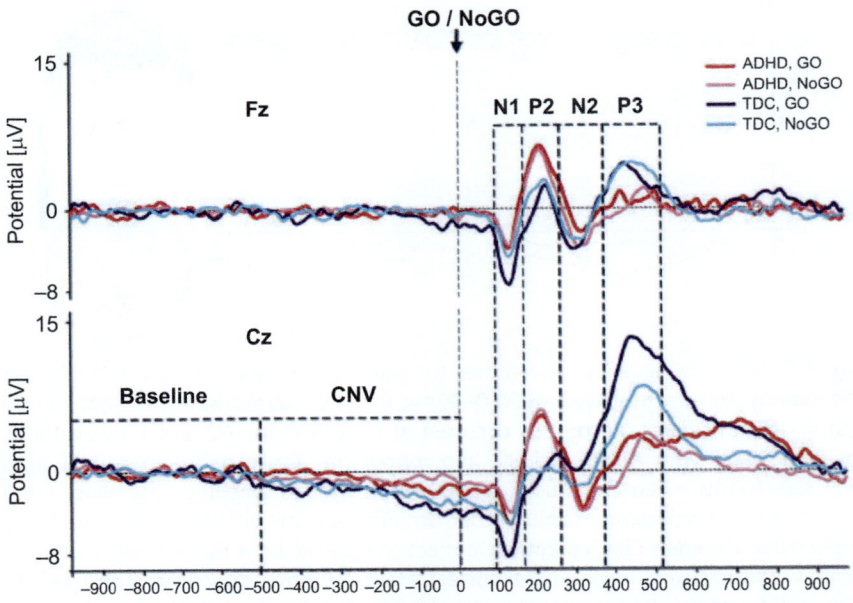

Fig. 8.8 Descriptive event-related potential (ERP) trace and components labeling. For each illustrated channel (Fz, Cz) the ERP traces (grand averaged) corresponding to the attention-deficit/hyperactivity disorder *(red)* and typically developing children *(blue)* groups in GO and NoGO conditions are superimposed. *(From Zarka, D., Cebolla, A.M., Cevallos, C., Palmero-Soler, E., Dan, B., Cheron, G., 2021. Caudate and cerebellar involvement in altered P2 and P3 components of GO/NoGO evoked potentials in children with attention-deficit/hyperactivity disorder. Eur. J. Neurosci. 53, 3447–3462. © 2021 Federation of European Neuroscience Societies and John Wiley & Sons Ltd., with permission from John Wiley and Sons.)*

concomitant to the reduced contribution of the dorsolateral prefrontal cortex, the insula, and the cerebellum.

8.4.2.4 Brain rhythms

Theta/beta amplitude ratios have been suggested as an index of arousal, but a number of studies (e.g., Barry et al., 2009) failed to find any association thereof in ADHD. Picken et al. (2020) tested again whether this could be confirmed in a sample of adults with the ADHD. They demonstrated that the theta/beta ratio correlated significantly with P300 latency, whereas alpha power did not correlate significantly with P300 amplitude or latency. Thus the theta/beta ratio may be a marker of cognitive processing capacity in both the general population and in participants with ADHD, and that the normal alpha/arousal linkage is anomalous in ADHD. Tombor et al. (2021) investigated the relationship of resting gamma activity with age, and found that resting gamma1 (30–39 Hz) increased with age and was significantly lower in ADHD than in control subjects from early adulthood. Tombor et al. (2021) also found no significant association between gamma2 (39–48 Hz) activity and age. Thus abnormal gamma power is present at all ages, highlighting the lifelong nature of ADHD. Cowley et al. (2022) analyzed neural correlates of attention—timing sensitivity and phase synchrony of response activations, and event-related (de)synchronization (ERS/ERD, respectively, increase and decrease in amplitude) of alpha and theta frequency band activity—in frontal and parietal scalp regions. The ADHD group showed significantly weaker target- and response-locked amplitudes that were strongly right lateralized at the N2 and weaker phase synchrony (longer reset poststimulus). They also manifested significantly less parietal prestimulus 8-Hz theta ERS, less frontal and parietal poststimulus 4-Hz theta ERS, and more frontal and parietal prestimulus alpha ERS during correct trials. Cowley et al. (2022) noted that these differences may reflect excessive modulation of endogenous activity by strong entrainment to stimulus (alpha), combined with deficient modulation by neural entrainment to task (theta).

8.4.2.5 MRI studies

Postema et al. (2021) performed the "largest ever" analysis of brain left–right asymmetry in ADHD in up to 1933 people with ADHD and 1829 unaffected controls. Asymmetry Indexes were calculated per participant for each bilaterally paired measure, and linear mixed effects modeling was applied separately in children, adolescents, adults, and the total sample, to test exhaustively for potential associations of ADHD with structural brain asymmetries. However, they found no evidence for altered caudate nucleus asymmetry in ADHD, in contrast to prior literature. In children, there was less rightward asymmetry of the total hemispheric surface area compared to controls. Lower rightward asymmetry of medial orbitofrontal cortex surface area in ADHD was similar to a recent finding for autism spectrum disorder. There were also some differences in cortical thickness asymmetry across age groups. In adults with ADHD, globus pallidus asymmetry was altered compared to those without ADHD. However, all effects were small (Cohen's d from -0.18 to 0.18) and would not survive study-wide correction for multiple testing.

To investigate the quantitative profiles of brain gray matter (GM) in pediatric drug-naïve ADHD patients using MRI, Su et al. (2022) enrolled drug-naïve pediatric ADHD and age- and gender-matched healthy controls. Quantitative parameters, T1 and T2 maps, were extracted from the MRI data. In general, T2 relaxation times are mainly influenced by tissue myelin content, while T1 relaxation times are related mainly (but not only) to the microstructural tissue integrity, including myelin content and iron content. Between-group quantitative maps were compared using a general linear model analysis. Pearson correlation analysis showed that compared with the healthy control group, altered T1 and T2 relaxation times in the ADHD group were mainly distributed in GM regions of the cerebellum, attention and execution control network, default mode network, and limbic areas. These altered T1, T2 values may reveal widespread micromorphology changes, i.e., brain iron deficiency, low myelin content, and enlarged vascular interstitial space in ADHD patients.

8.5 Do ERPs diagnose between ASD and ADHD?

Based on the data in Table 8.1, which show far more entries for ASD compared to ADHD, I can only draw some putative conclusions. Increased ABR I-V intervals are similar in both groups. Amplitude decrease in N2 is similar as is MMN amplitude reduction and latency increase. P3a and P3b amplitude decrease is also similar, but with more data in ADHD. Based on these data the putative conclusion is that ERPs in ASD and ADHD are not distinguishing. Let's look at some studies that include potentially more specific mismatch—particularly temporal processing—tests and compare ASD and ADHD participants in the same study.

Since children with ASD (Wallace and Stevenson, 2014; Boets et al., 2015; Foss-Feig et al., 2017, 2018) and those with ADHD (Mioni et al.,

2019; Anobile et al., 2022) show different temporal processing deficits, one would expect that duration-discrimination MMNs would have been used more often. However (Table 8.2), duration MMNs were only measured for ASD and not for ADHD, but much less as for frequency, and did not differ much from those based on frequency discrimination. Again, consistent findings are shown in bold font. Notably, Boets et al. (2015) observed impaired frequency discrimination in the ASD group and suggestive evidence of poorer temporal resolution as indexed by gap–in–noise detection thresholds (see also Eggermont, 2015). Foss-Feig et al. (2017) observed that children with ASD had substantially higher auditory gap detection thresholds compared to children with TD, and auditory gap detection thresholds were correlated significantly with several measures of language processing in this population.

Lim et al. (2015) scanned predominantly medication-naïve male adolescents with ADHD, medication-naïve male adolescents with ASD, and age-matched healthy male controls using high-resolution T1-weighted volumetric imaging in a 3-T MRI scanner. Voxel-based morphometry was used to test for group-level differences in structural gray matter and white matter volumes. They found a significant group difference in the gray matter of the right posterior cerebellum and left middle/superior temporal gyrus (MTG/STG). Post hoc analyses revealed that this was due to ADHD boys having a significantly smaller right posterior cerebellar gray matter volume compared to healthy controls and ASD boys, who did not differ from each other. ASD boys had a larger left MTG/STG gray matter volume relative to healthy controls and at a more lenient threshold relative to ADHD boys. They found no white matter differences between ASD and ADHD. Lukito et al. (2018) used fMRI to compare the functional correlates of duration discrimination between young adult males with ASD, ADHD, the comorbid condition of ASD+ADHD, and typical development using both region of interest (ROI) and whole-brain analyses. Both the ROI and the whole-brain analyses showed that compared to controls and for the ROI analysis relative to the other patient groups, the comorbid ASD+ADHD group had significant underactivation in right IFG—a key region for duration discrimination that is typically underactivated in boys with ADHD (Fig. 8.9).

8.6 Does predictive coding contribute?

Because there are only a few predictive coding (PC) studies applied to ADHD, we will consider them only in comparison to those for ASD.

Table 8.2 ASD and ADHD children.

	Autism			ADHD		
	Frequency	Duration	Phoneme	Frequency	Duration	Phoneme
MMN	↓1,2,8,9,13,16,26 ↑3,4,10,23,25 ≈7 ←1,26 →5,16,22	↓8,16,23 ≈7 →16	↓6,11,17 ↑3 ≈4,7,9,23 →12,14,15,22	↓43,44,45 ≈31,33,40,41 →43,44		
P3a	↓1,2,18,19,20,24,27,28 ↑4,8,25,26 ≈9,13		↓4,9,11,24 ↑12	↓31,34,35,36,37 ↓38,39,41,45 →39		
P3b	↓21			↓29,30,42 ↓32 ↑29 ≈29		

ERPs split by oddball stimulus type.

↓↑≈ amplitude changes; ←→ latency changes.

ASD: [1]Gomot et al. (2002), [2]Ferri et al. (2003), [3]Lepistö et al. (2008), [4]Whitehouse and Bishop (2008), [5]Roberts et al. (2011), [6]Ludlow et al. (2014), [7]Weismüller et al. (2015), [8]Vlaskamp et al. (2017), [9]Huang et al. (2018), [10]Leno et al. (2018), [11]Lindström et al. (2018), [12]Riva et al. (2018), [13]Goris et al. (2018), [14]Matsuzaki et al. (2019a), [15]Matsuzaki et al. (2019b), [16]Di Lorenzo et al. (2020), [17]Ruiz-Martínez et al. (2020), [18]Novick et al. (1980), [19]Niwa et al. (1983), [20]Oades et al. (1988), [21]Lincoln et al. (1993), [22]Oram Cardy et al. (2005), [23]Lepistö et al. (2005), [24]Irwin et al. (2017), [25]Hudac et al. (2018), [26]Charpentier et al. (2018), [27]Donkers et al. (2020), [28]Loiselle et al. (1980).

ADHD: [29]Holcomb et al. (1986), [30]Johnstone and Barry (1996), [31]Kemner et al. (1996), [32]Jonkman et al. (1997), [33]Oades et al. (1996), [34]Barry et al. (2003), [35]Brown et al. (2005), [36]Gumenyuk et al. (2005), [37]Alexander et al. (2008), [38]Senderecka et al. (2012), [39]Tsai et al. (2012), [40]Gomes et al. (2013), [41]Yang et al. (2015), [42]Janssen et al. (2016), [43]Yamamuro et al. (2016), [44]Lee et al. (2020), [45]Zhang et al. (2020).

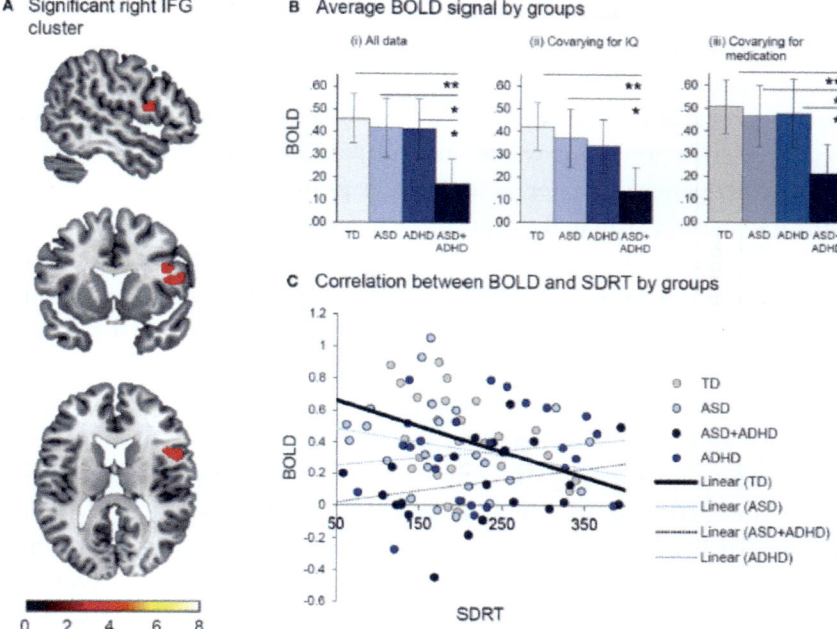

Fig. 8.9 Between-group effects of duration discrimination vs temporal order judgment. (A) Significant right IFG cluster where group effect was found. (B) Average BOLD signal by group is displayed (i) for all data, (ii) after covarying for IQ, and (iii) covarying for medication. (C) Correlation between BOLD and SDRT by group indicates that reduced BOLD is associated with increase in standard deviation of response time (SDRT). *P < .05, **P < .01. *(From Lukito, S.D., O'Daly, O.G., Lythgoe, D.J., Whitwell, S., Debnam, A., Murphy, C.M., et al., 2018. Neural correlates of duration discrimination in young adults with autism spectrum disorder, attention-deficit/hyperactivity disorder and their comorbid presentation. Front. Psychiatry 9, 569. https://doi.org/10.3389/fpsyt.2018. 00569. Open Access.)*

8.6.1 Predictive coding and ASD

Lawson et al. (2014) discussed "A recent and thought-provoking article presented a normative explanation for the perceptual symptoms of autism in terms of a failure of Bayesian inference" (Pellicano and Burr, 2012). Lawson et al. (2014) suggested that when Bayesian inference is grounded in predictive coding, many features of autistic perception can be attributed to aberrant precision (or beliefs about precision) within the context of hierarchical message passing in the brain (Chapters 1 and 7). A key belief underlying this explanation is that high sensory precision in autism, relative to prior precision, may be caused by a failure of sensory attenuation. In other words, an inability to contextualize sensory information renders sensory prediction errors too precise and context insensitive. Other predictive coding accounts (Van de Cruys et al., 2014) suggest that a key deficit in ASD concerns the inflexibility in modulating local prediction errors as a function of global top-down expectations. Goris et al. (2018) used EEG to investigate whether local prediction error processing was less modulated by global context (i.e., global stimulus frequency; Chapter 7) in ASD. A group of adults with ASD was compared with a matched group on a hierarchical auditory oddball task in which participants listened to short sequences of either five identical sounds (local standard) or four identical sounds and a fifth deviant sound (local deviant). The latter condition is known to generate the MMN, believed to reflect early sensory prediction error processing. Both groups showed an MMN that was modulated by global context (cf. Fig. 2.11). However, this effect was smaller in the ASD group as compared with the typically developed group. In contrast, the P3b, as a marker of conscious expectation processes, did not differ across groups. Grisoni et al. (2019) investigated ERP indicators of sound processing, semantic understanding, and predictive coding and their correlation with clinical symptoms of high-functioning ASD engaged in an auditory, passive listening, task. Early MMN-like responses to words (latency ~120 ms) were differentially modulated across groups: controls showed larger amplitudes for words in action sound compared to nonaction contexts, whereas ASD participants demonstrated enlarged early MMN-like responses only in a pure-tone context, with no other modulation dependent on action sound context. Late MMN-like responses around 560-ms poststimulus onset revealed body-part-congruent action-semantic priming for words in control participants, but not in the ASD group. The data suggest that high-functioning adults with ASD show a specific deficit in semantic processing and predictive coding of sounds and words related to action, which is absent for neutral, non-action, sounds.

Knight et al. (2020) explored the relationship between MMN and the predictive coding of pattern complexity in ASD. They tested the hypotheses that (1) individuals with ASD have reduced neural MMN responses to auditory stimuli that deviate in presentation timing from expected patterns, particularly as pattern complexity increases and shows higher baseline entropy and (2) MMN amplitude is inversely correlated with the level of impairment in social communication and repetitive behaviors in individuals with ASD. EEG was acquired as ASD and TD individuals (age 6–21 years) listened to repeated five–rhythm tones that varied in the Shannon entropy of the rhythm across three conditions (zero, medium 1-bit, and high 2-bit entropy). The majority of the tones conformed to the established rhythm (standard tones); occasionally the fourth tone was temporally shifted relative to its expected time of occurrence (deviant tones). Social communication and repetitive behaviors were measured using the Social Responsiveness Scale and Repetitive Behavior Scale-Revised. Both neurotypical controls and individuals with ASD showed stepwise decreases in MMN as a function of increasing entropy. "Contrary to the result forecasted by a predictive coding hypothesis, individuals with ASD do not differ from controls in these neural mechanisms of prediction error to auditory rhythms of varied temporal complexity, and there is no relationship between these signals and social communication or repetitive behavior measures" (Knight et al., 2020).

8.6.2 Comparing ASD and ADHD patients in the same study

8.6.2.1 Brain rhythms

Shephard et al. (2019) examined similarities and differences in large-scale oscillatory neural networks between boys aged 8–13 years with ASD, ADHD, ASD+ADHD, and TD controls. Oscillatory neural networks were computed using graph-theoretical methods from EEG data collected during an eyes-open resting state and attentional control and social cognition tasks. They found that children with ASD showed significant *hypoconnectivity* in large-scale networks during all three task conditions compared to children without ASD. In contrast, children with ADHD showed significant *hyperconnectivity* in large-scale networks during the attentional control and social cognition tasks, but not during the resting state, compared to children without ADHD. Children with cooccurring ASD+ADHD did not differ from children with ASD when paired with this group and vice versa when paired with the ADHD group, indicating that these children showed both ASD-like hypoconnectivity and ADHD-like hyperconnectivity. Shephard et al.'s (2019) findings suggest that ASD and ADHD are associated with distinct alterations in large-scale oscillatory networks, and these present together in children with both disorders. These alterations appear to be task independent in ASD but task related in ADHD and may underlie other neurocognitive aspects in these disorders.

8.6.2.2 Predictive coding

Gonzalez-Gadea et al. (2015) stated that "In ADHD, although no empirical or theoretical arguments based on predictive coding have been stated, deficits in top-down expectation could explain the observed symptoms of inattention and distractibility. Specifically, difficulties in generating predictions would increase reliance on novel sensory evidence. Accordingly, ADHD individuals (and contrary to ASD subjects) exhibit higher or even exaggerated neural responses to novel/unexpected stimuli (Gumenyuk et al., 2005) and lower responses to expected cues (Marzinzik et al., 2012). Additionally, abnormal PFC activation and executive dysfunction (Hart et al., 2013) could be related to difficulties in top-down expectation."

To provide further evidence, Gonzalez-Gadea et al. (2015) monitored brain dynamics of children (8–15 yr old) who had ASD or ADHD or who were control participants via high-density EEG. They performed analysis at the scalp and source-space levels while participants listened to standard and deviant tone sequences. Through task instructions, they manipulated top-down expectation by presenting expected and unexpected deviant sequences. Children with ASD showed reduced superior frontal cortex (SFC) responses to unexpected events but increased dorsolateral prefrontal cortex (DLPFC) activation to expected events. In contrast, children with ADHD exhibited reduced cortical responses in SFC to expected events but strong PFC activation to unexpected events. Based on the predictive coding account, top-down expectation abnormalities could be attributed to a disproportionate reliance (precision) allocated to prior beliefs in ASD and to sensory input in ADHD.

8.7 Developmental dyslexia

Developmental dyslexia affects around 5% of schoolchildren and is a specific learning disability of reading and spelling, which cannot be attributed to low intellectual ability or inadequate schooling (Lehongre et al., 2011). Intact phonological processing is crucial for successful literacy acquisition. Consequently, we will emphasize the auditory aspects of dyslexia. Three phonological symptoms of the deficit can be distinguished: (1) poor ability to pay attention to, and mentally manipulate individual speech sounds; (2) poor verbal short-term memory, i.e., the ability to repeat, for instance, pseudo-words or digit series; and (3) slow performance in naming a series of pictures, colors, or digits (Lehongre et al., 2011).

Díaz et al. (2012) noted: "There are various theories about the underlying cause of dyslexia. One of the most influential is the phonological deficit hypothesis, which posits that dyslexic persons have difficulties in processing and manipulating small units of speech, i.e., speech sounds or phonemes. The relevance of task-induced difficulties is also the hallmark of the sluggish attention shifting hypothesis, which proposes that dyslexics are impaired in shifting attention in tasks that have a fast rate of stimulus presentation (Hari and Renvall, 2001). In contrast, the magnocellular theory, another influential view, posits that abnormal medial geniculate body (MGB) function is the origin of poor phonological processing in dyslexic persons (Stein and Walsh, 1997). This theory is largely based on findings that the left MGB is altered in post-mortem brains of dyslexics. The magnocellular theory originally proposed that sensory deficits in fast-conducting visual [magnocellular] thalamic neurons are at the basis of dyslexia and predicts that dyslexics' deficits are not specific to linguistic processing, but that they extend to sensory processing in general (Livingstone et al., 1991)."

8.7.1 MMN and developmental dyslexia

8.7.1.1 Single-study findings

While MMN was not affected for oddball-tone stimuli, a significantly attenuated MMN in the dyslexic group for speech stimuli was found, specifically MMN amplitude was smaller in the time window of 225–600 ms (Schulte-Körne et al., 1998, 1999). In contrast, Baldeweg et al. (1999) found abnormal MMNs to changes in tone frequency but not to changes in tone duration in dyslexic subjects. Kujala et al. (2003) evaluated cortical auditory discrimination in dyslexia and found that the MMN elicited by pitch change was diminished over the left hemisphere of the dyslexic individuals suggesting left hemisphere auditory dysfunction. Kujala et al. (2006) compared MMNs in a 5-deviant paradigm, and found diminished pitch MMN and an enhanced location MMN in dyslexic individuals. Comparing children with reading difficulties (RD) with controls, Alonso-Búa et al. (2006) found significant longer MMN latency only in the task involving linguistic stimuli in the dyslexics. The LDN component (Appendix) showed lower amplitudes and delayed latencies in the dyslexics group during the processing of complex tones and linguistic stimuli.

In concert with temporal processing difficulties (Eggermont, 2015), MMN amplitudes in dyslexic children (mean age 11.8 ± 1.8 yrs) differed significantly from controls in the duration condition, whereas no differences were found in the frequency and phoneme conditions. In addition, the dyslexic children had delayed MMN in the all contrast conditions (Corbera et al., 2006). Similar findings were reported by Huttunen et al. (2007). However, for dyslectics with and without attention deficit (AD), Huttunen–Scott et al. (2008) found significant differences in the MMN in the right hemisphere in the reading disorder (RD) group, in all frontal and central channels in the RD+AD group, and the MMN peaks appeared earlier in frontal channels in the AD group. Hämäläinen et al. (2008) observed that control children produced larger responses than children with RD to pitch change in the P3a component but only when the sounds in the pair were close to each other. Compared to children with reading disabilities, MMN was smaller and LDN larger in control children in response to rise time change when the sounds in the pair were further apart. The nonoverlap in ERP measures between the groups was 40%–50%. Chobert et al. (2012) found no between-group differences for frequency deviants. However, whereas normal-reading children showed larger MMNs to large than to small deviants in vowel duration and voice onset time (VOT), no such deviance size effect was found in children with dyslexia.

In a three-deviant passive "oddball" paradigm and a corresponding blocked "deviant-alone" control condition, Stefanics et al. (2011) recorded ERPs to tones varying in rise time, duration, and intensity in children with dyslexia and typically developing children longitudinally. They reported results from test Phases 1 and 2, when participants were aged 8–10 years, and found an MMN to duration, but not to rise time or intensity deviants, at both time points for both groups. There was a slower frontocentral P1 response in the dyslexic group compared to controls. The amplitude of the P1 frontocentrally to tones with slower rise times and lower intensity was smaller compared to tones with sharper rise times and higher intensity in the oddball condition, for children with dyslexia only. The latency of this ERP component for all three stimuli was shorter on the right compared to the left hemisphere, but only for the blocked condition. Furthermore, Stefanics et al. (2011) found decreased N1 amplitude to tones with longer rise times compared to tones with shorter rise times for children with dyslexia, but only in the oddball condition. They concluded that "Although learning to read is sometimes considered a visual task, reading is parasitic on the spoken language system, and the data reported here suggest that clear instruction in the acoustic similarities and differences between spoken words is likely to be a useful aspect of reading instruction."

8.7.1.2 Systematic reviews

From a literature search, Bishop (2007) found 26 studies of the MMN in individuals with dyslexia or specific language impairment and 4 studies of infants or children at familial risk of these disorders. Findings were highly inconsistent. Overall, attenuation of the MMN and atypical lateralization in the clinical group were most likely to be found in studies using rapidly presented stimuli, including nonverbal sounds. "The MMN literature offers tentative support for the hypothesis that auditory temporal processing is impaired in language and literacy disorders, but the field is plagued by methodological inconsistencies, low reliability of measures, and low statistical power." Volkmer and Schulte-Körne (2018) conducted a systematic literature search that resulted in 17 studies reporting MMR to speech or nonspeech stimuli in children at risk of dyslexia. The results of the studies were inconsistent. Studies measuring speech MMR often found attenuated amplitudes in the at-risk group, but mainly in very young children. The results for older children (6–7 years) and for nonspeech stimuli were more heterogeneous. A moderate positive correlation of MMR amplitude size with later reading and spelling abilities was consistently found. Gu and Bi (2020) used a metaanalysis to quantitatively identify auditory processing deficits in individuals with dyslexia. They analyzed 81 results within 25 publications that employed passive oddball paradigms and identified MMN impairment in speech processing in children (Cohen's $d = 0.296$) and adults with dyslexia (Cohen's $d = 0.486$). Adults with dyslexia also showed atypical auditory processing of nonspeech (Cohen's $d = 0.409$). These moderate effect sizes ($d < 0.6$) limit the use for clinical application. [Cohen's d is an effect size used to indicate the standardized difference between two means.] These findings suggest for dyslexics that the auditory processing deficit in speech will persist into adulthood, combined with an auditory processing deficit is general in adults with dyslexia.

8.7.2 Role of the thalamus and potential of predictive coding in dyslexia

Using fMRI, Díaz et al. (2012) investigated whether phonological difficulties in dyslexia are associated with a dysfunction of the MGB in the thalamus. They found that, in dyslexic adults, the MGB responded abnormally when the task required attending to phonemes compared with other speech features, such as rapid automatic naming. Task-dependent modulation of left MGB activity correlated with dyslexia diagnostic scores, indicating that the task—rapid automatized naming, RAN—modulation of the MGB is critical for performance in dyslexics (Fig. 8.10).

Díaz et al. (2012) observed that "No other structure in the auditory pathway showed distinct functional neural patterns between the two tasks for dyslexic and control participants. These results suggest that deficits in dyslexia are associated with a failure of the neural mechanism that dynamically tunes MGB according to predictions from cortical areas to optimize speech

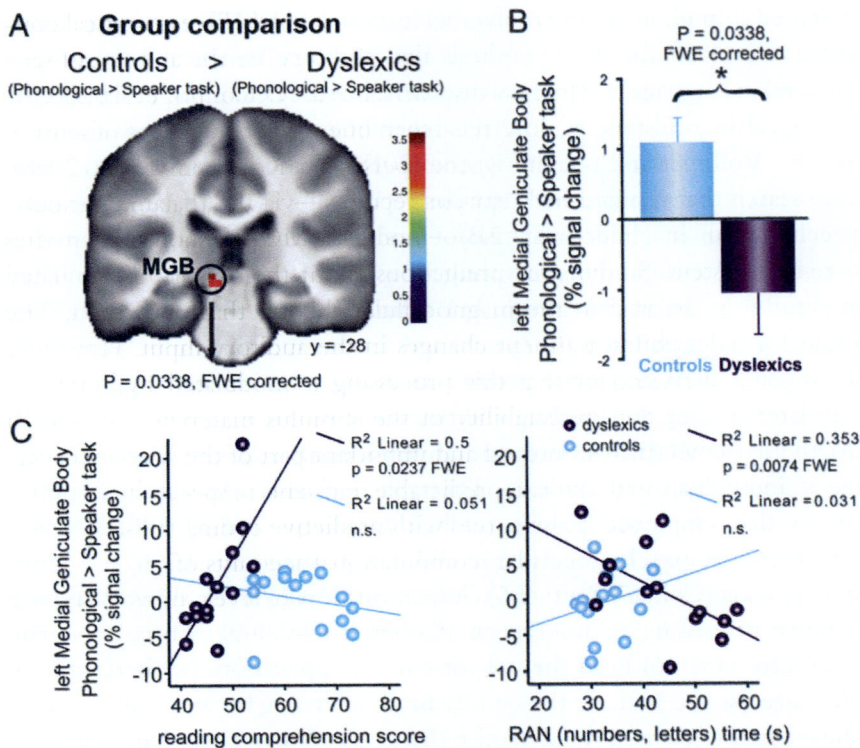

Fig. 8.10 Differences between normal readers and dyslexics in MGB activity during speech processing. (A) The between-group comparison reveals greater task-dependent modulation in the left MGB for control vs dyslexic participants ($P = .0338$, FWE corrected for MGB; local activation maxima, -9, -28, -8, MNI coordinates). The *color bar* represents t-values. (B) Estimated signal changes (as percentages) extracted from the left MGB for the contrast "phonological task – speaker task" for controls and dyslexics. Error bars represent SE. (C) Task-dependent modulation of the left MGB correlates positively with reading comprehension and negatively with time required for phonological access (i.e., rapid automatized naming, RAN) in dyslexic but not control participants. *(From Díaz, B., Hintz, F., Kiebel, S.J., von Kriegstein, K., 2012. Dysfunction of the auditory thalamus in developmental dyslexia. Proc. Natl. Acad. Sci. U. S. A. 109 (34), 13841–13846.)*

processing. This view on task–related MGB dysfunction in dyslexics has the potential to reconcile influential theories of dyslexia within a predictive coding framework of brain function." In that regard, they proposed that "sensory thalamic structures receive massive cortical backward connections, which are believed to play an important role in the processing of sensory information. Currently, there are two views, which are not necessarily in opposition, on the functional role of these corticothalamic connections.

One view is that the sensory thalamus is modulated by efferent cortical connections depending on attentional demands to regulate the amount of sensory information that is forwarded to cortical areas (O'Connor et al., 2002)."

The findings of Díaz et al. (2012) are in line with the concept of sensory thalamic structures as dynamic gatekeepers (von Kriegstein et al., 2008). They assume that cortical efferent connections—via the thalamic reticular nucleus (Chapter 4, Section 4.2.3)—modulate MGB response properties by providing specific dynamic predictions about the sensory input. They propose, as postulated in the magnocellular theory, that one role of the MGB is to process fast transient changes in the auditory input. However, in addition, they suggest that this processing is modulated by cognitive requirements and the predictability of the stimulus material. This would mean that the MGB is an integral and important part of the speech processing hierarchy, dealing with fast, predictable transients in speech signals. This adaptive role of the MGB also agrees with predictive coding (Friston, 2005; Kiebel et al., 2008). In particular, computational accounts of speech recognition suggest that, for optimal recognition, lower levels of the auditory hierarchy (e.g., the MGB) must process faster dynamics than higher levels (e.g., auditory cortex; Balaguer-Ballester et al., 2009; Kiebel et al., 2009). For evidence, see also Wehr and Zador (2005) and Chapter 3. This computational scheme of recognition posits that cortical predictions should optimize auditory processing at early, subcortical stages of the auditory processing hierarchy, especially for fast, complex, and highly predictable stimuli such as speech sounds (Díaz et al., 2012).

Jaffe-Dax et al. (2015) hypothesized that the perceptual deficits associated with dyslexia can be understood as poorly integrating prior information with noisy observations. They analyzed the performance of human participants in an auditory discrimination task using a two-parameter computational model. One parameter captures the internal noise in representing the current event, and the other captures the impact of recently acquired prior information. The model suggests that dyslexics' perceptual deficit can be accounted for by low weighting of their implicit memory of past trials relative to their internal noise. Underweighting the stimulus statistics decreased dyslexics' ability to compensate for noisy observations. ERP measurements (P2 component) while participants watched a silent movie indicated that dyslexics' perceptual deficiency may stem from poor automatic integration of stimulus statistics. Jaffe-Dax et al.'s (2015) computational model is tightly related to hypotheses that associate dyslexics' difficulties with a failure to make effective predictions that can facilitate task performance ("predictive coding";

Díaz et al., 2012). However, it is also compatible with hypotheses that dys-lexics are less resilient to external noise (Conlon et al., 2012; Partanen et al., 2012; the "noise exclusion hypothesis"). According to the Bayesian frame-work underpinning the "Implicit Memory Model," the prior information is used to compensate for the noise in the representation of the stimuli. Jaffe–Dax et al. (2015) found that dyslexics do not properly adjust the weight of previous trials to the level of internal noise. Subsequently, Jaffe–Dax et al. (2017) hypothesized that implicit memory decays faster among dyslexics (Fig. 8.11). They tested this by increasing the temporal intervals between consecutive trials, and by measuring the behavioral impact and ERP responses from the auditory cortex. Dyslexics showed a faster decay of implicit memory effects on both measures. For instance, N1 amplitude decreased 3–4 times faster in DD compared to controls, whereas for P2

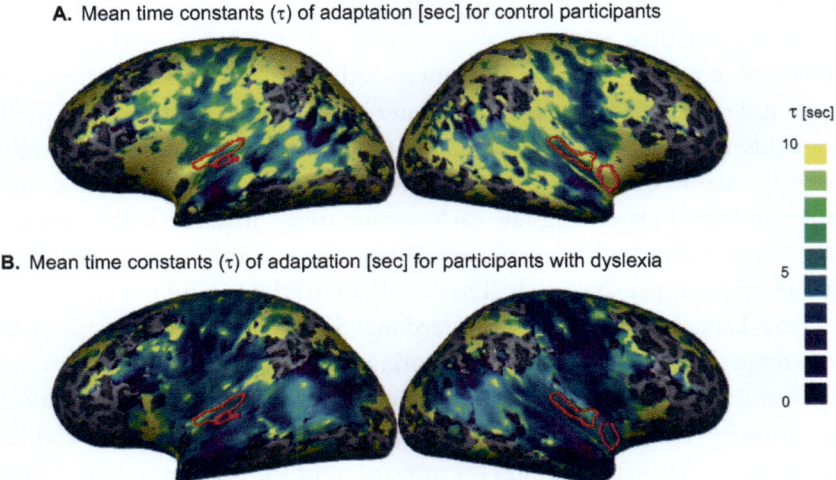

A. Mean time constants (τ) of adaptation [sec] for control participants

B. Mean time constants (τ) of adaptation [sec] for participants with dyslexia

Fig. 8.11 Cortical distribution of the groups' mean estimated time constants (τ) of adaptation, calculated separately for each of the responding voxels. (A) Control participants. (B) Participants with dyslexia. The estimated τs for participants with dyslexia were consistently shorter than those estimated for the control group. Significant group differences in the whole-brain analysis (Monte-Carlo cluster-level corrected: cluster threshold of 44 voxels) are outlined in *magenta*. The left and right primary auditory cortices, which were estimated as a source of P2 (ERP) component, are outlined in *orange*. An ROI analysis (see text) revealed a significant group difference in the left primary auditory cortex. *(From Jaffe-Dax, S., Kimel, E., Ahissar, M., 2018. Shorter cortical adaptation in dyslexia is broadly distributed in the superior temporal lobe and includes the primary auditory cortex. eLife 7, e30018. https://doi.org/10.7554/eLife.30018. Open Access.)*

the decrease was about 1.75 times faster. They found that shorter time constants in dyslexia are widespread across neocortex, although the most significant effects are found in the left superior temporal lobe, including the auditory cortex. This broad distribution suggests that the faster decay of implicit memory of individuals with dyslexia is a general characteristic of their cortical dynamics, which also affects sensory cortices. Jaffe-Dax et al. (2017) proposed "that dyslexics' shorter time constant of adaptation reflects a shorter time constant of implicit sound integration. Dyslexics' faster decay decreases the time constant of their integration of sound statistics, and consequently reduces the reliability of their implicitly calculated priors. Owing to their noisier priors, attaining the same level of processing reliability requires the extraction of more on-line information, which requires more processing time, leading to the stereotypic 'slower processing time' in dyslexia."

8.8 Summary

Although comorbidity between ASD and ADHD is high, differences in network connectivity are found. Children with ASD show significant *hypoconnectivity* in large-scale networks for attentional control, social cognition tasks, and resting state compared to children without ASD. In contrast, children with ADHD showed significant *hyperconnectivity* in large-scale networks of these attentional control and social cognition tasks, but not during the resting state, compared to children without ADHD. Children with ASD showed reduced superior frontal cortex (SFC) responses to unexpected events but increased dorsolateral prefrontal cortex (DLPFC) activation to expected events. In contrast, children with ADHD exhibited reduced cortical responses in SFC to expected events but strong PFC activation to unexpected events. In fMRI studies, a comparison of the functional correlates of duration discrimination between young adult males with ASD, ADHD, the comorbid condition of ASD+ADHD, and typical development showed that compared to controls the comorbid ASD+ADHD group had significant underactivation in right IFG—a key region for duration discrimination that is typically underactivated in boys with ADHD.

Based on the predictive coding account, top-down expectation abnormalities could be attributed to a disproportionate reliance (precision) allocated to prior beliefs in ASD and to sensory input in ADHD. Predictive coding also suggests that a key deficit in ASD concerns the inflexibility in modulating local prediction errors as a function of global top-down expectations. However, contrary to the result forecasted by a predictive coding hypothesis, individuals with ASD do not differ from controls in these neural mechanisms of prediction error to auditory rhythms of varied temporal complexity, and there is no relationship between these signals and social communication or repetitive behavior measures.

In concert with temporal processing difficulties, MMN amplitudes in dyslexic children differ significantly from controls in the duration-oddball condition, whereas no differences are found in the frequency and phoneme conditions. Phonological difficulties in dyslexia are associated with a dysfunction of the auditory thalamus. Even in dyslexic adults, the MGB responds abnormally when the task required attending to phonemes compared with other speech features, such as rapid automatic naming. MGB activity correlated with dyslexia diagnostic scores, indicating that the task modulation of the MGB is critical for performance in dyslexics. Deficits in dyslexia are associated with a failure of the neural mechanism that dynamically tunes MGB according to predictions from cortical areas to optimize speech processing. This view on task-related MGB dysfunction in dyslexics has the potential to reconcile influential theories of dyslexia within a predictive coding framework of brain function.

References

Alcántara, José, Weissblatt, Emma, Moore, Brian, Bolton, Patrick, 2004. Speech-in-noise perception in high-functioning individuals with autism orAsperger's syndrome. Journal of Child Psychology and Psychiatry 45 (6), 1107–1114.

Alexander, D.M., Hermens, D.F., Keage, H.A.D., Clark, C.R., Williams, L.M., Kohn, M.-R., et al., 2008. Event-related wave activity in the EEG provides new marker of ADHD. Clin. Neurophysiol. 119, 163–179.

Alonso-Búa, B., Díaz, F., Ferraces, M.J., 2006. The contribution of AERPs (MMN and LDN) to studying temporal vs. linguistic processing deficits in children with reading difficulties. Int. J. Psychophysiol. 59, 159–167.

Anobile, G., Bartoli, M., Pfanner, C., Masi, G., Cioni, G., Tinelli, F., 2022. Auditory time thresholds in the range of milliseconds but not seconds are impaired in ADHD. Sci. Rep. 12, 1352. https://doi.org/10.1038/s41598-022-05425-2.

Aykan, S., Gürses, E., Tokoz-Yilmas, S., Kalaycioglu, C., 2020. Auditory processin differences correlate withautisyic traint in males. Font. Hum. Neurosci. 14, 584784. https://doi.org/10.3389/fhum.2020.584704.

Balaguer-Ballester, E., Clark, N.R., Coath, M., Krumbholz, K., Denham, S.L., 2009. Understanding pitch perception as a hierarchical process with top-down modulation. PLoS Comput. Biol. 5, e1000301.

Baldeweg, T., Richardson, A., Watkins, S., Foale, C., Gruzelier, J., 1999. Impaired auditory frequency discrimination in dyslexia detected with mismatch evoked potentials. Ann. Neurol. 45, 495–503.

Barry, R.J., Clarke, A.R., Johnstone, S.J., 2003. A review of electrophysiology in attention-deficit/hyperactivity disorder: I. Qualitative and quantitative electroencephalography. Clin. Neurophysiol. 114, 171–183.

Barry, R.J., Clarke, A.R., McCarthy, R., Selikowitz, M., Brown, C.R., Heaven, P.C.L., 2009. Event-related potentials in adults with Attention-Deficit/Hyperactivity Disorder: an investigation using an inter-modal auditory/visual oddball task. Int. J. Psychophysiol. 71, 124–131.

Ben-Sasson, A., Hen, L., Fluss, R., Cermak, S.A., Engel-Yeger, B., Gal, E., 2009. A meta-analysis of sensory modulation symptoms in individuals with autism spectrum disorders. J. Autism Dev. Disord. 39, 1–11. https://doi.org/10.1007/s10803-008-0593-3.

Bishop, D.V.M., 2007. Using mismatch negativity to study central auditory processing in developmental language and literacy impairments: where are we, and where should we be going? Psychol. Bull. 133 (4), 651–672.

Boets, B., Verhoeven, J., Wouters, J., Steyaert, J., 2015. Fragile spectral and temporal auditory processing in adolescents with autism spectrum disorder and early language delay. J. Autism Dev. Disord. 45, 1845–1857. https://doi.org/10.1007/s10803-014-2341-1.

Brandwein, A.B., Foxe, J.J., Butler, J.S., Frey, H.-P., Bates, J.C., Shulman, L.H., et al., 2015. Neurophysiological indices of atypical auditory processing and multisensory integration are associated with symptom severity in autism. J. Autism Dev. Disord. 45, 230–244. https://doi.org/10.1007/s10803-014-2212-9.

Brown, C.R., Clarke, A.R., Barry, R.J., McCarthy, R., Selikowitz, M., Magee, C., 2005. Event-related potentials in attention-deficit/hyperactivity disorder of the predominantly inattentive type: an investigation of EEG-defined subtypes. Int. J. Psychophysiol. 58, 94–107.

Bruneau, N., Roux, S., Adrien, J.L., Barthélémy, C., 1999. Auditory associative cortex dysfunction in children with autism: evidence from late auditory evoked potentials (N1 wave–T complex). Clin. Neurophysiol. 110, 1927–1934.

Bruneau, N., Bonnet-Brilhault, F., Gomot, M., Adrien, J.-L., Barthélémy, C., 2003. Cortical auditory processing and communication in children with autism: electrophysiological/behavioral relations. Int. J. Psychophysiol. 51, 17–25.

Buchwald, J.S., Erwin, R., Van Lancker, D., Guthrie, D., Schwafel, J., Tanguay, P., 1992. Midlatency auditory evoked responses: P1 abnormalities in adult autistic subjects. Electroencephalogr. Clin. Neurophysiol. 84, 164–171.

Charpentier, J., Kovarski, K., Houy-Durand, E., Malvy, J., Saby, A., Bonnet-Brilhault, F., et al., 2018. Emotional prosodic change detection in autism Spectrum disorder: an electrophysiological investigation in children and adults. J. Neurodev. Disord. 10, 28. https://doi.org/10.1186/s11689-018-9246-9.

Chen, T.-C., Hsieh, M.H., Lin, Y.-T., Chan, P.-Y.S., Cheng, C.-H., 2020. Mismatch negativity to different deviant changes in autism spectrum disorders: a meta-analysis. Clin. Neurophysiol. 131, 766–777.

Chobert, J., François, C., Habib, M., Besson, M., 2012. Deficit in the preattentive processing of syllabic duration and VOT in children with dyslexia. Neuropsychologia 50, 2044–2055.

Ciesielski, K.T., Courchesne, E., Elmasian, R., 1990. Effects of focused selective attention tasks on event-related potentials in autistic and normal individuals. Electroencephalogr. Clin. Neurophysiol. 75, 207–220.

Clarke, A.R., Barry, R.J., Indraratna, A., Dupuy, F.E., McCarthy, R., Selikowitz, M., 2016. EEG activity in children with Asperger's Syndrome. Clin. Neurophysiol. 127, 442–451.

Conlon, E.G., Lilleskaret, G., Wright, C.M., Power, G.F., 2012. The influence of contrast on coherent motion processing in dyslexia. Neuropsychologia 50, 1672–1681.

Corbera, S., Escera, C., Artigas, J., 2006. Impaired duration mismatch negativity in developmental dyslexia. Neuroreport 17, 1051–1055.

Courchesne, E., Kilman, B.A., Galambos, R., Lincoln, A.J., 1984. Autism: processing of novel auditory information assessed by event-related brain potentials. Electroencephalogr. Clin. Neurophysiol. 59, 238–248.

Cowley, B.U., Juurmaa, K., Palomäki, J., 2022. Reduced power in fronto-parietal theta EEG linked to impaired attention-sampling in adult ADHD. eNeuro 9 (1), 1–20. ENEURO.0028-21.2021.

Dabbous, A.O., 2012. Characteristics of auditory brainstem response latencies in children with autism spectrum disorders. Audiol. Med. 10, 122–231.

Dawson, G., Finley, C., Phillips, S., Galpert, L., Lewy, A., 1988. Reduced P3 amplitude of the event-related brain potential: its relationship to language ability in autism. J. Autism Dev. Disord. 18 (4), 493–503.

Deiber, M.P., Ibañez, V., Fischer, C., Perrin, F., Maugière, P., 1988. Sequential mapping favours the hypothesis of distinct generators for Na and Pa middle latency auditory evoked potentials. Electroencephalogr. Clin. Neurophysiol. 71, 187–197.

Di Lorenzo, G., Riccioni, A., Ribolsi, M., Siracusano, M., Curatolo, P., Mazzone, L., 2020. Auditory mismatch negativity in youth affected by autism spectrum disorder with and without attenuated psychosis syndrome. Front. Psychiatry 11, 555340. https://doi.org/10.3389/fpsyt.2020.555340.

Díaz, B., Hintz, F., Kiebel, S.J., von Kriegstein, K., 2012. Dysfunction of the auditory thalamus in developmental dyslexia. Proc. Natl. Acad. Sci. U. S. A. 109 (34), 13841–13846.

Donkers, F.C.L., Schipul, S.E., Baranek, G.T., Cleary, K.M., Willoughby, M.T., Evans, A.-M., et al., 2015. Attenuated auditory event-related potentials and associations with atypical sensory response patterns in children with autism. J. Autism Dev. Disord. 45, 506–523. https://doi.org/10.1007/s10803-013-1948-y.

Donkers, F.C.L., Carlson, M., Schipul, S.E., Belger, A., Baranek, G.T., 2020. Auditory event-related potentials and associations with sensory patterns in children with autism spectrum disorder, developmental delay, and typical development. Autism 24 (5), 1093–1110. https://doi.org/10.1177/1362361319893196.

Dunn, M.A., Gomes, H., Gravel, J., 2008. Mismatch negativity in children with autism and typical development. J. Autism Dev. Disord. 38, 52–71.

Dwyer, P., De Meo-Monteil, R., Saron, C.D., Rivera, S.M., 2021. Effects of age on loudness-dependent auditory ERPs in young autistic and typically-developing children. Neuropsychologia 156, 107837.

Edgar, J.C., Fisk 4th, C.L., Liu, S., Pandey, J., Herrington, J.D., Schultz, R.T., et al., 2016. Translating adult electrophysiology findings to younger patient populations: difficulty measuring 40-Hz auditory steady-state responses in typically developing children and children with autism spectrum disorder. Dev. Neurosci. 38, 1–14. https://doi.org/10.1159/000441943.

Eggermont, J.J., 2015. Auditory Temporal Processing and Its Disorders. Oxford University Press, Oxford.

Eggermont, J.J., 2017. Hearing Loss: Causes, Prevention and Treatment. Academic Press, London.

Eggermont, J.J., 2019. Auditory brainstem response, audition, vestibular, and language testing. In: Levin, K.H., Chauvel, P. (Eds.), Handbook of Clinical Neurology. Clinical Neurophysiology: Basis and Technical Aspects, vol. 160 (3rd series), pp. 451–464, https://doi.org/10.1016/B978-0-444-64032-1.00030-8 (Chapter 2).

Eggermont, J.J., 2022. Tinnitus and Hyperacusis. Facts, Theories and Clinical Implications. Academic Press.

Erwin, R., Van Lancker, D., Guthrie, D., Schwafel, J., Tanguay, P., Buchwald, J.S., 1991. P3 responses to prosodic stimuli in adult autistic subjects. Electroencephalogr. Clin. Neurophysiol. 80, 561–571.

Ferri, R., Elia, M., Agarwal, N., Lanuzza, B., Musumeci, S.A., Pennisi, G., 2003. The mismatch negativity and the P3a components of the auditory event-related potentials in autistic low-functioning subjects. Clin. Neurophysiol. 114, 1671–1680.

Fisher, T., Aharon-Peretz, J., Pratt, H., 2011. Dis-regulation of response inhibition in adult Attention Deficit Hyperactivity Disorder (ADHD): an ERP study. Clin. Neurophysiol. 122, 2390–2399.

Ford, T.C., Woods, W., Crewther, D.P., 2017. Mismatch field latency, but not power, may mark a shared autistic and schizotypal trait phenotype. Int. J. Psychophysiol. 116, 60–67.

Foss-Feig, J.H., Schauder, K.B., Key, A.P., Wallace, M.T., Stone, W.L., 2017. Audition-specific temporal processing deficits associated with language function in children with autism spectrum disorder. Autism Res. 10, 1845–1856.

Foss-Feig, J.H., Stavropoulos, K.K.M., McPartland, J.C., Wallace, M.T., Stone, W.L., Key, A.P., 2018. Electrophysiological response during auditory gap detection: biomarker for sensory and communication alterations in autism spectrum disorder? Dev. Neuropsychol. 43 (2), 109–122. https://doi.org/10.1080/87565641.2017.13.

Friston, K., 2005. A theory of cortical responses. Philos. Trans. R. Soc. B 360, 815–836. https://doi.org/10.1098/rstb.2005.1622.

Galilee, A., Stefanidou, C., McCleery, J.P., 2017. Atypical speech versus non-speech detection and discrimination in 4- to 6-yr old children with autism spectrum disorder: an ERP study. PLoS One 12 (7), e0181354. https://doi.org/10.1371/journal.pone.018135.

Gandal, M.J., Edgar, J.C., Ehrlichman, R.S., Mehta, M., Roberts, T.P.L., Siegel, S.J., 2010. Validating γ oscillations and delayed auditory responses as translational biomarkers of autism. Biol. Psychiatry 68, 1100–1106.

Gomes, H., Duff, M., Ramos, M., Molholm, S., Foxe, J.J., Halperin, J., 2012. Auditory selective attention and processing in children with attention-deficit/hyperactivity disorder. Clin. Neurophysiol. 123, 293–302.

Gomes, H., Duff, M., Flores, A., Halperin, J.M., 2013. Automatic processing of duration in children with Attention-Deficit/Hyperactivity Disorder. J. Int. Neuropsychol. Soc. 19, 686–694.

Gomot, M., Wicker, B., 2012. A challenging, unpredictable world for people with Autism Spectrum Disorder. Int. J. Psychophysiol. 83, 240–247.

Gomot, M., Giard, M.-H., Adrien, J.-L., Barthelemy, C., Bruneau, N., 2002. Hypersensitivity to acoustic change in children with autism: electrophysiological evidence of left frontal cortex dysfunctioning. Psychophysiology 39, 577–584.

Gomot, M., Bernard, F.A., Davis, M.H., Belmonte, M.K., Ashwin, C., Bullmore, E.T., et al., 2006. Change detection in children with autism: an auditory event-related fMRI study. Neuroimage 29, 475–484.

Gomot, M., Blanc, R., Clery, H., Roux, S., Barthelemy, C., Bruneau, N., 2011. Candidate electrophysiological endophenotypes of hyper-reactivity to change in autism. J. Autism Dev. Disord. 41, 705–714.

Gonzalez-Gadea, M.L., Chennu, S., Bekinschtein, T.A., Rattazzi, A., Beraudi, A., Tripicchio, P., et al., 2015. Predictive coding in autism spectrum disorder and attention deficit hyperactivity disorder. J. Neurophysiol. 114, 2625–2636.

Goris, J., Braem, S., Nijhof, A.D., Rigoni, D., Deschrijver, E., Van de Cruys, S., et al., 2018. Sensory prediction errors are less modulated by global context in autism spectrum disorder. Biol. Psychiatry Cogn. Neurosci. Neuroimaging 3, 667–674.

Green, H.L., Shuffrey, L.S., Levinson, L., Shen, G., Avery, T., Randazzo Wagner, M., et al., 2020. Evaluation of mismatch negativity as a marker for language impairment in autism spectrum disorder. J. Commun. Disord. 87, 105997.

Grillon, C., Courchesne, E., Akshoomoff, N., 1989. Brainstem and middle latency auditory evoked potentials in autism and developmental language disorder. J. Autism Dev. Disord. 19 (2), 255–269.

Grisoni, L., Moseley, R.L., Motlagh, S., Kandia, D., Sener, N., Pulvermüller, F., et al., 2019. Prediction and mismatch negativity responses reflect impairments in action semantic processing in adults with autism spectrum disorders. Front. Hum. Neurosci. 13, 395. https://doi.org/10.3389/fnhum.2019.00395.

Groom, M.J., Bates, A.T., Jackson, G.M., Calton, T.G., Liddle, P.F., Hollis, C., 2008. Event-related potentials in adolescents with schizophrenia and their siblings: a comparison with Attention-Deficit/Hyperactivity Disorder. Biol. Psychiatry 63, 784–792.

Gu, C., Bi, H.-Y., 2020. Auditory processing deficit in individuals with dyslexia: a meta-analysis of mismatch negativity. Neurosci. Biobehav. Rev. 116, 396–405.

Gumenyuk, V., Korzyukov, O., Escera, C., Hämäläinen, M., Huotilainen, M., Häyrinen, T., et al., 2005. Electrophysiological evidence of enhanced distractibility in ADHD children. Neurosci. Lett. 374, 212–217.

Hämäläinen, J.A., Leppänen, P.H.T., Guttorm, T.K., Lyytinen, H., 2008. Event-related potentials to pitch and rise time change in children with reading disabilities and typically reading children. Clin. Neurophysiol. 119, 100–115.

Hari, R., Renvall, H., 2001. Impaired processing of rapid stimulus sequences in dyslexia. Trends Cogn. Sci. 5, 525–532.

Hart, H., Radua, J., Nakao, T., Mataix-Cols, D., Rubia, K., 2013. Meta-analysis of functional magnetic resonance imaging studies of inhibition and attention in Attention-deficit/Hyperactivity Disorder. Exploring task-specific, stimulant medication, and age effects. JAMA Psychiatry 70 (2), 185–198. https://doi.org/10.1001/jamapsychiatry.2013.277.

Holcomb, P.J., Ackerman, P.T., Dykman, R.A., 1986. Auditory event-related potentials in attention and reading disabled boys. Int. J. Psychophysiol. 3, 263–273.

Holstein, D.H., Vollenweider, F.X., Geyer, M.A., Csomor, P.A., Belser, N., Eich, D., 2013. Sensory and sensorimotor gating in adult attention-deficit/hyperactivity disorder (ADHD). Psychiatry Res. 205, 117–126.

Hours, C., Recasens, C., Baleyte, J.-M., 2022. ASD and ADHD comorbidity: what are we talking about? Front. Psychiatry 13, 837424. https://doi.org/10.3389/fpsyt.2022.837424.

Hsieh, M.H., Chien, Y.-L., Gau, S.S.F., 2021. Mismatch negativity and P3a in drug-naive adults with attention-deficit hyperactivity disorder. Psychol. Med. 12, 1–11. https://doi.org/10.1017/S0033291720005516.

Huang, D., Yu, L., Wang, X., Fan, Y., Wang, S., Zhang, Y., 2018. Distinct patterns of discrimination and orienting for temporal processing of speech and nonspeech in Chinese children with autism: an event-related potential study. Eur. J. Neurosci. 47, 662–668.

Hudac, C.M., DesChamps, T.D., Arnett, A.B., Cairney, B.E., Ma, R., Webb, S.J., et al., 2018. Early enhanced processing and delayed habituation to deviance sounds in autism spectrum disorder. Brain Cogn. 123, 110–119.

Huttunen, T., Halonen, A., Kaartinen, J., Lyytinen, H., 2007. Does mismatch negativity show differences in reading-disabled children compared to normal children and children with attention deficit? Dev. Neuropsychol. 31 (3), 453–470. https://doi.org/10.1080/87565640701229656.

Huttunen-Scott, T., Kaartinen, J., Tolvanen, A., Lyytinen, H., 2008. Mismatch negativity (MMN) elicited by duration deviations in children with reading disorder, attention deficit or both. Int. J. Psychophysiol. 69, 69–77.

Irwin, J., Avery, T., Turcios, J., Brancazio, L., Cook, B., Landi, N., 2017. Electrophysiological indices of audiovisual speech perception in the broader autism phenotype. Brain Sci. 7, 60. https://doi.org/10.3390/brainsci7060060.

Itagaki, S., Yabe, H., Mori, Y., Ishikawa, H., Takanashi, Y., Shin-Ichi Niwa, S.-i., 2011. Event-related potentials in patients with adult attention-deficit/hyperactivity disorder versus schizophrenia. Psychiatry Res. 189, 288–291.

Jafari, Z., Malayeri, S., Rostami, R., 2015. Subcortical encoding of speech cues in children with attention deficit hyperactivity disorder. Clin. Neurophysiol. 126, 325–332.

Jaffe-Dax, S., Raviv, O., Jacoby, N., Loewenstein, Y., Ahissar, M., 2015. A computational model of implicit memory captures dyslexics' perceptual deficits. J. Neurosci. 35 (35), 12116–12126.

Jaffe-Dax, S., Frenkel, O., Ahissar, M., 2017. Dyslexics' faster decay of implicit memory for sounds and words is manifested in their shorter neural adaptation. eLife 6, e20557. https://doi.org/10.7554/eLife.20557.

Janssen, T.W.P., Geladé, K., van Mourik, R., Maras, A., Oosterlaan, J., 2016. An ERP source imaging study of the oddball task in children with Attention Deficit/Hyperactivity Disorder. Clin. Neurophysiol. 127, 1351–1357.

Johnstone, S.J., Barry, R.J., 1996. Auditory event-related potentials to a two-tone discrimination paradigm in attention deficit hyperactivity disorder. Psychiatry Res. 64, 179–192.

Jonkman, L.M., Kemner, C., Verbaten, M.N., Koelega, H.S., Camfferman, G., Gaag, R.-J. v.d., et al., 1997. Event-related potentials and performance of attention-deficit hyperactivity disorder: children and normal controls in auditory and visual selective attention tasks. Biol. Psychiatry 41, 595–611.

Jorgensen, A.R., Whitehouse, A.J.Q., Fox, A.M., Maybery, M.T., 2021. Delayed cortical processing of auditory stimuli in children with autism spectrum disorder: a meta-analysis of electrophysiological studies. Brain Cogn. 150, 105709.

Kaiser, A., Aggensteiner, P.-M., Baumeister, S., Holz, N.E., Banaschewski, T., Brandeis, D., 2020. Earlier versus later cognitive event-related potentials (ERPs) in attention-deficit/ hyperactivity disorder (ADHD): a meta-analysis. Neurosci. Biobehav. Rev. 112, 117–134.

Kamita, M., Silva, L., Magliaro, F., Kawai, R., Fernandes, F., Matas, C., 2019. Brainstem auditory evoked potentials in children with autism spectrum disorder. J. Pediatr. (Rio. J.) 96 (3), 386–392. https://doi.org/10.1016/j.jped.2018.12.010.

Kemner, C., Verbaten, M.N., Cuperus, J.M., Camfferman, G., van Engeland, H., 1995. Auditory event-related brain potentials in autistic children and three different control groups. Biol. Psychiatry 38, 150–165.

Kemner, C., Verbaten, M.N., Koelega, H.S., Buitelaar, J.K., van der Gaag, R.J., Camfferman, G., van Engeland, H., 1996. Event-related brain potentials in children with attention-deficit and hyperactivity disorder: effects of stimulus deviancy and task relevance in the visual and auditory modality. Biol. Psychiatry 40, 522–534.

Kiebel, S.J., Daunizeau, J., Friston, K.J., 2008. A hierarchy of time-scales and the brain. PLoS Comput. Biol. 4, e1000209.

Kiebel, S.J., von Kriegstein, K., Daunizeau, J., Friston, K.J., 2009. Recognizing sequences of sequences. PLoS Comput. Biol. 5, e1000464.

Kikuchi, M., Shitamichi, K., Yoshimura, Y., Ueno, S., Hiraishi, H., Hirosawa, T., et al., 2013. Altered brain connectivity in 3-to 7-year-old children with autism spectrum disorder. Neuroimage Clin. 2, 394–401.

Kim, S., Baek, J.H., Kwon, Y.J., Lee, H.Y., Yoo, J.H., Shim, S.-h., Kim, J.S., 2021. Machine-learning-based diagnosis of drug-naive adult patients with attention-deficit hyperactivity disorder using mismatch negativity. Transl. Psychiatry 11, 484. https://doi.org/10.1038/s41398-021-01604-3.

Knight, E.J., Oakes, L., Hyman, S.L., Freedman, E.G., Foxe, J.J., 2020. Individuals with autism have no detectable deficit in neural markers of prediction error when presented with auditory rhythms of varied temporal complexity. Autism Res. 13, 2058–2072.

Kujala, T., Belitz, S., Tervaniemi, M., Näätänen, R., 2003. Auditory sensory memory disorder in dyslexic adults as indexed by the mismatch negativity. Eur. J. Neurosci. 17, 1323–1327.

Kujala, J., Pammer, K., Cornelissen, P., Roebroeck, A., Formisano, E., Salmelin, R., 2006. Phase coupling in a cerebro-cerebellar network at 8–13 Hz during reading. Cereb. Cortex 17 (6), 1476–1485. https://doi.org/10.1093/cercor/bhl059.

Kujala, T., Lepistö, T., Näätänen, R., 2013. The neural basis of aberrant speech and audition in autism spectrum disorders. Neurosci. Biobehav. Rev. 37, 697–704.

Kwon, S., Kim, J., Choe, B.-H., Ko, C., Park, S., 2007. Electrophysiologic assessment of central auditory processing by auditory brainstem responses in children with autism spectrum disorders. J. Korean Med. Sci. 22, 656–659.

Lawson, R.P., Rees, G., Friston, K.J., 2014. An aberrant precision account of autism. Front. Hum. Neurosci. 8, 302. https://doi.org/10.3389/fnhum.2014.00302.

Lee, D.H., Lee, P., Seo, S.W., Roh, J.H., Oh, M., Oh, J.S., et al., 2019. Neural substrates of cognitive reserve in Alzheimer's disease spectrum and normal aging. Neuroimage 186, 690–702.

Lee, Y.J., Jeong, M.Y., Kim, J.H., Kim, J.-S., 2020. Associations between the mismatch-negativity potential and symptom severity in medication-naïve children and adolescents with symptoms of attention-deficit/hyperactivity disorder. Clin. Psychopharmacol. Neurosci. 18 (2), 249–260. https://doi.org/10.9758/cpn.2020.18.2.249.

Lehongre, K., Ramus, F., Villiermet, N., Schwartz, D., Giraud, A.-L., 2011. Altered low-gamma sampling in auditory cortex accounts for the three main facets of dyslexia. Neuron 72, 1080–1090.

Lepistö, T., Kujala, T., Vanhala, R., Alku, P., Huotilainen, M., Näätänen, R., 2005. The discrimination of and orienting to speech and non-speech sounds in children with autism. Brain Res. 1066, 147–157.

Lepistö, T., Silokallio, S., Nieminen-von Wendt, T., Alku, P., Näätänen, R., Kujala, T., 2006. Auditory perception and attention as reflected by the brain event-related potentials in children with Asperger syndrome. Clin. Neurophysiol. 117, 2161–2171.

Leno, C., Chandler, S., White, P., Pickles, A., Baird, G., Hobson, C., et al., 2018. Testing the specificity of executive functioning impairments in adolescents with ADHD, ODD/CD and ASD. Eur. Child Adolesc. Psychiatry 27 (7), 899–908. https://doi.org/10.1007/s00787-017-1089-5.

Lepistö, T., Kajander, M., Vanhala, R., Alku, P., Huotilainen, M., Näätänen, R., et al., 2008. The perception of invariant speech features in children with autism. Biol. Psychol. 77, 25–31.

Li, A., Gao, G., Fu, T., Pang, W., Zhang, X., Qin, Z., Ge, R., 2020. Continued development of auditory ability in autism spectrum disorder children: a clinical study on click-evoked auditory brainstem response. Int. J. Pediatr. Otorhinolaryngol. 138, 110305.

Lim, L., Chantiluke, K., Cubillo, A.I., Smith, A.B., Simmons, A., Mehta, M.A., Rubia, K., 2015. Disorder-specific grey matter deficits in attention deficit hyperactivity disorder relative to autism spectrum disorder. Psychol. Med. 45, 965–976. https://doi.org/10.1017/S0033291714001974.

Lincoln, A.J., Courchesne, E., Harms, L., Allen, M., 1993. Contextual probability evaluation in autistic, receptive developmental language disorder, and control children: event-related brain potential evidence. J. Autism Dev. Disord. 23 (1), 37–58.

Lindström, R., Lepistö-Paisley, T., Makkonen, T., Reinvall, O., Nieminen-von Wendt, T., Alén, R., Kujala, T., 2018. Atypical perceptual and neural processing of emotional prosodic changes in children with autism spectrum disorders. Clin. Neurophysiol. 129, 2411–2420.

Liu, Y., Hanna, G.L., Hanna, B.S., Rough, H.E., Arnold, P.D., Gehring, W.J., 2020. Behavioral and electrophysiological correlates of performance monitoring and development in children and adolescents with Attention-Deficit/Hyperactivity Disorder. Brain Sci. 10, 79. https://doi.org/10.3390/brainsci10020079.

Livingstone, M.S., Rosen, G.D., Drislane, F.W., Galaburda, A.M., 1991. Physiological and anatomical evidence for a magnocellular defect in developmental dyslexia. Proc. Natl. Acad. Sci. U. S. A. 88 (18), 7943–7947.

Loiselle, D.L., Stamm, J.S., Maitinsky, S., Whipple, S.C., 1980. Evoked potential and behavioral signs of attentive dysfunctions in hyperactive boys. Psychophysiology 17 (2), 193–201.

Ludlow, A., Mohr, B., Whitmore, A., Garagnani, M., Pulvermüller, F., Gutierrez, R., 2014. Auditory processing and sensory behaviours in children with autism spectrum disorders as revealed by mismatch negativity. Brain Cogn. 86, 55–63.

Lukito, S.D., O'Daly, O.G., Lythgoe, D.J., Whitwell, S., Debnam, A., Murphy, C.M., et al., 2018. Neural correlates of duration discrimination in young adults with autism spectrum disorder, attention-deficit/hyperactivity disorder and their comorbid presentation. Front. Psychiatry 9, 569. https://doi.org/10.3389/fpsyt.2018.00569.

Mamashli, F., Khan, S., Bharadwaj, H., Michmizos, K., Ganesan, S., Garel, K.-L.A., et al., 2017. Auditory processing in noise is associated with complex patterns of disrupted functional connectivity in autism spectrum disorder. Autism Res. 10, 631–647.

Marzinzik, F., Wahl, M., Krüger, D., Gentschow, L., Colla, M., Klostermann, F., 2012. Abnormal distracter processing in adults with attention-deficit-hyperactivity disorder. PLoS One 7 (3), e33691. https://doi.org/10.1371/journal.pone.0033691.

Matsuzaki, J., Kagitani-Shimono, K., Goto, T., Sanefuji, W., Yamamoto, T., Sakai, S., et al., 2012. Differential responses of primary auditory cortex in autistic spectrum disorder with auditory hypersensitivity. Neuroreport 23 (2), 113–118.

Matsuzaki, J., Kagitani-Shimono, K., Sugata, H., Hanaie, R., Nagatani, T., Yamamoto, T., et al., 2017. Delayed mismatch field latencies in autism spectrum disorder with abnormal auditory sensitivity: a magnetoencephalographic study. Front. Hum. Neurosci. 11 (446). https://doi.org/10.3389/fnhum.2017.00446.

Matsuzaki, J., Ku, M., Berman, J.I., Blaskey, L., Bloy, L., Chen, Y.-h., et al., 2019a. Abnormal auditory mismatch fields in adults with autism spectrum disorder. Neurosci. Lett. 698, 140–145.

Matsuzaki, J., Kuschner, E.S., Blaskey, L., Bloy, L., Kim, M., Ku, M., et al., 2019b. Abnormal auditory mismatch fields are associated with communication impairment in both verbal and minimally verbal/nonverbal children who have autism spectrum disorder. Autism Res. 12, 1225–1235.

Maziade, M., Mérette, C., Cayer, M., Roy, M.-A., Szatmari, P., Coté, R., et al., 2000. Prolongation of brainstem auditory-evoked responses in autistic probands and their unaffected relatives. Arch. Gen. Psychiatry 57 (11), 1077–1083.

McClelland, R.J., Eyre, D.G., Watson, D., Calvert, G.J., Sherrard, E., 1992. Central conduction time in childhood autism. Br. J. Psychiatry 160, 659–663.

Mereu, M., Contarini, G., Buonaguro, E.F., Latte, G., Managò, F., Iasevoli, F., et al., 2017. Dopamine transporter (DAT) genetic hypofunction in mice produces alterations consistent with ADHD but not schizophrenia or bipolar disorder. Neuropharmacology 121, 179–194.

Mioni, G., Capodieci, A., Biffi, V., Porcelli, F., Cornoldi, C., 2019. Difficulties of children with symptoms of attention-deficit/hyperactivity disorder in processing temporal information concerning everyday life events. J. Exp. Child Psychol. 182, 86–101.

Miron, O., Roth, D.A.-E., Gabis, L.V., Henkin, Y., Shefer, S., Dinstein, I., Geva, R., 2016. Prolonged auditory brainstem responses in infants with autism. Autism Res. 9, 689–695.

Näätänen, R., Picton, T.W., 1986. N2 and automatic versus controlled processes. Electroencephalogr. Clin. Neurophysiol. Suppl. 38, 169–186.

Niwa, S.-i., Ohta, M., Yamazaki, K., 1983. P300 and stimulus evaluation process in autistic subjects. J. Autism Dev. Disord. 13 (1), 33–42.

Novick, B., Vaughan Jr., H.G., Kurtzberg, D., Simson, R., 1980. An electrophysiologic indication of auditory processing defects in autism. Psychiatry Res. 3, 107–114.

O'Connor, D.H., Fukui, M.M., Pinsk, M.A., Kastner, S., 2002. Attention modulates responses in the human lateral geniculate nucleus. Nat. Neurosci. 5, 1203–1209.

O'Neill, B.V., Croft, R.J., Nathan, P.J., 2008. The loudness dependence of the auditory evoked potential (LDAEP) as an in vivo biomarker of central serotonergic function in humans: rationale, evaluation and review of findings. Hum. Psychopharmacol. 23, 355–370.

Oades, R.D., Walker, M.K., Geffen, L.B., Stern, L.M., 1988. Event-related potentials in autistic and healthy children on an auditory choice reaction time task. Int. J. Psychophysiol. 6, 25–37.

Oades, R.D., Dittmann-Balcar, A., Schepker, R., Eggers, C., Zerbin, D., 1996. Auditory event-related potentials (ERPs) and mismatch negativity (MMN) in healthy children and those with attention-deficit or Tourette/tic symptoms. Biol. Psychol. 43, 163–185.

Oram Cardy, J.E., Flagg, E.J., Roberts, W., Roberts, T.P.L., 2005. Delayed mismatch field for speech and non-speech sounds in children with autism. Neuroreport 16, 521–525.

Orekhova, E.V., Stroganova, T.A., Nygren, G., Tsetlin, M.M., Posikera, I.N., Gillberg, C., et al., 2007. Excess of high frequency electroencephalogram oscillations in boys with autism. Biol. Psychiatry 62, 1022–1029.

Orekhova, E.V., Stroganova, T.A., Prokofiev, A.O., Nygren, G., Gillberg, C., Elam, M., 2009. The right hemisphere fails to respond to temporal novelty in autism: evidence from an ERP study. Clin. Neurophysiol. 120, 520–529.

Panagiotidi, M., Overton, P.G., Stafford, T., 2017. Multisensory integration and ADHD-like traits: evidence for an abnormal temporal integration window in ADHD. Acta Psychol. 181, 10–17.

Park, E.J., Park, Y.-M., Lee, S.-H., Kim, B., 2022. The loudness dependence of auditory evoked potentials is associated with the symptom severity and treatment in boys with attention deficit hyperactivity disorder. Clin. Psychopharmacol. Neurosci. 20 (3), 514–525. https://doi.org/10.9758/cpn.2022.20.3.514.

Partanen, M., Fitzpatrick, K., Mädler, B., Edgell, D., Bjornson, B., Giaschi, D.E., 2012. Cortical basis for dichotic pitch perception in developmental dyslexia. Brain Lang. 123, 104–112.

Pellicano, E., Burr, D., 2012. When the world becomes 'too real': a Bayesian explanation of autistic perception. Trends Cogn. Sci. 16 (10), 504–510.

Pfeiffer, B., Stein Duker, L., Murphy, A., Shui, C., 2019. Effectiveness of noise-attenuating headphones on physiological responses for children with autism spectrum disorders. Front. Integr. Neurosci. 13, 65. https://doi.org/10.3389/fnint.2019.00065.

Picken, C., Clarke, A.R., Barry, R.J., McCarthy, R., Selikowitz, M., 2020. The theta/beta ratio as an index of cognitive processing in adults with the combined type of attention deficit hyperactivity disorder. Clin. EEG Neurosci. 51 (3), 167–173. https://doi.org/10.1177/1550059419895142.

Picton, T.W., 2011. Human Auditory Evoked Potentials. Plural Publishing, San Diego, CA.

Picton, T.W., Hillyard, S.A., Krausz, H.I., Galambos, R., 1974. Human auditory evoked potentials. I. Evaluation of components. Electroencephalogr. Clin. Neurophysiol. 36 (2), 179–190. https://doi.org/10.1016/0013-4694(74)90155-2.

Pillion, J.P., Boatman-Reich, D., Gordon, B., 2018. Auditory brainstem pathology in autism spectrum disorder: a review. Cogn. Behav. Neurol. 31 (2), 53–78.

Ponton, C.W., Eggermont, J.J., Kwong, B., Don, M., 2000. Maturation of human central auditory system activity: evidence from multi-channel evoked potentials. Clin. Neurophysiol. 111, 220–236.

Ponton, C.W., Eggermont, J.J., Khosla, D., Kwong, B., Don, M., 2002. Maturation of human central auditory system activity: separating auditory evoked potentials by dipole source modeling. Clin. Neurophysiol. 113, 407–420.

Port, R.G., Edgar, J.C., Ku, M., Bloy, L., Murray, R., Blaskey, L., et al., 2016. Maturation of auditory neural processes in autism spectrum disorder—a longitudinal MEG study. Neuroimage Clin. 11, 566–577.

Postema, M.C., Hoogman, M., Ambrosino, S., Asherson, P., Banaschewski, T., Bandeira, C.E., et al., 2021. Analysis of structural brain asymmetries in attention-deficit/hyperactivity disorder in 39 datasets. J. Child Psychol. Psychiatry 62 (10), 1202–1219. https://doi.org/10.1111/jcpp.13396.

Puente, A., Ysunza, A., Pamplona, M., Silva-Rojas, A., Lara, C., 2002. Short latency and long latency auditory evoked responses in children with attention deficit disorder. Int. J. Pediatr. Otorhinolaryngol. 62, 45–51.

Ralli, M., Romani, M., Zodda, A., Yoshie Russo, F., Altissimi, G., Patrizia, M., et al., 2020. Hyperacusis in children with attention deficit hyperactivity disorder: a preliminary study. Int. J. Environ. Res. Public Health 17, 3045. https://doi.org/10.3390/ijerph17093045.

Ramezani, M., Lotfi, Y., Moossavi, A., Bakhshi, E., 2019. Auditory brainstem response to speech in children with high functional autism spectrum disorder. Neurol. Sci. 40, 121–125. https://doi.org/10.1007/s10072-018-3594-9.

Retz, W., González-Trejo, E., Römer, K.D., Philipp-Wiegmann, F., Reinert, P., Low, Y.-F., et al., 2012. Assessment of post-excitatory long-interval cortical inhibition in adult attention-deficit/hyperactivity disorder. Eur. Arch. Psychiatry Clin. Neurosci. 262, 507–517. https://doi.org/10.1007/s00406-012-0299-6.

Riva, V., Cantiani, C., Mornati, G., Gallo, M., Villa, L., Mani, E., et al., 2018. Distinct ERP profiles for auditory processing in infants at-risk for autism and language impairment. Sci. Rep. 8, 715. https://doi.org/10.1038/s41598-017-19009-y.

Roberts, T.P.L., Cannon, K.M., Tavabi, K., Blaskey, L., Khan, S.Y., Monroe, J.F., et al., 2011. Auditory magnetic mismatch field latency: a biomarker for language impairment in autism. Biol. Psychiatry 70, 263–269.

Roberts, T.P.L., Lanza, M.R., Dell, J., Qasmieh, S., Hines, K., Blaskey, L., et al., 2013. Maturational differences in thalamocortical white matter microstructure and auditory evoked response latencies in autism spectrum disorders. Brain Res. 1538, 79–85.

Roberts, T., Masuzaki, J., Blaskey, L., Bloy, L., Edgar, J., Kim, M., et al., 2019. Delayed M50/M100 evoked response component latency in minimally verbal/nonverbal children who have autism spectrum disorder. Mol. Autism 10 (34). https://doi.org/10.1186/s13229-019-0283-3.

Rosenhall, U., Nordin, V., Brantberg, K., Gillberg, C., 2003. Autism and auditory brain stem responses. Ear Hear. 24, 206–214.

Roth, D.A.-E., Muchnik, C., Shabtai, E., Hildesheimer, M., Henkin, Y., 2011. Evidence for atypical auditory brainstem responses in young children with suspected autism spectrum disorders. Dev. Med. Child Neurol. 54 (1), 23–29. https://doi.org/10.1111/j.1469-8749.2011.04149.x.

Rubenstein, J.L.R., Merzenich, M.M., 2003. Model of autism: increased ratio of excitation/inhibition in key neural systems. Genes Brain Behav. 2, 255–267.

Ruiz-Martínez, F., Rodeiguez-Martinez, E., Wilson, C., Yau, S., Saldaña, D., Gómez, C., 2020. Impaired P1 habituation and mismatch negativity in children with autism spectrum disorder. J. Autism Dev. Disord. 50 (2), 603–616. https://doi.org/10.1007/s10803-019-04299-0.

Sadaghiani, S., D'Esposito, M., 2015. Functional characterization of the cingulo-opercular network in the maintenance of tonic alertness. Cereb. Cortex 25, 2763–2773. https://doi.org/10.1093/cercor/bhu072.

Santos, M., Marques, C., Nóbrega Pinto, A., Fernandes, R., Coutinho, B.M., Almeida e Sousa, C., 2017. Autism spectrum disorders and the amplitude of auditory brainstem response wave I. Autism Res. 10, 1300–1305.

Schulte-Körne, G., Deimel, W., Bartling, J., Remschmidt, H., 1998. Auditory processing and dyslexia: evidence for a specific speech processing deficit. Neuroreport 9, 337–340.

Schulte-Körne, G., Deimel, W., Bartling, J., Remschmidt, H., 1999. Pre-attentive processing of auditory patterns in dyslexic human subjects. Neurosci. Lett. 276, 41–44.

Senderecka, M., Grabowska, A., Gerc, K., Szewczyk, J., Chmylak, R., 2012. Event-related potentials in children with attention deficit hyperactivity disorder: an investigation using an auditory oddball task. Int. J. Psychophysiol. 85, 106–115.

Sergeant, J., 2000. The cognitive-energetic model: an empirical approach to Attention-Deficit Hyperactivity Disorder. Neurosci. Biobehav. Rev. 24, 7–12.

Shephard, E., Tye, C., Ashwood, K.L., Azadi, B., Johnson, M.H., Charman, T., et al., 2019. Oscillatory neural networks underlying resting state, attentional control and social cognition task conditions in children with ASD, ADHD and ASD+ADHD. Cortex 117, 96–110.

Skoff, B.F., Mirsky, A.F., Turner, D., 1980. Prolonged brainstem transmission time in autism. Psychiatry Res. 2, 157–166.

Stefanics, G., Fosker, T., Huss, M., Mead, N., Szucs, D., Goswami, U., 2011. Auditory sensory deficits in developmental dyslexia: a longitudinal ERP study. Neuroimage 57, 723–732.

Stein, J., Walsh, V., 1997. To see but not to read; the magnocellular theory of dyslexia. Trends Neurosci. 20 (4), 147–152.

Stoodley, C.J., 2014. Distinct regions of the cerebellum show gray matter decreases in autism, ADHD, and developmental dyslexia. Front. Syst. Neurosci. 8, 92. https://doi.org/10.3389/fnsys.2014.00092.

Stroganova, T.A., Nygren, G., Tsetlin, M.M., Posikera, I.N., Gillberg, C., Elam, M., et al., 2007. Abnormal EEG lateralization in boys with autism. Clin. Neurophysiol. 118, 1842–1854.

Su, S., Chen, Y., Dai, Y., Lin, L., Qian, L., Zhou, Q., et al., 2022. Quantitative synthetic MRI reveals grey matter abnormalities in children with drug-naïve attention-deficit/hyperactivity disorder. Brain Imaging Behav. 16, 406–414. https://doi.org/10.1007/s11682-021-00514-8.

Talge, N.M., Tudor, B.M., Kileny, P.R., 2018. Click-evoked auditory brainstem responses and autism spectrum disorder: a meta-analytic review. Autism Res. 11, 916–927.

Talge, N.M., Adkins, M., Kileny, P.R., Frownfelter, I., 2022. Click-evoked auditory brainstem responses and autism spectrum disorder: a meta-analytic investigation of disorder specificity. Pediatr. Res. 92, 40–46. https://doi.org/10.1038/s41390-021-01730-0.

Tanguay, P.E., Edwards, R.M., Buchwald, J., Schwafel, J., Allen, V., 1982. Auditory brainstem evoked responses in autistic children. Arch. Gen. Psychiatry 39 (2), 174–180.

Tas, A., Yagiz, R., Tas, M., Esme, M., Uzun, C., Karasalihoglu, A.R., 2007. Evaluation of hearing in children with autism by using TEOAE and ABR. Autism 11 (1), 73–79. https://doi.org/10.1177/1362361307070908.

Taylor, M.J., Rosenblatt, B., Linschoten, L., 1982. Auditory brainstem response abnormalities in autistic children. Can. J. Neurol. Sci. 9 (4), 429–433. https://doi.org/10.1017/s0317167100044346.

Testo, A.A., Felicione, J.M., Ellard, K.K., Peters, A.T., Chou, T., Gosai, A., et al., 2020. Neural correlates of the ADHD self-report scale. J. Affect. Disord. 263, 141–146.

Tombor, L., Kakuszi, B., Papp, S., Réthelyi, J., Bitter, I., Czobor, P., 2021. Atypical resting-state gamma band trajectory in adult attention deficit/hyperactivity disorder. J. Neural Transm. 128, 1239–1248. https://doi.org/10.1007/s00702-021-02368-2.

Tonnquist-Uhlen, I., Ponton, C.W., Eggermont, J.J., Kwong, B., Don, M., 2003. Maturation of human central auditory system activity: the T-complex. Clin. Neurophysiol. 114, 685–701.

Tsai, M.-L., Hung, K.-L., Lu, H.-H., 2012. Auditory event-related potentials in children with attention deficit hyperactivity disorder. Pediatr. Neonatol. 53, 118–124.

Van de Cruys, S., Evers, K., Van der Hallen, R., Van Eylen, L., Boets, B., De Wit, L., et al., 2014. Precise minds in uncertain worlds: predictive coding in autism. Psychol. Rev. 121 (4), 649–675. https://doi.org/10.1037/a0037665.

Ververi, A., Vargiami, E., Papadopoulou, V., Tryfonas, D., Zafeiriou, D.I., 2015. Brainstem auditory evoked potentials in boys with autism: still searching for the hidden truth. Iran. J. Child. Neurol. 9 (2), 21–28.

Vlaskamp, C., Oranje, B., Falcher Madsen, G., Møllegaard Jepsen, J.R., Sarah Durston, S., Cantio, C., et al., 2017. Auditory processing in autism spectrum disorder: mismatch negativity deficits. Autism Res. 10, 1857–1865.

Volkmer, S., Schulte-Körne, G., 2018. Cortical responses to tone and phoneme mismatch as a predictor of dyslexia? A systematic review. Schizophr. Res. 191, 148–160.

von Kriegstein, K., Patterson, R.D., Griffiths, T.D., 2008. Task-dependent modulation of medial geniculate body is behaviorally relevant for speech recognition. Curr. Biol. 18, 1855–1859.

Wallace, M.T., Stevenson, R.A., 2014. The construct of the multisensory temporal binding window and its dysregulation in developmental disabilities. Neuropsychologia 64, 105–123.

Wehr, M., Zador, A.M., 2005. Synaptic mechanisms of forward suppression in rat auditory cortex. Neuron 47, 437–445.

Weismüller, B., Thienel, R., Youlden, A.-M., Fulham, R., Koch, M., Schall, U., 2015. Psychophysiological correlates of developmental changes in healthy and autistic boys. J. Autism Dev. Disord. 45, 2168–2175.

Whitehouse, A.J.O., Bishop, D.V.M., 2008. Do children with autism 'switch off' to speech sounds? An investigation using event-related potentials. Dev. Sci. 11 (4), 516–524.

Williams, Z.J., Abdelmessih, P.G., Key, A.P., Woynaroski, T.G., 2021. Cortical auditory processing of simple stimuli is altered in autism: a meta-analysis of auditory evoked responses. Biol. Psychiatry Cogn. Neurosci. Neuroimaging 6, 767–781. www.sobp.org/BPCNNI.

Wilson, T.W., Rojas, D.C., Reite, M.L., Teale, P.D., Rogers, S.J., 2007. Children and adolescents with autism exhibit reduced MEG steady-state gamma responses. Biol. Psychiatry 62, 192–197.

Yamamuro, K., Ota, T., Iida, J., Nakanishi, Y., Kishimoto, N., Kishimoto, T., 2016. Associations between the mismatch-negativity component and symptom severity in children and adolescents with attention deficit/hyperactivity disorder. Neuropsychiatr. Dis. Treat. 12, 3183–3190.

Yang, M.T., Hsu, C.-H., Yeh, P.-W., Lee, W.-T., Liang, J.-S., Fu, W.-M., et al., 2015. Attention deficits revealed by passive auditory change detection for pure tones and lexical tones in ADHD children. Front. Hum. Neurosci. 9, 470. https://doi.org/10.3389/fnhum.2015.00470.

Yoshimura, Y., Kikuchi, M., Hayashi, N., Hiraishi, H., Hasegawa, C., Takahashi, T., et al., 2017. Altered human voice processing in the frontal cortex and a developmental language delay in 3- to 5-year-old children with autism spectrum disorder. Sci. Rep. 7, 17116. https://doi.org/10.1038/s41598-017-17058-x.

Yoshimura, Y., Ikeda, T., Hasegawa, C., An, K.-M., Tanaka, S., Yaoi, K., et al., 2021. Shorter P1m response in children with autism spectrum disorder without intellectual disabilities. Int. J. Mol. Sci. 22, 2611. https://doi.org/10.3390/ijms22052611.

Yu, L., Wang, S., Huang, D., Wu, X., Zhang, Y., 2018. Role of inter-trial phase coherence in atypical auditory evoked potentials to speech and nonspeech stimuli in children with autism. Clin. Neurophysiol. 129, 1374–1382.

Zarka, D., Cebolla, A.M., Cevallos, C., Palmero-Soler, E., Dan, B., Cheron, G., 2021. Caudate and cerebellar involvement in altered P2 and P3 components of GO/NoGO evoked potentials in children with attention-deficit/hyperactivity disorder. Eur. J. Neurosci. 53, 3447–3462. https://doi.org/10.1111/ejn.15198.

Zhang, J., Qiu, M., Pan, J., Zhao, L., 2020. The preattentive change detection in preschool children with attention deficit hyperactivity disorder: a mismatch negativity study. Neuroreport 31, 776–779.

CHAPTER 9

Predictive coding in aging, tinnitus, MCI, and Alzheimer's disease

In this chapter we will discuss tinnitus, for which recently exciting progress is made in the application of predictive coding, and findings in patients with cognitive impairment, mild as well of the Alzheimer type. We will however start with the cognitive aspects of aging as they cooccur with the previously mentioned disorders and their effects on event-related potentials (ERPs) interact. "Converging evidence from structural, metabolic and functional connectivity MRI suggests that neurodegenerative diseases, such as Alzheimer's disease (AD), target specific neural networks. However, age-related network changes commonly co-occur with neuropathological ones, limiting efforts to disentangle disease-specific alterations in network function from those associated with normal ageing" (Chhatwal et al., 2018). They observed the specific degradation of cognitive—especially the default mode (DMN) and dorsal attention network (DAN)—compared to motor and sensory networks in early AD and found that this distinctive degradation pattern was magnified in more advanced stages of disease. Chhatwal et al. (2018) found that aging in the absence of AD biomarkers was characterized by network degradation across cognitive and sensory networks, with between- and within-network connections. Combining the contrasting patterns of connectivity in AD and aging argues against AD as a form of accelerated aging. Somewhat contrasting this, Dennis and Thompson (2014) had also argued that "connectivity and network integrity appear to decrease in healthy aging, but this decrease is accelerated in AD, with specific systems hit hardest, such as the DMN."

9.1 Auditory aspects of aging

This aging section is largely based on Eggermont (2019a). Changes associated with age-related hearing impairment occur in three levels, comprising

Brain Responses to Auditory Mismatch and Novelty Detection
https://doi.org/10.1016/B978-0-443-15548-2.00009-0

the peripheral auditory system, central auditory system, and cognitive systems. An important aspect to consider is that auditory stimuli, especially speech, activate neurons in the prefrontal and other nonauditory cognition-related areas. Using speech stimuli and functional magnetic resonance imaging (fMRI)-based connectivity analyses, Langers and Melcher (2011) demonstrated coordinated activity between a wide range of brain structures (Fig. 9.1). Resting-state fMRI revealed largely similar networks meaning that although their activity is not auditory stimulus dependent, they may still be driven by the high spontaneous activity in the auditory system. These 'nonauditory' areas are often implicated in declining speech discrimination. Thus the auditory brain extends well beyond the classical auditory areas in the temporal lobe (see also Fig. 1.7). In hearing–impaired elderly

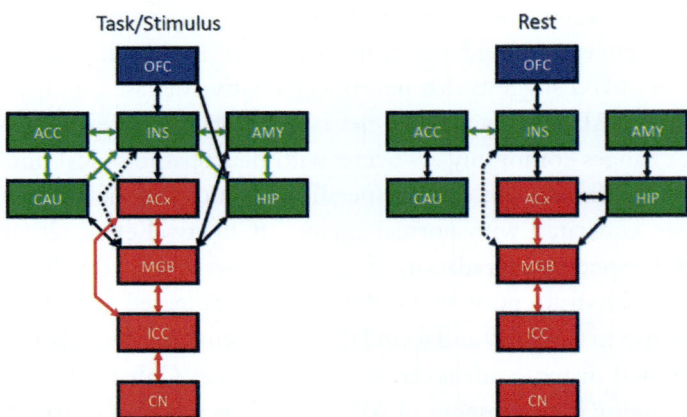

Fig. 9.1 Effective connectivity between auditory and nonauditory ROIs. The ROIs correspond to local maxima in the spatial maps of independent components. Color of the ROIs: red, auditory system; green, limbic system; blue, OFC. The connectivities shown for (A) the task- and stimulus-driven condition and (B) the resting-state condition are based on separate analyses. Only highly significant connections ($P < 0.001$, corrected for multiple comparisons) are indicated (lines with arrows, labeled with the corresponding partial correlation coefficients; solid and dotted lines indicate positive and negative correlations, respectively). Significance was assessed by means of a bootstrap resampling technique. Many of the connections apparent during the sound-stimulated condition were also apparent during the resting state. For both datasets the MGB and INS were nexus of connectivity in that they showed significant effective connectivity with the greatest number of other ROIs. ACC, anterior cingulate cortex; ACx, auditory cortex; AMY, amygdala; CAU, caudate; CN, cochlear nucleus; HIP, hippocampus; INS, insula; OFC, orbitofrontal cortex. Based on Langers and Melcher (2011).

patients, the age-related declines of peripheral and central auditory processing interact with the diminished cognitive functions and support, leading to reduced auditory perception of speech.

9.1.1 ABR changes and aging

We will start with a short overview of the effects of aging on the auditory brainstem response (ABR). ABR latencies reflect two aspects of brainstem changes, loss of input from the cochlea and changes in brainstem conduction velocity. Hearing loss is reflected in increased latencies for all ABR waves, whereas decreased brainstem conduction velocity results in increased interpeak latencies, predominantly measured as the wave I-wave V latency difference. Which, if any, of these aspects are characteristic for aging? Rowe (1978) had studied ABRs in young and older adults and found that wave peak latencies increased with age, but that interpeak latencies were not affected by age. Rosenhall et al. (1985) recorded ABRs in healthy individuals ranging in age from 5 to 75 years. Their pure-tone thresholds were 20-dB HL or better at the frequencies 125 Hz to 2000 Hz and 35 dB or better at the frequencies 4000–8000 Hz. They found that the latencies of waves I, III, and V increased slightly with increasing age, but that the I–V interpeak latency was the same in all age groups. Rosenhall et al. (1986) then reported results in 209 elderly participants with hearing loss, and compared to young or middle-aged adults also with a cochlear hearing loss. They found that the older individuals had generally longer ABR wave latencies than the young and middle-aged participants. The I–V interpeak latency was significantly prolonged in the older age groups compared with the group of younger individuals, except for subjects with pronounced hearing loss. This suggested to Rosenhall et al. (1986) that presbycusis may cause an age-related dysfunction in the auditory brainstem. Martini et al. (1991) reported a latency increase of all principal waves of ABR, reflecting the hearing loss, but without a significant change in the I–V delay in healthy males and females with a mean age of 67.2 years. Ottaviani et al. (1991) compared ABRs in older adults aged 60–80 years with presbycusis with those from normal-hearing older (and young adults with a sloping hearing loss). They showed no significant differences between presbycusic and young cochlear hearing loss adults concerning I–III and I–V interpeak latencies. More recently, Konrad-Martin et al. (2012) performed a cross-sectional study in Veteran participants, aged 26–71 years. Aging increased ABR peak latencies, whereas the I–V interpeak latency did not change.

These studies all suggest largely an effect of the age-related hearing loss causing absolute latency increase, without brainstem abnormalities—except in one study—as reflected in interpeak latencies. A recent review on the ABR changes in auditory and neurological disorders is found in Eggermont (2019b). Comparisons with mild cognitive impairment (MCI) and Alzheimer's are presented in Table 9.1.

9.1.2 Event-related potential changes caused by aging

Typically, one finds an increasing influence of aging on late cognitive processes reflected in preattentive mismatch negativity (MNN) and task-related (P3) than in the obligatory perceptual (e.g., P1–N1–P2) ones, albeit that some of these are affected by attention. P3a changes often accompany those in the MMN (Chapter 5). Studies using the label P3 often imply P3b. In particular, a reduced MMN response to duration- and frequency-deviants is a robust feature among the elderly, which is due to a decline in frontal-based control mechanisms, with alterations in connectivity between temporal and frontal regions. An overview of age-related ERP changes is presented in Table 9.1. We start with effects of aging on the obligatory P1–N1–P2 complex.

9.1.2.1 Obligatory ERP components

Alain et al. (2022) measured neuroelectric brain activity in young (18–30 years) and older adults (60–88 years) in response to a rapid randomized sequence of lateralized auditory, visual, and somatosensory stimuli. They found that older adults' early sensory evoked responses were greater in amplitude than those of young adults in all three modalities, which coincided with enhanced source activity in auditory, visual, and somatosensory cortices (Fig. 9.2). By comparing phase-locking values (PLV), a measure of phase synchrony between two time series attributed to different cortical regions, Alain et al. (2022) found that older adults also showed stronger neural synchrony than young adults between superior prefrontal and sensory cortices, implying increased functional connectivity. Furthermore, in older adults, the degree of phase synchrony was positively correlated with the magnitude of source activity in sensory areas. For visual stimuli, older adults also showed stronger PLV between the right and left visual cortices, between the right visual cortex and the mid-parietal area, and between the right visual cortex and the right anterior temporal cortex. In contrast, Alain et al. (2022) found that the synchrony for somatosensory stimuli only trended toward

Table 9.1 Amplitude and latency changes in ERPs.

	Aging	MCI	Alzheimer's
ABR (I–V)	~75,77,78,79,80,81 →76	~21	~2,5,9,95 →94
P1	↑71	↑20,21,24 ~34 →20	~2 ↑29,33
N1	≈68,93 ~68,93 →86	~26 ↑40 ≈26	→10,16
P2	↓63 →63,86,93 ↓93	~26,34 ↑34,35 ≈26 ↓40	~10
N2	→85,86,89	≈26,32 →26,38 ↑27 ↓28,32	→1,3,10,11,14,17,38
MMN passive	↓63,64,65,66,67,68,69,72,89,90,91 →63,65,66,69,90 ≈68,73 ~68,73	↓30,31 ←37 →41	↓29,36 →29
MMN active			↓15
P3a passive	→69,70,84,85,86,87,88,89,90 ~68,74 ↓69,72,84,88,89 ≈68,74	→20,24,26,28,38 ≈26,27 ↑34,39	→1,3,4,8,10,12,14,17,22,38 ~7 ↓3,13,23 ≈12
P3a active		→25 ≈32	→6,18,19 ≈11 ↓6,18
P3b	→82,83,87,92,93 ↓74,82,83,92,93	↓25	

↓↑≈ amplitude decrease, increase, no change; ← → ~ latency decrease, increase, no change.

AD and MCI: [1]Goodin and Asimov (1986), [2]Grimes et al. (1987), [3]St Clair et al. (1988), [4]Ball et al. (1989), [5]Tachibana et al. (1989), [6]Polich et al. (1990), [7]Kraiuhin et al. (1990), [8]Marsh et al. (1990), [9]Kuskowski et al. (1991), [10]Williams et al. (1991), [11]Verleger et al. (1992), [12]Ortiz et al. (1994), [13]Holt et al. (1995), [14]O'Mahony et al. (1996), [15]Kazmerski et al. (1997), [16]Pekkonen et al. (1999), [17]Sumi et al. (2000), [18]Yamaguchi et al. (2000), [19]Golob and Starr (2000), [20]Golob et al. (2001), [21]Irimajiri et al. (2005), [22]Gironell et al. (2005), [23]Ally et al. (2006), [24]Golob et al. (2007), [25]Juckel et al. (2008), [26]Caravaglios et al. (2008), [27]Papaliagkas et al. (2008), [28]Papaliagkas et al. (2011), [29]Cheng et al. (2012), [30]Mowszowski et al. (2012), [31]Lindin et al. (2013), [32]Cid-Fernández et al. (2014), [33]Green et al. (2015), [34]Li et al. (2016), [35]Lister et al. (2016), [36]Jiang et al. (2017), [37]Gao et al. (2018), [38]Cintra et al. (2018), [39]Correa-Jaraba et al. (2018), [40]Buján et al. (2019), [41]Chan et al. (2021), [94]Tachibana et al. (1989), [95]Kuskowski et al. (1991), *Aging:* [63]Bertoli et al. (2002), [64]Kisley et al. (2005), [65]Cooper et al. (2006), [66]Näätänen et al. (2012), [67]Cheng et al. (2013), [68]Getzmann and Näätänen (2015), [69]Tsolaki et al. (2015), [70]Correa-Jaraba et al. (2016), [71]Nowak et al. (2016), [72]Morrison et al. (2019), [73]Ruohonen et al. (2020), [74]Hsu et al. (2021), [75]Otto and McCandles (1982), [76]Rosenhall et al. (1986), [77]Martini et al. (1991), [78]Ottaviani et al. (1991), [79]Rowe (1978), [80]Rosenhall et al. (1985), [81]Konrad-Martin et al. (2012), [82]Pfefferbaum et al. (1980a), [83]Picton et al. (1984), [84]Pratt et al. (1989), [85]Coyle et al. (1991), [86]Iragui et al. (1993), [87]Gaál et al. (2007), [88]Fjell and Walhovd (2001), [89]Cooper et al. (2006), [90]Schiff et al. (2008), [91]Rimmele et al. (2012), [92]Ford et al. (1979), [93]Pfefferbaum et al. (1980b).

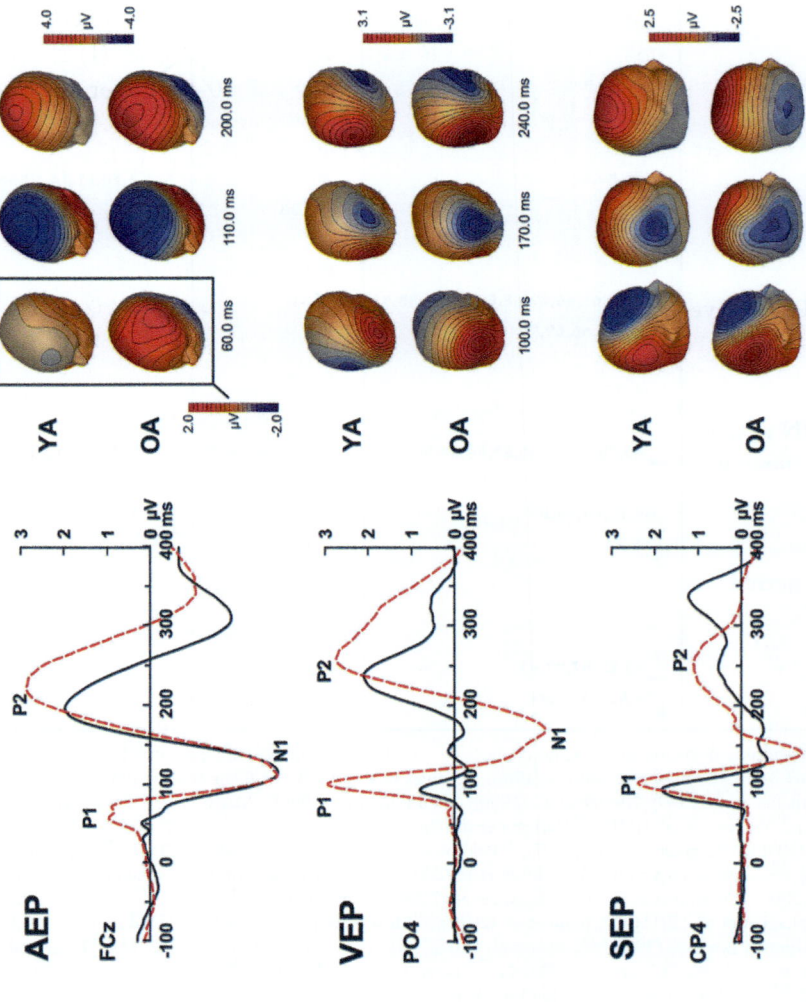

Fig. 9.2 Left. Group mean AEPs, VEPs, and SEPs in young (YA, black solid lines) and older (OA red dashed lines) adults. Right, the isocontour maps highlighting the P1, N1, and P2 deflections in all three sensory modalities. FCz = midline frontal central; PO4 = right parietal-occipital; CP4 = right central-parietal. *From Alain et al. (2022).*

significance, with weaker synchrony between the mid–PFC and the right somatosensory cortex.

9.1.2.2 Endogenous potentials

Because often hearing loss is not tested—or not reported—in psychiatric and neurological studies, we start with a note of caution pertaining to isolating the effects of hearing loss and cognition on the P3. Pollock and Schneider (1992) evoked the P3 by auditory stimuli under two conditions: (1) 250-Hz standard, 500-Hz target and (2) 1000-Hz standard, 2000-Hz target. A group with normal hearing (mean age = 52.7 years), defined as less than 16-dB loss at each tone frequency, was compared to a group with hearing impairment (mean age = 69.9 years), defined with a loss of 16 dB or more at each tone frequency. At 500 Hz the mean loss in this group was 26 dB, and 44.4 dB at 2000 Hz. The hearing-impaired group showed significant P3 latency prolongations compared to those with normal hearing for the 2000 Hz, but not for the 500-Hz target, even though subject age was used as a covariate. Under both conditions, hearing-impaired subjects showed reduced P3 amplitudes compared to those with normal hearing. This raises the possibility that some of the latency prolongations ascribed to age or cognitive processing deficits might instead be attributed to hearing impairment.

We proceed with aging studies that focused on the task-related P3b component. Ford et al. (1979) used N1 and P3b components to test participants' abilities to decrease attention to irrelevant stimuli in order to better detect target stimuli. Twelve healthy old (mean age = 80.3 years) and 12 healthy young adults (mean age = 22.0 years) listened to 1500-Hz tones in one ear and 800-Hz tones in the other ear. Note that audiometry was performed but only to exclude subjects with threshold levels greater than 30-dB SPL at 500 Hz. Infrequently, the frequency of either tone was raised to function as the target. During one run, infrequent tones in the right ear were targets, and in the other run those in the left ear were targets. The participants were asked to count the targets. For both groups, the early N1 component was larger to tones in the attended ear than in the unattended ear, and the P3b component was largest to the targets. P3 latency was longer for old subjects and this indicated to Ford et al. (1979) that these took longer to decide stimulus relevance. However, in the light of our caution, they likely had a greater hearing loss in the test frequencies compared to the controls. This caution applies to several of the following studies. The same group (Pfefferbaum et al., 1980b) tested elderly (mean age = 78.6 years) and young women in a reaction–time task designed to elicit middle and late ERPs.

Auditory thresholds were obtained only at 1000 Hz. Tones of 1000 Hz (72% of the time), 500 Hz (14%), and 2000 Hz (14%) were presented randomly. The subject was instructed to press a button when one of the two types of infrequent tones occurred. The aged subjects differed from the young with respect to the later occurring ERP components: P2 was larger and had a longer latency; P3b had a longer latency and had a different scalp distribution. In contrast, no age-related differences were found for N1 amplitude or latency.

To investigate a correlation between ERPs and reaction time (RT), Picton et al. (1984) recorded ERPs from healthy subjects aged 20–79 years. All subjects were screened for normal hearing at 20 dB above normal threshold for 1000- and 2000-Hz tones. The ERP to a detected improbable signal (2000 Hz, probability = 10%) in a series of 1000-Hz tones (90%) contained a P3b. The latency of the P3b wave in the response to the auditory signal increased regularly with increasing age at a rate of 1.36 ms per year, and its amplitude decreased at a rate of 0.18 µV per year. This occurred independently of any change in the reaction time, which showed no significant age-related change. This indicates different mechanisms for reaction time and P3b. Bertoli et al. (2002) found no significant differences in the psychoacoustic gap detection—at 1 kHz—thresholds between young and elderly subjects. However, longer gaps were needed to elicit MMN in elderly subjects. Pure-tone thresholds were ≤20 dB from 0.25 to 3 kHz. They also had significantly reduced MMN peak amplitudes, increased MMN peak latencies, a significantly smaller P2 amplitude, and longer P2 latency in their responses to the standard stimulus when compared to the same measures in young subjects. Changes in N1 were not significant. The MMN to the difficult-to-detect deviants, but not to the larger deviants, was enhanced in amplitude with auditory attention when compared to the unattended condition.

Schiff et al. (2008) recorded ERPs in normal subjects, aged 20–80 years, with 10–12 subjects per age decade. No audiograms were taken. Four blocks of short tones were delivered (20% rare 2000 Hz and 80% frequent 1000 Hz, presented at 110-dB SPL). In the first two blocks, subjects performed a distracting visual search task (distracted condition); in the latter two blocks, they had to attend the occurrence of the rare tones (active condition). There were no changes in N1, whereas MMN amplitude decreased with age. N2b and P3 latencies increased with age, while their amplitudes decreased. Females had larger amplitude P3 than males. In the elderly, P3 latency was longer in the second block than in the first one. Thus Schiff et al. (2008) "detected a higher influence of aging on late cognitive processes (P3) than on the perceptual (N1) and pre-attentive (MNN) ones." Morrison et al. (2019) determined whether healthy older adults (65+ years) involuntarily detect unattended auditory stimuli as efficiently as younger adults. To test this, younger and older adults were presented with to-be-ignored auditory sequences consisting of frequently presented 80-dB SPL standards and rarely presented increments (+10 dB) and decrements (−20 dB). The deviance-related negativity (representing the summation of the N1 changes and MMN) to the decrement did not differ between the two groups. On the other hand, the deviance-related negativity to the increment was significantly reduced in the older adults. Importantly, the P3a was also significantly reduced in the older adults. This reduced P3a may reflect a deficit in the involuntary shift of attention from current cognitive demands to a potentially more critical event (Fig. 9.3).

9.1.3 Cortical changes in aging

Vaden et al. (2015) used word recognition in multitalker babble to evaluate if the cingulo-opercular network (CON) engagement provides performance benefit for older adults. The CON is composed of anterior insula/operculum, dorsal anterior cingulate cortex, and thalamus (Sadaghiani and D'Esposito, 2015). Healthy older adults (50–81 years of age; mean pure-tone thresholds <32-dB HL from 0.25 to 8 kHz, best ear) performed word

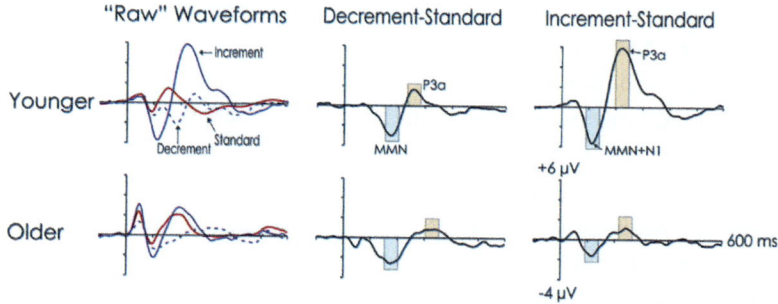

Fig. 9.3 Raw standard, decrement, and increment ERPs (left column) in younger and older adults. Data are from the Cz electrode site. Positivity in this and all other figures is indicated by an upward deflection. A small negative deflection occurring at about 100 ms is apparent after the presentation of the standard (N1). Its amplitude was not significantly different between younger and older adults. The decrement standard and increment standard are illustrated in the center and right columns, respectively. The MMN and P3a are better observed in the difference wave. Abbreviations: ERP, event-related potential; MMN, mismatch negativity. *Reprinted from Event-related potentials associated with auditory attention capture in younger and older adults. Neurobiol. Aging 77, 20–25, Morrison, C., Kamal, F., Campbell, K., Taler, V., 2019, with permission from Elsevier.*

recognition in multitalker babble at two signal-to-noise ratios (SNR, +3 or +10 dB) during a sparse sampling fMRI experiment. Vaden et al. (2015) found that elevated CON activity was associated with an increased likelihood of correct recognition on the following trial independently of SNR and performance on the preceding trial. The CON effect increased for participants with the best overall performance. These effects were lower for older adults compared with a younger, normal-hearing adult sample.

9.1.4 Predictive coding in aging

From a Bayesian perspective (Chapters 1 and 7), the brain represents a model of its environment and offers predictions about the environment, while responding, through changing synaptic strengths to novel interactions and experiences. Moran et al. (2014) hypothesized that these predictive and updating processes are modified as we age, representing an optimization of neuronal architecture. Using magnetoencephalogram (MEG) recordings, Moran et al. (2014) elicited the MMN in a large cohort of humans in their third to ninth decade. Following previous dynamic causal modeling (DCM) studies of the MMN (cf. Chapters 1 and 7), they "used a six–source model— Heschl's gyrus (HG), superior temporal gyrus (STG) and inferior frontal gyrus (IFG) all bilateral—to characterize age effects within the MMN network. In this dynamic causal model, auditory input enters bilaterally at HG; these primary auditory sources are connected via forward connections to STG sources, which in turn sent forward connections to the IFG. Reciprocal backward connections are included to allow signal propagation down the hierarchy from IFG to STG and from STG to HG (Fig. 9.4C). Each MMN source was modeled with a neural mass model comprising three neuronal populations, with distinct receptor types and intrinsic connectivity." Specifically, the model contains synaptic parameters that encode the contribution of AMPA, NMDA, and $GABA_A$ receptor-mediated currents in three populations: comprising pyramidal cells, inhibitory interneurons, and granular-layer spiny-stellate cells (Chapter 1; Fig. 1.12). These populations are connected intrinsically and receive extrinsic inputs according to their laminar disposition: forward connections drive spiny stellate cells and backward connections drive pyramidal cells and inhibitory interneurons. Moran et al. (2014) found no evidence for age-dependent differences in the model fit; however, they found a selective age-related attenuation of synaptic connectivity changes that drive rapid sensory learning. In contrast, baseline synaptic connectivity strengths stayed consistently strong over the decades.

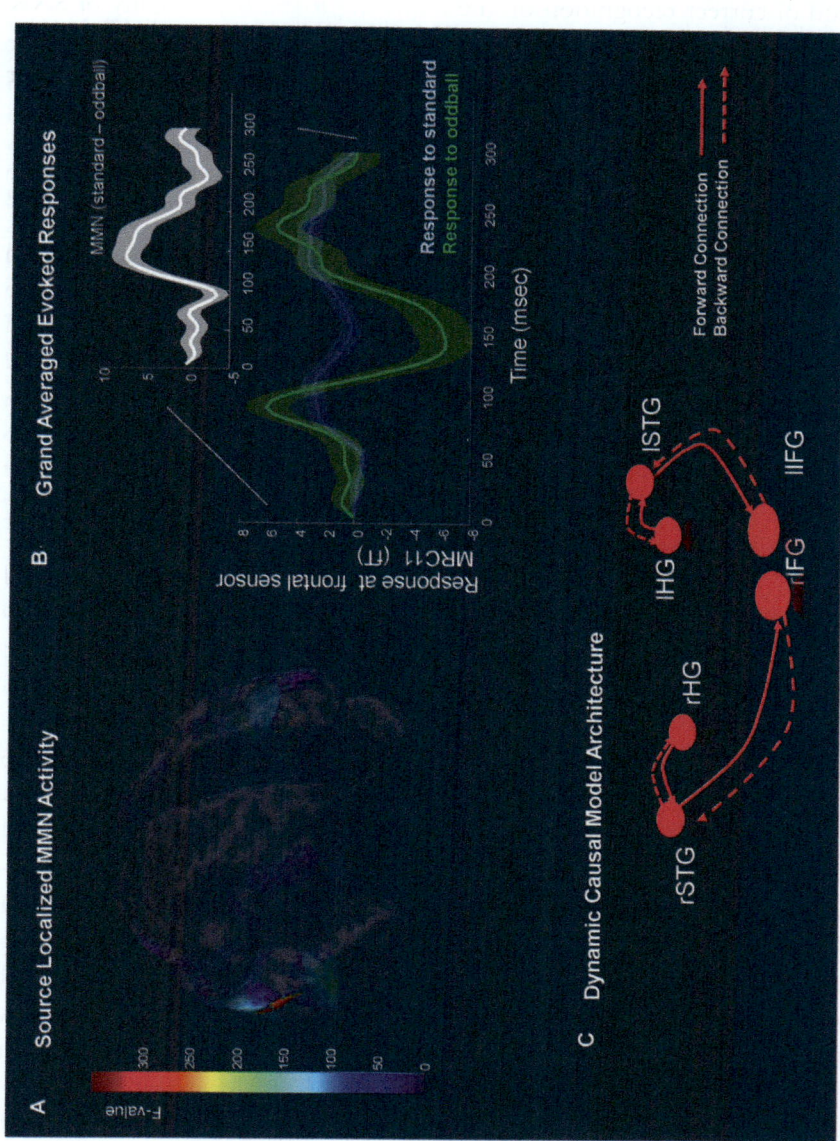

A Source Localized MMN Activity

B Grand Averaged Evoked Responses

C Dynamic Causal Model Architecture

Fig. 9.4 See legend on next page.

(Continued)

Fig. 9.4 Mismatch Network. **(A)** Statistical parametric mapping (SPM) of mismatch (standard–oddball) effect across subjects ($P < 0.05$ FWE corrected) sharing a color-coded F-statistic on a semitransparent canonical cortical inflated mesh. This SPM compares the power (in frequencies from 0 to 30 Hz, over 60–300 ms of peristimulus time), evoked by oddball stimuli with the equivalent power evoked by standard stimuli. **(B)** Auditory evoked responses recorded at one magnetoencephalogram (MEG) sensor over right frontal cortex. Plotted are the grand-averaged evoked measurements across all sessions (shaded areas represent their standard deviation) in response to standard tones (blue) and oddball tones (green). The difference in these responses constitutes the mismatch negativity (MMN); seen here as the negative differences from 100 to 200 ms (white inset)—as predicted from the literature. Both types of trials were fitted for each subject in the dynamic causal modeling (DCM) analysis. **(C)** In the DCM, we modeled the transmission of neuronal activity from primary sensory to frontal regions using three sources reciprocally connected in each hemisphere; source location priors were as follows: left Heschl's gyrus (HG): $x = 242$, $y = 222$, $z = 7$; right HG: $x = 46$, $y = 214$, $z = 8$; left STG: $x = 261$, $y = 232$, $z = 8$; right STG: $x = 59$, $y = 225$, $z = 8$; left IFG: $x = 246$, $y = 20$, $z = 8$; right IFG: $x = 46$, $y = 20$, $z = 8$. Inputs entered HG bilaterally and were passed via forward connections to superior temporal gyrus (STG) within each hemisphere. STG sent top-down backward connections to HG. STG also sent forward connections up to inferior frontal gyrus (IFG) and received backward connections from IFG. Each source is modeled in the DCM with a neural mass model. The parameters of synaptic interactions within each source, as well as the extrinsic connections between sources, were optimized during model inversion. The extrinsic connectivity was equipped with an additional parameter that allowed for different connection strengths during standard or oddball stimulus processing. *From Moran et al. (2014).*

Cooray et al. (2014) also used DCM to study connectivity in healthy young and old subjects. MMN was elicited with an auditory oddball paradigm in young (mean age = 26) and older (mean age = 76) healthy subjects. They showed that the MMN-generating network consisted of five nodes—missing the left IFG compared to Moran et al. (2014)—that could modulate all intra- and internodal connections. This model showed that old subjects had significantly increased input from right STG to the right IFG (red arrow) together with increased inhibition of IFG (pyramidal cells; blue feedback loop). Furthermore, there was reduced modulation of activity within right IFG on stimulus change (Fig. 9.5). In this model (compare with the maturation model in Fig. 6.13 and generic models in Chapter 1), the age–related change in MMN is due to a decline in frontal-based control mechanisms, with alterations in connectivity between temporal and frontal regions together with a dysregulation of the excitatory-inhibitory balance in the right IFG (Cooray et al., 2014).

Model 3

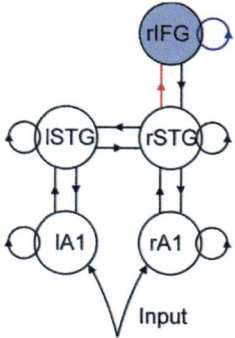

Input

Fig. 9.5 DCM analysis revealed that there was a significant reduction in modulation of intrinsic connectivity at rIFG for old subjects ($B=-0.08\pm0.5$, coupling constant) compared to young subjects ($B=0.3\pm0.2$). Furthermore, there was increased extrinsic connectivity in the network between rSTG and rIFG ($P<0.01$, red line in figure) and increased inhibitory activity at rIFG ($P<0.03$, node colored blue) when generating the standard tone ERP. *Reprinted from 'A mechanistic model of mismatch negativity in the aging brain. Clin. Neurophysiol. 125, 1774–1782. Cooray, G., Garrido, M.I., Hyllienmark, L., Brismar, T., 2014. ©2014 International Federation of Clinical Neurophysiology, with permission from Elsevier.*

Predictive processing is fundamental to audition. This is because the auditory system essentially deals with sequential inputs, which contains relevant information for the imminent future. It seems that older brains are less predisposed to updating the prior probability estimate—but see Fig. 9.5 for increased updating of rIFG—leading to a perception of the environment increasingly dominated by top-down information (i.e., from IFG to STG; Moran et al., 2014). In other words, age may turn our brain into a one-sided prediction machine where the sensory input is weighted less to optimize predictions. Hsu et al. (2021) investigated whether this age-related shift from emphasis on sensation to prediction occurs at all levels of hierarchical message passing. They recorded ERPs with an auditory local-global mismatch paradigm (cf. Fig. 2.11) in a cohort of 108 healthy participants from 3 groups: seniors (55–82 years), adults (19–27 years), and adolescents (15–18 years). The detection of local deviancy was largely preserved in older individuals at earlier latency (including the MMN followed by the P3a but not the reorienting negativity, RON; cf. Appendix). In contrast, attenuation of P3b and worse task performance showed that the detection of global deviancy is clearly compromised in older individuals. Hsu et al.'s (2021) findings demonstrate that older brains show little decline in sensory (i.e., first-order) prediction errors but significant diminution in contextual (i.e., second-order) prediction errors.

Aging typically degrades the precision of peripheral and central processing, which leads to decreased weighting of sensory inputs and increased reliance on predictions (Chan et al., 2021). Age-related attenuation of learning-dependent increase in forward connectivity from primary auditory cortex suggests a reduced sensitivity to the ascending prediction errors (Moran et al., 2014). In addition, there is an age-related increase of the inhibitory effect at the right IFG, indicating increased firing rate of the inhibitory neurons (Cooray et al., 2014) over the lifespan. It seems that older brains are less predisposed to updating the prior probability estimate, leading to a perception of the environment increasingly dominated by top-down information.

9.2 Tinnitus

Tinnitus is an auditory percept occurring without a related sound source in- or out-side the body. This defines tinnitus as a phantom, akin to phantom pain. Tinnitus, in the form of ringing, buzzing, roaring, or hissing, is perceived by ~15% of the population and population wise ~2.5% experiences a severely bothersome tinnitus. Tinnitus is mostly, but not exclusively, a result of hearing loss and is more frequent in the older age groups (Eggermont, 2012, 2022).

9.2.1 Tinnitus and event-related potentials
9.2.1.1 MMN
Using EEG-based MMN, Weisz et al. (2004) demonstrated abnormalities in tinnitus sufferers that are specific to frequencies located at the audiometrically normal lesion edge as compared to normal-hearing controls. Groups also differed with respect to the cortical locations underlying the MMN. Sources in the 90–135 ms latency window were generated in more anterior brain regions in the tinnitus group. Both measures of abnormality correlated significantly with emotional–cognitive distress related to tinnitus. Using an oddball paradigm with frequency, duration, and silent gap deviants, Mohebbi et al. (2019) showed lower MMN amplitude for the higher-frequency deviant and for the silent gap deviant in decompensated tinnitus group compared to normal control and compensated tinnitus group. Decompensated tinnitus is defined as a complex psychosomatic process in which the person suffers considerably from tinnitus and does not habituate to it. In this context, mental and emotional factors affect the perception of

tinnitus. Tinnitus is considered compensated when the person hears this phantom sound but habituates to it and does not essentially feel affected by it, or only complains in specific situations such as quiet environments, stressful situations, and physical tension. Mohebbi et al. (2019) suggested a deficit in sensory memory and change detection processing in decompensated tinnitus subjects. This causes persistent prediction errors, consequently the tinnitus signal is consistently detected as a new signal and activates the brain salience network and consequently prevents habituation to tinnitus. Chen et al. (2022) recorded EEGs under an oddball paradigm in healthy control subjects, tinnitus patients, patients with sensorineural hearing loss (SNHL), and patients with both SNHL and tinnitus. Compared with controls, tinnitus patients with or without SNHL had decreased MMN amplitudes, and SNHL patients had longer MMN latencies.

9.2.1.2 P300

Using an auditory oddball paradigm (for top-down analyses) and a passive listening paradigm (for bottom-up analyses) while recording EEGs, Hong et al. (2016) explored whether top-down or bottom-up components were more critical in the neuropathology of tinnitus, independent of peripheral hearing loss. They observed significantly reduced P300 amplitudes (reflecting fundamental cognitive processes such as attention) and evoked theta power (reflecting top-down regulation in memory systems) for target stimuli at the tinnitus frequency of patients with tinnitus but without hearing loss. The contingent negative variation (CNV), reflecting top-down expectation of a subsequent event prior to stimulation (Appendix), and N1 (reflecting auditory bottom-up selective attention) were different between the healthy controls and tinnitus groups. Interestingly, when tinnitus patients were divided into two subgroups based on their P300 amplitudes—a higher attentional resourcing group (larger P300; T1) and a lower attentional resourcing group (smaller P300; T2)—their P170 and N200 components to tinnitus-frequency sound were different (Fig. 9.6).

From a literature review, Cardon et al. (2020) determined which ERP components differed systematically between tinnitus patients and controls. They found that P300 amplitude was significantly smaller in tinnitus patients, while its latency was significantly prolonged. No other investigated ERP components were found to differ between tinnitus and nontinnitus subjects. Using an oddball paradigm, Cardon et al. (2022) confirmed that the P300 was significantly reduced in the tinnitus group. Source estimation revealed that the response of tinnitus patients was characterized by a

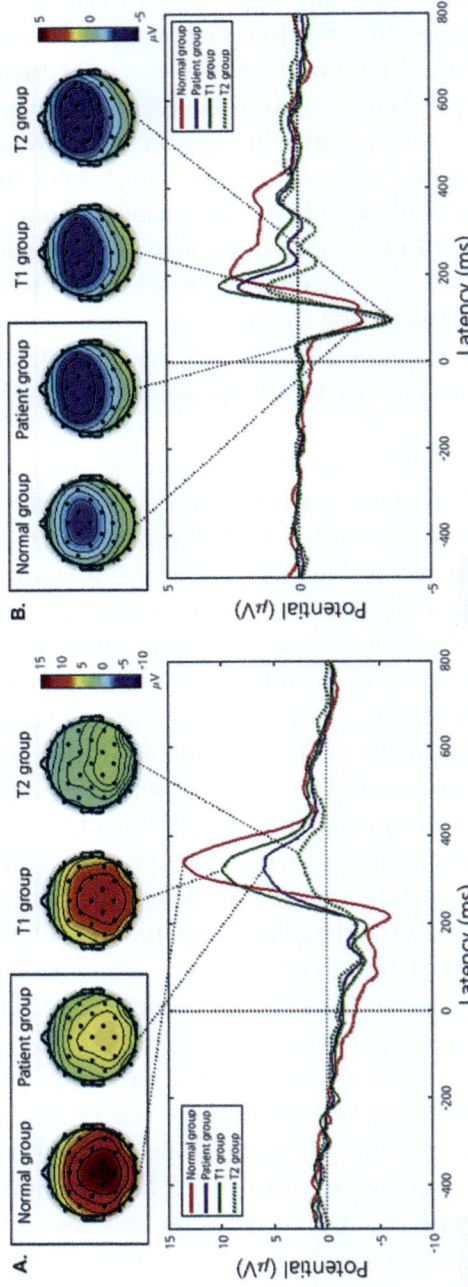

Fig. 9.6 Grand-averaged ERP time courses for the normal healthy controls (red solid line), patients with tinnitus (blue solid line), T1 (green solid line), and T2 (green dotted line) groups at the Cz electrode and the topographies of their maxima (A) during the processing of target stimuli in the oddball task and (B) the processing of standard stimuli. Note the decreased CNV and systematically reduced P300 amplitudes in the patient groups in (A), and the enhanced N1 amplitudes and shorter latencies of P170 and N200 in the patient groups compared to the healthy group in (B). Time zero indicates stimulus onset. The color bar indicates the amplitude (mV). All topographies are shown from the vertex view, with the nose at the top. *From Hong et al. (2016). Open Access.*

decreased activity in temporal cortex, parahippocampus, and insula. They concluded that tinnitus was associated with a decreased cognitive performance, especially on tasks measuring delayed memory.

9.2.1.3 Brain rhythms

Song et al. (2015) correlated the tinnitus awareness percentage with source-localized cortical oscillatory activity and functional connectivity in tinnitus patients. The activity of bilateral rostral anterior cingulate cortices (ACCs), left dorsal- and pregenual ACCs for the delta band, bilateral rostral/pregenual/subgenual ACCs for the theta band, and left rostral/pregenual ACC for the beta1 band displayed significantly negative correlations with tinnitus awareness percentage. Also, the connectivity between the left primary auditory cortex (A1) and the rostral ACC, as well as between the left A1 and the subgenual ACC for the beta 1 band, was negatively correlated with tinnitus awareness percentage. To put this in a predictive-coding perspective, Sedley et al. (2016b) used recordings from the human auditory cortex surface in medically refractory epilepsy and found that surprise due to prediction violations is encoded by oscillations in the gamma band (>30 Hz), whereas changes to predictions occur in the beta band (12–30 Hz), and the precision of predictions appears to quantitatively relate to alpha-band oscillations (8–12 Hz). These results confirm oscillatory codes for critical aspects of generative models of perception.

9.2.2 Predictive coding and tinnitus

In one of the first studies mentioning the potential of predictive coding to understand tinnitus (Roberts et al., 2013), we presented a qualitative model for a role of attention in tinnitus (Fig. 9.7) and discussed evidence in the light of predictive coding. "A key assumption of the model is that in tinnitus there is a disparity between what the brain predicts it should be hearing (this expectation influenced by neural activity underlying the tinnitus percept) and the acoustic information that is delivered by the ear to the brain, when cochlear damage indexed by the audiogram or more sensitive measures is present. The disparity between the predicted and obtained input activates mechanisms of auditory attention that may contribute to the establishment and persistence of tinnitus and to its modulation by competing tasks. The model is intended to illustrate how predicted auditory events might be evaluated at the level of primary auditory cortex. Although input from auditory pathways to the cortex is presumed to be impaired in tinnitus, leading to prediction failure, the output of cortical neurons representing the tinnitus

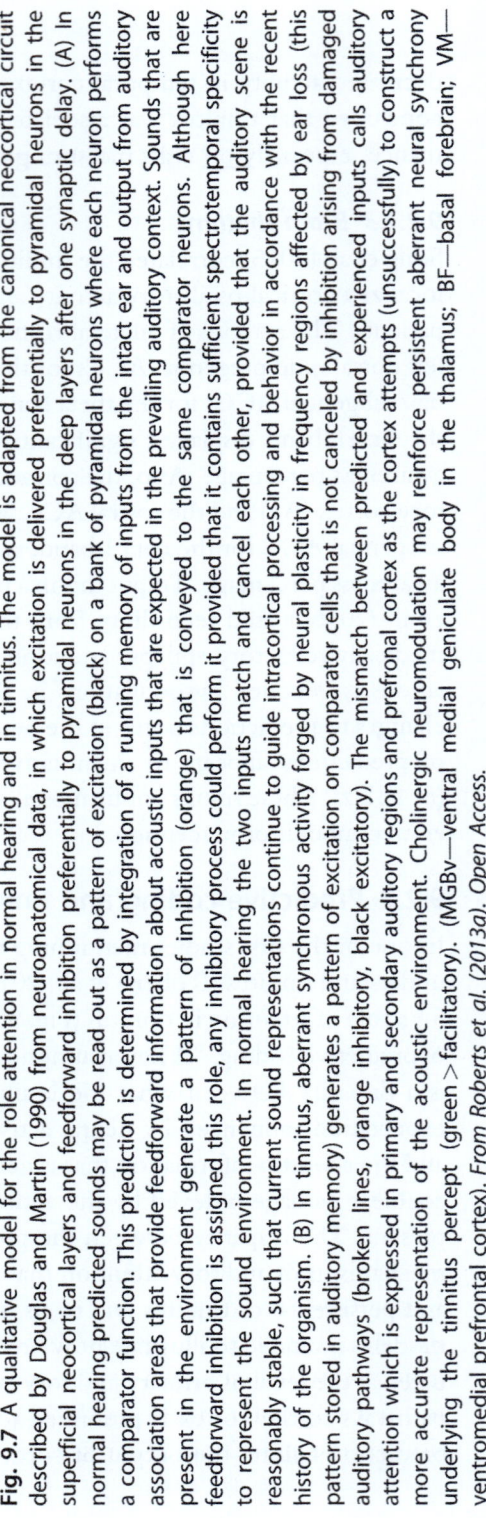

Fig. 9.7 A qualitative model for the role attention in normal hearing and in tinnitus. The model is adapted from the canonical neocortical circuit described by Douglas and Martin (1990) from neuroanatomical data, in which excitation is delivered preferentially to pyramidal neurons in the superficial neocortical layers and feedforward inhibition preferentially to pyramidal neurons in the deep layers after one synaptic delay. (A) In normal hearing predicted sounds may be read out as a pattern of excitation (black) on a bank of pyramidal neurons where each neuron performs a comparator function. This prediction is determined by integration of a running memory of inputs from the intact ear and output from auditory association areas that provide feedforward information about acoustic inputs that are expected in the prevailing auditory context. Sounds that are present in the environment generate a pattern of inhibition (orange) that is conveyed to the same comparator neurons. Although here feedforward inhibition is assigned this role, any inhibitory process could perform it provided that it contains sufficient spectrotemporal specificity to represent the sound environment. In normal hearing the two inputs match and cancel each other, provided that the auditory scene is reasonably stable, such that current sound representations continue to guide intracortical processing and behavior in accordance with the recent history of the organism. (B) In tinnitus, aberrant synchronous activity forged by neural plasticity in frequency regions affected by ear loss (this pattern stored in auditory memory) generates a pattern of excitation on comparator cells that is not canceled by inhibition arising from damaged auditory pathways (broken lines, orange inhibitory, black excitatory). The mismatch between predicted and experienced inputs calls auditory attention which is expressed in primary and secondary auditory regions and prefrontal cortex as the cortex attempts (unsuccessfully) to construct a more accurate representation of the acoustic environment. Cholinergic neuromodulation may reinforce persistent aberrant neural synchrony underlying the tinnitus percept (green > facilitatory). (MGBv—ventral medial geniculate body in the thalamus; BF—basal forebrain; VM—ventromedial prefrontal cortex). *From Roberts et al. (2013a). Open Access.*

sound remains intact, such that the predicted pattern of auditory activity may be conveyed top-down to subcortical structures as well as bottom-up to higher association areas, enabling the prediction to be assessed at multiple levels of the auditory projection pathway. Prediction has been proposed by several authors as an operating principle in the auditory system (Chapter 1). [...] A final aspect of the model concerns the nature of the attention system that is activated when prediction failure occurs. The basal forebrain cholinergic system (possibly accompanied by activation of tegmental cholinergic projections) may be a key component of this system. However, the basal forebrain cholinergic system receives top-down input from the PFC and in turn projects to all primary and secondary cortical sensory regions facilitating neural processing in these regions. In principle top-down inputs to the basal forebrain system could reflect the outcome of processing in other sensory domains, so that involvement of this system is not unique to auditory attention" (Roberts et al., 2013).

As a note on interaction between attention and predictive coding, Schröger et al. (2015) wrote: "Predictive coding theories (Chapter 1) postulate a predictive model that generates (feedback) inferences on the content of and the confidence in the sensory evidence. The difference between input and prediction is expressed in a prediction error, which is a (feed forward) signal that updates the model and, thus, helps to reduce the ambiguity with respect to how we interpret the world. Attention affects the precision of the prediction for the attended content via the model by modulating the gain of the prediction error. This improves the model and—as a consequence—solves the perceptual problem. If we consider perception as an active inference process, and accept that attention modulates the neural signals involved in this predictive process, it is clear that attention and prediction should interact. This interaction can take two forms: (i) attention affecting predictive processes and (ii) predictive processes driving attention."

Sedley et al. (2016a) proposed a framework, based on predictive coding, in which spontaneous neural activity in the subcortical auditory pathway constitutes a 'tinnitus precursor' which is normally ignored as imprecise evidence against the prevailing percept of 'silence.' However, direct evidence for such a system, and the physiological basis of its computations, is lacking. Current models feature as contributory mechanisms acting to increase either the intensity of the precursor or its precision. If precision (i.e., postsynaptic gain) rises sufficiently then tinnitus is perceived. Perpetuation arises through focused attention (Roberts et al., 2013), which further increases the precision of the precursor, and resetting of the default prediction to expect tinnitus. Sedley et al. (2016a) proposed a multiplicative effect, between changes in four categories, on the ultimate impact of the tinnitus precursor on perceptual inference. These can be summarized based on how they affect precision: (1) The strength of afferent input, for instance spontaneous firing rate (SFR). (2) Increasing gain within auditory cortex affords greater precision to subcortical input of a given intensity. (3) Synchronous firing of error-coding neurons allows temporal summation of excitatory postsynaptic potentials on their targets, thereby increasing their influence on postsynaptic responses, and in effect their precision. (4) Low-frequency oscillations are likely to play an important role in modulating the precision of the tinnitus precursor and/or its transmission to higher perceptual areas (Sedley et al., 2016a).

A recent study (Eggermont and Kral, 2016), employed a cat model of congenital single-sided deafness (SSD) and reported that SFR in the auditory cortex—an often-used index for tinnitus in animal experiments (Eggermont, 2012, 2022)—was significantly reduced relative to normal-hearing controls, which suggests the absence of tinnitus in congenitally deaf ears, which have no reference or memory of hearing. Lee et al. (2017) noted that missing auditory information due to hearing loss may cause auditory phantom percepts, i.e., tinnitus. This type of deafferentation-induced auditory phantom percept should be preceded by auditory experience because the fill-in phenomenon, namely tinnitus, is based upon auditory prediction and the resultant prediction error. Thus auditory perception requires previous auditory experience. In line with the findings of our animal study, Lee et al. (2017) found that none of human subjects with congenital SSD who had never had any auditory experiences in the affected ear perceived tinnitus on the SSD side, whereas 30 of 44 subjects with acquired SSD experienced tinnitus on the affected side. This suggests that the lack of "auditory memory" is hemisphere specific and likely so is the tinnitus-prediction signal.

Sedley et al. (2019) tested the theory that tinnitus is caused by resetting of auditory predictions toward a persistent low-intensity sound. EEG-based MMN, which quantify the violation of sensory predictions, to unattended tinnitus-like sounds was greater in response to upward than downward intensity deviants (cf. Fig. 9.3) in unselected chronic tinnitus subjects with normal to severely impaired hearing, and in acute tinnitus subjects, but not in hearing and age-matched controls, or in healthy and hearing-impaired controls presented with simulated tinnitus. They proposed that, once the brain has recognized the tinnitus signal as a sound source, it forms a default prediction of that sound continuing, which prevents the spontaneous activity being ignored as noise, and that prediction ensures the persistence of tinnitus once present for a sufficient length of time. Hullfish et al. (2019) also made the case for modeling tinnitus from the perspective of predictive coding. They emphasize two key claims: (1) acute tinnitus reflects an increase in sensory precision in related frequency channels and (2) chronic tinnitus reflects a change in the brain's default prediction. Under predictive coding, learning and inference are equivalent (Friston, 2003), which fits the emerging consensus that the transition from acute to chronic phantom perception is a learning process, albeit a maladaptive one. Mohan et al. (2022) hypothesized that specifically, in patients with minimal to no hearing loss, there is a more top-down subtype of tinnitus that may be driven by maladaptive changes in an auditory predictive coding network. They tested this by using an auditory oddball paradigm with omission of global deviants, a measure that is previously shown to empirically characterize hierarchical prediction errors (PEs; Fig. 9.8).

Mohan et al. (2022) observed: (1) increased predictions characterized by increased prestimulus response and increased alpha connectivity between the parahippocampus (PHC), dorsal ACC and PHC, pregenual ACC, and posterior cingulate cortex (PCC); (2) increased PEs characterized by increased P300 amplitude and gamma activity and increased theta connectivity between auditory cortices, PHC, and dorsal ACC in the tinnitus group; (3) increased overall feedforward connectivity in theta from the auditory cortex and PHC to the dorsal ACC (Fig. 9.9); (4) correlations of prestimulus theta activity to tinnitus loudness and alpha activity to tinnitus distress.

Mohan et al. (2022) concluded that tinnitus patients show increased PEs by stimuli that do not possess the audiological characteristics of the phantom sound. This suggests a predisposition of these patients to PEs generated throughout the auditory domain. This idea expands on recent evidence showing neural changes to tinnitus-related stimuli and the theory that chronically the brain updates tinnitus as the new prediction (Hullfish et al., 2019; Sedley et al., 2019, 2016a). This suggests that changes in predictive coding may not be tinnitus specific but a systemic change in auditory predictive coding (Mohan et al., 2022a).

In addition to effects in auditory processing areas (Heschl's gyrus, superior temporal gyrus), Schlossmacher et al. (2022) found several distinct clusters that indexed deviance processing in frontal and parietal regions (anterior cingulate cortex/supplementary motor area (ACC/SMA), inferior parietal

xxxxx protocol

xxxxY protocol

Omission protocol

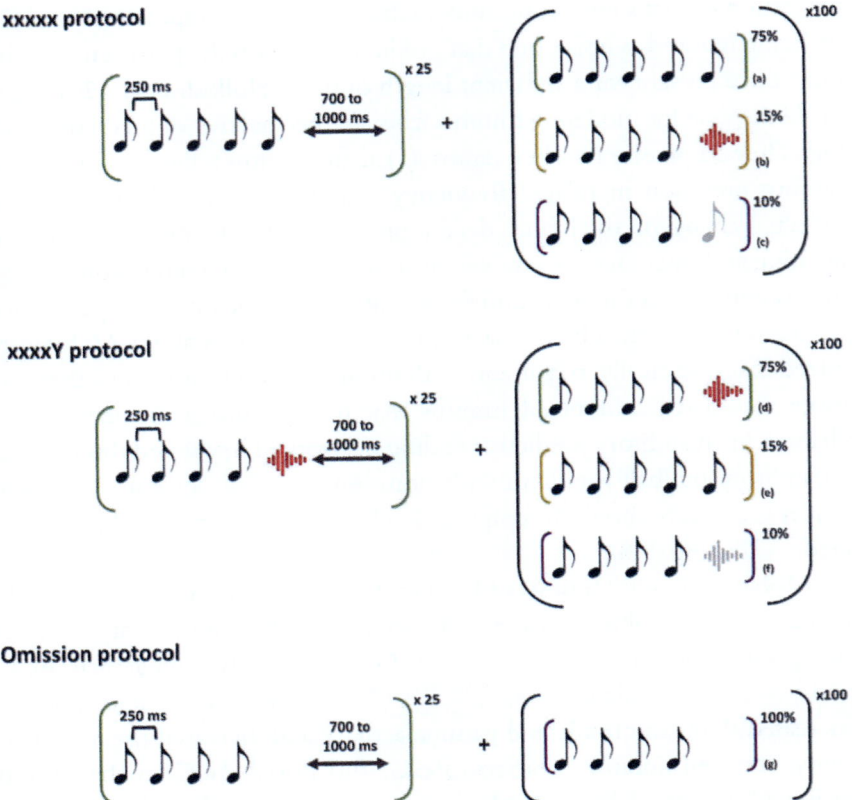

Fig. 9.8 Summary of the auditory oddball paradigm used in this study. Each black note represents the 500-Hz tone, the red burst represents the noise burst, and the light gray sound represents the missing tone/noise burst as per the context. In the xxxxX condition the five-tone sequence is presented with 75% probability, the four tones followed by a noise burst is presented with 15% probability, and the four tones followed by a light gray note representing the omission to the fifth tone is presented with 10% probability. In the xxxxY condition, the four tones followed by a noise burst is presented with 75% probability, the five-tone sequence is presented with 15% probability, and the four tones followed by the light gray burst representing the omission to fifth noise burst is presented with 10% probability. A separate omission sequence consisting of four tones followed by a silence is presented with 100% probability. *From Mahon et al. (2022a). Open Access.*

lobule (IPL), anterior insula (AI), inferior frontal junction (IFJ)). They also observed significant effects of adaptation in Heschl's gyrus, STG, and ACC/SMA, while prediction error–related activity was observed in STG, IPL, AI, and IFJ. Additional dynamic causal modeling confirmed the superiority of a hierarchical processing structure compared to a flat structure (Fig. 9.10).

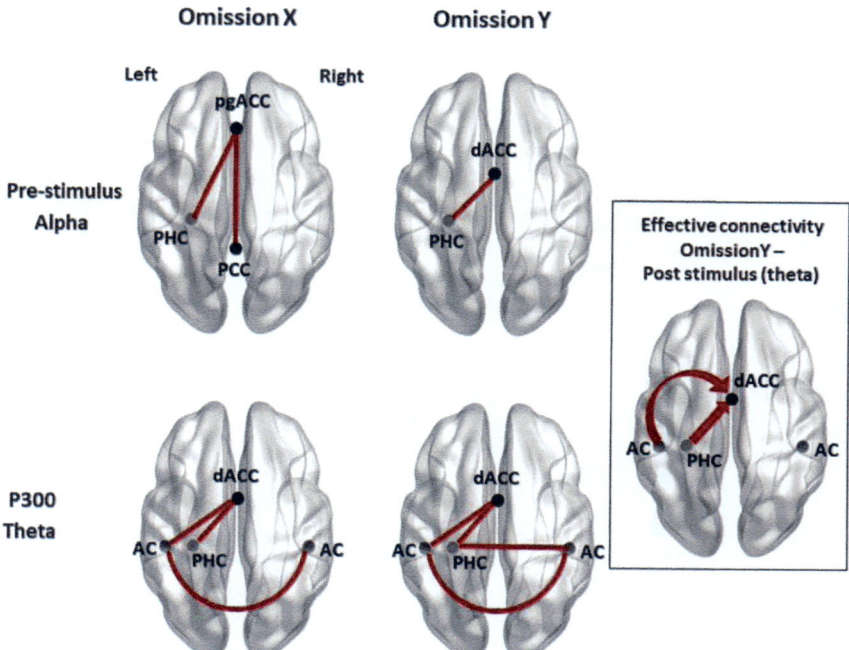

Fig. 9.9 Summary of functional connectivity changes of the deviant-standard responses in for the two omission conditions in the prestimulus (top-row) and P300 (bottom-row) timeframes. The figure shows the significantly increased functional connectivity in the tinnitus group. It also shows the direction of significantly increased flow of information in the tinnitus group highlighted by the boxed image. *From Mahon et al. (2022a). Open access.*

9.3 ERPs in aging, mild cognitive impairment, and Alzheimer's disease

Here our aim will be to understand the transition between mild cognitive impairment (MCI) to Alzheimer's disease (AD). It will therefore be proper to present first Table 9.1 (in Section 9.1.2.1) with about 100 appropriate references—excluding meta-analyses and systematic reviews, which will be presented in the main text—that combines findings from aging, MCI, and AD. This will give a measured overview and also indicates differences and communalities at a glance. There are no exhaustive recordings of the P1–N1–P2 complex changes in the three groups of subjects, and except for aging hardly any data on the ABR. N2 findings are slightly more occurring, and typically reflect increased latencies. MMN changes include increased latencies and decreased amplitudes in aging and MCI, but have

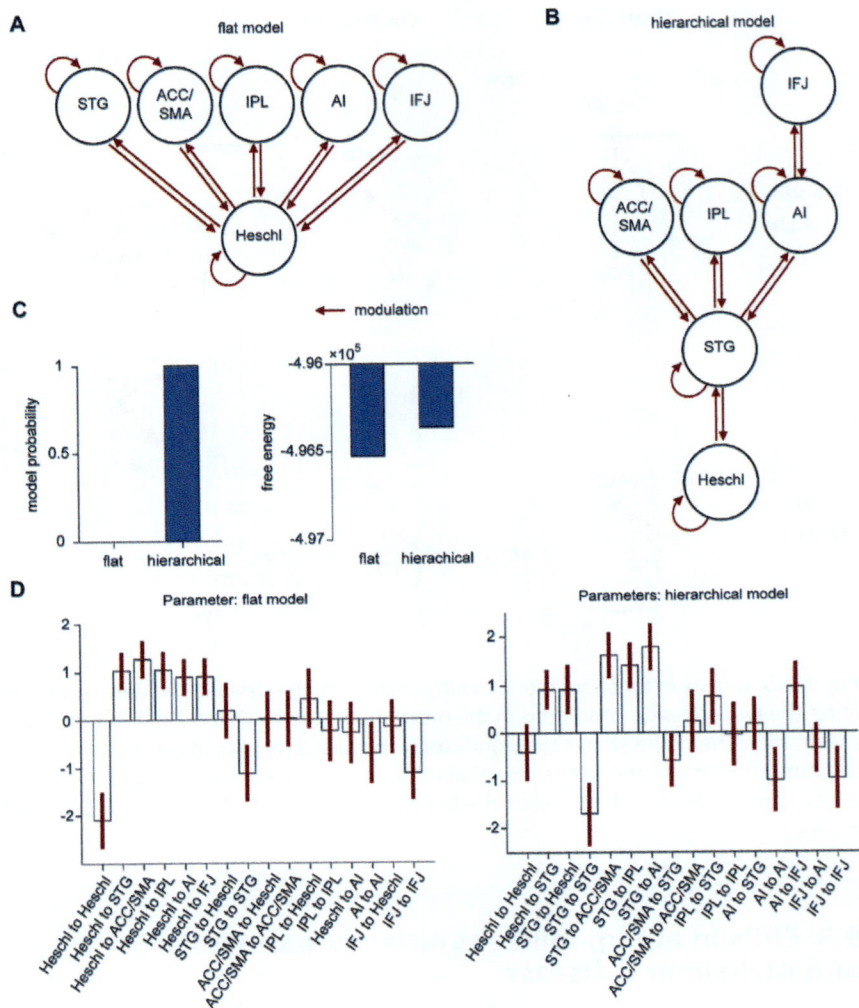

Fig. 9.10 DCM analysis. (B) Model architecture of the hierarchical model. (C) Results of Bayesian model comparison including model probabilities of the second-level and free energy. (D) Parameter estimates for the flat and the hierarchical model. Error bars represent 95% credible intervals. ACC/SMA, anterior cingulate cortex/supplementary motor area; IFJ, inferior frontal junction; IPL, inferior parietal lobe. *From Schlossmacher et al. (2022). Open Access.*

barely been quantified in AD. A much-used measure is found in the P3 complex, which universally shows increased latencies and reduced amplitude in aging as well as in the neurological disorders. Here, the effect of hearing loss cannot be excluded (recall Section 9.1.2.2). Consequently, a tentative first conclusion is that simple parameters of the ERPs do not distinguish between the two disorders and aging. The prediction error amplitudes—reflected in

N2, MMN, and P3—are all reduced and so do not trigger the presence of newness, do not result in corrective top-down reactions, and maintain the various cognitive disabilities.

9.3.1 ABR and MLR studies

Grimes et al. (1987) studied ABRs and middle-latency responses (MLRs) in a group of patients with AD to determine whether a correlate of dementia existed with these potentials. Comparison of absolute and interwave latencies on ABR, and absolute latency and amplitude of the MLR in patients with AD and normal aged controls, showed no significant differences between groups for any measure. Further, no relationship with degree of dementia or temporal lobe involvement, as assessed through dichotic speech recognition studies, and MLRs could be demonstrated. It was concluded that the temporal lobe atrophy and hypometabolism seen in AD are not generally sufficient to disrupt the generating of ABR and MLR potentials. Tachibana et al. (1989) recorded ABRs in patients with multiinfarct dementia (MID; mean age 71.2 years), patients with AD (mean age 70.6 years), and normal subjects (mean age 69.1 years). Both MID and AD patients showed significant prolonged I–V interpeak latencies compared to normal subjects. In patients with MID, both I–III and III–V latencies were significantly longer than those of normal subjects. There were no significant differences between MID and AD with regard to any of the interpeak latencies. Results suggest that brainstem lesions are located in the auditory pathways in patients with MID and AD. Kuskowski et al. (1991) recorded ABRs from patients with mild-to-moderate AD dementia and age-matched normal controls. There were no significant differences between the two groups on any latency measure (wave V latency, I–V and III–V interpeak latencies). Within the AD group, there was no association between ABR latency measures and cognitive status, indexed by scores on the Mini-Mental Status Exam (MMSE). Eight of the AD subjects were retested 6 months and 1 year later. Despite significant evidence of continued cognitive decline, there were no changes in ABR latency measures over time.

9.3.2 Early ERP studies

Goodin and Aminoff (1986) recorded cortical auditory evoked potentials in demented patients with AD, Parkinson's disease, and clinically definite Huntington's disease. There was a significant prolongation in the latency of both the N2 and P3 component of the response for all three demented groups compared with normal controls, with 60% of the combined demented patients having either N2 or P3 prolongations that exceeded the normal regression line by two or more standard errors. In an additional 17% of the patients neither the N2 nor the P3 component could be identified in the evoked potential waveforms. Ball et al. (1989) undertook a longitudinal study of the changes in latency of the P3 wave of the auditory ERP in a group of thoroughly screened and diagnosed possible and probable AD patients and normal controls. For ERP recording, individuals listened to a series of moderate intensity, low-pitched tones of 250 Hz, 70-dB sound pressure level, and 100 ms in duration, with a 1.5-s interstimulus interval. Occasionally and randomly these tones were replaced by a higher pitched, louder tone (450 Hz, 84 dB) occurring on 20% of trials. Subjects were asked to keep a cumulative count of the target tones (32 in all). Subjects who were unable to keep an accurate internal count of the target stimuli were allowed to simply lift a finger in recognition of the target stimuli. On initial record-ing, P3 latency was significantly prolonged in the probable AD group by more than 1.5 standard deviations (40 ms) beyond the normal group. Over the course of the next 3 years, the rate of increase in P3 latency was signif-icantly greater for the patient group than for the controls. The rate of change in P3 latency may reflect accelerated senescence in AD. The same group (Marsh et al., 1990) also recorded P3 in patients with probable AD and found that P3 latency was prolonged by more than 1.5 standard deviations from age expectancy in 14 of 18 patients, but none of 17 controls. In these subjects P3 latency was shown to be inversely correlated with relative metabolic rates (assessed with positron emission tomography, PET) of parietal and, to a lesser extent, temporal and frontal association areas, but not with subcortical areas. Polich et al. (1990) recorded P3 in demented patients presumed to be in the early stages of AD and normal controls matched for age, sex, educa-tion, and occupational level. All subjects performed a simple auditory dis-crimination task in which a target tone was presented randomly on 20% of the trials. P3 amplitude was smaller and peak latency longer for the AD patients compared to controls. In another ERP session, the target tone occurred 50% of the time and AD patients demonstrated less amplitude dif-ference between the target and standard sequences and longer overall laten-cies compared to the control group.

9.3.3 More recent studies

9.3.3.1 ERP studies

In AD and controls, Golob and Starr (2000) measured event-related poten-tials (P50, N100, P200, N200, and P300), the prestimulus readiness potential

(RP), and reaction time as a function of the stimulus sequence. The RP is a manifestation of cortical contribution to the premotor planning of volitional movement and develops before the presentation of both targets and frequents. The target detection task was an "oddball" paradigm containing a sequence of 300 tones with a constant ISI of 2.5 s. Frequent stimuli were 1000-Hz pure tones delivered with a probability of 0.80 (240 tones/sequence). Target tones were 2000-Hz pure tones presented with a probability of 0.20 (60 tones/sequence). Subjects were instructed to quickly press a response button with the thumb of their dominant hand in response to the target tones. Grand-averaged potentials in AD showed a significant reduction of the amplitude of the RP and an increase of P300 latency (Fig. 9.11). Both controls and AD showed faster reaction time, increases in RP amplitude, and decreases in P300 latency as a function of the number of frequents preceding the target.

Golob et al. (2001) then examined ERPs and reaction time in elderly controls and MCI. Subjects listened to a sequence of tones and responded to high-pitched target tones ($P=0.20$) that were randomly mixed with low-pitched tones ($P=0.80$). Two potentials differed between groups; the P50 amplitude and latency were significantly increased in MCI, and P300 latency was significantly longer in MCI. They found that ERPs in MCI subjects during target detection have certain features similar to healthy aging (RP, N100, P200, N200), and other features similar to Alzheimer's disease (delayed P300 latency, Fig. 9.12, and slower reaction time). P50 differences in MCI may reflect pathophysiological changes in the modulation of auditory cortex by association cortical regions having neuropathological changes in early AD.

The same group, Irimajiri et al. (2005), tested whether increased P50 amplitudes in MCI were accompanied by changes of MLRs occurring around 50 ms and/or ABRs. Relative to controls, MCI subjects had larger P50 amplitudes at all stimulus rates. Amplitudes of the MLR components (Pa, Nb, Pb peaking at approximately 30, 40, and 50 ms, respectively; note Pb is part of the P1 complex) did not differ between groups, but a slow wave between 30 and 49 ms on which the middle-latency components arose was significantly increased in MCI. ABR Wave V latency and amplitude did not differ significantly between groups. Golob et al. (2007) then studied subjects with amnestic MCI, AD, and both younger and age-matched older controls. Baseline auditory sensory (P50, N100) and cognitive potentials (P300) were recorded during an auditory discrimination task. MCI patients were followed for up to 5 years, and outcomes were classified as continued diagnosis of MCI (MCI-stable), probable Alzheimer's disease (MCI-convert), or

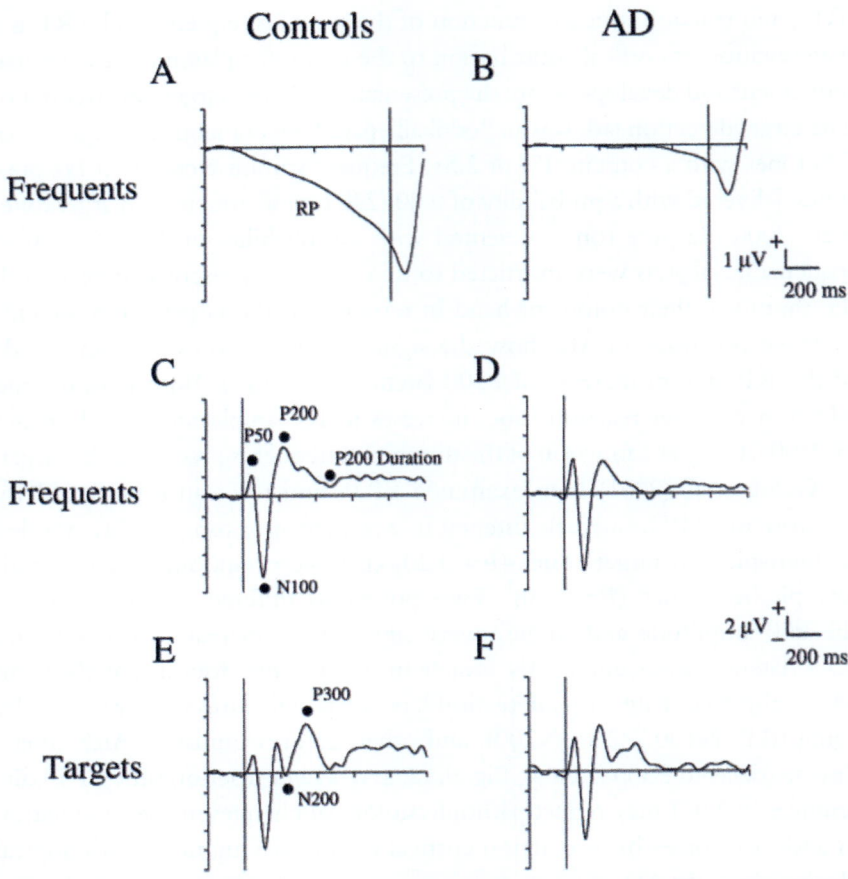

Fig. 9.11 Filtered grand-average tracings divided according to group and stimulus type. Stimulus onset is shown by the vertical line through each tracing. (A,B) RPs preceding frequent tones in controls and AD. Potentials were low-pass filtered (DC-3 Hz). (C,D) Evoked potentials to frequent tones in controls (C) and AD (D). Potentials were band-pass filtered (1–16 Hz) to attenuate slow shifts. (E,F) Evoked potentials to target tones in controls (E) and AD (F). As in (C,D) tracings were band-pass filtered (1–16 Hz). *Reprinted from 'Effects of stimulus sequence on event-related potentials and reaction time during target detection in Alzheimer's disease. Clin. Neurophysiol. 111, 1438–1449. Golob, E.J., Starr, A., 2000. © 2000 Elsevier Science Ireland Ltd., with permission from Elsevier.*

other outcomes. P50 amplitude increased with normal aging and had additional increases in MCI as a function of both initial diagnosis and outcome (MCI-convert > MCI-stable). P300 latency increased with normal aging and had additional increases in MCI but did not differ among outcomes. Golob et al. (2007) concluded that ERPs differ among amnestic MCI subtypes and outcomes occurring up to 5 years later.

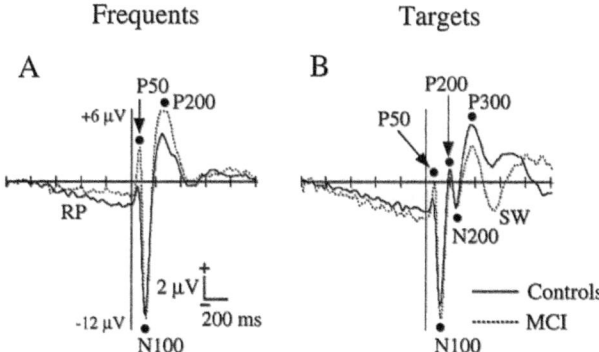

Fig. 9.12 Grand-average potentials for MCI and control groups to frequent (A) and target (B) tones at the Cz electrode site. Significant group differences were observed for P50 amplitude and latency and P300 latency. Two second epochs are shown (1 s before to 1 s after stimulus onset). Vertical line indicates stimulus onset, and potentials were low-pass filtered (DC–16 Hz). *Reprinted from Auditory event-related potentials during target detection are abnormal in mild cognitive impairment. Clin. Neurophysiol. 113, 151–161. Golob, E.J., Johnson, J.K., Starr, A., 2001b. © 2002 Elsevier Science Ireland Ltd., with permission from Elsevier.*

Fruehwirt et al. (2019) examined which ERP component correlates the most with AD severity, as measured by the Mini-Mental Status Examination (MMSE), and analyzed the temporal change of this component as AD progressed. P3 latency exhibited the strongest association with AD severity but there were also significant correlations for N2 latency. P1 amplitude, as measured by different detection methods and at various scalp sites, did not significantly correlate with disease severity—neither in probable AD, possible AD, nor in both subgroups of patients combined. Shad et al. (2022) discussed the notion of functional deficits presenting prior to structural abnormalities in AD. Imaging of the inferior colliculi using magnetic resonance spectroscopy (MRS) shows significant decrease in the *N*-acetyl aspartate/creatinine ratio and increase in the glial marker myo-inositol in subjects with MMSE scores greater than 24 and with no signs of atrophy in their MRI of the medial temporal lobe. They observed a significant decrease in amplitude and increase in latency during the first 10 ms of CAEPs measured on EEG indicating slowing of the ABR.

9.3.3.2 Studies including brain rhythms

Hsiao et al. (2014) investigated "the alterations of functional connectivity in AD during auditory change detection processing. MEG responses to deviant and standard sounds were recorded in AD patients, young and elderly controls. Larger source amplitudes and shorter peak latencies were found in the right temporal mismatch responses of young controls compared with elderly controls and AD patients. During deviant stimuli, the right temporal-frontal theta–phase synchrony was significantly smaller in AD than in young controls and elderly controls. Moreover, the left temporal-frontal synchronization at theta and alpha bands was reduced in AD and elderly controls compared with young controls. Hsiao et al. (2014) concluded that the loss in temporo-frontal theta synchronization might be an electrophysiological hallmark of AD." Jovicich et al. (2019) tested auditory "oddball" ERPs, resting-state fMRI (rs-fMRI) connectivity, and rs-EEG rhythms as longitudinal (2 year) functional biomarkers of prodromal AD. Functional biomarkers that showed significant group effects ("positive" vs "negative") regardless of time were (1) reduced rs-fMRI connectivity in both the DMN and the PCC, both also giving significant time effects (connectivity decay regardless of Group); (2) increased rs-EEG source activity at delta (<4 Hz) and theta (4–8 Hz) rhythms and decreased source activity at low-frequency alpha (8–10.5 Hz) rhythms; and (3) reduced parietal and posterior cingulate source activities of oddball ERPs.

9.3.3.3 Effects of genetic risk factors

Apolipoprotein E (ApoE) status and gender are risk factors for the development of AD, which is more prevalent in female than in male carriers of the ApoE ε4 gene. Irimajiri et al. (2010) examined P50, N100, and P300 in an auditory target detection task in females for ApoE ε4 carriers and ApoE ε4 noncarriers to define the incidence of abnormalities prior to the clinical expression of cognitive impairments. Both neuropsychological test scores and sensory cortical potentials (P50, N100) did not differ between the two ApoE groups. In contrast, the cognitive P3 complex was significantly decreased in amplitude and delayed in latency for ApoE ε4 carriers compared to noncarriers. Four out of the 10 ApoE ε4 carriers had abnormally (>2SD) delayed P3 latency compared to one out of 20 noncarriers (Fig. 9.13). Abnormal cognitive processes reflected by P300 latency delays are expressed at significantly higher incidence in normal older females who are carriers of the ε4 allele than in noncarriers of this allele.

Grothe and Teipel (2016) assessed regional amyloid load using AV45–PET, neuronal metabolism using FDG-PET, and gray matter volume using structural MRI in 473 participants from the Alzheimer's Disease Neuroimaging Initiative, including preclinical, predementia, and clinically manifest AD stages. They found that "amyloid deposition in AD dementia showed

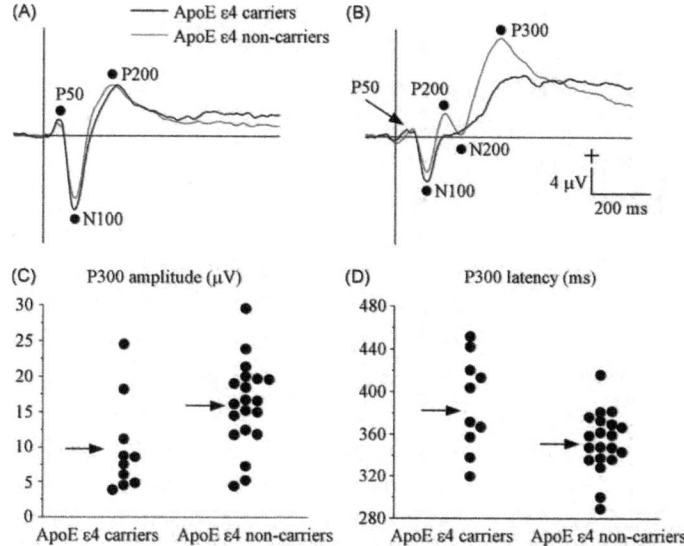

Fig. 9.13 Event-related potentials to (A) nontargets and (B) targets for healthy older female ApoE ε4 carriers and ApoE ε4 noncarriers and plots from individual 10 ApoE ε4 carriers and 20 ApoE ε4 noncarriers for (C) P300 amplitude and (D) P300 latency to targets. Averaged potentials were measured from Cz for P50, N100, and P200 components and Pz for N200 and P300 components and were low-pass filtered (DC—30 Hz). The vertical lines for (A) and (B) indicate stimulus onset, and the epoch lasts from −100 ms to 800 ms relative to the stimulus onset. Arrows in (C) and (D) indicate mean values for ApoE ε4 carriers and noncarriers. *Reprinted from ApoE genotype and abnormal auditory cortical potentials in healthy older females. Neurobiol. Aging 31, 1799–1804. Irimajiri, R., Golob, E.J., Starr, A., 2010.* © *2008 Elsevier Science Ireland Ltd., with permission from Elsevier.*

a preference for the default mode network, but high effect sizes were also observed for other neocortical networks, most notably the frontoparietal-control network (FPN). Atrophic changes were most specific for an anterior limbic network, followed by the DMN, whereas other neocortical networks were relatively spared. Hypometabolism appeared to be a mixture of both amyloid- and atrophy-related profiles. Similar patterns of modality-dependent network specificity were also observed in the predementia and, for amyloid deposition, in the preclinical stage." Grothe and Teipel (2016) "confirmed a high vulnerability of the DMN for multimodal-imaging abnormalities in AD. However, rather than being selective for the DMN, imaging abnormalities more generally affect higher order cognitive networks and, importantly, the vulnerability profiles of these networks markedly differ for distinct aspects of AD pathology."

Gao et al. (2018) investigated amnestic MCI (aMCI) subjects and healthy controls (HCs) that underwent neuropsychological assessment and ApoE genotyping. Novelty MMN and P3a components were recorded during an auditory oddball task. Novelty MMN latency was significantly shorter in aMCI than in HCs. Novelty MMN latency was negatively correlated with episodic memory in aMCI, but not in HC. Novelty P3a latency was negatively correlated with information processing speed in all subjects. Novelty MMN latency was shorter in aMCI females than in HC females. Moreover, novelty P3a amplitudes were lower in males than in females in both aMCI and HC. Novelty MMN latency was shorter in aMCI ApoE ε4 than HC ApoE ε4. The shorter MMN latency may reflect a compensation mechanism for novelty. Cintra et al. (2018) compared the results of neuropsychological tests, N200 and P300, and polymorphisms of ApoE and BDNF rs6265 between patients with normal cognition and those with MCI and AD. They found that patients with cognitive impairment (MCI or AD) showed increase in the latencies of P300 and N200. BNDF gene was not associated with cognitive impairment. Latencies of N200 and P300 increased in cognitively impaired patients with the presence of ApoE ε4 allele. Hojjati et al. (2021) provided preliminary evidence that topographically overlapping Aβ and tau within the DMN are more important for the underlying pathophysiology of AD than each of the tau and/or Aβ pathologies alone. They analyzed publicly available MRI-neuroimaging data from 303 individuals and showed that the probability of observing overlapping Aβ and tau is significantly higher within than outside the DMN. They then showed that using Aβ and tau overlap can increase the reliability of the prediction of healthy individuals converting to mild cognitive impairment and to a lesser degree converting from MCI to AD.

Cheng et al. (2021) recorded MMNm activity in HC and individuals with subjective cognitive decline (SCD). They also analyzed gray matter (GM) volumes in the MMNm-related regions through voxel-based morphometry and performed APOE ε4 genotyping for all the participants. There were no significant differences in GM volume and proportions of APOE ε4 carriers between HC and SCD groups. However, individuals with SCD exhibited weakened z-corrected MMNm responses in the left IPL and right IFG as compared to HC. Based on the regions showing significant between-group differences, z-corrected MMNm amplitudes of the right IFG significantly correlated with the memory performance among the SCD participant (Fig. 9.14). Cheng et al.'s (2021) data suggest that neurophysiological changes of the brain, as indexed by MMNm, precede structural atrophy in the individuals with SCD compared to those without SCD.

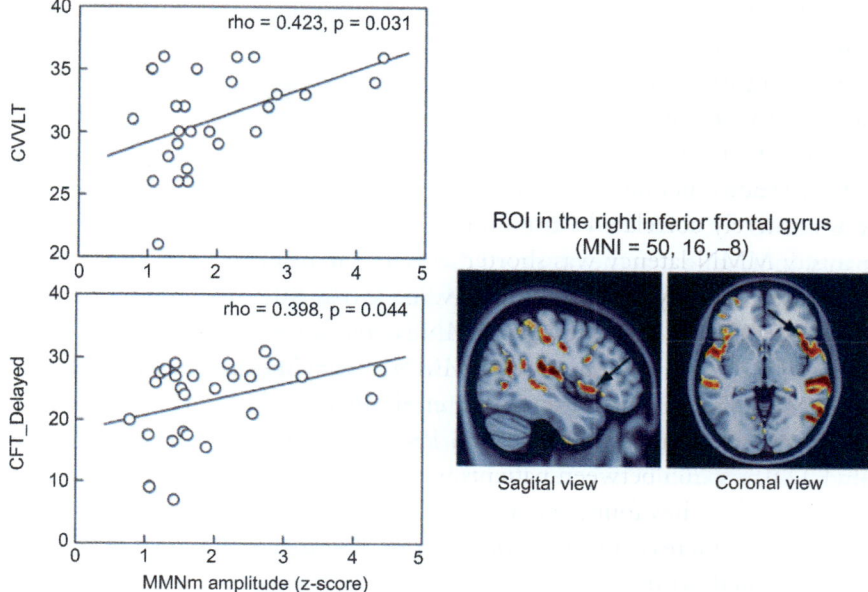

Fig. 9.14 Since significant between-group differences of *z*-corrected MMNm amplitudes were found in the left IPL and right IFG (black arrows), we further examined whether *z*-corrected MMNm amplitudes in these two regions would be associated with the cognitive tests that were related to memory function. Among the individuals with SCD, *z*-corrected MMNm amplitudes of the right IFG were significantly and positively correlated with the performance of total scores of the Chinese Version Verbal Learning Test (CVVLT) and delayed recall scores of Rey–Osterrieth Complex Figure Test (CFT). *From Cheng, C.-H., Chang, C.-C., Chao, Y.-P., Lu, H., Peng, S.-W., Wang, P.-N., 2021c. Altered mismatch response precedes gray matter atrophy in subjective cognitive decline. Psychophysiol. 58, e13820. © 2021 Society for Psychophysiological Research, with permission from John Wiley and Sons.*

9.3.4 Predictive coding and cognitive impairment

Altered predictive coding may underlie the reduced auditory MMN amplitude observed in patients with dementia. Gjini et al. (2022) hypothesized that accumulating dementia-associated pathologies, including amyloid-β and tau, lead to disturbed predictions of our sensory environment. They studied a cross-sectional cohort of participants who underwent PET imaging and high-density EEG during an oddball paradigm and used dynamic casual modeling and Bayesian statistics to make inferences about the neuronal architectures (generators) and mechanisms (effective connectivity) underlying the observed auditory evoked responses. Amyloid-β imaging with [C-11] Pittsburgh Compound-B PET was qualitatively rated using established criteria. Tau-positive PET scans, with [18F]MK-6240, were defined by MK–6240 standardized uptake value ratio positivity threshold at 2 standard deviations above the mean of the Amyloid(−) group in the entorhinal cortex (entorhinal MK–6240 standardized uptake value ratio > 1.27). DCM showed that tau pathology was associated with increased feedforward connectivity and decreased feedback connectivity, with increased excitability of STG but not IFG (Fig. 9.15). Gjini et al. (2022) found that this effect on STG was consistent with the distribution of tau disease on PET in these participants, indicating that the observed differences in mismatch negativity reflect pathological changes evolving in preclinical dementia. Exclusion of participants with diagnosed MCI or dementia did not affect the results. These observational data provide proof of concept that abnormalities in predictive coding may be detected in the preclinical phase of Alzheimer's disease.

Shaw et al. (2021) used behavioral variant frontotemporal dementia (bvFTD) as a model disease. To probe neural circuits in bvFTD, they used a passive auditory oddball paradigm. Auditory stimuli were either standard tones or deviations in 1 of 5 dimensions (frequency, loudness, laterality,

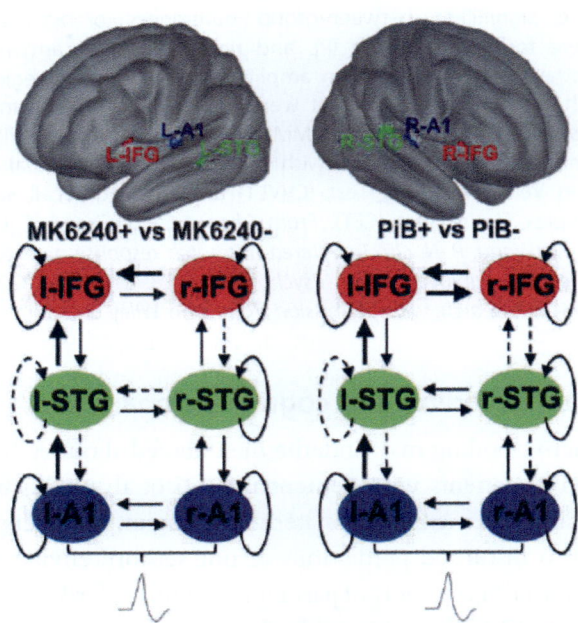

Fig. 9.15 Tau pathology (MK6240+) is associated with increased excitability of superior temporal gyrus (STG) as well as increased feedforward connectivity (from auditory cortex (A1) to STG and from STG to inferior frontal gyrus (IFG)) and decreased feedback connectivity (from IFG to STG). Similar changes in dynamics were associated with amyloid disease (PiB+). *From Gjini et al. (2022). Open Access.*

duration, or a central silent period). Evoked responses to deviant tones and large-scale cortical interactions (Hughes and Rowe, 2013) during such auditory oddball paradigms are grossly abnormal in bvFTD and related disorders. DCM parameters were optimized by inverting to the whole time series of the initial MMN (300 ms), not merely the peak amplitude and latency. Shaw et al. (2021) built a moderately complex model that includes 6 principal generators that have been most extensively studied by MEG, EEG, and direct ECoG (cf. Fig. 9.4; Garrido et al., 2009; Moran et al., 2014). Crucially, analysis of human MEG and ECoG confirmed similar hierarchical network features. Shaw et al. (2021) found that the model included right but not left IFG, just as in models for related MMN in young healthy adults (Garrido et al., 2009), which suggests that the IFG asymmetry is not a result of aging or frontotemporal dementia. However, the Shaw et al. (2021) model also had a feature whose importance was identified by Phillips et al. (2015) (Fig. 7.6), in terms of an expectancy input—putatively from PFC—onto IFG.

Importantly, Pérez-González et al. (2022) reviewed the relation between auditory aging and AD with respect to predictive coding. They observed that a very important and critical implication for predictive coding is directly related to hearing loss and its restoration, but it is unclear how cognitive processes from higher levels may affect purely sensory processes at the lowest levels. They continued with "The reliance on auditory predictions can be significantly disrupted in mild cognitive impairment and dementia. These abilities use temporo-parietal areas that are affected by AD (Golden et al., 2015), and accordingly, patients have difficulty using top-down information to follow conversations in the presence of background noise. Patients with AD show impairments in segregating, tracking, and grouping auditory objects that evolve over time (Goll et al., 2012), and in perceiving sound location and motion (Golden et al., 2015). They are also worse at adapting to expected auditory stimuli as they show reduced auditory MMN responses (Pekkonen et al., 2001; Laptinskaya et al., 2018). Even otherwise healthy APOE4 carriers (i.e., elevated risk of AD) show impairments in detecting auditory targets using contextual information (Zimmermann et al., 2019). Patients with amnestic and logopenic phenotypes of AD are impaired in processing a melodic contour, which depends on working memory to predict the upcoming sounds (Golden et al., 2017). All these ideas are also highlighted by Kocagoncu et al. (2020) and reviewed by Swords et al. (2018). Taken together, the reviewed studies suggested to Pérez-González et al. (2022) that "a core cognitive deficit in AD may involve a deficiency of predictive coding, and this deficiency may be related to the loss of cholinergic adjustment of weighting of bottom–up and top–down inputs in conjunction with loss of auditory sensory input."

9.4 Summary

There is an increasing influence of aging on cognitive processes reflected in preattentive mismatch negativity (MNN) and task–related (P3) compared to the obligatory perceptual (e.g., P1–N1–P2) ERPs, albeit that some of these are affected by attention. In particular, a reduced MMN response to duration– and frequency–deviants is a robust feature among the elderly, which is due to a decline in frontal–based control mechanisms, with alterations in connectivity between temporal and frontal regions. Predictive processing is fundamental to audition, which deals with sequential inputs that contain relevant information for the imminent future. It seems that older brains are less predisposed to updating prior probability estimates, leading to a perception of the environment increasingly dominated by top–down information (e.g., from IFG to STG). It also appears that older brains show little decline in sensory (i.e., first–order) prediction errors but significant diminution in contextual (i.e., second–order) prediction errors.

Using an auditory oddball paradigm (for top–down analyses) and a passive listening paradigm (for bottom–up analyses) while recording EEGs, one may investigate whether top–down or bottom–up components were more critical in the neuropathology of tinnitus. Significantly reduced P300 amplitudes (reflecting fundamental cognitive processes such as attention) and evoked theta power (reflecting top–down regulation in memory systems) for target stimuli at the tinnitus frequency of patients with tinnitus but without hearing loss are found. Modeling tinnitus from the perspective of predictive coding showed (1) increased predictions characterized by increased prestimulus response and increased alpha connectivity between the parahippocampus (PHC), dorsal ACC and PHC, pregenual ACC, and posterior cingulate cortex (PCC); (2) increased prediction errors characterized by increased P300 amplitude and gamma activity and increased theta connectivity between auditory cortices, PHC, and dorsal ACC in the tinnitus group. This suggests increased overall feedforward connectivity in theta from the auditory cortex and PHC to the dorsal ACC. Furthermore, there are correlations of prestimulus theta activity to tinnitus loudness and alpha activity to tinnitus distress.

In aging and MCI, MMN changes include increased latencies and decreased amplitudes, but MMN recordings have barely been used in AD. P3 complex analyses, which are more universally used, show increased latencies and reduced amplitude in aging as well as in MCI and AD. A tentative conclusion is that simple parameters of the ERPs do not distinguish between MCI, AD, and aging. Prediction error amplitudes—reflected in N2, MMN, and P3—are all reduced and so do not trigger the presence of newness, do not result in corrective top-down reactions, and maintain the various cognitive disabilities. Altered predictive coding may underlie this reduced auditory MMN amplitude observed in patients with dementia, attributed to pathologies such as amyloid and tau deposits. DCM shows that tau pathology is associated with increased feedforward connectivity and decreased feedback connectivity, with increased excitability of STG but not IFG. The effect on STG was consistent with the distribution of tau disease on PET, indicating that the observed differences in MMN may already reflect pathological changes evolving in preclinical dementia. Core cognitive deficits in AD may involve a deficiency of predictive coding, and this deficiency may be related to the loss of cholinergic adjustment of weighting of bottom-up and top-down inputs in conjunction with loss of auditory sensory input.

References

Alain, C., Chow, R., Lu, J., Rabi, R., Sharma, V.V., Shen, D., et al., 2022. Aging enhances neural activity in auditory, visual, and somatosensory cortices: the common cause revisited. J. Neurosci. 42 (2), 264–275.

Ally, B.A., Jones, G.E., Cole, J.A., Budson, A.E., 2006. The P300 component in patients with Alzheimer's disease and their biological children. Biol. Psychol. 72, 180–187.

Ball, S.D., Marsh, J.T., Schubarth, G., Brown, W.S., Strandburg, R., 1989. Longitudinal P300 latency changes in Alzheimer's disease. J. Gerontol. 44 (6), M195–M200.

Bertoli, S., Smurzynski, J., Probst, R., 2002. Temporal resolution in young and elderly subjects as measured by mismatch negativity and a psychoacoustic gap detection task. Clin. Neurophysiol. 113, 396–406.

Buján, A., Lister, J.L., O'Brien, J.L., Edwards, J.D., 2019. Cortical auditory evoked potentials in mild cognitive impairment: evidence from a temporal-spatial principal component analysis. Psychophysiology 56, e13466.

Caravaglios, G., Costanzo, E., Palermo, F., Muscoso, E.G., 2008. Decreased amplitude of auditory event-related delta responses in Alzheimer's disease. Int. J. Psychophysiol. 70, 23–32.

Cardon, E., Joossen, I., Vermeersch, H., Jacquemin, L., Mertens, G., Vanderveken, O.M., et al., 2020. Systematic review and meta-analysis of late auditory evoked potentials as a candidate biomarker in the assessment of tinnitus. PLoS One 15 (12), e0243785. https://doi.org/10.1371/journal.pone.0243785.

Cardon, E., Vermeersch, H., Joossen, I., Jacquemin, L., Mertens, G., Vanderveken, O.M., et al., 2022. Cortical auditory evoked potentials, brain signal variability and cognition as biomarkers to detect the presence of chronic tinnitus. Hear. Res. 420, 108489.

Chan, J.S., Wibral, M., Stawowsky, C., Brandl, M., Helbling, S., Naumer, M.J., et al., 2021. Predictive coding over the lifespan: increased reliance on perceptual priors in older adults—a magnetoencephalography and dynamic causal modeling study. Front. Aging Neurosci. 13, 631599. https://doi.org/10.3389/fnagi.2021.631599.

Chen, J., Zhao, Y., Zou, T., Wen, X., Zhou, X., Yu, Y., et al., 2022. Sensorineural hearing loss affects functional connectivity of the auditory cortex, parahippocampal gyrus and inferior prefrontal gyrus in tinnitus patients. Front. Neurosci. 16, 816712. https://doi.org/10.3389/fnins.2022.816712.

Cheng, C.-H., Baillet, S., Hsiao, F.-J., Lin, Y.-Y., 2013. Effects of aging on neuromagnetic mismatch responses to pitch changes. Neurosci. Lett. 544, 20–24.

Cheng, C.-H., Chang, C.-C., Chao, Y.-P., Lu, H., Peng, S.-W., Wang, P.-N., 2021. Altered mismatch response precedes gray matter atrophy in subjective cognitive decline. Psychophysiology 58, e13820. https://doi.org/10.1111/psyp.13820.

Cheng, C.-H., Wang, P.-N., Hsu, W.-Y., Lin, Y.-Y., 2012. Inadequate inhibition of redundant auditory inputs in Alzheimer's disease: an MEG study. Biol. Psychol. 89, 365–373.

Chhatwal, J.P., Schultz, A.P., Johnson, K.A., Hedden, T., Jaimes, S., Benzinger, T.L.S., et al., 2018. Preferential degradation of cognitive networks differentiates Alzheimer's disease from ageing. Brain 141, 1486–1500.

Cid-Fernández, S., Lindín, M., Diaz, F., 2014. Effects of amnestic mild cognitive impairment on N2 and P3 Go/NoGo ERP components. J. Alzheimer's Dis. 38, 295–306. https://doi.org/10.3233/JAD-130677.

Cintra, M.T.G., Ávila, R.T., Soares, T.O., Cunha, L.C.M., Silveira, K.D., de Moraes, E.N., et al., 2018. Increased N200 and P300 latencies in cognitively impaired elderly carrying ApoE ε-4 allele. Int. J. Geriatr. Psychiatry 33, e221–e227.

Cooper, R.J., Todd, J., McGill, K., Michie, P.T., 2006. Auditory sensory memory and the aging brain: a mismatch negativity study. Neurobiol. Aging 27, 752–762.

Cooray, G., Garrido, M.I., Hyllienmark, L., Brismar, T., 2014. A mechanistic model of mismatch negativity in the ageing brain. Clin. Neurophysiol. 125, 1774–1782.

Correa-Jaraba, K.S., Cid-Fernández, S., Lindín, M., Díaz, F., 2016. Involuntary capture and voluntary reorienting of attention decline in middle-aged and old participants. Front. Hum. Neurosci. 10, 129. https://doi.org/10.3389/fnhum.2016.00129.

Correa-Jaraba, K.S., Lindín, M., Díaz, F., 2018. Increased amplitude of the P3a ERP component as a neurocognitive marker for differentiating amnestic subtypes of mild cognitive impairment. Front. Aging Neurosci. 10, 19. https://doi.org/10.3389/fnagi.2018.00019.

Coyle, S., Gordon, E., Howson, A., Meares, R., 1991. The effects of age on auditory event-related potentials. Exp. Aging Res. 17 (2), 103–111.

Dennis, E.L., Thompson, P.M., 2014. Functional brain connectivity using fMRI in aging and Alzheimer's disease. Neuropsychol. Rev. 24, 49–62. https://doi.org/10.1007/s11065-014-9249-6.

Eggermont, J.J., 2012. The Neuroscience of Tinnitus. Oxford University Press, Oxford, UK.

Eggermont, J.J., 2019a. The Auditory Brain and Age-Related Hearing Impairment, Academic Press, London, UK. ISBN 978-0-12-815304-8.

Eggermont, J.J., 2019b. Auditory brainstem response. In: Levin, K.H., Chauvel, P. (Eds.), Handbook of Clinical Neurology. Clinical Neurophysiology: Basis and Technical Aspects, vol. 160, pp. 451–464, https://doi.org/10.1016/B978-0-444-64032-1.00030-8 (3rd series).

Eggermont, J.J., 2022. Tinnitus and Hyperacusis. Facts, Theories and Clinical Implications. Academic Press, UK.

Eggermont, J.J., Kral, A., 2016. Somatic memory and gain increase as preconditions for tinnitus: insights from congenital deafness. Hear. Res. 333, 37–48.

Fjell, A.M., Walhovd, K.B., 2001. P300 and neuropsychological tests as measures of aging: scalp topography and cognitive changes. Brain Topogr. 14 (1), 25–40.

Ford, J.M., Hink, R.F., Hopkins, W.F., Roth, W.T., Pfefferbaum, A., Kopell, B.S., 1979. Age effects on event-related potentials in a selective attention task. J. Gerontol. 34 (3), 388–395.

Friston, K., 2003. Learning and inference in the brain. Neural Netw. 16, 1325–1352. https://doi.org/10.1016/j.neunet.2003.06.005.

Fruehwirt, W., Dorffner, G., Roberts, S., Gerstgrasser, M., Grossegger, D., Schmidt, R., et al., 2019. Associations of event-related brain potentials and Alzheimer's disease severity: a longitudinal study. Progr. Neuropsychopharmacol. Biol. Psychiatry 92, 31–38.

Gaál, Z., Csuhaj, R., Molnár, M., 2007. Age-dependent changes of auditory evoked potentials—effect of task difficulty. Biol. Psychol. 76, 196–208. https://doi.org/10.1016/j.biopsycho.2007.07.009.

Gao, L., Chen, J., Gu, L., Shu, H., Wang, Z., Liu, D., et al., 2018. Effects of gender and apolipoprotein E on novelty MMN and P3a in healthy elderly and amnestic mild cognitive impairment. Front. Aging Neurosci. 10, 256. https://doi.org/10.3389/fnagi.2018.00256.

Garrido, M.I., Kilner, J.M., Stephan, K.E., Friston, K.J., 2009. The mismatch negativity: a review of underlying mechanisms. Clin. Neurophysiol. 120 (3), 453–463. https://doi.org/10.1016/j.clinph.2008.11.029.

Getzmann, S., Näätänen, R., 2015. The mismatch negativity as a measure of auditory stream segregation in a simulated "cocktail-party" scenario: effect of age. Neurobiol. Aging 36, 3029–3037.

Gironell, A., García-Sánchez, C., Estévez-González, A., Boltes, A., Kulisevsky, J., 2005. Usefulness of P300 in subjective memory complaints. A prospective study. J. Clin. Neurophysiol. 22, 279–284.

Gjini, K., Casey, C., Tanabe, S., Bo, A., Parker, M., White, M., et al., 2022. Greater tau pathology is associated with altered predictive coding. Brain Comm. https://doi.org/10.1093/braincomms/fcac209.

Golden, H.L., Clark, C.N., Nicholas, J.M., Cohen, M.H., Slattery, C.F., Paterson, R.W., et al., 2017. Music perception in dementia. J. Alzheimers Dis. 55, 933–949. https://doi.org/10.3233/JAD-160359.

Golden, H.L., Nicholas, J.M., Yong, K.X.X., Downey, L.E., Schott, J.M., Mummery, C.J., et al., 2015. Auditory spatial processing in Alzheimer's disease. Brain 138, 189–202. https://doi.org/10.1093/brain/awu337.

Goll, J.C., Kim, L.G., Ridgway, G.R., Hailstone, J.C., Lehmann, M., Buckley, A.H., et al., 2012. Impairments of auditory scene analysis in Alzheimer's disease. Brain 135, 190–200. https://doi.org/10.1093/brain/awr260.

Golob, E.J., Irimajiri, R., Starr, A., 2007. Auditory cortical activity in amnestic mild cognitive impairment: relationship to subtype and conversion to dementia. Brain 130, 740–752.

Golob, E.J., Johnson, J.K., Starr, A., 2001. Auditory event-related potentials during target detection are abnormal in mild cognitive impairment. Clin. Neurophysiol. 113, 151–161.

Golob, E.J., Starr, A., 2000. Effects of stimulus sequence on event-related potentials and reaction time during target detection in Alzheimer's disease. Clin. Neurophysiol. 111, 1438–1449.

Goodin, D.S., Aminoff, M.J., 1986. Electrophysiological differences between subtypes of dementia. Brain 109, 1103–1113.

Green, D.L., Payne, L., Polikar, R., Moberg, P.J., Wolk, D.A., Kounios, J., 2015. P50: a candidate ERP biomarker of prodromal Alzheimer's disease. Brain Res. 1624, 390–397.

Grimes, A.M., Grady, C.L., Pikus, A., 1987. Auditory evoked potentials in patients with dementia of the Alzheimer type. Ear Hear. 8 (3), 157–161.

Grothe, M., Teipel, S., Alzheimer's Disease Neuroimaging Initiative, 2016. Spatial patterns of atrophy, hypometabolism, and amyloid deposition in Alzheimer'sdisease correspond to dissociable functional brain networks. Hum. Brain Mapp. 37, 35–53. https://doi.org/10.1002/hbm.23018.

Hojjati, S.H., Feiz, F., Ozoria, S., Razlighi, Q.R., The Alzheimer's Disease Neuroimaging Initiative, 2021. Topographical overlapping of the amyloid-β and tau pathologies in the default mode network predicts Alzheimer's disease with higher specificity. J. Alzheimers Dis. 83, 407–421. https://doi.org/10.3233/JAD-210419.

Holt, L.E., Raine, A., Pa, G., Schneider, L.S., Henderson, V.W., Pollock, V.E., 1995. P300 topography in Alzheimer's disease. Psychophysiology 32, 257–265.

Hong, S.K., Park, S., Ahn, M.-H., Min, B.-K., 2016. Top-down and bottom-up neurodynamic evidence in patients with tinnitus. Hear. Res. 342, 86–100.

Hsiao, F.-J., Chen, W.-T., Wang, P.-N., Cheng, C.-H., Lin, Y.-Y., 2014. Temporo-frontal functional connectivity during auditory change detection is altered in Alzheimer's disease. Hum. Brain Mapp. 35, 5565–5577.

Hsu, Y.-F., Waszak, F., Strömmer, J., Hämäläinen, J.A., 2021. Human brain ages with hierarchy-selective attenuation of prediction errors. Cereb. Cortex 31, 2156–2168.

Hughes, L.E., Rowe, J.B., 2013. The impact of neurodegeneration on network connectivity: a study of change detection in frontotemporal dementia. J. Cogn. Neurosci. 25, 802–813.

Hullfish, J., Sedley, W., Vanneste, S., 2019. Prediction and perception: insights for (and from) tinnitus. Neurosci. Biobehav. Rev. 102, 1–12.

Iragui, V.J., Kutas, M., Mitchiner, M.R., Hlllyard, S.A., 1993. Effects of aging on event-related brain potentials and reaction times in an auditory oddball task. Psychophysiology 30, 10–22.

Irimajiri, R., Golob, E.J., Starr, A., 2005. Auditory brain-stem, middle- and long-latency evoked potentials in mild cognitive impairment. Clin. Neurophysiol. 116, 1918–1929.

Irimajiri, R., Golob, E.J., Starr, A., 2010. ApoE genotype and abnormal auditory cortical potentials in healthy older females. Neurobiol. Aging 31, 1799–1804.

Jiang, S., Yan, C., Qiao, Z., Yao, H., Jiang, S., Qiu, X., et al., 2017. Mismatch negativity as a potential neurobiological marker of early-stage Alzheimer disease and vascular dementia. Neurosci. Lett. 647, 26–31.

Jovicich, J., Babiloni, C., Ferrari, C., Marizzoni, M., Moretti, D.V., Del Percio, C., et al., 2019. Two-year longitudinal monitoring of amnestic mild cognitive impairment patients with prodromal Alzheimer's disease using topographical biomarkers derived from functional magnetic resonance imaging and electroencephalographic activity. J. Alzheimers Dis. 69, 15–35. https://doi.org/10.3233/JAD-180158.

Juckel, G., Clotz, F., Frodl, T., Kawohl, W., Hampel, H., Pogarell, O., et al., 2008. Diagnostic usefulness of cognitive auditory event-related P300 subcomponents in patients with Alzheimers disease? J. Clin. Neurophysiol. 25, 147–152.

Kazmerski, V.A., Friedman, D., Ritter, W., 1997. Mismatch negativity during attend and ignore conditions in Alzheimer's disease. Biol. Psychiatry 42, 382–402.

Kisley, M.A., Davalos, D.B., Engleman, L.L., Guinther, P.M., Davis, H.P., 2005. Age-related change in neural processing of time-dependent stimulus features. Cogn. Brain Res. 25, 913–925.

Kocagoncu, E., Quinn, A., Firouzian, A., Cooper, E., Greve, A., Gunn, R., et al., 2020. Tau pathology in early Alzheimer's disease is linked to selective disruptions in neurophysiological network dynamics. Neurobiol. Aging 92, 141–152. https://doi.org/10.1016/j.neurobiolaging.2020.03.009.

Konrad-Martin, D., Dille, M.F., McMillan, G., Griest, S., McDermott, D., Stephen, A., et al., 2012. Age-related changes in the auditory brainstem response. J. Am. Acad. Audiol. 23 (1), 18–35. https://doi.org/10.3766/jaaa.23.1.3.

Kraiuhin, C., Gordon, E., Coyle, S., Sara, G., Rennie, C., Howson, A., et al., 1990. Normal latency of the P300 event-related potential in mild-to-moderate Alzheimer's disease and depression. Biol. Psychiatry 28, 372–386.

Kuskowski, M.A., Morley, G.K., Malone, S.M., Okaya, A.J., 1991. Longitudinal measurements of brainstem auditory evoked potentials in patients with dementia of the Alzheimer type. Int. J. Neurosci. 60 (1), 79–84. https://doi.org/10.3109/00207459109082039.

Langers, D.R.M., Melcher, J.R., 2011. Hearing without listening: functional connectivity reveals the engagement of multiple nonauditory networks during basic sound processing. Brain Connect. 1 (3), 233–244. https://doi.org/10.1089/brain.2011.0023.

Laptinskaya, D., Thurm, F., Küster, O.C., Fissler, P., Schlee, W., Kolassa, S., et al., 2018. Auditory memory decay as reflected by a new mismatch negativity score is associated with episodic memory in older adults at risk of dementia. Front. Aging Neurosci. 10, 5. https://doi.org/10.3389/fnagi.2018.00005.

Lee, S.-Y., Nam, D.W., Koo, J.-W., De Ridder, D., Vanneste, S., Song, J.-J., 2017. No auditory experience, no tinnitus: lessons from subjects with congenital- and acquired single-sided deafness. Hear. Res. 354, 9–15.

Li, B.-Y., Tang, H.-D., Chen, S.-D., 2016. Retrieval deficiency in brain activity of working memory in amnesic mild cognitive impairment patients: a brain event-related potentials study. Front. Aging Neurosci. 8, 54. https://doi.org/10.3389/fnagi.2016.00054.

Lindín, M., Correa, K., Zurrón, M., Díaz, F., 2013. Mismatch negativity (MMN) amplitude as a biomarker of sensory memory deficit in amnestic mild cognitive impairment. Front. Aging Neurosci. 5, 79. https://doi.org/10.3389/fnagi.2013.00079.

Lister, J.L., Harrison Bush, A.L., Andel, R., Matthews, C., Morgan, D., Edwards, J.D., 2016. Cortical auditory evoked responses of older adults with and without probable mild cognitive impairment. Clin. Neurophysiol. 127, 1279–1287.

Marsh, J.T., Schubarth, G., Brown, W.S., Riege, W., Strandburg, R., Dorsey, D., et al., 1990. PET and P300 relationships in early Alzheimer's disease. Neurobiol. Aging 11 (4), 471–476.

Martini, A., Comacchio, F., Magnavita, V., 1991. Auditory evoked responses (ABR, MLR, SVR) and brain mapping in the elderly. Acta Otolaryngol. 111 (sup476), 97–104.

Mohan, A., Luckey, A., Weisz, N., Vanneste, S., 2022. Predisposition to domain-wide maladaptive changes in predictive coding in auditory phantom perception. NeuroImage 248, 118813.

Mohebbi, M., Daneshi, A., Asadpour, A., Mohsen, S., Farhadi, M., Mahmoudian, S., 2019. The potential role of auditory prediction error in decompensated tinnitus: an auditory mismatch negativity study. Brain Behav. 2019, e01242. https://doi.org/10.1002/brb3.1242.

Moran, R.J., Symmonds, M., Dolan, R.J., Friston, K.J., 2014. The brain ages optimally to model its environment: evidence from sensory learning over the adult lifespan. PLoS Comput. Biol. 10 (1), e1003422. https://doi.org/10.1371/journal.pcbi.1003422.

Morrison, C., Kamal, F., Campbell, K., Taler, V., 2019. Event-related potentials associated with auditory attention capture in younger and older adults. Neurobiol. Aging 77, 20–25.

Mowszowski, L., Hermens, D.F., Diamond, K., Norrie, L., Hickie, I.B., Lewis, S.J.G., et al., 2012. Reduced mismatch negativity in mild cognitive impairment: associations with neuropsychological performance. J. Alzheimers Dis. 30, 209–219. https://doi.org/10.3233/JAD-2012-111868.

Näätänen, R., Kujala, T., Escera, C., Baldeweg, T., Kreegipuu, K., Carlson, S., Ponton, C., 2012. The mismatch negativity (MMN)—a unique window to disturbed central auditory processing in ageing and different clinical conditions. Clin. Neurophysiol. 123, 424–458.

Nowak, K., Oron, A., Szymaszek, A., Leminen, M., Näätänen, R., Szelag, E., 2016. Electrophysiological indicators of the age-related deterioration in the sensitivity to auditory duration deviance. Front. Aging Neurosci. 8, 2. https://doi.org/10.3389/fnagi.2016.00002.

O'Mahony, D., Coffey, J., Murphy, J., O'Hare, N., Hamilton, D., Rowan, M., 1996. Event-related potential prolongation in Alzheimer's disease signifies frontal lobe impairment: evidence from SPECT imaging. J. Gerontol. Med. Sci. 51A (3), M102–M107.

Ortiz, T., Loeches, M.M., Miguel, F., Abdad, E.V., Puente, A.E., 1994. P300 latency and amplitude in the diagnosis of dementia. J. Clin. Psychol. 50 (3), 381–388.

Ottaviani, F., Maurizi, M., D'alatri, L., Almadori, G., 1991. Auditory brainstem responses in the aged. Acta Otolaryngol. 111 (sup476), 110–113.

Otto, W.C., McCandless, G.A., 1982. Aging and auditory site of lesion. Ear Hear. 3 (3), 110–117.

Papaliagkas, V., Kimiskidis, V., Tsolaki, M., Anogianakis, G., 2008. Usefulness of event-related potentials in the assessment of mild cognitive impairment. BMC Neurosci. 9, 107. https://doi.org/10.1186/1471-2202-9-107.

Papaliagkas, V.T., Kimiskidis, V.K., Tsolaki, M.N., Anogianakis, G., 2011. Cognitive event-related potentials: longitudinal changes in mild cognitive impairment. Clin. Neurophysiol. 122, 1322–1326.

Pekkonen, E., Hirvonen, J., Jääskeläinen, I.P., Kaakkola, S., Huttunen, J., 2001. Auditory sensory memory and the cholinergic system: implications for Alzheimer's disease. NeuroImage 14, 376–382. https://doi.org/10.1006/nimg.2001.0805.

Pekkonen, E., Jääskeläinen, I.P., Hietanen, M., Huotilainen, M., Näätänen, R., Ilmoniemi, R.J., et al., 1999. Impaired preconscious auditory processing and cognitive functions in Alzheimer's disease. Clin. Neurophysiol. 110, 1942–1947.

Pérez-González, D., Schreiner, T.G., Llano, D.A., Malmierca, M.S., 2022. Alzheimer's disease, hearing loss, and deviance detection. Front. Neurosci. 16, 879480. https://doi.org/10.3389/fnins.2022.879480.

Pfefferbaum, A., Ford, J.M., Roth, W.T., Kopell, B.S., 1980a. Age differences in P3-reaction time associations. Electroencephalogr. Clin. Neurophysiol. 49, 257–265.

Pfefferbaum, A., Ford, J.M., Roth, W.T., Kopell, B.S., 1980b. Age-related changes in auditory event-related potentials. Electroencephalogr. Clin. Neurophysiol. 49, 266–276.

Phillips, H.N., Blenkmann, A., Hughes, L.E., Bekinschtein, T.A., Rowe, J.B., 2015. Hierarchical organization of frontotemporal networks for the prediction of stimuli across multiple dimensions. J. Neurosci. 35 (25), 9255–9264.

Picton, T.W., Stuss, D.T., Champagne, S.C., Nelson, R.F., 1984. The effects of age on human event-related potentials. Psychophysiology 21 (3), 312–325.

Polich, J., Ladish, C., Bloom, F.E., 1990. P300 assessment of early Alzheimer's disease. Electroencephalogr. Clin. Neurophysiol. 77, 179–189.

Pollock, V.E., Schneider, L.S., 1992. P3 from auditory stimuli in healthy elderly subjects: hearing threshold and tone stimulus frequency. Int. J. Psychophysiol. 12, 237–241.

Pratt, H., Michalewski, H.J., Patterson, J.V., Starr, A., 1989. Brain potentials in a memory-scanning task. II. Effects of aging on potentials to the probes. Electroencephalogr. Clin. Neurophysiol. 72, 507–517.

Rimmele, J., Sussman, E., Keitel, C., Jacobsen, T., Schröger, E., 2012. Electrophysiological evidence for age effects on sensory memory processing of tonal patterns. Psychol. Aging 27 (2), 384–398.

Roberts, L.E., Husain, F.T., Eggermont, J.J., 2013. Role of attention in the generation and modulation of tinnitus. Neurosci. Biobehav. Rev. 37, 1754–1773.

Rosenhall, U., Bjorkman, G., Pedersen, K., Kall, A., 1985. Brain-stem auditory evoked potentials in different age groups. Electroencephalogr. Clin. Neurophysiol. 162, 426–430.

Rosenhall, U., Pedersen, K., Dotevall, M., 1986. Effects of presbycusis and other types of hearing loss on auditory brainstem responses. Scand. Audiol. 15 (4), 179–185.

Rowe III, M.J., 1978. Normal variability of the brain-stem auditory evoked response in young and old adult subjects. Electroencephalogr. Clin. Neurophysiol. 144, 459–470.

Ruohonen, E.M., Kattainen, S., Li, X., Taskila, A.-E., Ye, C., Astikainen, P., 2020. Event-related potentials to changes in sound intensity demonstrate alterations in brain function related to depression and aging. Front. Hum. Neurosci. 14, 98. https://doi.org/10.3389/fnhum.2020.00098.

Sadaghiani, S., D'Esposito, M., 2015. Functional characterization of the cingulo-opercular network in the maintenance of tonic alertness. Cereb. Cortex 25, 2763–2773. https://doi.org/10.1093/cercor/bhu072.

Schiff, S., Valenti, P., Andrea, P., Lot, M., Bisiacchi, P., Gatta, A., Amodio, P., 2008. The effect of aging on auditory components of event-related brain potentials. Clin. Neurophysiol. 119, 1795–1802.

Schlossmacher, I., Dilly, J., Protmann, I., Hofmann, D., Dellert, T., Roth-Paysen, M.-L., et al., 2022. Differential effects of prediction error and adaptation along the auditory cortical hierarchy during deviance processing. NeuroImage 259, 119445.

Schröger, E., Kotz, S.A., San Miguel, I., 2015. Bridging prediction and attention in current research on perception and action. Brain Res. 1626, 1–13.

Sedley, W., Alter, K., Gander, P.E., Berger, J., Griffiths, T.D., 2019. Exposing pathological sensory predictions in tinnitus using auditory intensity deviant evoked responses. J. Neurosci. 39 (50), 10096–10103.

Sedley, W., Friston, K.J., Gander, P.E., Kumar, S., Griffiths, T.D., 2016a. An integrative tinnitus model based on sensory precision. Trends Neurosci. 39 (12), 799–812. https://doi.org/10.1016/j.tins.2016.10.004.

Sedley, W., Gander, P.E., Kumar, S., Kovach, C.K., Oya, H., Kawasaki, H., et al., 2016b. Neural signatures of perceptual inference. eLife 5, e11476. https://doi.org/10.7554/eLife.11476.

Shad, K., Soubra, W., Cordato, D., 2022. The auditory afferent pathway as a clinical marker of Alzheimer's disease. J. Alzheimer's Dis. 85, 47–53. https://doi.org/10.3233/JAD-215206.

Shaw, A.D., Hughes, L.E., Moran, R., Coyle-Gilchrist, I., Rittman, T., Rowe, J.B., 2021. In vivo assay of cortical microcircuitry in frontotemporal dementia: a platform for experimental medicine studies. Cereb. Cortex 31 (3), 1837–1847.

Song, J.J., Vanneste, S., De Ridder, D., 2015. Dysfunctional noise cancelling of the rostral anterior cingulate cortex in tinnitus patients. PLoS One 10 (4), e0123538. https://doi.org/10.1371/journal.pone.0123538.

St Clair, D., Blackburn, I., Blackwood, D., Tyrer, G., 1988. Measuring the course of Alzheimer's disease. A longitudinal study of neuropsychological function and changes in P3 event-related potential. Br. J. Psychiatry 152, 48–54.

Sumi, N., Nan'no, H., Fujimoto, O., Ohta, Y., Takeda, M., 2000. Interpeak latency of auditory event-related potentials (P300) in senile depression and dementia of the Alzheimer type. Psychiatry Clin. Neurosci. 54, 679–684.

Swords, G.M., Nguyen, L.T., Mudar, R.A., Llano, D.A., 2018. Auditory system dysfunction in Alzheimer disease and its prodromal states: a review. Ageing Res. Rev. 44, 49–59. https://doi.org/10.1016/j.arr.2018.04.001.

Tachibana, H., Takeda, M., Sugita, M., 1989. Brainstem auditory evoked potentials in patients with multi-infarct dementia and dementia of the Alzheimer type. Int. J. Neurosci. 48 (3–4), 325–331. https://doi.org/10.3109/002074589090021.

Tsolaki, A., Kosmidou, V., Hadjileontiadis, L., Kompatsiaris, I., Tsolaki, M., 2015. Brain source localization of MMN, P300 and N400: aging and gender differences. Brain Res. 1603, 32–49.

Vaden Jr., K.I., Kuchinsky, S.E., Ahlstrom, J.B., Dubno, J.R., Eckert, M.A., 2015. Cortical activity predicts which older adults recognize speech in noise and when. J. Neurosci. 35 (9), 3929–3937.

Verleger, R., Kompf, D., Neukater, W., 1992. Event-related EEG potentials in mild dementia of the Alzheimer type. Electroencephalogr. Clin. Neurophysiol. 84, 332–343.

Weisz, N., Voss, S., Berg, P., Elbert, T., 2004. Abnormal auditory mismatch response in tinnitus sufferers with high-frequency hearing loss is associated with subjective distress level. BMC Neurosci. 5, 8.

Williams, P.A., Jones, G.H., Briscoe, M., Thomas, R., Cronin, P., 1991. P300 and reaction-time measures in senile dementia of the Alzheimer type. Br. J. Psychiatry 159, 410–414.

Yamaguchi, S., Tsuchiya, H., Yamagata, S., Toyoda, G., Kobayashi, S., 2000. Event-related brain potentials in response to novel sounds in dementia. Clin. Neurophysiol. 111, 195–203.

Zimmermann, J., Alain, C., Butler, C., 2019. Impaired memory-guided attention in asymptomatic APOE4 carriers. Sci. Rep. 9, 8138. https://doi.org/10.1038/s41598-019-44471-1.

CHAPTER 10

Brain networks involved in deviance and novelty detection: Are they sensory modality specific?

10.1 Introduction

Most approaches to connectivity in the neuroimaging literature use functional measures, such as phase synchronization and temporal correlations or spectral coherence, to establish statistical dependencies between activities in two sources (Eggermont, 2021). Although functional connectivity can be used to establish a statistical dependency it does not provide information about the causal architecture of the interactions. Several techniques (detailed in Chapter 1) such as Granger causality and transfer entropy (Gourévitch and Eggermont, 2007), structural equation modeling (McIntosh and Gonzalez-Lima, 1994), and dynamic causal modeling (DCM; Friston et al., 2003; Garrido et al., 2007) use the concept of effective connectivity. Magnetic resonance imaging (MRI) based diffusion tensor imaging (DTI) assesses physical connections via neural tracts.

Effective connectivity refers explicitly to the influence one neuronal system exerts over another. In DCM, the brain is regarded as a deterministic nonlinear dynamic system (Friston et al., 2003; Chapter 1). DCM provides an account of the interactions among cortical regions and allows one to make inferences about system parameters—intrinsic excitation and inhibition; extrinsic connection strength and direction—and investigate how these parameters are influenced by experimental factors. Furthermore, by taking the marginal likelihood over the conditional density of the model parameters, one can estimate the probability of the input data, given a particular putative model (Garrido et al., 2007).

We have seen that the mismatch negativity (MMN) and potentially the P3 serve as error signals in the predictive coding (PC) theory of brain function (Friston, 2005; Chapter 7). As we will see in this chapter, the estimated

Brain Responses to Auditory Mismatch and Novelty Detection
https://doi.org/10.1016/B978-0-443-15548-2.00010-7

networks underlying both auditory MMN (aMMN) and visual MMN (vMMN) often have the right inferior frontal gyrus (IFG) as its "endpoint," albeit that the aMMN network starts in primary auditory cortex (A1) and for vMMN in primary visual cortex (V1). This happens regardless of the presence of a neurological disorder, thus the network for the predictive coding based on decreased aMMN amplitudes and increased latencies is basically the same for autism, attention deficit/hyperactivity disorder (ADHD), aging, mild cognitive impairment (MCI), and Alzheimer's disease.

First of all, we will briefly review evidence that the vMMN and aMMN share similar "mismatch" properties. Although less is known about somatosensory MMN (sMMN), we will also find that it is very similar to aMMN and vMMN. Then we recall how DCM explores the evidence for similar endpoints of these predictive coding networks. We then review fMRI and electroencephalography/magnetoencephalography (EEG/MEG) based network estimates underlying several neurological and psychiatric disorders. In Chapter 12, we will investigate how this MMN network may be embedded with disease-specific network changes.

10.2 Sensory modality-dependent mismatch negativity

Until now, we have limited our study to the auditory-oddball MMN, but for a generalization of the mismatch network to provide estimates of predictive coding errors it is appropriate to also investigate the vMMN and potentially the sMMN. Both of these are not as frequently studied as the aMMN.

10.2.1 Visual MMN
10.2.1.1 Similarities of auditory and visual MMN
Stefanics et al. (2014) reviewed the theoretical underpinnings of vMMN and discussed this in a predictive coding framework: (1) experimental protocols and procedures to control refractoriness effects, such as repetition suppression (Chapter 4); (2) methods to control attention; and (3) relation of vMMN to the direct perception of stimuli. The predictive coding account suggests that repetition suppression depends on the probability structure of the environment (Chapters 4 and 7) and involves an active process which generates models of the causes of the sensory input. These generative models can be thought of as hierarchical memory representations of stimulus characteristics (Chapter 1). A stimulus that does not match this memory representation elicits a "mismatch process." This process is manifested as an

aMMN/vMMN, depending on the sensory stimuli. Predictive coding uses precision-weighted prediction errors (pwPEs) to continuously update this model. Stefanics et al. (2018) used a visual "roving standard" paradigm to elicit vMMN in humans by unexpected changes in either color or emotional expression of faces. They simulated pwPE trajectories of a Bayes-optimal observer and used these to conduct a trial-by-trial analysis. They found significant modulation of brain activity by both color (strong effect) and emotion (weak effect) pwPEs (Fig. 10.1). Stefanics et al. (2018) suggest that vMMN responses reflect trial-wise pwPEs, as postulated by predictive coding.

In a study by Hedge et al. (2015), healthy young adults completed a visual oddball task in separate EEG and fMRI sessions. A region of interest (ROI) analysis was conducted on left and right middle frontal (MFG) and inferior frontal (IFG) gyri, of which the latter has also been identified as a potential aMMN generator (Chapters 1 and 7). A significant increase in activation was observed in the left IFG and MFG in response to blocks containing visually deviant stimuli. These findings suggest that, like the aMMN, a frontal mechanism underlies change detection in the vMMN paradigm, in this case localized to the left IFG. This raises the possibility of a common frontal change detection mechanism.

Rosburg et al. (2019) designed an event-related potential (ERP) study to investigate simultaneously the aMMN and vMMN, as well as the impact of facial emotional stimuli on the aMMN. A vMMN was clearly elicited by fearful-face deviants, but hardly by neutral-face deviants. Neither the aMMN nor the processing of visual targets was modulated by facial emotion. Tse et al. (2021) postulated a generic frontosensory cortical network underlying the prediction violation mechanism: the IFG is responsible for non–modality-specific prediction processes while the sensory cortices are responsible for a modality-specific error signal generation process. They examined the involvement of the IFG-occipital cortex network (OCN) in visual preattentive change detection. Modulations of the IFG–OCN mismatch response patterns by abstractness and stimulus-train length reflect the processing demands on the prediction processes and are similar to that of the IFG–superior temporal gyrus (STG) network in auditory change detection. These findings demonstrated that the frontosensory cortical structure—i.e., IFG—is not unique to auditory preattentive change detection and provided supports for a universal neural mechanism across sensory modalities as suggested by the prediction violation hypothesis.

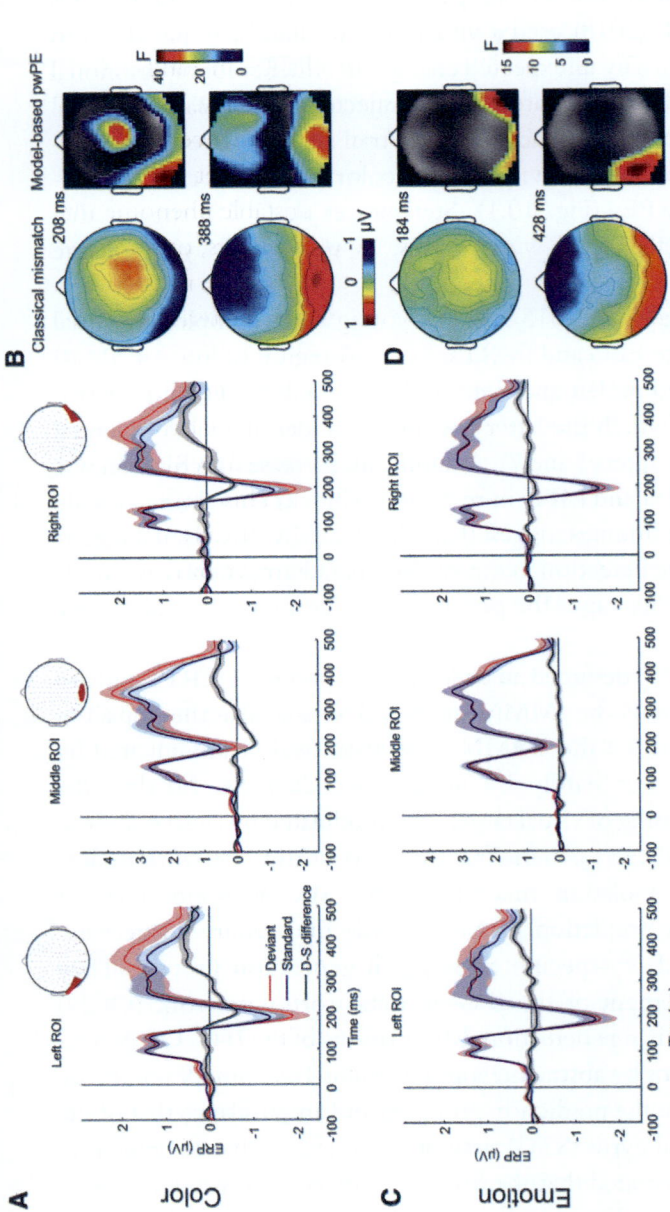

Fig. 10.1 ERP waveforms, scalp voltage maps, and topographic statistical parametric maps. (A) ERPs with 95% confidence interval for changes in color obtained with traditional averaging deviant-minus-standard subtraction. Red areas in channel layout plots show scalp regions where electrodes were used for plotting the ERP waveforms. (B) Scalp potential plots of deviant-minus standard difference waveform (left) at two time points of cluster maxima where SPM analysis yielded significant results. Statistical parametric maps (right) for model-based color pwPE estimates (pooled across emotions) of the F test. Note the high similarity of topographic distributions for the traditionally obtained mismatch responses (with negative and positive posterior scalp distributions) and the SPM obtained with computational model-based analyses. (C, D) Data for the emotion changes, plotted similarly as for color. *From Stefanics, G., Heinzle, J., Horváth, A.A., Stephan, K.E., 2018. Visual mismatch and predictive coding: a computational single-trial ERP study. J. Neurosci. 38(16), 4020–4030.*

10.2.1.2 Limitations of the visual MMN

In two experiments, Male et al. (2020) examined whether the vMMN follows similar principles to the aMMN. They searched for a genuine vMMN from simple, physiologically plausible stimuli that change in fundamental dimensions: orientation, contrast, phase, and spatial frequency. They controlled for attention and eye movements and found no evidence for a genuine vMMN, despite adequate statistical power. Male et al. (2020) concluded that either the genuine vMMN is a rather unstable phenomenon that depends on still-to-be-identified experimental parameters, or it is confined to visual stimuli for which monitoring across time is more natural than monitoring over space. Beck et al. (2021) assessed the characteristics of the vMMN elicited during the categorization of previously unknown visual stimuli. To examine this, they used five-dot patterns that allowed for the formation of categories through rotation and reflection. They observed that both between-category and within-category deviants elicited a vMMN, and that both vMMNs were comparable in magnitude. This suggests that abstract categorical representations are not always automatically processed at early visual stages and demonstrate limitations of generalization from the auditory domain to visual domain.

10.2.1.3 Unconscious prediction errors

Rowe et al. (2020) used a visual oddball paradigm in which participants engaged in a central letter task during EEG recordings while presented with task-irrelevant high- or low-coherence background random-dot motion. Critically, once in a while, the motion direction of the dots changed. They then applied DCM to the EEG data to assess evidence for top–down modulations for PE to visible or invisible change. When an unexpected stimulus occurs, the PE signal is propagated from lower- to higher-brain areas, resulting in upregulation of forward connectivity. This will cause a revised prediction from high- to low-level brain areas, resulting in increased feedback connectivity. Finding of increased forward connectivity from left medial temporal cortex (MTC) to inferior temporal gyrus (ITG) for visible PE is consistent with the theoretical prediction. Surprisingly, unconscious PE related to unseen changes was accompanied by significant decreases in top–down feedback connectivity from right MTC to right V1 (Fig. 10.2). "This means that the neural prediction from MTC to V1 decreased when the motion direction was unexpected, which appears inconsistent with the general framework of predictive coding. One possible explanation is that when the prediction is violated, the system suspends prediction,

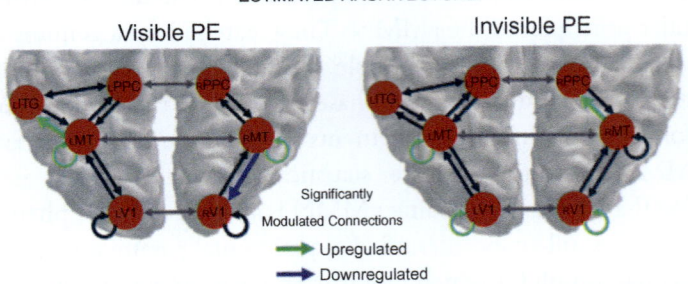

Fig. 10.2 Dynamic causal modeling (DCM) for visible and invisible PE responses. Model evidence results from our random-effects Bayesian Model Selection analyses, displayed as the exceedance probability of each model (i.e., the probability that a particular model is more likely than any other model given the group data). Using Bayesian Model Averaging we were able to obtain weighted parameter estimates using the model evidence across all models and participants. Significantly modulated connections (false discovery rate corrected) are shown in green (upregulated) and blue (downregulated). ITG, inferior temporal gyrus; MT, middle temporal cortex; PPC, posterior parietal cortex. *From Rowe, E.G., Tsuchiya, N., Garrido, M.I., 2020. Detecting (un)seen change: the neural underpinnings of (un)conscious prediction errors. Front. Syst. Neurosci. 14:541670. doi:10.3389/fnsys.2020.541670. Open access.*

corresponding to the down regulated prediction from the high- to low-level area. Invisible PE, on the other hand, only induced enhanced feedforward connectivity from right MTC to PPC" (Rowe et al., 2020).

10.2.2 Somatosensory MMN

Naeije et al. (2018) investigated the spatiotemporal dynamics and the neural mechanisms of the MEG-based sMMN. They identified cortical sources of evoked responses using equivalent current dipole modeling and found that early tactile change detection involves mainly contralateral secondary somatosensory cortex. Shen et al. (2018) examined two factors potentially modulating the amplitude of the sMMN and P300 responses elicited by touch to pairs of body parts, i.e., hand, neck, and lip, varying in (a) the distance between the representation of these body parts in somatosensory cortex, and (b) the distances between the stimulated points on the body surface. The sMMN and the P300 response were elicited by tactile stimulation in two oddball protocols. The neck-lip pairing resulted in significantly larger sMMN responses (with shorter latencies) than the hand-lip pairing, whereas the reverse was true for the amplitude of the P300. Mean sMMN amplitude and latency did not differ between finger pairings. Shen et al. (2018) suggested that, for certain combinations of body parts, early automatic sMMN may be influenced by distance between the cortical representations of these body parts (cf. frequency in aMMN), whereas the later P300 response may be more influenced by the distance between stimulated body parts on the body surface (cf. sound location in aMMN).

10.2.3 Sensory modality-dependent MMN sources

Butler et al. (2011) investigated whether small duration discriminations, represented by the MMN, were generated in the same cortical regions regardless of the sensory modality. Scalp recordings pointed to statistically distinct MMN topographies across auditory and tactile senses, implying differential underlying cortical generator configurations. Intracranial recordings confirmed these noninvasive findings, showing generators of the auditory MMN along the STG with no evidence of a somatosensory MMN in this region, whereas a robust somatosensory MMN was recorded from postcentral gyrus in the absence of an auditory MMN (Fig. 10.3). This indicates stimulus specificity in the more peripheral areas of the MMN-generating networks. Butler et al. (2011) noted that: "Despite a clear dissociation of the areas involved in auditory and somatosensory duration processing as indexed by the MMN, these data do not rule out the possibility that amodal duration representations come online during the active performance of a duration task, or that there might be amodal processing that feeds into sensory-specific areas. It is also possible that there are other metrics of unisensory duration processing that would reveal an amodal representation. However, the MMN provides a solid metric of duration processing for which there are no alternatives that we are aware of, and the loci of MMN generation is commonly considered to reflect where the deviant feature is processed (Molholm et al., 2005)."

10.3 Network changes underlying cognition

We will first present an overview of resting-state networks (RSNs), and then focus our attention on the default mode network (DMN) and the inferior frontal gyrus.

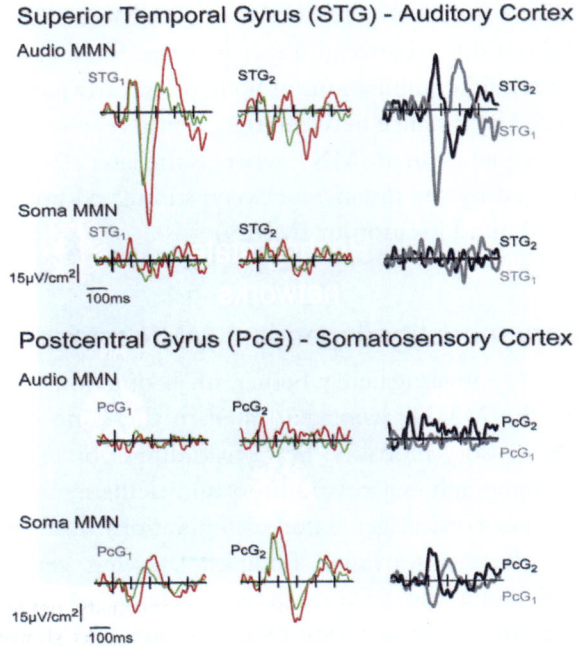

Fig. 10.3 Grand-mean CSD, at intracranial electrodes sites deviant (red line), standard (green line), and the subtraction waveform, MMN (black and gray lines). The yellow and magenta dots represent the electrode locations on the MRI-CTs. The yellow and magenta arrows indicate the Sylvian fissure and the postcentral gyrus, respectively. *From Butler, J.S., Molholm, S., Fiebelkorn, I.C., Mercier, M.R., Schwartz, T.H., Foxe, J.J., 2011a. Common or redundant neural circuits for duration processing across audition and touch. J. Neurosci. 31(9), 3400 –3406.*

10.3.1 Overview of cortical networks

The following is based on a previous review largely discussing the role of resting-state brain rhythms underlying intrinsic and extrinsic network-connectivity changes (Eggermont, 2021). Most agreed–upon resting-state networks are illustrated in Fig. 10.4. These networks can be characterized (Tessitore et al., 2019) as follows: (1) The DMN is involved in introspection, mind–wandering, active episodic memory, and becomes deactivated during specific goal-directed behavior. It encompasses mainly precuneus and posterior cingulate (PCC), bilateral inferior-lateral-parietal, and ventromedial prefrontal cortices (VMPFC). Parts of the IFG are also considered within the DMN. (2) The sensorimotor network (SMN) has a central role in detection and processing of sensory input and preparation and execution of motor functions. It comprises the primary sensorimotor cortex (SI), supplementary

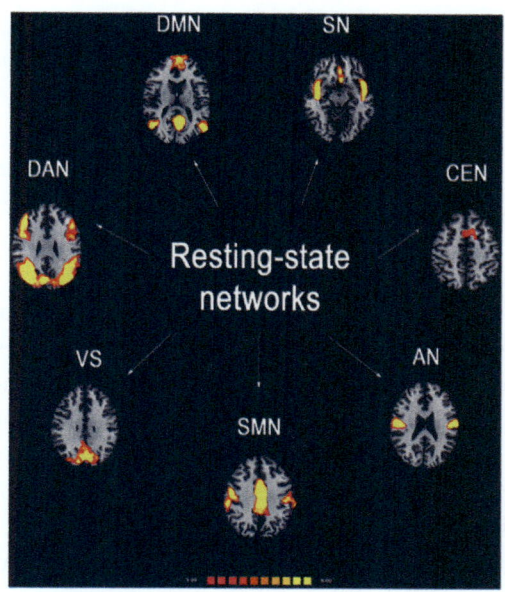

Fig. 10.4 Most reported resting-state functional connectivity networks in healthy controls. Mean resting-state functional MRI imaging networks shown in axial view and three-dimensional reconstructions ($P < 0.05$ corrected). Colors represent percentage BOLD signal change, overlaid on the average anatomic images in standard space. DMN, default mode network; SN, salience network; CEN, central executive network; DAN, dorsal attention network; SMN, sensorimotor network; VS, visual network; AN, auditory network. *From Tessitore, A., Cirillo, M., De Miccoa, R., 2019. Functional connectivity signatures of Parkinson's disease. J. Parkinsons Dis. 9, 637–652. DOI 10.3233/JPD-191592. Open Access.*

motor area (SMA), and secondary somatosensory cortices (SII). (3) The central executive network (CEN) is involved in executive control and working memory function and includes anterior cingulate (ACC) and para–cingulate cortices. This is often considered part of a task-positive network (TPN). (4) The salience network (SAL; SN in the Figure) detects and responds to behaviorally salient events and encompasses mainly the dorsal ACC and bilateral insulae. (5) The dorsal attention network (DAN) is involved in voluntary (top–down) orienting and selective attention. Superior parietal and superior frontal areas, including intraparietal sulcus and frontal eye fields, are the most involved cortical areas. (6) The auditory network (AN) comprises the right and left primary auditory cortex (A1), Heschl's gyrus, lateral superior temporal gyrus (STG), and posterior insular cortex. (7) The visual network (VS) comprises mainly the lateral and superior occipital gyrus as well as the lingual gyrus (Tessitore et al., 2019).

In a meta-analytic approach, Ficco et al. (2021) used the Activation Likelihood Estimation (ALE) algorithm to detect spatial convergence across studies, related to prediction error and encoding. Overall, the ALE results suggest the ultimate role of the left IFG and left insula in both processes. Moreover, they employed a meta-analytic connectivity method (Seed-Voxel Correlations Consensus). This technique reveals a large, bilateral predictive network, which resembles large-scale networks involved in task-driven attention and execution. A key feature of this network is its remarkable similarity to the TPN. The TPN is a set of areas involved in task execution and is usually divided into three large-scale brain networks: salience (SAL), DAN, and ventral attention network (VAN). Activity patterns of the DMN and those of the TPN are anticorrelated (Fig. 10.5 left) and possibly involved in different forms of cognition. In sum, Ficco et al. (2021) find that: (1) predictive processing seems to occur more in certain brain regions than others, when considering different sensory modalities at a time; (2) there is no evidence, at the network level, for a distinction between error and prediction processing, implying that these processes are based on distinct neuron populations within the same networks.

Two studies have detailed the anticorrelation of the DMN and the TPN. Hu et al. (2017) noted that the DMN, a task–negative network, is temporally anticorrelated with the TPN (Fig. 10.5 left), which is, in turn, associated with task-induced increased alertness. The TPN is generally considered to be linked to externally oriented attention, whereas the DMN is associated with introspectively oriented cognitive processes such as self-referential and reflective activity. The inverse correlation of activity in the two types of networks may reflect a competition for attentional resources between external and internal orientation, competitively allocating attentional resources to ready the brain itself, or to keep the brain alert. Magnuson et al. (2015) showed group incidence maps representing the DMN and TPN. Anatomical regions associated with high levels of connectivity in the DMN are shown in Fig. 10.5 top right and include the PCC/precuneus, angular gyri, medial prefrontal cortex, middle frontal gyrus, and the superior frontal gyrus. The highest correlation values within the TPN mask are associated with the dorsal prefrontal cortex, premotor cortex, inferior parietal lobules, and the medial frontal gyrus (Fig. 10.5 bottom right).

10.3.2 The default mode network

The default mode network underlies the baseline mental activity in humans. Its higher energy consumption compared to other brain networks and its coupling with conscious awareness are both pointing to an overarching

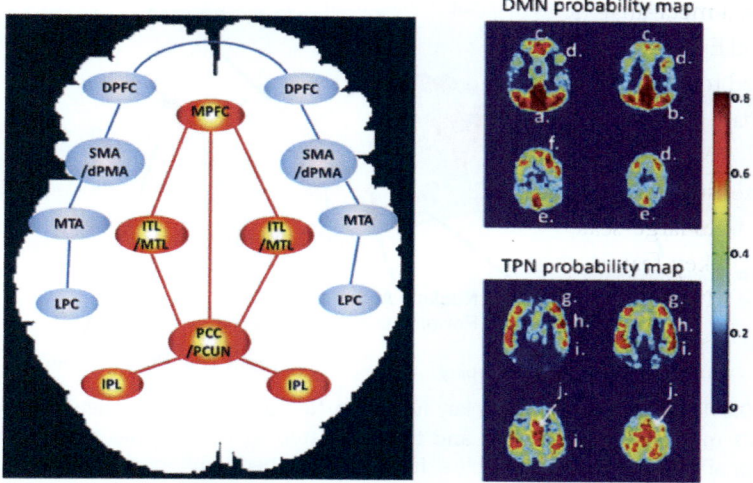

Fig. 10.5 (Left) Simplified illustration of the principal regions for the two anticorrelated networks, the DMN (red) and TPN (blue). Abbreviations: DMN, default mode network; TPN, task positive network; PCC/PCUN, posterior cingulate cortex and adjacent precuneus; MPFC, medial prefrontal cortex; ITL/MTL, inferior/mesial temporal lobe; IPL, inferior parietal lobe; DPFC, dorsolateral PFC; MTAs, middle temporal area; SMA, supplementary motor area; dPMA, dorsal premotor areas; LPC, lateral parietal cortex. (Right) DMN and TPN group incidence maps. Network maps are highly reproducible and localized. Anatomical areas indicated on the incidence maps specified by the nearest areas composed of red/yellow voxels: (A) precuneus/posterior cingulate cortex, (B) angular gyrus, (C) medial prefrontal cortex, (D) middle frontal gyrus, (E) precuneus, (F) superior frontal gyrus, (G) dorsolateral prefrontal cortex, (H) premotor cortex, (I) inferior parietal lobe, (J) medial frontal gyrus. *Left Panel From Hu, M.-L., Zong, X.-F., Mann, J.J., Zheng, J.-J., Liao, Y-H., Li, Z.-C., et al., 2017. A review of the functional and anatomical default mode network in schizophrenia. Neurosci. Bull. 33(1), 73–84.with permission from Springer Nature; Right Panel From Magnuson, M.E., Thompson, G.J., Schwarb, H., Pan, W.-J., McKinley, A., Schumacher, E.H., Keilholz, S.D., 2015. Errors on interrupter tasks presented during spatial and verbal working memory performance are linearly linked to large-scale functional network connectivity in high temporal resolution resting state fMRI. Brain Imag. Beh. 9, 854–867, with permission from Springer Nature.*

function (Smallwood et al., 2021). Understanding the DMN in the context of processing hierarchies shows how these shape brain activity (through abstract representation, stable dynamics, or prediction error, for example) as in the mechanistic role the DMN plays in human cognition.

Studies of reinforcement learning, which can be characterized by prediction–error models (Bayer and Glimcher, 2005), identify activity within medial prefrontal regions of the DMN (Fig. 10.6). Dohmatob et al. (2020) note that "Analogous to the posteromedial cortex (PMC),

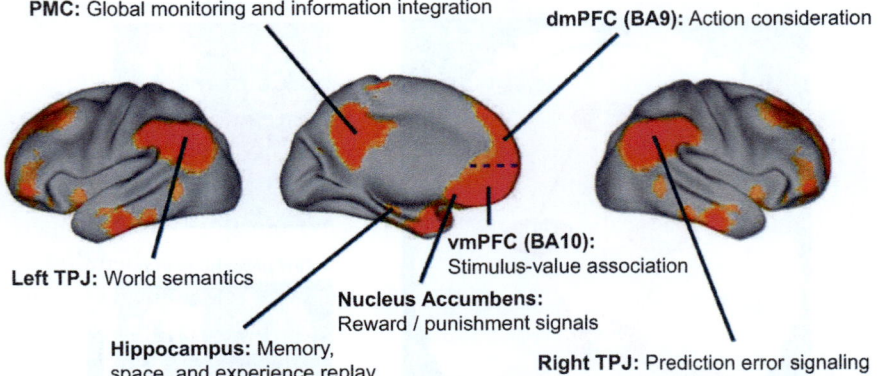

Fig. 10.6 Default mode network: key functions. Neurobiological overview of the DMN with its major constituent parts and the associated functional roles relevant in our functional interpretation. The blue horizontal dashed line in the middle image indicates the cytoarchitectonic border between the more dorsal BA9 and the more ventral BA10. Axonal tracing in monkeys and diffusion tractography in humans suggested that the NAc of the reward circuitry has monosynaptic fiber connections to the vmPFC. Evaluation of propagated value information and triggered affective states encoded in the vmPFC may then feed into the functionally connected partner nodes of the DMN, such as the dmPFC and PMC. *From Dohmatob, E., Dumas, G., Bzdok, D., 2020. Dark control: the default mode network as a reinforcement learning agent. Hum. Brain Mapp. 41, 3318–3341. Open access.*

the dorsomedial PFC (DMPFC, BA9) of the DMN is believed to underlie multi-sensory processes across time, space, and types of information processing to exert top-level control on behavior. […] This DMN part potentially promotes the de-novo construction and manipulation of meaning representations instructed by stored semantics and memories. The DMPFC may underlie representation and assessment of one's own and other individuals' action considerations. […] Comparing to the DMPFC, the ventromedial PFC (VMPFC, BA10) is probably more specifically devoted to subjective value evaluation and risk estimation of relevant environmental stimuli" (Fig. 10.6). The ventromedial prefrontal DMN underlies adaptive behavior by bottom-up-driven processing of "what matters now," drawing on value representations. The temporo-parietal junction (TPJ) of the DMN has distinctly different functions in left and right hemisphere. Dohmatob et al. (2020) note that the left TPJ may have a functional relationship to Wernicke's area involved in semantic processes. In contrast, the right TPJ is closely related to multisensory event representation and prediction error signaling. Thus (Fig. 10.6), DMN function naturally accommodates as special cases (a) predictive coding (right TPJ), (b) semantic associations (left TPJ), and (c) global monitoring and information integration (PFC).

Smallwood et al. (2021) elaborating on the functional role of the DMN in the way cognitive processes emerge within the temporal structure of a task, note "During such tasks, DMN regions in the parietal cortex (posteromedial cortex, PMC and angular gyrus, AG), medial temporal cortex (MTC) and IFG showed greater activity during decision making when the decisions were based on information from a prior trial than when similar decisions were made based on immediate sensory input." Such hierarchies in processing shape the temporal dynamics of complex systems and are a core premise of accounts of predictive coding (Chapter 1).

10.3.3 The inferior frontal gyrus

As we have seen in previous chapters (Chapters 1 and 7 in particular), aMMN-based networks derived by DCM more often include the right IFG (cf. Fig. 10.11), which is part of the DMN but also participates in nodes of the TPN, such as executive control, dorsal and ventral attention networks, and provides top-down "guidance" to minimize prediction errors.

10.3.3.1 Structure and subdivisions

The IFG neural circuitry includes functional connections within and between the three major IFG subgyri: the pars orbitalis, pars triangularis, and pars opercularis (Greenlee et al., 2007; Fig. 10.7). These different subregions of ventrolateral prefrontal cortex (VLPFC) participate in distinct cortical networks (Barredo et al., 2016). IFG subnetworks support separable cognitive functions: anterior VLPFC (IFG pars orbitalis) functionally correlates with a ventral frontotemporal network associated with top-down influences on memory retrieval, while mid-VLPFC (IFG pars triangularis) functionally correlates with a dorsal frontoparietal network associated with postretrieval control processes. Barredo et al. (2016) demonstrated a functional ventral-dorsal division within IFG. Ventral IFG as a whole connects broadly to lateral temporal cortex. Although several different individual white matter tracts form connections between ventral IFG and lateral temporal cortex, functional connectivity analysis of fMRI data indicates that these are part of the same ventral functional network. By contrast, across subdivisions, dorsal IFG was connected with the midfrontal gyrus and correlated as a separate dorsal functional network.

Resulting in even more subdivisions, Hartwigsen et al. (2019) identified 5 clusters in the right IFG that differed with respect to their coactivation

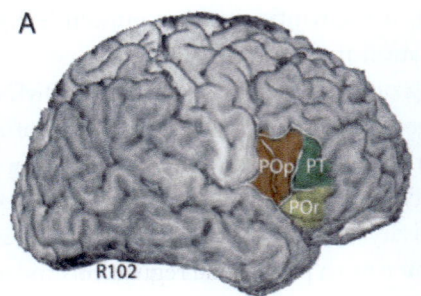
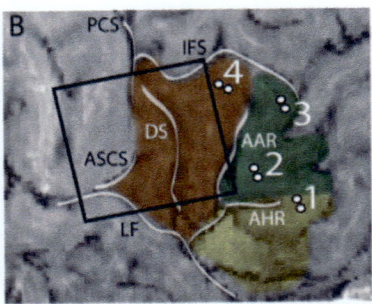

Fig. 10.7 (Top A) Lateral MRI surface rendering from subject R102 with IFG subgyri labeled and sulci traced. In this and all subsequent figures colors denote the POr (yellow), the PT (green), and the POp (brown). (B) Expanded view of the IFG with major sulci (white lines) labeled, four sites of stimulation (paired open circles, labeled 1–4), and position of the recording array (black box, 25-mm square) indicated. POp, pars opercularis; POr, pars orbitalis; PT, pars triangularis. *From Greenlee, J.D.W., Oya, H., Kawasaki, H., Volkov, I.O., Severson III, M.A., Howard III, M.A., Brugge, J.F., 2007. Functional connections within the human inferior frontal gyrus. J. Comp. Neurol. 503, 550–559.* © 2007 WILEY-LISS, INC., with permission from John Wiley and Sons.

patterns. Two clusters in the right posterior IFG were functionally associated with action inhibition and execution, while two anterior clusters were related to reasoning and social cognitive processes. A fifth cluster was associated with spatial attention (DAN). Strikingly, the functional organization of the right IFG can thus be characterized by a posterior-to-anterior axis with action-related functions on the posterior and cognition-related functions on the anterior end. They observed further subdivisions along a dorsal-to-ventral axis in posterior IFG between action execution and inhibition, and in anterior IFG between reasoning and social cognition.

10.3.3.2 Effective connectivity

The IFG is also involved in the evaluation of linguistic, interoceptive, and emotional information. Briggs et al. (2019) identified four major connections of the IFG: the frontal aslant tract connecting to the superior frontal gyrus; the superior longitudinal fasciculus connecting to the inferior parietal lobule, lateral occipital area, posterior temporal areas, and the temporal pole; the inferior fronto-occipital fasciculus connecting to the cuneus and lingual gyrus; and the uncinate fasciculus connecting to the temporal pole. These play a dominant role in language processing as described in Chapter 11 (cf. Fig. 11.15). A callosal fiber bundle connecting the inferior frontal gyri bilaterally was also identified. Hsu et al. (2020) reported that "the lateral orbitofrontal cortex (LOFC) and its closely connected right IFGs (rIFG) have direct connections with the supracallosal ACC; all of which are punishment or non-reward related areas. The LOFC and rIFG also have direct connections with the right supramarginal gyrus and inferior parietal cortex, and with some premotor cortical areas, which may provide outputs for the LOFC and rIFG. Direct connections of the OFC and IFG were with especially the temporal pole part of the temporal lobe. The left IFG, which includes Broca's area, has direct connections with the left angular and supramarginal gyri." The IFG, OFC, and ventromedial PFC have been linked to the regulation of anxiety during threat exposure. Gold et al. (2015) showed greater functional connectivity between the right amygdala and bilateral IFG, OFC, VMPFC, ACC, and frontopolar cortex that was associated with threat exposure.

On the basis of the nucleus accumbens (NAc) as a seed region, Xu et al. (2019) conducted a Granger causality analysis to investigate the directional connectivity and the relationship with tinnitus duration or distress. Compared with healthy controls, tinnitus patients exhibited abnormal directional connectivity between the NAc and the prefrontal cortex, principally the middle frontal gyrus (MFG), OFC, and IFG (Fig. 10.8).

Nakae et al. (2020) integrated spatiotemporal profiles of cortico-cortical evoked potentials (CCEPs) recorded intraoperatively in 14 surgical patients. Visualization of the combined CCEP data showed that the left IFG pars opercularis (Broca's area) is connected to the posterior temporal cortices and the supramarginal gyrus, whereas the IFG pars orbitalis is connected to the anterior lateral temporal cortices and angular gyrus (Fig. 10.9). Quantitative topographical analysis of CCEP connectivity confirmed an anterior-posterior gradient of connectivity from IFG stimulus sites to the temporal response sites. Reciprocity analysis indicated that the anterior part of the IFG is bidirectionally connected to the temporal or parietal area (Fig. 10.9C). This study shows that "each IFG subdivision has different connectivity to the temporal lobe with an anterior–posterior gradient and supports the classical connectivity concept of Dejerine (Fig. 10.9D) that is, that the frontal lobe is connected to the temporal lobe through the arcuate fasciculus and also a double fan-shaped structure anchored at the limen insulae."

10.4 Networks underlying prediction-error coding

10.4.1 General

From an auditory EEG/MEG-based DCM study, Dürschmid et al. (2019) identified a hierarchical feedforward-feedback cascade in which the IFG sits at the top, providing top-down predictions to—and receiving prediction-

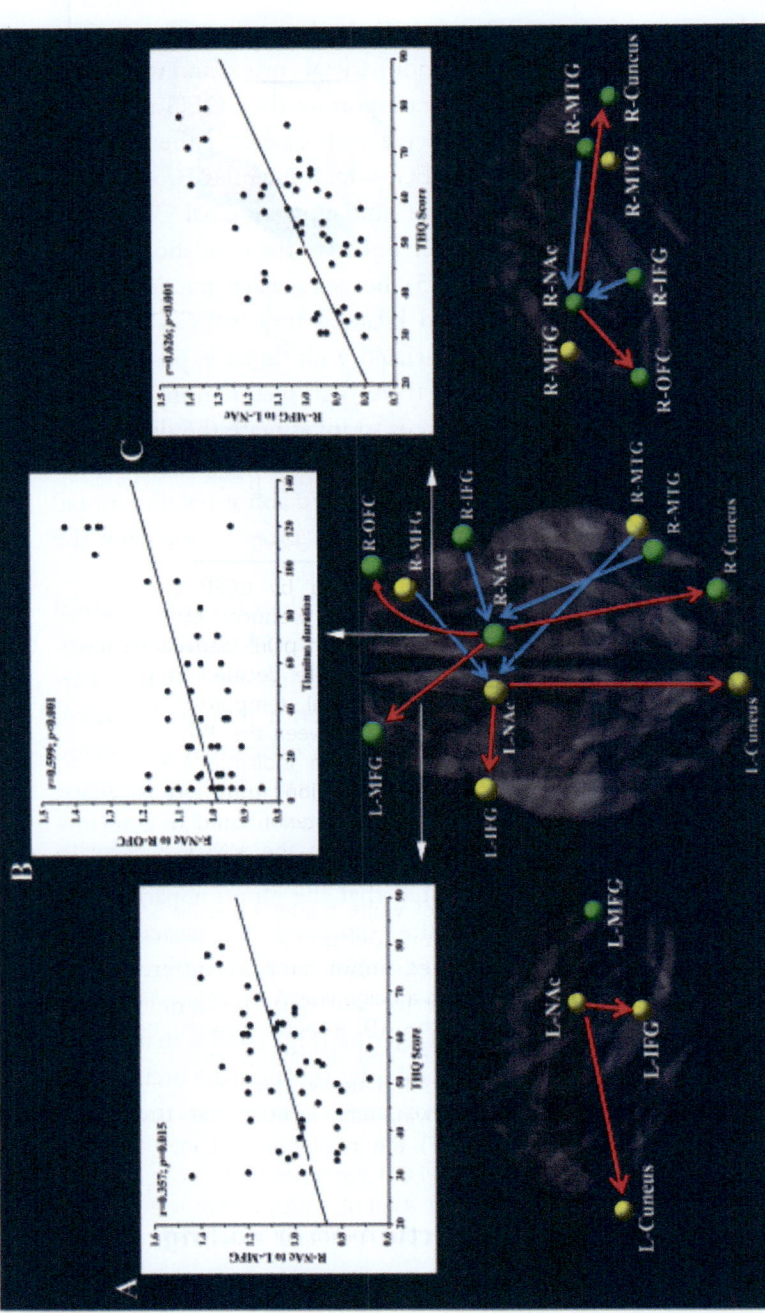

Fig. 10.8 The directional functional connectivity networks of the bilateral NAc. The red line represents the directional functional connectivity from the bilateral NAc to the other brain regions; the blue line represents the directional functional connectivity from the other brain regions to the bilateral NAc. (A) Positive correlation between the increased directional connectivity from the right NAc to left MFG and the THQ score ($r = 0.357$, $P = 0.015$). (B) Positive correlation between the increased directional connectivity from the right NAc to right OFC and the tinnitus duration ($r = 0.599$, $P < 0.001$). (C) Positive correlation between the enhanced directional connectivity from the right MFG to left NAc and the THQ score ($r = 0.626$, $P < 0.001$). *From Xu, J.-J., Cui, J., Feng, Y., Yong, W., Chen, H., Chen, Y.-C., et al., 2019. Chronic tinnitus exhibits bidirectional functional dysconnectivity in frontostriatal circuit. Front. Neurosci. 13, 1299. doi:10.3389/fnins.2019.01299. Open access.*

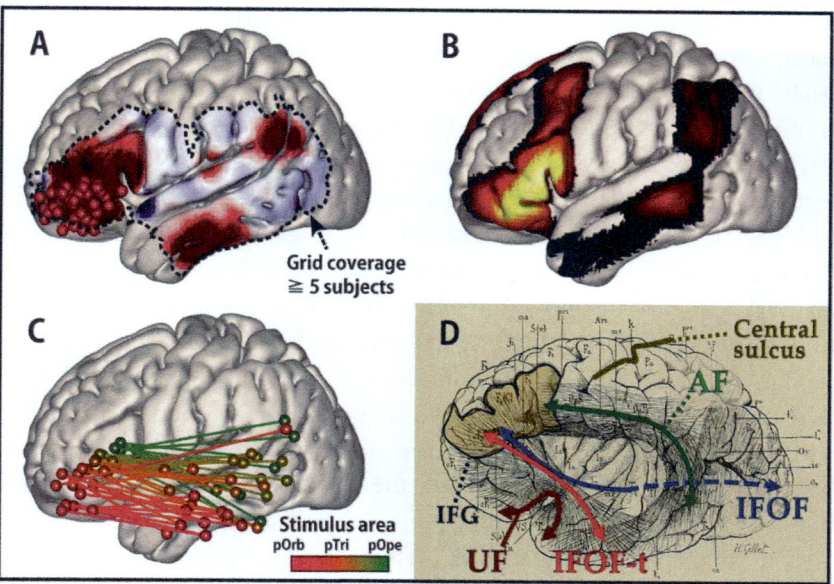

Fig. 10.9 Comparison of connectivity values discovered by CCEP and rs-fMRI: A connectivity gradient. (A) The averaged response map produced by stimulating pOrb, delayed phase. (B) The functional connectivity from pOrb derived from the NeuroSynth database. In this study, the seed voxels were decided as the pOrb stimulus sites. The connectivity pattern in the lateral temporo-parietal area resembles that of CCEP. (C) The connectivity gradient between the IFG and MTG as assessed using the CCEP database. The color gradation indicates the anterior-posterior coordinate of the stimulus sites. The gradation from red to green corresponds to the transition from the anterior to the posterior stimulation site. All pairs of stimulus and max response sites are plotted in the MNI *y–z* plane to illustrate the connectivity gradient. (D) An illustration of the long tracts in the vicinity of the IFG, overlaid on the reprinted schema of white matter dissection from the classical textbook "Anatomie des centres nerveux" (Dejerine and Dejerine-Klumpke, 1895). Major pathways are annotated with colored arrows, the IFG is outlined with a black line, and the central sulcus is outlined with a beige line. A fan-shaped structure can be seen connecting the frontal lobe and the ATL through the temporal stem. While the superior longitudinal fasciculus (SLF) and arcuate fasciculus (AF) are established as the main pathway of the dorsal network, additional studies are required to identify the connectivity underlying the ventral network, which presumably includes the uncinate fasciculus (UF), the inferior longitudinal fasciculus (ILF), and the inferior fronto-occipital fasciculus (IFOF). *From Nakae, T., Matsumoto, R., Kunieda, T., Arakawa, Y., Kobayashi, K., Shimotake, A., et al., 2020. Connectivity gradient in the human left inferior frontal gyrus: intraoperative cortico-cortical evoked potential study. Cereb. Cortex 30, 4633–4650. Open access.*

error (PE) signals from—the STG, which in turn provides top–down predictions to (and receives PE signals from) the A1. Dürschmid et al.'s (2019) findings showed clear effects of predictability in the ventral PFC for pitch deviations. Previously, Phillips et al. (2015) and Phillips et al. (2016) validated these models, originally tested on EEG/MEG data, with electrocorticography (ECoG) data from two patients. However, Phillips et al.'s (2015, 2016) models suggested a PFC-generated prediction signal affecting the IFG that was limited to temporal deviations (duration and gap deviations in their study), but not with pitch, intensity, or location deviations. Dürschmid et al. (2019) speculated on the "functional advantage of maintaining predictions that account for global regularities allows the prefrontal cortex to efficiently direct attention only to unexpected events (Sussman et al. 2003), whereas for the auditory cortex, detecting all local changes is advantageous for parsing the auditory input into meaningful chunks (e.g., in speech perception)."

Critically, Heilbron and Chait (2018) noted that "even if we fully accept the network modulations suggested by DCM, this doesn't mean that these changes necessarily reflect predictive coding, or even a single underlying mechanism. Indeed, it is difficult to see why changes in A1 excitability and STG/IFG connectivity should be uniquely characteristic of predictive coding. This problem is reinforced by the fact that the discussed studies have mostly used designs in which expectation and adaptation are confounded, which makes arbitrating between predictive and nonpredictive interpretations difficult. As such, while the discussed studies constitute exciting methodological developments in the analysis of noninvasive electrophysiological data, their strength as empirical support for predictive coding theory seems limited" (Chapter 1).

What is needed in addition to the DCM would be actually "recording" from the commonly used putative sources; IFG, STG, and A1; and potentially additional ones. This can be done using fMRI, but that is less sensitive to synchronization changes in processing than EEG/MEG. A good approximation was provided by Cope et al. (2022) who used combined sLORETA-based source reconstructed MEG responses using individual head models based on T1-weighted MRI images. They based their source selection on nodes from the multiple demand (MD) system: a network of brain regions engaged in flexible organization and control of cognitive operations across diverse mental activities. The MD system comprises domain general regions in middle frontal gyrus, inferior frontal sulcus, anterior insula, and intraparietal sulcus, as well as supplementary motor area and anterior cingulate cortex (Duncan, 2010; Duncan et al., 2020). Cope et al.'s (2022) selection of sources assumed that "atrophy of any multiple demand node is sufficient to impair auditory neurophysiological responses to change in frequency, location, intensity, continuity or duration. There was no similar association with atrophy of the cingulo-opercular, salience or language networks, or with global atrophy."

Cope et al. (2022) investigated four neurodegenerative syndromes, of which we only present here the Alzheimer's disease with mild cognitive impairment (AD/MCI) one. To establish the functional role of IFG and inferior parietal cortex (IPC) in generating the MMN response observed in temporal cortex (STG and A1), Cope et al. (2022) assessed effective connectivity with DCM. A whole-brain contrast confirmed modulation of all of their ROIs by the overall contrast between standard and deviant tones: the strongest mismatch response was around A1 and STG bilaterally (Fig. 10.10). Specifically, the DCM result for AD/MCI compared to control is shown in Fig. 10.11F. Cope et al. (2022) concluded that "dementia-related reductions in the neurophysiological amplitude of the auditory mismatch response are not a simple reflection of global atrophy, but are rather a specific manifestation of neurodegeneration in these frontal and parietal regions. Neurodegeneration reduced top-down effective connectivity from affected fronto-parietal nodes and increased effective connectivity from unaffected fronto-parietal nodes. This compensation was only partially effective, as overall top-down influence on STG fell with the involvement of any frontal or parietal node set. All degenerative nodes showed increased functional connectivity, again consistent with compensatory upregulation."

10.4.2 DCM variability in predictive coding

David et al. (2006) used a conventional reconstruction algorithm, showing that cortical sources were localized symmetrically along the medial part of the upper bank of the Sylvian fissure, in the right MTG, left medial and posterior cingulate, and bilateral OFC. A1 has major interhemispheric connections through the corpus callosum. In addition, these areas project to temporal and frontal lobes following different streams (Fig. 1.7; Medalla and Barbas, 2014). Finally, cingulate activations are often found in relation to oddball tasks, either auditory or visual (Linden et al., 1999). Using these sources and prior knowledge about the functional anatomy of the auditory system, David et al. (2006) constructed the DCM shown in Fig. 10.11A: an extrinsic (thalamic) input entered bilateral A1 which was connected to

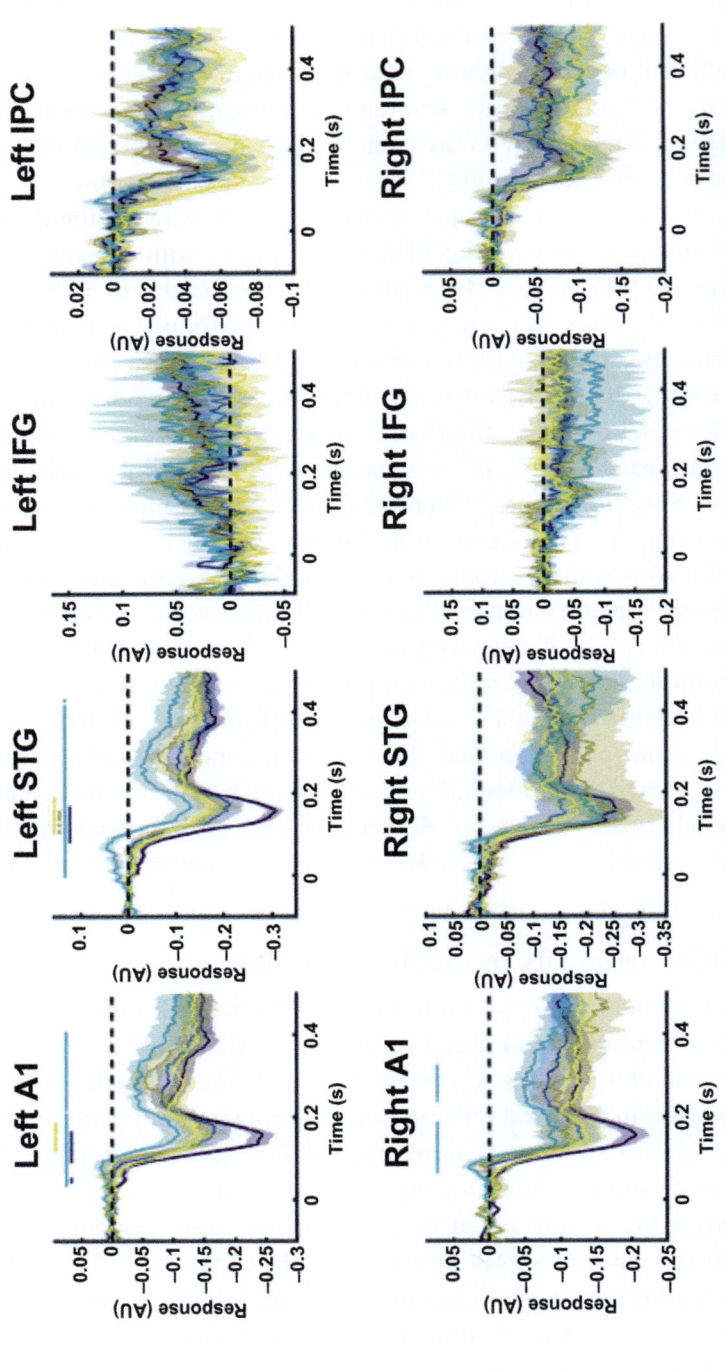

Fig. 10.10 Source-localized evoked responses for MMN. Means and SEs across subjects are plotted, along with an illustrative significance bar, showing time points at which patient group responses significantly differed from control responses, $P < 0.05$ FDR corrected. A1, primary auditory cortex; STG, superior temporal gyrus; IFG, inferior frontal gyrus; IPC, inferior parietal cortex. Control, blue; AD/MCI, yellow. Other colors, different disorders. *From Cope, T.E., Hughes, L.E., Phillips, H.N., Adams, N.E., Jafariana, A., Nesbittb, D., et al., 2022. Causal evidence for the multiple demand network in change detection: auditory mismatch magnetoencephalography across focal neurodegenerative diseases. J. Neurosci. 42(15), 3197–3215. Open access.*

ipsilateral OFC (OF in the Figure). In the right hemisphere, an indirect forward pathway was specified from A1 to OF through the STG. All these connections were reciprocal. At the highest level in the hierarchy, OFC and left PCC (PC in the Figure) were laterally and reciprocally connected.

In Fig. 10.11B, Garrido et al. (2007) present the quantitative results of inversion of a five-source model, using the grand-mean response over subjects. It illustrates the reconstructed source activity generating the ERPs to standards and deviants for all three neuronal populations in each source. These differences are caused by, and only by, differences in coupling between and within sources. In contrast to the David et al. (2006) model, they feature bilateral STGs and unilateral IFG and did not involve the OFC. Using the Garrido et al. (2007) model to study effects of aging, Cooray et al. (2014) performed DCM analysis, which revealed a significant reduction in modulation of intrinsic connectivity at rIFG for older subjects compared to young subjects. Furthermore, there was increased extrinsic connectivity in the network between rSTG and rIFG ($P < 0.01$, red line in Fig. 10.11C) and increased inhibitory activity at rIFG ($P < 0.03$, node colored blue) when generating the standard tone ERP. Quiroga-Martinez et al. (2021) noted that their network (Fig. 10.11D) is very similar to classical DCM studies of MMN responses (i.e., Garrido et al., 2009) with two important differences. First, they included anterior STG (instead of posterior STG or planum temporale), which has been related to the processing of pitch sequences (Gander et al., 2019; Patterson et al., 2002). Second, they included the frontal operculum (FOP), which is part of IFG but posterior to the IFG node used more commonly (cf. Fig. 10.7). Results of the DCM analysis of fMRI data for general deviant detection by Schlossmacher et al. (2022) can be found in Fig. 10.11E. To identify potential nodes in the network analysis, they first compared oddball and standard stimuli, confirmed deviance-related activation bilaterally in Heschl's gyrus and STG. They also found several distinct mismatch response clusters in frontal and parietal areas as well as the anterior insula (AI). Activations were mainly detected in the ventral attention network (VAN), which is involved in orienting attention to salient events and thus alerting the organism to environmental changes. This network includes the AI and ACC/SMA, which are also strongly involved in detecting salient events. Furthermore, there was strong bilateral inferior frontal junction (IFJ) activity. The IFJ is involved in the dorsal frontoparietal network modulating goal-directed attention in sensory areas in a top-down fashion, but also in a network activated by unexpected salient environmental changes. The IFJ presents a dynamic region

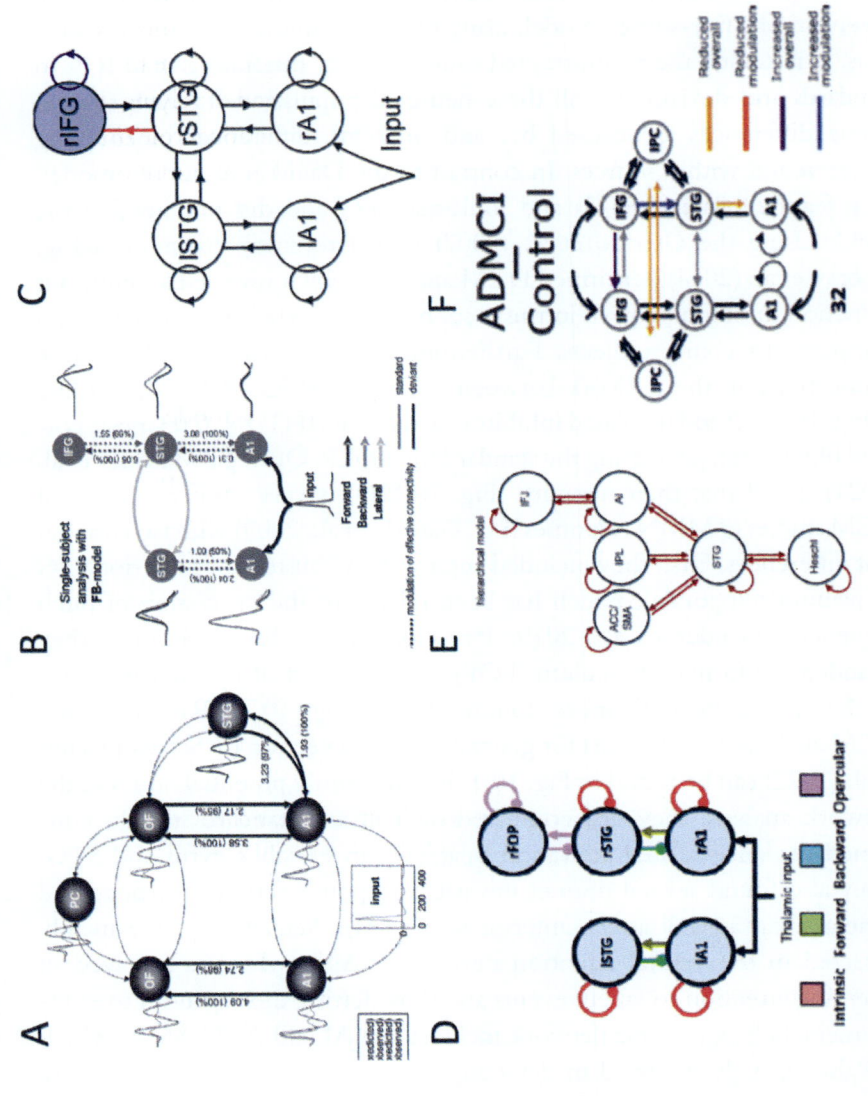

Fig. 10.11 See legend on next page.

integrating information from both dorsal and ventral networks. Schlossmacher et al. (2022) also found deviance-related effects in the left IPL, which is part of a frontoparietal control network. In the AI, IFJ, and IPL, a significant prediction error effect was found. Surprisingly, there was no significant prediction error-related activity in ACC/SMA, probably due to the task-irrelevant oddball paradigm chosen here. In contrast to just the IFG, the many structures feeding into the STG in this study might be due to actually investigating active regions during an oddball test, instead of relying on previous EEG/MEG-based studies. To understand the functional role of frontal (IFG) and parietal (IPC) cortex in generating the MMN response observed in temporal cortex, Cope et al. (2022) assessed effective connectivity with dynamic causal modeling. Fig. 10.11F shows results for a specific group: Alzheimer's disease with MCI.

10.4.3 Taking neural delays into account

Despite the enormous interest in predictive coding as a model of cortical processing, most models of predictive coding do not consider that neural processing itself takes time. In discussing this missing item in PC, Hogendoorn and Burkitt (2019) show that there is an important distinction to be made regarding the sense in which PC models predict. PC models are typically considered to reflect some kind of expectation about the future.

Fig. 10.11 Some representative DCM results. *Panel A Reprinted from David, O., Kiebel, S.J., Harrison, L.M., Mattout, J., Kilner, J.M., Friston. K.J., 2006. Dynamic causal modeling of evoked responses in EEG and MEG. NeuroImage 30, 1255–1272. with permission from Elsevier; Panel B Reprinted from Garrido, M.I., Kilner, J.M., Kiebel, S.J., Stephan, K.E., Friston, K.J., 2007a. Dynamic causal modeling of evoked potentials: a reproducibility study. NeuroImage 36, 571–580, with permission from Elsevier; Panel C Reprinted from Cooray, G., Garrido, M.I., Hyllienmark, L., Brismar, T., 2014. A mechanistic model of mismatch negativity in the aging brain. Clin. Neurophysiol. 125, 1774–1782, with permission from Elsevier; Panel D From Quiroga-Martinez, D.R., Hansen, N.C., Højlund, A., Pearce, M., Brattico, E., Holmes, E., et al., 2021. Musicianship and melodic predictability enhance neural gain in auditory cortex during pitch deviance detection. Hum. Brain Mapp. 42, 5595–5608. DOI:10.1002/hbm.25638. Open Access; Panel E From Schlossmacher, I., Dilly, J., Protmann, I., Hofmann, D., Dellert, T., Roth-Paysen, M.-L., et al., 2022. Differential effects of prediction error and adaptation along the auditory cortical hierarchy during deviance processing. Neuroimage 259, 119445. Open Access; Panel F From Cope, T.E., Hughes, L.E., Phillips, H.N., Adams, N.E., Jafariana, A., Nesbittb, D., et al., 2022. Causal evidence for the multiple demand network in change detection: auditory mismatch magnetoencephalography across focal neurodegenerative diseases. J. Neurosci. 42(15), 3197–3215.*

For example, in perception, preactivation of the neural representation of an expected sensory event ahead of the actual occurrence of that event reflects the nervous system predicting a future event (Garrido et al., 2009; Kok et al., 2017; Hogendoorn and Burkitt, 2018). However, rather than predicting the future, conventional models of PC such as the one first proposed by Rao and Ballard (1999) are predictive in the hierarchical sense of higher areas predicting the activity of lower areas (Rao and Ballard, 1999; Bastos et al., 2012). As such, what is missing from conventional PC models is the fact that neural communication itself takes time. DCM models have neglected that neural transmission incurs significant delays. These delays mean that forward and backward signals are misaligned in time. Hogendoorn and Burkitt (2018) note: "Because of neural transmission delays, prediction error is minimized not when a backward signal represents the sensory information that originally generated it, but when it represents the sensory information that is going to be available at the lower level by the time the backward signal arrives. In other words, prediction error is minimized when the backward signal anticipates the future state of the lower hierarchical level."

Hogendoorn and Burkitt (2019) further show that "when the framework is extended to allow for neural delays, a model emerges that provides a more natural explanation for a wide range of experimental findings (Fig. 10.12). For stimuli that are changing at a constant rate, estimating that future state requires only rate-of-change information about the relevant feature, and it follows that if such information is available at a given level, it will be recruited to minimize prediction error. When allowing for transmission delays, hierarchical predictions therefore need to become temporal predictions: they need to predict the future" (Hogendoorn and Burkitt, 2019).

10.5 Summary

Comparing the auditory, visual, and somatosensory MMN shows that the estimated networks underlying both auditory MMN (aMMN) and visual MMN (vMMN) often have the inferior frontal gyrus (IFG) as its "endpoint," albeit that the aMMN network starts in primary auditory cortex (A1) and the vMMN in primary visual cortex (V1). As in the aMMN, a frontal mechanism underlies change detection in the vMMN paradigm localized to the left IFG, contrasting the right IFG in the aMMN network. Although less is known about somatosensory MMN (sMMN), it appears very similar to aMMN and vMMN. Intracranial recordings showed aMMN generators along the STG with no evidence of a somatosensory MMN in this region,

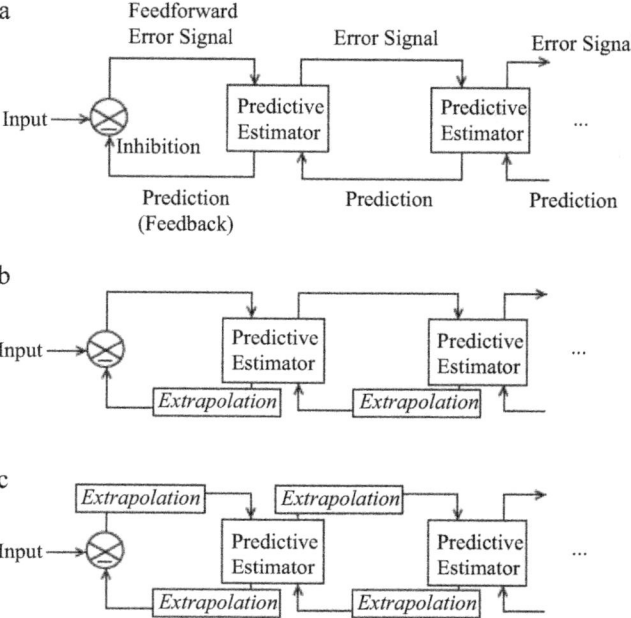

Fig. 10.12 The classical hierarchical predictive coding model and two possible extensions. (A) The classical predictive model (model A). This model consists of hierarchically organized loops of forward and backward connections. Backward signals carry predictions, and forward signals carry prediction errors (Rao and Ballard, 1999). (B) The predictive model with extrapolated feedback (model B). To handle time-varying stimuli such as motion, the classical model can be expanded to include an extrapolation mechanism on the backward step. This would be one way to minimize prediction error for time-varying stimuli. (C) The predictive model with real-time alignment (model C). In this model, extrapolation (time delay) mechanisms work on both forward and backward steps. Like model B, this would minimize prediction error, but has the additional consequence that it aligns the content of neural representations across the hierarchy. Diagram labels in B and C are as in A, but are omitted for clarity. *From Hogendoorn, H., Burkitt, A.N., 2019. Predictive coding with neural transmission delays: a real-time temporal alignment hypothesis. eNeuro, 6(2) e0412-18.2019 1–12. Open access.*

whereas a robust sMMN was recorded from postcentral gyrus in the absence of an auditory MMN. This indicates stimulus specificity in the intermediate areas of the aMMN- and sMMN-generating networks.

Compiling data across dynamic causal modeling (DCM) studies shows large variation in the higher-order cortex structures involved in the aMMN network. Whereas all investigations agree with the involvement of A1 and STG, the next stage while typically including only right IFG, in some cases bilateral IFG results. In some models the IFG is replaced with the orbito-frontal cortex (OFC), the frontal operculum, or a combination of ACC, inferior parietal lobe, and anterior insula. In some models there is a modulation of the "IFG-layer" from PFC or inferior frontal junction. Consequently, the nonunique choice of structures, potentially determined by the mode of data acquisition, by each investigator determines what the aMMN generation model will look like.

References

Barredo, J., Verstynen, T.D., Badre, D., 2016. Organization of corticocortical pathways supporting memory retrieval across subregions of the left ventrolateral prefrontal cortex. J. Neurophysiol. 116, 920–937.

Bastos, A.M., Usrey, W.M., Adams, R.A., Mangun, G.R., Fries, P., Friston, K.J., 2012. Canonical microcircuits for predictive coding. Neuron 21, 695–711.

Bayer, H.M., Glimcher, P.W., 2005. Midbrain dopamine neurons encode a quantitative reward prediction error signal. Neuron 47, 129–141.

Beck, A.-K., Berti, S., Czernochowski, D., Lachmann, T., 2021. Do categorical representations modulate early automatic visual processing? A visual mismatch-negativity study. Biol. Psychol. 163, 108139.

Briggs, R.G., Chakraborty, A.R., Anderson, C.D., Abraham, C.J., Palejwala, A.H., Conner, A.K., et al., 2019. Anatomy and white matter connections of the inferior frontal gyrus. Clin. Anat. 32, 546–556.

Butler, J.S., Molholm, S., Fiebelkorn, I.C., Mercier, M.R., Schwartz, T.H., Foxe, J.J., 2011. Common or redundant neural circuits for duration processing across audition and touch. J. Neurosci. 31 (9), 3400–3406.

Cooray, G., Garrido, M.I., Hyllienmark, L., Brismar, T., 2014. A mechanistic model of mismatch negativity in the ageing brain. Clin. Neurophysiol. 125, 1774–1782.

Cope, T.E., Hughes, L.E., Phillips, H.N., Adams, N.E., Jafariana, A., Nesbittb, D., et al., 2022. Causal evidence for the multiple demand network in change detection: auditory mismatch magnetoencephalography across focal neurodegenerative diseases. J. Neurosci. 42 (15), 3197–3215.

David, O., Kiebel, S.J., Harrison, L.M., Mattout, J., Kilner, J.M., Friston, K.J., 2006. Dynamic causal modeling of evoked responses in EEG and MEG. Neuroimage 30, 1255–1272. https://doi.org/10.1016/j.neuroimage.2005.10.045.

Dejerine, J., Dejerine-Klumpke, A., 1895. Anatomie des centres nerveux: Méthodes générales d'étude-embryologie-histogénèse et histologie. In: Rueff (Ed.), Anatomie du Cerveau. Rueff et Cie, Paris.

Dohmatob, E., Dumas, G., Bzdok, D., 2020. Dark control: the default mode network as a reinforcement learning agent. Hum. Brain Mapp. 41, 3318–3341.

Duncan, J., 2010. The multiple-demand (MD) system of the primate brain: mental programs for intelligent behaviour. Trends Cogn. Sci. 14, 172–179.

Duncan, J., Assem, M., Shashidhara, S., 2020. Integrated intelligence from distributed brain activity. Trends Cogn. Sci. 24, 838–852.

Dürschmid, S., Reichert, C., Hinrichs, H., Heinze, H.-J., Kirsch, H.E., Knight, R.T., 2019. Direct evidence for prediction signals in frontal cortex independent of prediction error. Cereb. Cortex 11, 4530–4538.

Eggermont, J.J., 2021. Brain Oscillations, Synchrony and Plasticity. Basic Principles and Application to Auditory-Related Disorders. Academic Press, London, UK, ISBN: 978-0-12-819818-6, pp. 1–250.

Ficco, L., Mancuso, L., Manuello, J., Teneggi, A., Liloia, D., Duca, S., et al., 2021. Disentangling predictive processing in the brain: a meta-analytic study in favour of a predictive network. Sci. Rep. 11, 16258. https://doi.org/10.1038/s41598-021-95603-5.

Friston, K., 2005. A theory of cortical responses. Philos. Trans. R. Soc. B 360, 815–836. https://doi.org/10.1098/rstb.2005.1622.

Friston, K.J., Harrison, L., Penny, W., 2003. Dynamic causal modelling. Neuroimage 19, 1273–1302.

Gander, P.E., Kumar, S., Sedley, W., Nourski, K.V., Oya, H., Kovach, C.K., et al., 2019. Direct electrophysiological mapping of human pitch-related processing in auditory cortex. Neuroimage 202, 116076. https://doi.org/10.1016/j.neuroimage.2019.116076.

Garrido, M.I., Kilner, J.M., Kiebel, S.J., Friston, K.J., 2009. Dynamic causal modeling of the response to frequency deviants. J. Neurophysiol. 101, 2620–2631.

Garrido, M.I., Kilner, J.M., Kiebel, S.J., Stephan, K.E., Friston, K.J., 2007. Dynamic causal modelling of evoked potentials: a reproducibility study. Neuroimage 36, 571–580.

Garrido, M.I., Kilner, J.M., Stephan, K.E., Friston, K.J., 2009. The mismatch negativity: a review of underlying mechanisms. Clin. Neurophysiol. 120 (3), 453–463. https://doi.org/10.1016/j.clinph.2008.11.029.

Gold, A.L., Morey, R.A., McCarthy, G., 2015. Amygdala–prefrontal cortex functional connectivity during threat-induced anxiety and goal distraction. Biol. Psychiatry 77, 394–403.

Gourévitch, B., Eggermont, J.J., 2007. Evaluating information transfer between auditory cortical neurons. J. Neurophysiol. 97, 2533–2543.

Greenlee, J.D.W., Oya, H., Kawasaki, H., Volkov, I.O., Severson III, M.A., Howard III, M.A., Brugge, J.F., 2007. Functional connections within the human inferior frontal gyrus. J. Comp. Neurol. 503, 550–559.

Hartwigsen, G., Neef, N.E., Camilleri, J.A., Margulies, D.S., Eickhoff, S.B., 2019. Functional segregation of the right inferior frontal gyrus: evidence from coactivation-based parcellation. Cereb. Cortex 29, 1532–1546.

Hedge, C., Stothart, G., Todd Jones, J., Rojas Frías, P., Magee, K.L., Brooks, J.C.W., 2015. A frontal attention mechanism in the visual mismatch negativity. Behav. Brain Res. 293, 173–181.

Heilbron, M., Chait, M., 2018. Great expectations: is there evidence for predictive coding in auditory cortex? Neuroscience 389, 54–73.

Hogendoorn, H., Burkitt, A.N., 2018. Predictive coding of visual object position ahead of moving objects revealed by time-resolved EEG decoding. Neuroimage 171, 55–61.

Hogendoorn, H., Burkitt, A.N., 2019. Predictive coding with neural transmission delays: a real-time temporal alignment hypothesis. eNeuro 6 (2), 1–12. e0412-18.2019.

Hsu, C.-C.H., Rolls, E.T., Huang, C.-C., Chong, S.T., Lo, C.Y.Z., Feng, J., Lin, C.-P., 2020. Connections of the human orbitofrontal cortex and inferior frontal gyrus. Cereb. Cortex 30, 5830–5843.

Hu, M.-L., Zong, X.-F., Mann, J.J., Zheng, J.-J., Liao, Y.-H., Li, Z.-C., et al., 2017. A review of the functional and anatomical default mode network in schizophrenia. Neurosci. Bull. 33 (1), 73–84.

Kok, P., Mostert, P., de Lange, F.P., 2017. Prior expectations induce prestimulus sensory templates. Proc. Natl. Acad. Sci. U. S. A. 114, 10473–10478.

Linden, J.F., Grunewald, A., Andersen, R.A., 1999. Responses to auditory stimuli in macaque lateral intraparietal area. II. Behavioral modulation. J. Neurophysiol. 82 (1), 343–358. https://doi.org/10.1152/jn.1999.82.1.343.

Magnuson, M.E., Thompson, G.J., Schwarb, H., Pan, W.-J., McKinley, A., Schumacher, E.H., Keilholz, S.D., 2015. Errors on interrupter tasks presented during spatial and verbal working memory performance are linearly linked to large-scale functional network connectivity in high temporal resolution resting state fMRI. Brain Imaging Behav. 9, 854–867. https://doi.org/10.1007/s11682-014-9347-3.

Male, A.G., O'Shea, R.P., Schröger, E., Müller, D., Roeber, U., Widmann, A., 2020. The quest for the genuine visual mismatch negativity (vMMN): event-related potential indications of deviance detection for low-level visual features. Psychophysiology 57, e13576. https://doi.org/10.1111/psyp.13576.

Mclntosh, A.R., Gonzalez-Lima, F., 1994. Structural equation modeling and its application to network analysis in functional brain imaging. Hum. Brain Mapp. 2, 2–22.

Medalla, M., Barbas, H., 2014. Specialized prefrontal "auditory fields": organization of primate prefrontal-temporal pathways. Front. Neurosci. 8, 77. https://doi.org/10.3389/fnins.2014.00077.

Molholm, S., Martinez, A., Ritter, W., Javitt, D.C., Foxe, J.J., 2005. The neural circuitry of pre-attentive auditory change-detection: an fMRI study of pitch and duration mismatch negativity generators. Cereb. Cortex 15, 545–551. https://doi.org/10.1093/cercor/bhh155.

Naeije, G., Vaulet, T., Wens, V., Marty, B., Goldman, S., De Tiège, X., 2018. Neural basis of early somatosensory change detection: a magnetoencephalography study. Brain Topogr. 31, 242–256. https://doi.org/10.1007/s10548-017-0591-x.

Nakae, T., Matsumoto, R., Kunieda, T., Arakawa, Y., Kobayashi, K., Shimotake, A., et al., 2020. Connectivity gradient in the human left inferior frontal gyrus: intraoperative cortico-cortical evoked potential study. Cereb. Cortex 30, 4633–4650.

Patterson, R.D., Uppenkamp, S., Johnsrude, I.S., Griffiths, T.D., 2002. The processing of temporal pitch and melody information in auditory cortex. Neuron 36 (4), 767–776. https://doi.org/10.1016/S0896-6273 (02)01060-7.

Phillips, H.N., Blenkmann, A., Hughes, L.E., Bekinschtein, T.A., Rowe, J.B., 2015. Hierarchical organization of frontotemporal networks for the prediction of stimuli across multiple dimensions. J. Neurosci. 35 (25), 9255–9264.

Phillips, H.N., Blenkmann, A., Hughes, L.E., Kochen, S., Bekinschtein, T.A., Cam-Can, Rowe, J.B., 2016. Convergent evidence for hierarchical prediction networks from human electrocorticography and magnetoencephalography. Cortex 82, 192–205.

Quiroga-Martinez, D.R., Hansen, N.C., Højlund, A., Pearce, M., Brattico, E., Holmes, E., et al., 2021. Musicianship and melodic predictability enhance neural gain in auditory cortex during pitch deviance detection. Hum. Brain Mapp. 42, 5595–5608. https://doi.org/10.1002/hbm.25638.

Rao, R.P.N., Ballard, D.H., 1999. Predictive coding in the visual cortex: a functional interpretation of some extra-classical receptive-field effects. Nat. Neurosci. 2, 79–87.

Rosburg, T., Weigl, M., Deuring, G., 2019. Enhanced processing of facial emotion for target stimuli. Int. J. Psychophysiol. 146, 190–200.

Rowe, E.G., Tsuchiya, N., Garrido, M.I., 2020. Detecting (un)seen change: the neural underpinnings of (un)conscious prediction errors. Front. Syst. Neurosci. 14:541670. doi:10.3389/fnsys.2020.541670.

Schlossmacher, I., Dilly, J., Protmann, I., Hofmann, D., Dellert, T., Roth-Paysen, M.-L., et al., 2022. Differential effects of prediction error and adaptation along the auditory cortical hierarchy during deviance processing. Neuroimage 259, 119445.

Shen, G., Smyk, N.J., Meltzoff, A.N., Marshall, P.J., 2018. Using somatosensory mismatch responses as a window into somatotopic processing of tactile stimulation. Psychophysiology 55, e13030. https://doi.org/10.1111/psyp.13030.

Smallwood, J., Bernhardt, B.C., Leech, R., Bzdok, D., Jefferies, E., Margulies, D.S., 2021. The default mode network in cognition: a topographical perspective. Nat. Rev. Neurosci. 22 (8), 503–513.

Stefanics, G., Heinzle, J., Horváth, A.A., Stephan, K.E., 2018. Visual mismatch and predictive coding: a computational single-trial ERP study. J. Neurosci. 38 (16), 4020–4030.

Stefanics, G., Kremláček, J., Czigler, I., 2014. Visual mismatch negativity: a predictive coding view. Front. Hum. Neurosci. 8, 666. https://doi.org/10.3389/fnhum.2014.00666.

Sussman, E., Winkler, I., Schröger, E., 2003. Top-down control over involuntary attention switching in the auditory modality. Psychon. Bull. Rev. 10 (3), 630–637.

Tessitore, A., Cirillo, M., De Miccoa, R., 2019. Functional connectivity signatures of Parkinson's disease. J. Parkinsons Dis. 9, 637–652. https://doi.org/10.3233/JPD-191592.

Tse, C.-Y., Shum, Y.-H., Xiao, X.-Z., Wang, Y., 2021. Fronto-occipital mismatch responses in pre-attentive detection of visual changes: implication on a generic brain network underlying Mismatch Negativity (MMN). Neuroimage 244, 118633.

Xu, J.-J., Cui, J., Feng, Y., Yong, W., Chen, H., Chen, Y.-C., et al., 2019. Chronic tinnitus exhibits bidirectional functional dysconnectivity in frontostriatal circuit. Front. Neurosci. 13, 1299. https://doi.org/10.3389/fnins.2019.01299.

CHAPTER 11

Predictive coding in music, speech, and language

"Traditionally, music and language have been treated as different psychological faculties. This duality is reflected in older theories noting that speech functions were thought to be localized in the left and music functions in the right-hemisphere of the brain. This view has been challenged in recent years mainly because of the advent of modern brain imaging techniques and the improvement in neurophysiological measures to investigate brain functions. The findings of these more recent studies show that music and speech functions have many aspects in common and that several neural modules are similarly involved in speech and music" (Jäncke, 2012).

11.1 Right-hemisphere dominance

Functional neuroimaging studies in humans indicate that the left and right auditory hemispaces are coded asymmetrically, with a rightward attentional bias that reflects spatial attention in vision. Based on human electroencephalographic (EEG) data acquired during an auditory-location oddball paradigm, Dietz et al. (2014) applied Dynamic Causal Modeling (DCM) of effective connectivity and Bayesian model comparison to assess this hemisphere dominance. Their results support a hemispheric asymmetry in a frontoparietal network (FPN; Chapter 10) that conforms to the right-hemisphere dominance model. Within the FPN, forward connectivity increases selectively in the hemisphere contralateral to the side of sensory stimulation. They interpreted this in light of hierarchical predictive coding (Chapter 1) as a selective increase in attentional gain mediated by feedforward connections that carry precision-weighted prediction errors (pwPE) during perceptual inference.

I introduce here three main connectivity networks for the temporal-parietal junction (TPJ), reviewed by Igelstrom and Graziano (2017) that play a role in further discussion of hemispheric dominance. These are: (a) the dorsal sector of the TPJ close to the intraparietal sulcus, that is

Brain Responses to Auditory Mismatch and Novelty Detection
https://doi.org/10.1016/B978-0-443-15548-2.00011-9
345

functionally connected to the lateral anterior prefrontal cortex and the anterior sectors of the cingulate and paracingulate cortex; (b) the caudal sector of the TPJ in the angular gyrus that is linked with the posterior cingulate and the hippocampus; (c) an anterior area in the supramarginal gyrus that is strongly connected with regions of the ventral attentional network like the ventrolateral prefrontal cortex, the middle frontal gyrus, and the frontal operculum (Fig. 11.1).

Dietz et al. (2021) evaluated changes in connection strengths within this hierarchical network during left and right salient stimuli. An increase in connection strength corresponds to an increase in postsynaptic efficacy or the influence of an axonal projection on its postsynaptic target population. This analysis revealed that (Fig. 11.2), when salient stimuli appeared on the left, there was an increase in feedforward connection strength from the right inferior frontal gyrus (IFG) to the inferior parietal lobe (IPL). This feedforward influence was reciprocated by an increase in feedback connection strength from right IPL to IFG and an increase in feedback connection strength from right IFG to Heschl's gyrus (HG). When salient stimuli appeared on the right, there was an increase in feedforward connection strength from left IFG to IPL. This increase in the feedforward influence was reciprocated by an increase in feedback connection strength from left IPL to IFG and an increase in feedback connection strength from left IFG to HG. Finally, there was an increase in the feedback connection strength from IFG to HG in the right hemisphere. In other words, when stimuli appeared on the left or the right side in ego-centric coordinates, there was an increase in the strength of feedforward and feedback connectivity in the hemisphere contralateral to the side of stimulation, with an additional feedback influence in the right hemisphere when stimuli appeared on the right. Finally, as a complement to Bayesian model comparison, Dietz et al. (2021) assessed the accuracy of the optimal model in explaining the observed data in terms of the difference between the model's predicted responses and the observed electrophysiological responses. This showed that the right-dominant DCM explained, on average, 59% of the variance.

As Igelstrom and Graziano (2017) remarked: "The right-dominance of attentional functions, and left-dominance of memory and language processing, are reflected in the localization of activations in task-based studies. Symmetrically located network nodes in the left and right IPL/TPJ may thus play distinct functional roles. ... The IPL/TPJ might be a site for

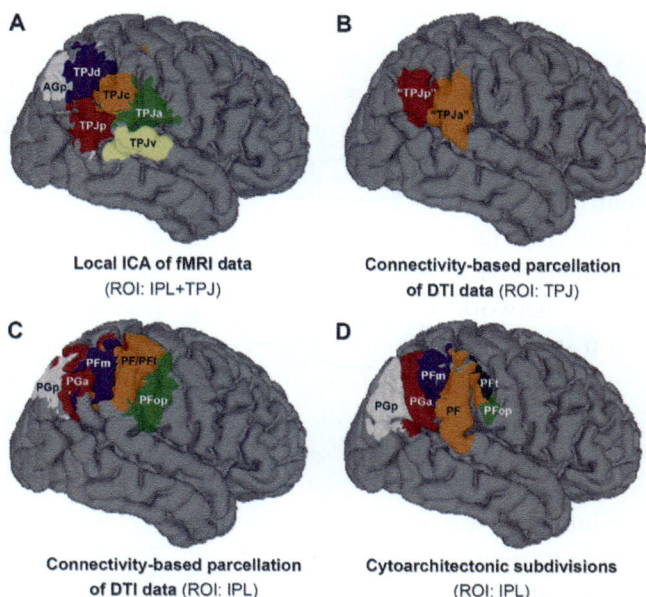

Fig. 11.1 Parcellations of the IPL/TPJ derived from different methods within different regions of interest (ROIs). (A) Subdivisions identified with independent component analysis applied within an ROI comprising the AG, SMG, posterior STS, and posterior superior temporal gyrus. (B) TPJ subdivisions identified with structural connectivity-based parcellation, applied within an ROI bounded by the intraparietal sulcus dorsally, and the dorsal bank of the STS ventrally. Shown is a winner-take-all map constructed from the TPJ parcellation atlas available in FSL, thresholded at 25% for illustration. (C) Subdivisions identified with structural connectivity-based parcellation of the IPL within a ROI in the IPL, reaching ventrally to the level of the ventral tip of the postcentral sulcus and the horizontal segment of the STS. Shown is a winner-take-all map constructed from the Parietal Cortex atlas available in FSL, thresholded at 25% for illustration. The labels reflect a comparison with the cytoarchitectonic atlas. (D) Cytoarchitectonic subdivisions of the IPL. All data are shown as colored overlays on a standard cortical surface in MNI space. Colors are matched to improve clarity and are based on similarities in locations, connectivity patterns, or task-related activations (see main text for discussion). Note that some differences in the spatial extent of the subdivisions (e.g., PGa versus TPJp, PF vs TPJc) are the result of major differences in the dorsoventral coverage of the ROIs chosen for the studies, indicated in the main text. *Reprinted from Igelstrom, K.M., Graziano, M.S.A., 2017. The inferior parietal lobule and temporoparietal junction: a network perspective. Neuropsychologia 105, 70–83., with permission from Elsevier.*

communication between neighboring, perhaps partially overlapping, network nodes, and thereby form a cognitive hub where multiple networks converge and interact."

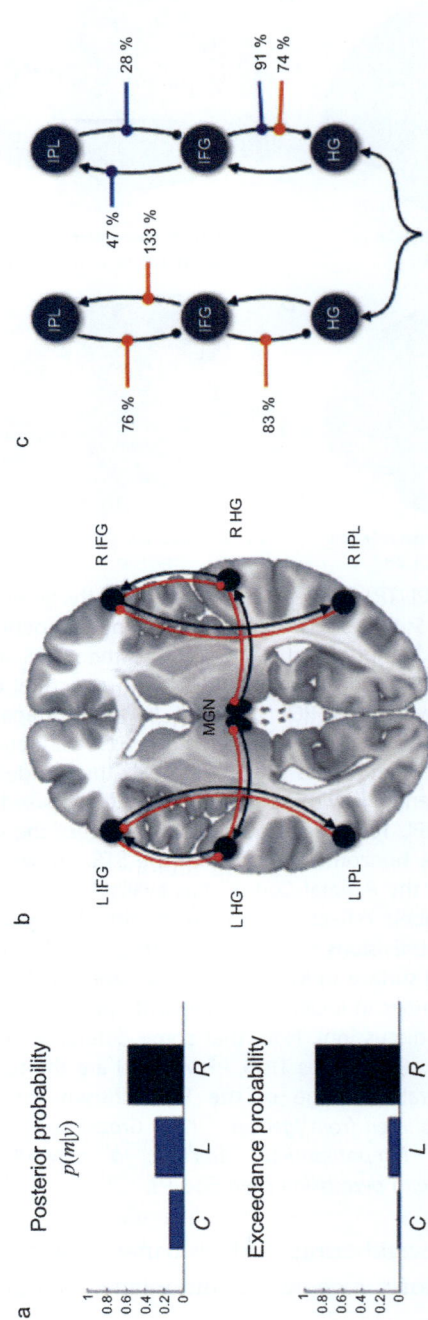

Fig. 11.2 Right-dominant network in healthy controls. (A) Posterior probability and protected exceedance probability of the alternative network models representing contralateral encoding (C), left-lateralized encoding (L), and right-dominant encoding (R) in healthy controls. (B) Architecture of the bilateral right-dominant cortical hierarchy with feedforward connections in black and feedback connections in red. (C) Network showing contralateral increases in connection strength (%) during left stimuli in blue and during right stimuli in orange. *From Dietz, M.J., Nielsen, J.F., Roepstorff, A., Garrido, M.I., 2021. Reduced effective connectivity between right parietal and inferior frontal cortex during audiospatial perception in neglect patients with a right-hemisphere lesion. Hear. Res. 399, 108052. Open Access.*

11.2 Similar brain areas are activated by music and language

In an fMRI study, Koelsch et al. (2002) presented musical chord sequences to participants, infrequently containing unexpected musical events. These events activated the areas of Broca and Wernicke, the superior temporal sulcus, Heschl's gyrus, both planum polare and planum temporale, as well as the anterior superior insular cortices. It is obvious that these structures activated by music are also involved in the processing of language, especially the areas of Broca and Wernicke. They found that the network activated by music appears to be very similar compared to the network that serves the understanding of language. This finding suggests that the cortical language network is less domain specific than previously believed.

A growing body of evidence has also highlighted behavioral connections between musical rhythm and linguistic syntax, suggesting that these abilities may be mediated by common neural resources. Heard and Lee (2020) performed a quantitative meta-analysis of neuroimaging studies using an activation likelihood estimate (ALE) to localize the shared neural structures engaged in a representative set of musical rhythm (rhythm, beat, and meter) and linguistic syntax (merge movement and reanalysis) operations. They found that "rhythm engaged a bilateral sensorimotor network throughout the brain consisting bilaterally of the IFG, supplementary motor area (SMA), superior temporal gyri (STG)/TPJ, insula, IPL, and putamen. By contrast, syntax mostly recruited the left sensorimotor network including the IFG, posterior STG, premotor cortex, and SMA. Intersections between rhythm and syntax maps yielded overlapping regions in the left IFG, left SMA, and bilateral insula—neural substrates involved in temporal hierarchy processing and predictive coding" (Fig. 11.3).

11.3 Cortical pathways in linguistic processing

11.3.1 Dual-stream speech processing

Hickok and Poeppel (2007) outlined a dual-stream model of speech processing wherein a ventral stream processes speech comprehension and a dorsal stream map speech acoustics to frontal lobe articulatory networks. The model assumes that the ventral stream is largely bilaterally organized—although there are important computational differences between the left- and right-hemisphere systems—and that the dorsal stream is strongly left-hemisphere dominant (Fig. 11.4). A critical component of this control

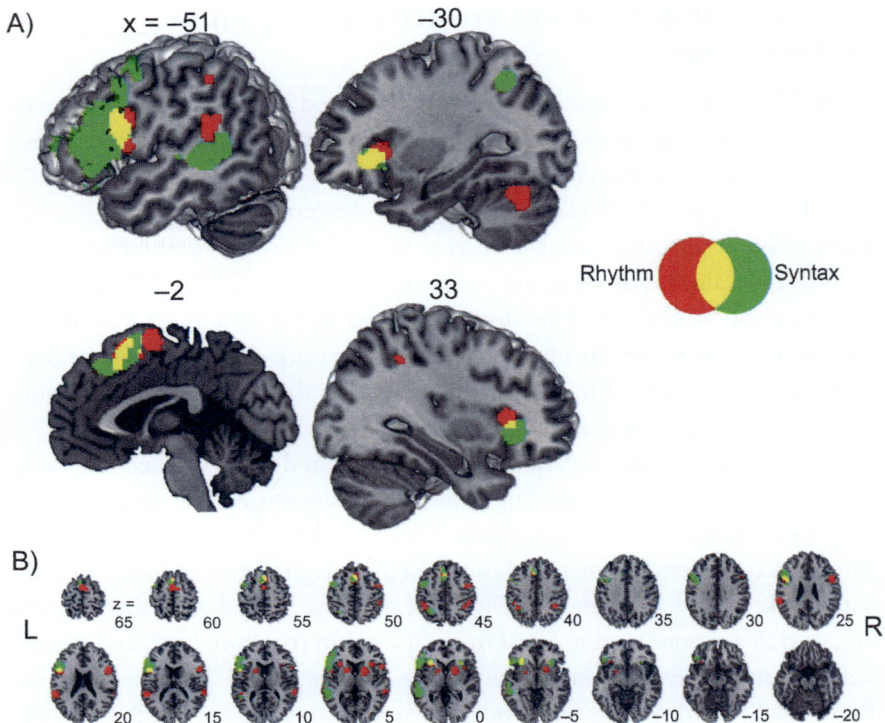

Fig. 11.3 Overlap between rhythm and syntax analyses. (A) Renders of rhythm and syntax ALE maps at slices (from left to right, top to bottom) $x = [-51, -30, -2, 33]$. (B) A series of axial slices for rhythm and syntax areas. Both the rhythm and syntax maps are thresholded at voxel-level $P < 0.001$ (uncorrected) in combination with cluster-level $P < 0.05$ corrected using family-wise error. *Reprinted from Heard, M., Lee, Y. S., 2020. Shared neural resources of rhythm and syntax: An ALE meta-analysis. Neuropsychologia 137, 107–284. with permission from Elsevier.*

system is forward sensory prediction, which affords a natural mechanism for limited motor influence on perception. As an alternative, Rauschecker and Scott (2009) proposed a dual-stream model where the auditory ventral processing stream runs parallel to the visual ventral stream in the infratemporal cortex (Fig. 11.5). In Hickok and Poeppel's (2007) model, phonological processes are assumed to take place largely in the bilateral STS. In Rauschecker and Scott's (2009) model, phonological-relevant encoding is observed through the middle and anterior STG.

Bidelman et al. (2019) measured EEGs in older adults with and without mild hearing loss during a signal-in-noise identification task. Using functional connectivity and graph-theoretic analyses, they showed that

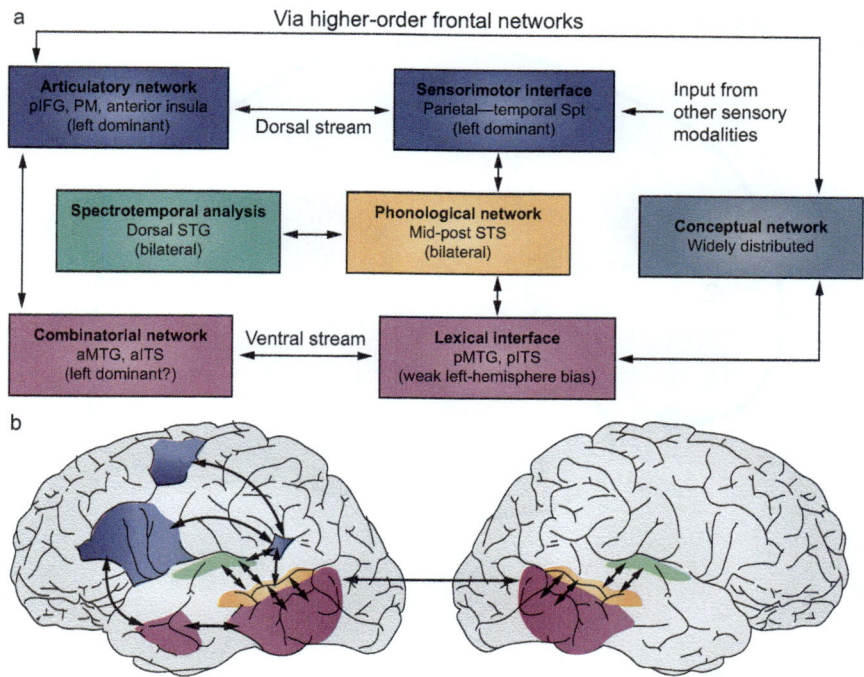

Fig. 11.4 Dual-stream model of speech processing. The dual-stream model holds that early stages of speech processing occur bilaterally in auditory regions on the dorsal STG (spectrotemporal analysis; green) and STS (phonological access/representation; yellow) and then diverges into two broad streams: a temporal lobe ventral stream supports speech comprehension (lexical access and combinatorial processes; pink) whereas a strongly left dominant dorsal stream supports sensorimotor integration and involves structures at the parietal–temporal junction (Spt) and frontal lobe. The conceptual-semantic network (gray box) is assumed to be widely distributed throughout cortex. IFG, inferior frontal gyrus; ITS, inferior temporal sulcus; MTG, middle temporal gyrus; PM, premotor; Spt, Sylvian parietaltemporal; STG, superior temporal gyrus; STS, superior temporal sulcus. *From Hickok, G., Poeppel, D., 2007a. The cortical organization of speech processing. Nat. Rev. Neurosci. 8, 393–402, with permission from Springer Nature.*

hearing-impaired (HI) listeners have more extended (less integrated) communication pathways and less efficient information exchange among widespread brain regions (larger network eccentricity) than their normal-hearing (NH) peers. They found a reversal in directed neural signaling in left hemisphere dependent on hearing status among specific connections within the dorsal–ventral speech pathways (Fig. 11.6). NH listeners showed an overall net "bottom-up" signaling directed from auditory cortex (A1) to inferior frontal gyrus (IFG; Broca's area), whereas the HI group showed the reverse signal (i.e., "top-down" Broca's → A1). A similar flow reversal was noted

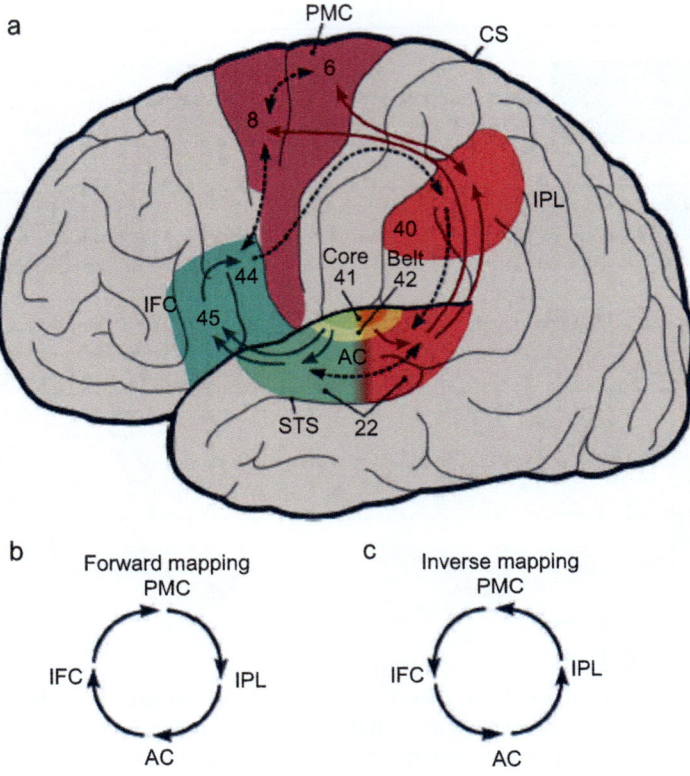

Fig. 11.5 Dual auditory processing scheme of the human brain and the role of internal models in sensory systems. This expanded scheme closes the loop between speech perception and production and proposes a common computational structure for space processing and speech control in the postero-dorsal auditory stream. (A) Antero-ventral (green) and postero-dorsal (red) streams originating from the auditory belt. The postero-dorsal stream interfaces with premotor areas and pivots around inferior parietal cortex, where a quick sketch of sensory event information is compared with a predictive efference copy of motor plans. (B) In one direction, the model performs a forward mapping: object information, such as speech, is decoded in the antero-ventral stream all the way to category-invariant inferior frontal cortex (area 45) and is transformed into motor-articulatory representations (area 44 and ventral PMC), whose activation is transmitted to the IPL (and posterior superior temporal cortex) as an efference copy. (C) In reverse direction, the model performs an inverse mapping, whereby attention- or intention-related changes in the IPL influence the selection of context-dependent action programs in PFC and PMC. Both types of dynamic model are testable using techniques with high temporal precision (e.g., magnetoencephalography in humans or single-unit studies in monkeys) that allow determination of the order of events in the respective neural systems. *From Rauschecker, J.P., Scott, S.K., 2009. Maps and streams in the auditory cortex: nonhuman primates illuminate human speech processing. Nat. Neurosci. 12(6), 718–724. With permission from Springer Nature.*

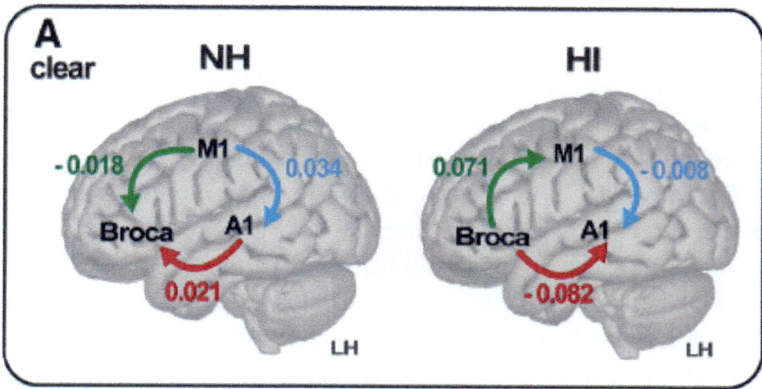

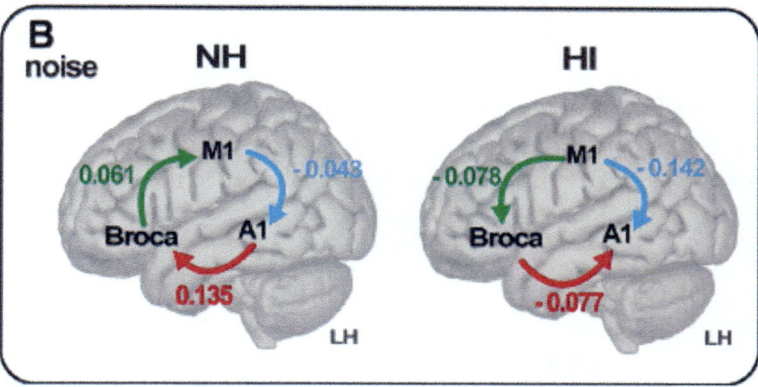

Fig. 11.6 Directional flow of neural signaling within the dorsal–ventral stream reverses with age-related hearing loss. A. Clear speech. , B. Noise-degraded speech. Values represent the strength of connectivity within LH via phase-transferentropy, whereas the direction (causality) of communication is determined by sign. Arrows denote flow from region A→B. The NH listeners show signaling directed from A1→Broca's (pars opercularis), whereas the HI group shows the reverse (Broca's→A1), suggesting bottom-up versus top-down configurations within the same pathway dependent on hearing status. During noise-degraded speech, communication between linguistic and motor areas reverses from Broca's driving M1 to M1 driving Broca's, but only in HI listeners (cf. green connection, A vs. B). A1, auditory cortex; M1, motor cortex. *From Bidelman, G.M., Mahmud, M.S., Yeasin, M., Shen, D., Arnott, S.R., Alain, C., 2019. Age-related hearing loss increases full-brain connectivity while reversing directed signaling within the dorsal–ventral pathway for speech. Brain Struct. Funct. 224, 2661–2676, with permission from Springer Nature.*

between left IFG and motor cortex (M1). This full–brain connectivity results demonstrate that even mild forms of hearing loss alter how the brain routes information within the auditory–linguistic–motor loop (Bidelman et al., 2019).

In a recent review, Bhaya-Grossman and Chang (2022) note that whereas in the models of Hickok and Poepel (2007) and Rauschecker and Scott (2009) information is assumed to travel from the primary auditory (A1) to the nonprimary auditory cortex (in STG), Hamilton et al. (2021) have found that transient functional lesioning of A1 via electrical stimulation or focal ablation of the A1 does not impair speech comprehension. Furthermore, electrical stimulation targeting the STG impairs speech comprehension without impairing tone discrimination (Boatman, 2004). "The double dissociation between A1 and nonprimary auditory cortices as well as the highly distributed nature of speech feature encoding throughout the STG pose significant challenges to the prevailing anatomically defined, hierarchical stream models of cortical processing" (Bhaya-Grossman and Chang, 2022). Specifically, Hamilton et al. (2021) used intracranial recordings across the entire human auditory cortex—636 electrode sites in the left temporal plane and STG in nine participants, including both grid and depth electrodes, electrocortical stimulation, and surgical ablation. They identified an onset-specific region indicated by 15 black electrodes in Fig. 11.7C

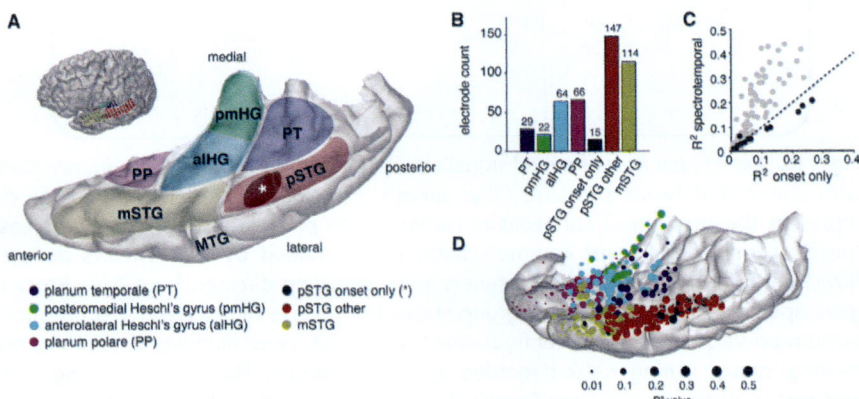

Fig. 11.7 Anatomical parcellations of temporal lobe regions of the human auditory cortex and electrode coverage. (A) Anatomical regions of interest on the left hemisphere temporal lobe of an example participant. STG = superior temporal gyrus, MTG = middle temporal gyrus. (B) Electrode counts across anatomical areas for all nine participants. (C) Comparison between onset-only and spectrotemporal models shows a population described by only a singular onset feature. (D) All participants' electrodes projected onto a Montreal Neurological Institute (MNI) atlas brain (cvs_avg35_inMNI152). Electrode size reflects the maximum amount of variance (R2) explained by the encoding models tested in our analyses. Electrode sites are colored according to their anatomical location. *Reprinted from Hamilton, L.S., Oganian, Y., Hall, J., Chang, E.F., 2021. Parallel and distributed encoding of speech across human auditory cortex. Cell 184, 4626–4639., with permission from Elsevier.*

and D (13/15 were located in anatomically defined posterior STG area). This onset-selective region, which exhibited strong transient responses at the onset of sentences followed by relative quiescence, was observed only in a localized region of the lateral STG and not on the superior temporal plane. Hamilton et al. (2021) also showed that cortical processing across auditory areas is not consistent with a serial hierarchical organization (i.e., from A1 to "higher" auditory areas). Instead, response latency and receptive field analyses demonstrate parallel and distinct information processing in the primary and nonprimary auditory cortices. This functional dissociation was also observed where electrical stimulation of A1 evokes auditory hallucination but does not distort or interfere with speech perception. Opposite effects were observed during stimulation of nonprimary cortex in STG. Ablation of A1 does not affect speech perception.

11.3.2 Network assessment, predictive coding, and the role of IFG

It is not surprising that the left IFG, which contains Broca's area, is involved in linguistic processing. Musso et al. (2015) examined the brain network involved in the recognition and integration of words and chords that were not hierarchically related to the preceding syntax, that is, those deviating from the universal principles of grammar and tonal relatedness. This kind of syntactic processing in both domains was found to rely on a shared network in the left hemisphere centered on the inferior part of the IFG, including pars opercularis and pars triangularis (cf. Fig. 10.7), and on dorsal and ventral long association tracts connecting this brain area with temporo-parietal regions. Direct partial correlation analysis within the activation nodes of the shared structural network in both domains (resulting from the conjunction analysis of structurally deviant > baseline conditions in music and language) showed that the functional connectivity within IFG is highly differentiated (Fig. 11.8): F3op (pars opercularis; blue triangle in Fig. 11.8) interacts with inferior parietal cortex (IPC) and middle temporal gyrus (MTG), whereas the pars triangularis (F3tri; red triangle in Fig. 11.8) interacts only with MTG. The correlation analysis also showed that the insula (INS) significantly interacts with MTG and superior frontal gyrus (SFG). The dorsal (blue) and ventral (red) pathways (Fig. 11.8; cf. Fig. 11.4) may support distinct aspects of language processing. Potentially, this is reflected in the dual-route model of reading, with the dorsal pathway more involved in grapho-phonological conversion during phonological tasks, and the ventral pathway performing lexico-semantic access during semantic tasks. Such a subdivision is also suggested at the level of the

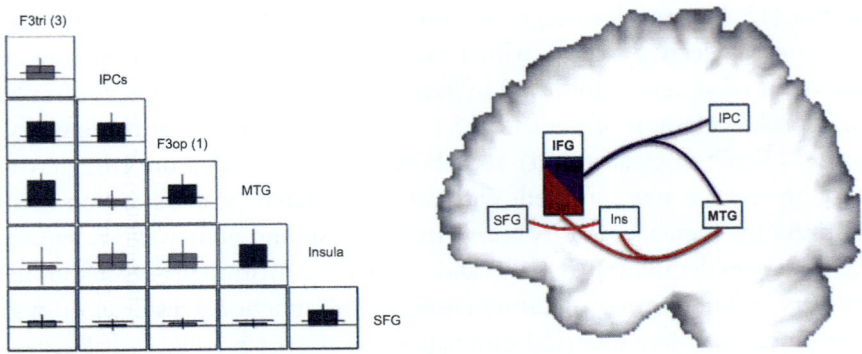

Fig. 11.8 Directed partial correlation (dPC) results within the overlapping syntactic network in music and language. Left: dPC values are plotted for each pairwise interaction within the overlapping syntactic network in music and language. Significant interactions are displayed as dark gray plots, not significant interactions in lighter gray. Right: Schematic view of the dPC results. Significant functional interactions are drawn as lines following the anatomical course as identified by tractography. The common structural network in music and language shows the involvement of local connections between both IFG activations and between insula and the superior frontal areas, ventral tracts along EmC (red) connecting F3tri and insula with MTG, and dorsal tracts along superior longitudinal fasciculus and arcuate fasciculus (SLF/AF) system (blue) connecting F3op with IPC and MTG. Thus it is clear that both ventral and dorsal tracts are functionally relevant for performing the structural tasks in both domains. F3op=pars opercularis; F3tr, pars triangularis; F3orb, pars orbitalis, IPC, inferior parietal cortex; MTG, middle temporal gyrus; SFG, superior frontal gyrus. *Reprinted from Musso, M., Weiller, C., Horn, A., Glauche, V., Umarova, R., Hennig, J., et al., 2015. A single dual-stream framework for syntactic computations in music and language. NeuroImage 117, 267–283., with permission from Elsevier.*

IFG, involving ventral and dorsal parts for lexico-semantic and phonological processing, respectively (Musso et al., 2015).

Ylinen et al. (2017) proposed that inferring predictions facilitate speech processing and word learning in the early stages of language development. Twelve- and 24-month-olds' event-related potential (ERP) responses to heard syllables are faster and more robust when the preceding word context predicts the ending of a familiar word (Fig. 11.9). For unfamiliar, novel word forms, however, word expectancy violation generates a prediction error response, the strength of which significantly correlates with children's vocabulary scores at 12 months. Ylinen et al. (2017) suggest "that predictive coding may accelerate word recognition and support early learning of novel words, including not only the learning of heard word forms but also their mapping to meanings. Prediction error may mediate learning via attention,

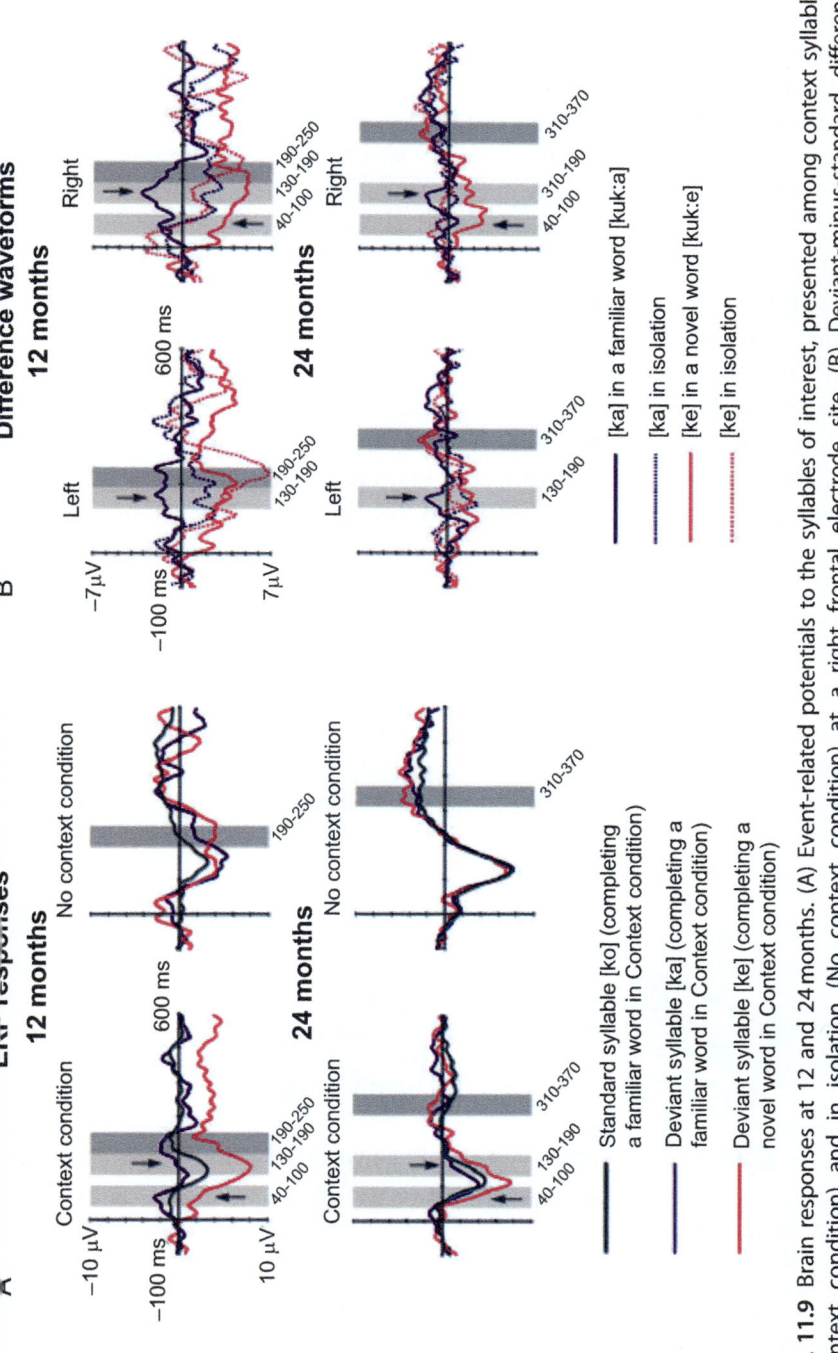

Fig. 11.9 Brain responses at 12 and 24 months. (A) Event-related potentials to the syllables of interest, presented among context syllables (Context condition) and in isolation (No context condition) at a right frontal electrode site. (B) Deviant-minus-standard difference waveforms at left and right frontal electrode sites. In both (A) and (B), time windows with significant word-level and phonetic effects have been marked with light gray and dark gray, respectively, and arrows point to those deviant responses that are significantly different compared with all other deviant responses across conditions in that time window. *From Ylinen, S., Bosseler, A., Junttila, K., Huotilainen, M., 2017. Predictive coding accelerates word recognition and learning in the early stages of language development. Dev. Sci. 20, e12472. ©2016 John Wiley & Sons Ltd., with permission.*

since infants' attention allocation to the entire learning situation in natural environments could account for the link between prediction error and the understanding of word meanings."

Della Rosa et al. (2018) conducted an fMRI study, using a lexical decision task involving both abstract and concrete words, whose imageability and context availability values were explicitly modeled in separate parametric analyses. "The left IFG [including Broca's area] was significantly more activated for abstract than for concrete words, and a conjunction analysis showed a common activation for words with low imageability or low context availability only in the left IFG, in the same area reported for abstract words. [...] Only the left middle temporal gyrus/angular gyrus, known to be involved in semantic processing, was a significant predictor of left IFG activity differentiating abstract from concrete words."

Using fMRI, Fei et al. (2020) compared brain responses to intelligible and unintelligible speech between older and young adults. They found reduced brain activation and lower regional pattern distinctions in response to intelligible versus unintelligible speech in the left anterior STG and the left IFG in the older group compared with young adults. In both the older group and young adults, they found bidirectional positive modulations between the anterior STG and IFG, forward modulation from the angular gyrus (AG) to the IFG and from the posterior MTG to the AG, and backward modulations from the anterior STG to the posterior MTG (Fig. 11.10). The strength of the effective connectivity from the IFG to AG was

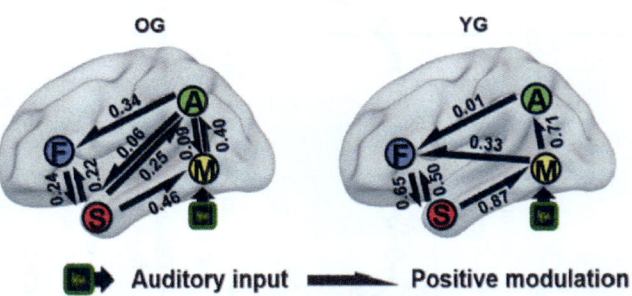

Fig. 11.10 Functional and effective connectivity analysis. (D) Positive connections significantly modulated by intelligible speech processing in the older group (OG; 62–80 years) and younger group (YG; 21–28 years) with FDR correction at $P < 0.05$ (pMTG-aSTG-IFG-AG, i.e., M-S-F-A in the Figure). Abbreviations: pMTG, posterior middle temporal gyrus; aSTG, anterior superior temporal gyrus; IFG, inferior frontal gyrus; AG, angular gyrus; OG, older group; YG, young group. *Reprinted from Fei, N., Ge, J., Wang, Y., Gao, J.-H., 2020. Aging-related differences in the cortical network subserving intelligible speech. Brain Lang. 201, 104–713. ©2019, with permission from Elsevier.*

significantly higher in the older group compared to young adults. Notably, the functional connectivity between the left IFG and the left AG was increased and a significantly enhanced bidirectional effective connectivity between the left anterior STG and the left AG was observed in the older adults for processing speech intelligibility (Fig. 11.10). Contrasting findings were also found between groups; in the older group, positive modulation from the AG to the posterior MTG and bidirectional positive modulations between the anterior STG and AG were observed; while in the young adults, forward modulation from the posterior MTG to the IFG was found.

Wang et al. (2022) used MEG and ERPs to track the time course and localization of evoked activity produced by expected, unexpected plausible, and implausible words during incremental language comprehension (Fig. 11.11). They "suggest that the full pattern of results can be explained within a hierarchical predictive coding framework in which increased evoked activity reflects the activation of residual information that was not already represented at a given level of the fronto-temporal hierarchy ("error" activity). Between 300 and 500 ms [the N400], the three conditions produced progressively larger responses within left temporal cortex (lexico-semantic prediction error), whereas implausible inputs produced a selectively enhanced response within inferior frontal cortex (prediction error at the level of the event model). Between 600 and 1000 ms, unexpected plausible words activated left inferior frontal and middle temporal cortices (feedback activity that produced top-down error), whereas highly implausible inputs activated left inferior frontal cortex, posterior fusiform (unsuppressed orthographic prediction error/reprocessing), and medial temporal cortex (possibly supporting new learning). Therefore, predictive coding may provide a unifying theory that links language comprehension to other domains of cognition." Of course, as Wang et al. (2022) surmised, no single study can provide definitive evidence for the predictive coding model of language comprehension. As they noted, several findings, including the graded increase in N400 activity within the temporal cortex, and reprocessing of anomalies within posterior fusiform cortex are also consistent with other psycholinguistic or neurobiological models.

11.4 Role of brain rhythms in music and language coding

11.4.1 Music

Using magnetoencephalography (MEG), Fujioka et al. (2015) examined beta-band oscillatory activities around 20 Hz while participants listened to metronome beats and imagined musical meters such as a march and waltz.

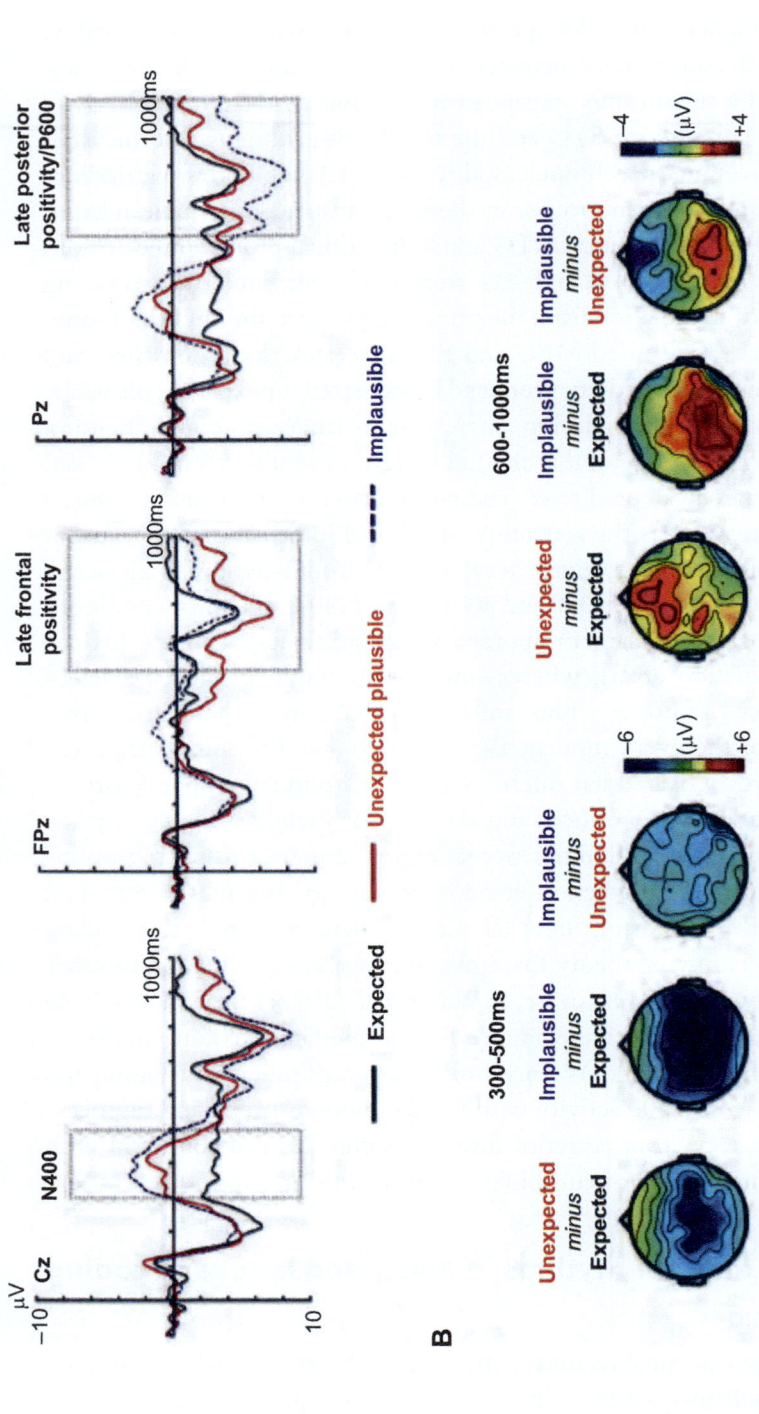

Fig. 11.11 ERP results and language comprehension. (A) Grand-averaged ERP waveforms elicited by critical words in each of the three conditions, shown at three representative electrode sites: Cz, FPz, and Pz. Expected: solid black line; implausible: dashed blue line. Negative voltage is plotted upwards. Dotted boxes are used to indicate the time windows corresponding to the N400 (300–500 ms), the late frontal positivity (600–1000 ms), and the late posterior positivity/P600 (600–1000 ms) ERP components. (B) Voltage maps show the topographic distributions of the ERP effects produced by contrasting *expected*, *unexpected plausible*, and *implausible* critical words between 300 and 500 ms (left panel) and between 600 and 1000 ms (right panel). Note that the N400 effects and the late positivity effects are shown at different voltage scales to better illustrate the scalp distribution of each effect. *From Wang, L., Schoot, L., Brothers, T., Alexander, E., Warnke, L., Kim, M., et al., 2022_Epub. Predictive coding across the left fronto-temporal hierarchy during language comprehension. Cerebral Cortex, © The Author(s) 2022. by permission of Oxford University Press.*

They demonstrated that beta-band event-related desynchronization—decreased amplitude—in the auditory cortex differentiates between beat positions, specifically between downbeats and the following beat. Thus beta-band oscillations are related to hierarchical and internalized timing information. Moreover, the meter representation in the beta oscillations was widespread across the brain, including sensorimotor and premotor cortices, parietal lobe, and cerebellum. The results also extend to the role of beta oscillations in neural processing of predictive timing. Edalati et al. (2021) explored the oscillatory properties of the neural response to rhythmic incongruence and the cross-frequency coupling between multiple frequencies to further investigate the mechanisms underlying rhythm perception. They "designed an experiment to investigate the neural response to rhythmic deviations in which the tone either arrived earlier than expected or the tone in the same metrical position was omitted. These two manipulations modulate the rhythmic structure differently, with the former creating a larger violation of the general structure of the musical stimulus than the latter. Both deviations resulted in an MMN response, whereas only the rhythmic deviant resulted in a subsequent P3a. Rhythmic deviants due to the early occurrence of a tone, but not omission deviants, seemed to elicit a late high gamma response (60–80 Hz) at the end of the P3a over the left frontal region, which correlated with the P3a amplitude over the same region and was also nested in theta oscillations" (Edalati et al., 2021).

11.4.2 Speech and language

From electrocorticograms (ECoGs) in 10 human subjects with intractable epilepsy, Potes et al. (2014) identified stimulus-related modulations in the alpha (8–12 Hz) and high-gamma (70–110 Hz) bands at neuroanatomical locations implicated in auditory processing of music. They found stimulus-related ECoG modulations in the alpha band in areas adjacent to tonotopically organized A1. Stimulus-related ECoG modulations in the high-gamma band in areas close to A1 and in other perisylvian areas were also found and preceded those in the alpha band by 280 ms. Consequently, high-gamma band activity predicted alpha activity, but not vice versa. Using Granger causality analysis they identified relationships of high-gamma activity between posterior STG (Wernicke's area) and inferior frontal gyrus (Broca's area), and between STG and premotor cortex (Fig. 11.12). This suggested to Potes et al. (2014) that these relationships reflect direct cortico-cortical connections rather than common driving input from the medial geniculate body.

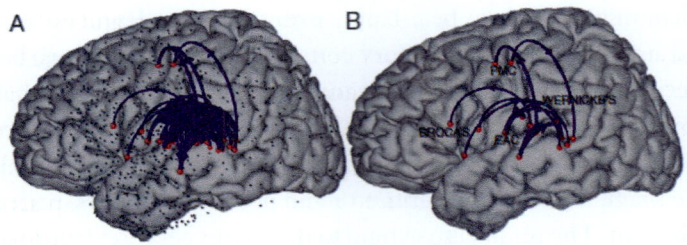

Fig. 11.12 Intersubject Granger causality analysis. Causal relationships of activity in the high-gamma band recorded during listening to music and between different pairs of electrode locations. Across all possible 630,800 electrode combinations, only 68 connections were significant at $\alpha = 1.58e-8$ after Bonferroni correction (A). These connections were further reduced to the ten most salient 10 connections (B). *Reprinted from Potes,C., Brunner, P., Gunduz, A., Knight, R.T., Schalk, G., 2014. Spatial and temporal relationships of electrocorticographic alpha and gamma activity during auditory processing. NeuroImage 97, 188–195., with permission from Elsevier.*

Lewis and Bastiaansen (2015) proposed that relating beta and gamma oscillations to sentence-level language comprehension might be unified under a predictive coding account. They suggested that "oscillatory activity in the beta frequency range may reflect both the active maintenance of the current network configuration responsible for representing the sentence-level meaning under construction, and the *top-down* propagation of predictions to hierarchically lower processing levels based on that representation. In addition, they suggest that oscillatory activity in the low and middle gamma range reflect the matching of top-down predictions with bottom-up linguistic input, while evoked high gamma might reflect the propagation of bottom-up prediction errors to higher levels of the processing hierarchy." Bidelman (2015) recorded EEG activity while participants rapidly classified sounds along a vowel continuum (/u/ to /a/). Time-frequency analyses revealed distinct temporal dynamics in induced (nonphase locked) oscillations. The found that "increased beta (15–30 Hz) coded prototypical vowel sounds carrying well-defined phonetic categories whereas increased gamma (50–70 Hz) accompanied ambiguous tokens near the categorical boundary."

Meyer (2018) synthesized a mapping from each linguistic processing domain to a unique set of oscillatory mechanisms. "Gamma-band oscillations synchronize with phonemes, theta-band oscillations synchronize with syllables, and delta-band oscillations synchronize with intonational phrase boundaries. The three bands form a hierarchical relationship, such that low-frequency   oscillations   can   top-down   align   higher-frequency oscillations, amplifying bottom-up information extraction and establishing a coherent percept." Also based on hierarchical oscillations, language comprehension can proceed along two processing streams: (1) A syntactic stream "employing delta-band cycles to group words into syntactic phrases, alpha-band power modulations to store syntactic phrases in verbal working memory and theta-band power modulations to retrieve information from sentences' hierarchical working memory representations, as well as from long-term memory." (2) A predictive stream "employing beta-band oscillations to top-down predict the semantics of upcoming words that fit the cumulative meaning of the prior word sequence, and gamma-band oscillations to bottom-up assess the contextual semantic fit of incoming words" (Meyer (2018).

Donhauser and Baillet (2020) trained a neural network that uses context to predict speech at the phoneme level. Using this model, they show that speech-related rhythmic activity is hierarchically organized into two time-scales: fast responses (theta: 4–10 Hz), restricted to early auditory regions, and slow responses (delta: 0.5–4 Hz), dominating in downstream auditory regions. Neural activity in these bands is selectively modulated by predictions: the gain of early theta responses varies according to the contextual uncertainty of speech, while later delta responses are selective to surprising speech inputs. Donhauser and Baillet (2020) conclude that theta sensory sampling is tuned to maximize expected information gain, while delta encodes only nonredundant information. Hovsepyan et al. (2020) explored the effects of theta–gamma coupling on bottom-up/top-down dynamics during on-line syllable identification. In human auditory cortex, bottom-up gamma activity is modulated at the low-beta rate, which could offer top-down integration constants that are intermediate between syllables and gamma phonemic-range chunks. Alternatively, the beta top-down rhythm could be related to expected precision, which weighs bottom-up prediction errors reflected in gamma activity. They designed a computational model (Precoss—predictive coding and oscillations for speech) that can recognize syllable sequences in continuous speech. "The model uses predictions from internal spectro-temporal representations of syllables and theta oscillations to signal syllable onsets and duration. Syllable recognition is best when theta-gamma coupling is used to temporally align spectro-temporal predictions with the acoustic input. This neurocomputational modelling work demonstrates that the notions of predictive coding and neural oscillations can be brought together to account for on-line dynamic sensory processing."

Summarizing, there is general agreement, both for music and language, that in terms of predictive coding, beta activity represents top–down predictions, whereas gamma represents bottom–up prediction error.

11.5 Predictive coding in more detail

11.5.1 General

To assess the neuroanatomical foundations of predictive coding, Siman-Tov et al. (2019) conducted an activation likelihood estimation (ALE) meta-analysis of 39 neuroimaging studies of three functional domains (action perception, language, and music) inherently involving prediction. They found that "anatomically, this network consists of regions within the IFG, anterior insula (AI), medial frontal gyrus (MFG), premotor cortex (PMC), pre-SMA, TPJ, striatum, thalamus/subthalamus and the cerebellum" (Fig. 11.13). Functionally, this network appears to be associated with high-level perception and action, as well as with aspects of attention, implicit learning, mirroring, and social cognition. In light of the evidence presented, Siman-Tov et al. (2019) "believe that this network is a fundamental and robust brain system, probably interacting with other large-scale brain networks, such as the salience network (SAL; possibly through the AI and ACC), mediating interoceptive inference aspects of emotional processing (Seth and Friston, 2016) and the dorsal attention network (DAN; possibly through the frontal eye field, FEF, and intraparietal sulcus), mediating attention allocation aimed at prediction error minimization." Siman-Tov et al. (2019) concluded that "the predictive coding model helps understand how these various brain functions, each independently being associated with analogous set of brain regions, can actually converge under one umbrella, namely implicit prediction."

11.5.2 Predictive coding in music

"Music offers a most illuminating paradigm to understand the fundaments of the predictive brain, largely because every type of music is based on predictable regularities. e.g., temporal, melodic, and, depending on the musical culture or musical tradition, harmonic, timbral, and textural structure" (Koelsch et al., 2019).

Vuust et al. (2009) employed MEG to test (1) that neuronal markers of rhythmic incongruities behave in accordance with a predictive coding framework, and (2) that musical competence affects the composition of the neuronal networks involved in the processing of rhythm by affecting

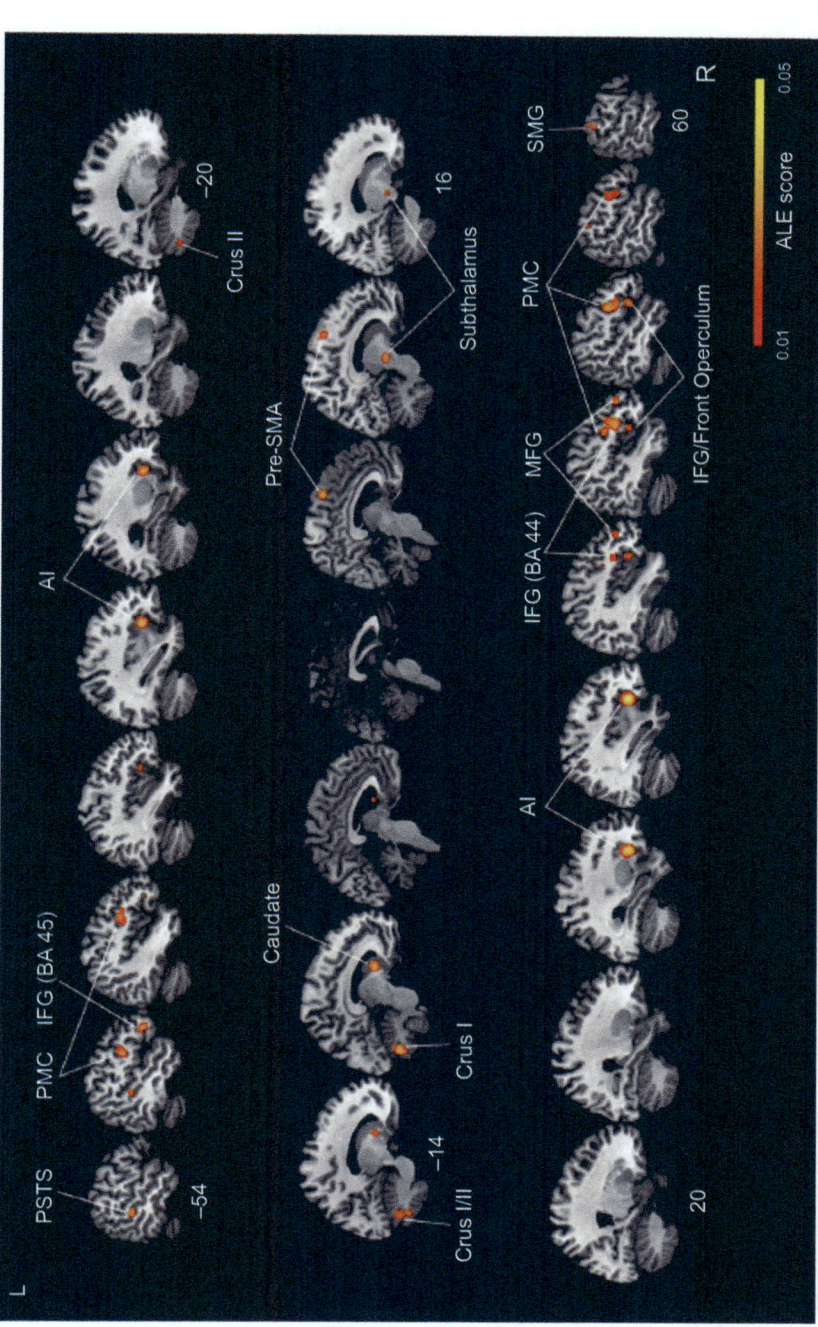

Fig. 11.13 ALE meta-analysis results of 39 neuroimaging studies involving prediction formation and/or violation in three functional domains: action perception, language, and music, presented on sagittal sections (FDR, $q = 0.05$, cluster size = 200 mm^3). AI, anterior insula, Crus I/II, cerebellar lobule VII, Crus I/II, Front. Operculum, frontal operculum, IFG, inferior frontal gyrus, MFG, middle frontal gyrus, PMC, premotor cortex, Pre-SMA, presupplementary motor area, PSTS, posterior superior temporal sulcus, SMG, supramarginal gyrus. *Reprinted from Siman-Tov, T., Granot, R.Y., Shany, O., Singer, N., Hendler, T., Gordon, C.R., 2019. Is there a prediction network? Meta-analytic evidence for a cortical-subcortical network likely subserving prediction. Neurosci. Biobehav. Rev. 105, 262–275., with permission from Elsevier.*

the neuronal integration. As a special case, the rhythmic regularity in music is generated by expectations created in different layers of the musical structure. Vuust et al. (2009) found event-related responses to strong rhythmic incongruence in all subjects, the MMN peaking at 110–130 ms and the P3a around 80 ms after the MMN in expert jazz musicians and some of the rhythmically unskilled subjects, as well as responses to subtle rhythmic incongruence in most of the expert musicians. Vuust et al. (2012) then showed that "performing musicians' characteristics of style and genre influence their perceptual skills and their brains' processing of sound features embedded in a musical context, as indexed by larger MMN. Such influences of training on low-level, pre-attentive neural processing exemplify the longer-term contextual, environmental and cultural aspects of predictive coding."

Koelsch et al. (2019) noted that MMNs of expert musicians were stronger in the left hemisphere than in the right hemisphere in contrast to P3a showing a slight, nonsignificant, right lateralization. They interpreted MMN and P3a as a prediction error generated in the auditory cortex and its subsequent evaluation in a broader network including generators in the auditory cortex as well as higher-level neuronal sources. This is in keeping with expectations based on the predictive coding theory. In line with the preattentive nature of the MMN, it seems that attention does not affect early sensory processing. Koelsch et al. (2019) considered the early latency right anterior negativity (ERAN, a response to music-syntactic irregularities; Appendix): "Crucially, the irregularities eliciting the ERAN rest upon (usually implicit) knowledge of musical syntax. Acquisition of such knowledge requires substantial exposure to music, whereas the events eliciting the MMN are inferred in real-time from the acoustic environment." They also showed that "the MMN and ERAN dissociate in two ways: in contrast to the MMN, which attenuates with repeated acoustical irregularities and which is relatively impervious to attention or predictions of predictability, the ERAN persists with repeated exposure to syntactical irregularities. Particularly, a typical ERAN did not differ in amplitude when participants knew or did not know how the sequence would end. However, later positive potentials differed between conditions: when participants had no knowledge about the sequence ending, the irregular chords elicited a P3a (but no P3b), and when participants were told about the sequence endings, irregular endings elicited a more parietal, P3b-like potential (but no P3a)" (Koelsch et al., 2019).

Lumaca et al. (2021) applied DCM and parametric empirical Bayes on fMRI data to identify modulation of effective brain connectivity that takes place during perceptual learning of complex tone patterns. They used a complex oddball paradigm (cf. Fig. 2.11) based on tone patterns as opposed to simple deviant tones as well as fMRI which allowed high spatial accuracy in the identification of involved cortical regions. These regions served then as empirical regions of interest for the analysis of effective connectivity. Deviant patterns induced an increased BOLD response, compared to standards, in early auditory (Heschl's gyrus [HG]) and association auditory areas (planum temporale [PT]) bilaterally. Within this network, Lumaca et al. (2021) found a left-lateralized increase in feedforward connectivity from HG to PT during deviant responses and an increase in excitation within left HG. In contrast to previous findings, including the fMRI studies of Della Rosa et al. (2018) and Schlossmacher et al. (2022), they did not find frontal activity such as in IFG, nor did they find modulations of backward connections in response to oddball sounds. These results suggested to them that complex auditory prediction errors are encoded by changes in feedforward and intrinsic connections, confined to the STG.

The same group (Quiroga-Martinez et al., 2021) employed MEG and DCM to investigate whether the MMNm—and its modulation by context predictability and musical expertise—are associated with enhanced neural gain of auditory areas, as a plausible mechanism for encoding precision-weighted prediction errors. Using Bayesian model comparison, they "asked whether models with intrinsic connections within A1 and STG—typically related to gain control—or extrinsic connections between A1 and STG—typically related to propagation of prediction and error signals—better explained MEG responses. They found that, compared to regular sounds, out-of-tune pitch deviations were associated with lower intrinsic (inhibitory) connectivity in A1 and STG, and lower backward (inhibitory) connectivity from STG to A1, consistent with disinhibition and enhanced neural gain in these auditory areas. More predictable melodies were associated with disinhibition in right A1, while musicianship was associated with disinhibition in left A1 and reduced connectivity from STG to left A1 (Fig. 11.14). These results indicate that musicianship and melodic predictability, as well as pitch deviations themselves, enhance neural gain in auditory cortex during deviance detection." Quiroga-Martinez et al. (2021) noted that their network is very similar to classical DCM studies of MMN responses (Dietz et al., 2014; Garrido et al., 2009; Schmidt et al., 2013) with two important differences. First, they included anterior STG (instead of posterior STG or planum temporale), which has been related to the processing of pitch sequences (Gander et al., 2019). Second,

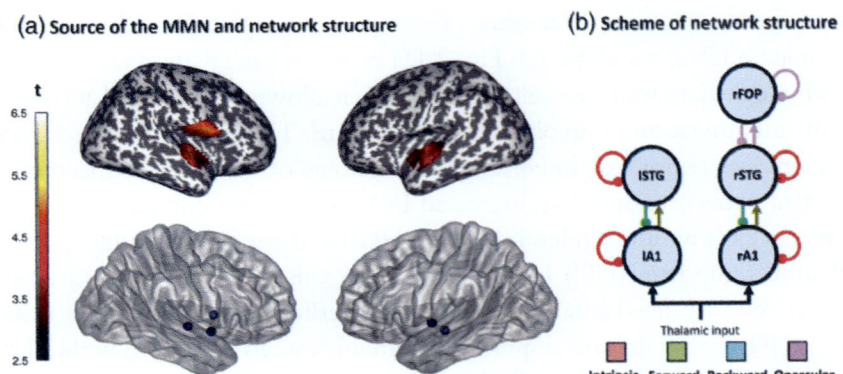

Fig. 11.14 Structure of the network. (A) Results of the source reconstruction (top) and prior location of the sources (bottom). (B) Scheme of the network. The connections modulated in each of the four model families are indicated with colors. The combination of these families yielded a total of 24 = 16 models, including a null model in which no connections were modulated. A1, primary auditory cortex; FOP, frontal operculum; STG, superior temporal gyrus. *From Quiroga-Martinez, D.R., Hansen, N.C., Højlund, A., Pearce, M., Brattico, E., Holmes, E., et al., 2021. Musicianship and melodic predictability enhance neural gain in auditory cortex during pitch deviance detection. Hum. Brain Mapp. 42, 5595–5608. https://doi.org/10.1002/hbm.25638. Open access.*

they included the frontal operculum (FOP), which is posterior to the IFG node used in most other models. Given that Quiroga-Martinez et al. (2021) would have had access to the Lumaca et al. (2021) study their finding of a downstream source (right FOP), influencing the STG, gives credence to my view that neural synchrony (reflected in the MEG, but not in the fMRI) changes play a significant role in the identification of network nodes in the MMN system.

11.5.3 Predictive coding in speech and language
11.5.3.1 *White matter pathways and mismatch responses*
Predictive coding models in the context of overt speech contain auditory white matter pathways such as the arcuate fasciculus and the frontal aslant. The arcuate and aslant fasciculi are two fiber bundles that are potentially involved in predictive coding in the context of speech. The arcuate fasciculus provides a direct connection between speech production (Broca's in left IFG) and speech perception (Wernicke's in left STG) areas. In addition to direct, long segment fibers connecting Broca's and Wernicke's area, the arcuate fasciculus also has shorter, indirect connections consisting of an anterior pathway which connects Broca's area to Geschwind's territory (in the

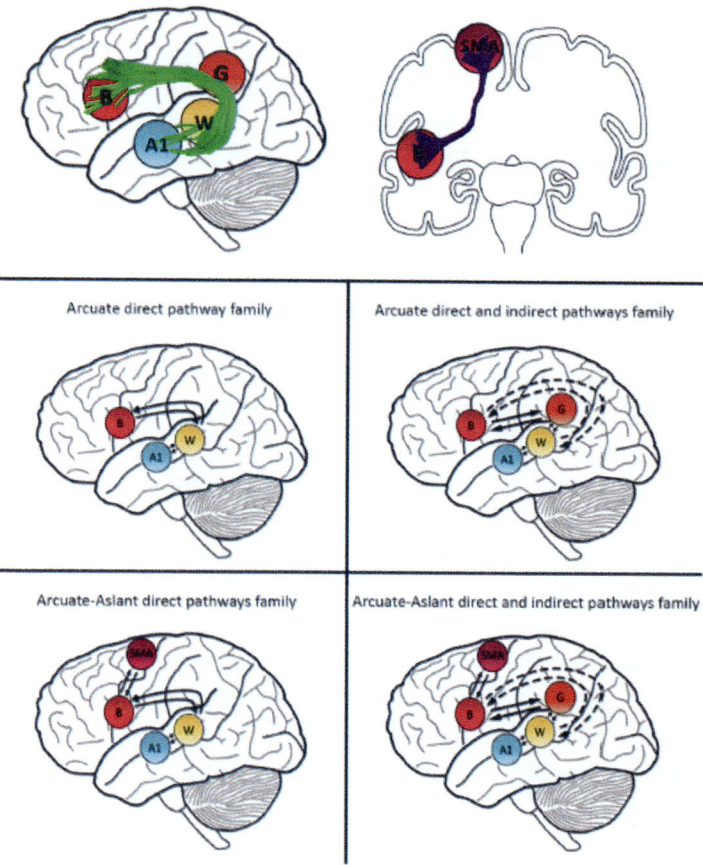

Fig. 11.15 Schematic representation of family definitions and anatomical white matter pathways. Primary auditory cortex (A1), Wernicke's area (W), Geschwind's territory (G), and Broca's area (B) are interconnected via the arcuate fasciculus (green). B and supplementary motor area (SMA) are interconnected by the frontal aslant (blue). *From Oestreich, L.K.L., Whitford, T.J., Garrido, M.I., 2018. Prediction of speech sounds is facilitated by a functional fronto-temporal network. Front. Neural Circuits 12, 43. https:// doi.org/10.3389/fncir.2018.00043. Open access.*

inferior parietal lobe, IPL), and a posterior pathway which connects Geschwind's territory and Wernicke's area (cf. Fig. 11.15; Catani et al., 2005; Oestreich et al., 2018).

In a follow-up study by Oestreich et al. (2019a), EEG recordings were obtained from participants listening to a classical two-tone duration deviant oddball paradigm and a stochastic oddball paradigm. A large subset of participants underwent diffusion-weighted MRI. Fractional anisotropy (FA) was extracted from the arcuate fasciculi and the auditory interhemispheric pathway. While prediction errors evoked by the classical oddball paradigm failed to reveal significant effects, the stochastic oddball paradigm elicited significant clusters at the typical mismatch negativity time window. Using the subset of this study for which both EEG and MRI were available, Oestreich et al. (2019b) investigate whether structural connectivity of auditory white matter pathways enables the effective connectivity underpinning auditory mismatch responses. Anatomically constrained tractography was used to extract auditory white matter pathways, namely the bilateral arcuate fasciculi, inferior fronto-occipital fasciculi (IFOF), and the auditory interhemispheric pathway, from which "Apparent Fibre Density" was calculated. For the EEG recorded during a stochastic oddball paradigm, DCM was used to investigate the effective connectivity underlying auditory mismatch responses generated in brain regions interconnected by the abovementioned auditory white matter pathways. Oestreich et al. (2019b) showed that brain areas interconnected by all auditory white matter pathways best explained the dynamics of auditory mismatch responses. Furthermore, 'Apparent Fiber Density' in the right arcuate fasciculus was significantly associated with the effective connectivity between the cortical regions that lie within it. "Taken together, these findings indicate that auditory prediction errors recruit a fronto-temporal network of brain regions that are effectively and structurally connected by auditory white matter pathways" (Oestreich et al., 2019b).

11.5.3.2 Predictive coding

At least parts of both crucial areas—STG and IFG—in the "common" auditory predictive coding network are involved in speech processing. This is clearly reflected in the study by Yvert et al. (2012) who used DCM of ERPs, combined with group source reconstruction, to estimate how those processes translate into context-dependent modulation of effective connectivity within the temporal-frontal language network. Fifteen healthy human subjects performed a phoneme detection task in pseudo-words (phoneme detection) and a semantic categorization task in words (presented visually). Cortical current densities revealed the sequential activation of temporal regions, from the occipital–temporal junction (OT in Fig. 11.16) toward the anterior temporal lobe (AT in Fig. 11.16), before reaching the IFG (Fig. 11.16). The comparison between models revealed robust and direct connection from occipital-temporal junction to IFG. It allows early activation of frontal cortex during visual language and object recognition tasks, possibly by the means of direct anatomical connections between early visual

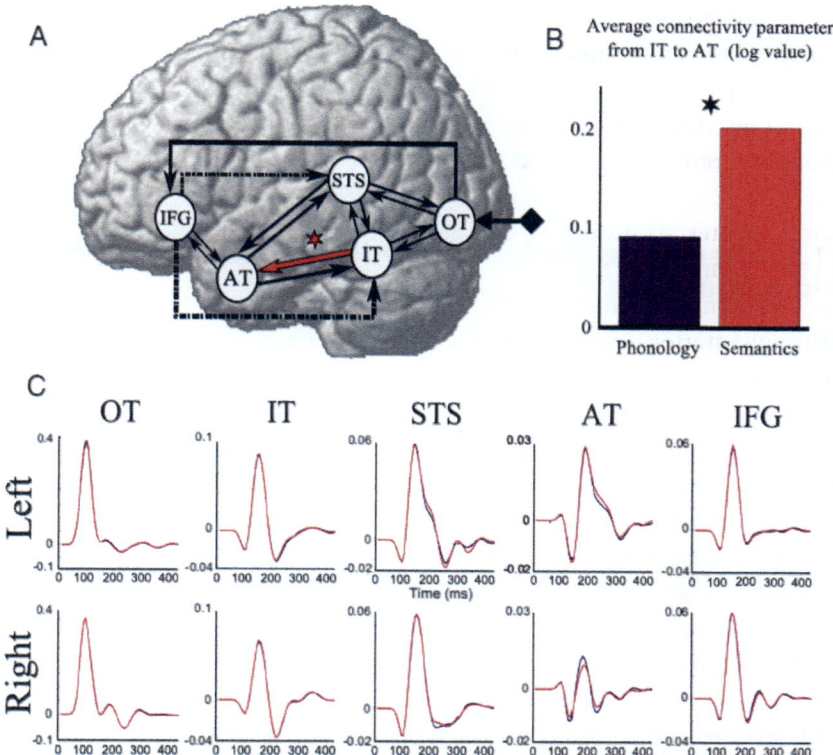

Fig. 11.16 (A) Winning model family composed of a direct connection from the occipito-temporal junction (OT) to inferior frontal gyrus (IFG), a late level of modulation and a feedback from IFG either to posterior superior temporal sulcus (STS) or posterior inferior temporal sulcus (IT) or both. AT, anterior temporal pole. The red arrow indicates higher connectivity for semantics than for phonology ($*P < 0.05$, Bonferroni corrected). (B) Corresponding connectivity parameters obtained after Bayesian model averaging over subjects; they were significantly different between conditions. (C) Group cortical time series of the same model family (blue, phonology; red, semantics). *From Yvert, G., Perrone-Bertolotti, M., Baciu, M., David, O., 2012. Dynamic causal modeling of spatiotemporal integration of phonological and semantic processes: an electroencephalographic study. J. Neurosci. 32(12), 4297–4306.*

regions and frontal cortex. Functionally, visual stimuli may activate very early IFG to prime the language network for the upcoming cross–modal interaction between visual and language systems for word recognition (Yvert et al., 2012).

Yvert et al. (2012) also identified a difference of activation between phonology and semantics in the anterior temporal lobe within the 240–300 ms peristimulus time window. DCM indicated this increase of activation of the

AT in the semantic (visual) condition as a consequence of an increase of forward connectivity from the posterior inferior temporal lobe (IT in Figure) to the AT. In addition, fast activation of the IFG, which allowed a feedback control of frontal regions on the superior temporal and posterior inferior temporal cortices, was found to be likely.

11.6 Alternative models to predictive coding in speech and language

While the notion of the brain as a prediction machine has been extremely influential and productive in cognitive science, there are competing accounts of how best to model and understand the predictive capabilities of brains. Commonly used, a "Bayesian brain" framework explicitly generates predictions and uses resultant errors to guide adaptation. Falandays et al. (2021) suggest that the prediction-generation component of this framework may involve little more than a pattern completion process. They applied this reasoning to language, where explicitly predictive models are perhaps most popular. They demonstrate how a connectionist model, TRACE, exhibits predictive processing without any representations of predictions or errors. TRACE (McClelland and Elman, 1986) is an interactive activation model with feedforward connections (from features to phonemes to words) as well as excitatory feedback from its word layer to its phoneme layer. These findings may indicate that interactive activation is functionally equivalent or approximant to predictive coding, or that caution is warranted in interpreting neural signal reduction as diagnostic of predictive coding. Falandays et al. (2021) then "present the TRACE model that is entirely unsupervised and memoryless, but nonetheless exhibits prediction-like behavior in its pursuit of homeostasis. These explorations demonstrate that brain-like systems can get prediction 'for free,' without the need to posit formal logical representations with Bayesian probabilities or an inference machine that holds them in working memory." Luthra et al. (2021) also showed that "despite not explicitly implementing prediction, the TRACE model of speech perception exhibits this putative hallmark of predictive coding, with reductions in total lexical activation, total lexical feedback, and total phoneme activation when the input conforms to expectations."

Statistical learning refers to the implicit mechanism of extracting regularities in our environment. Tsogli et al. (2022) used fMRI to investigate the neural correlates of processing events that are irregular based on learned local dependencies. A stream of consecutive sound triplets was presented.

Unbeknown to the subjects, triplets were either (a) standard, namely triplets ending with a high probability sound or (b) statistical deviants, namely triplets ending with a low probability sound. Participants underwent a learning phase outside the scanner followed by an fMRI session. They found that "processing of statistical deviants activated a set of regions encompassing the STG bilaterally, the right deep frontal operculum including lateral orbitofrontal cortex, and the right premotor cortex. Results demonstrate that the violation of local dependencies within a statistical learning paradigm does not only engage sensory processes, but is instead reminiscent of the activation pattern during the processing of local syntactic structures in music and language, reflecting the online adaptations required for predictive coding in the context of statistical learning."

11.7 Summary

One of the dominant areas for the application of predictive coding is language and music. Influential dual-stream pathway models of speech processing have been proposed, wherein typically a ventral stream processes speech signals for comprehension, and a dorsal stream maps acoustic speech signals to frontal lobe articulatory networks. In these models, information is assumed to travel from the primary auditory (A1) to the nonprimary auditory cortex (in STG). Intracortical recordings in humans during neurosurgery, however, have found that transient functional lesioning of A1 via electrical stimulation or focal ablation does not impair speech comprehension. Furthermore, electrical stimulation targeting the STG impairs speech comprehension without impairing tone discrimination. These findings are not consistent with a serial hierarchical organization (i.e., from A1 to "higher" auditory areas). Instead, response latency and receptive field analyses demonstrate parallel and distinct information processing in the primary and nonprimary auditory cortices. In support of this, Granger causality analysis identified relationships of high-gamma activity—the prediction-error signal—between distinct locations in early auditory pathways within the superior temporal gyrus (STG), between posterior STG (Wernicke's area) and inferior frontal gyrus (Broca's area), and between STG and premotor cortex. This suggests that these relationships reflect direct cortico-cortical connections rather than common driving input from the medial geniculate body via the A1.

Predictive coding models in the context of overt speech contain auditory white matter pathways such as the arcuate fasciculus and the frontal aslant.

The arcuate fasciculus provides a direct connection between speech production (Broca's in left IFG) and speech perception (Wernicke's in left STG) areas. In addition to direct, long segment fibers connecting Broca's and Wernicke's area, the arcuate fasciculus also has shorter, indirect connections consisting of an anterior pathway which connects Broca's area to Geschwind's territory (in the inferior parietal lobe, IPL), and a posterior pathway which connects Geschwind's territory and Wernicke's area. DCM showed that brain areas interconnected by all auditory white matter pathways best explained the dynamics of auditory mismatch responses.

Besides that brain oscillations play a role in predictive coding—where beta activity represents top-down predictions, and gamma represent bottom-up prediction error—the various brain rhythms have also been implicated in segmentation and identification of linguistically meaningful units, on three timescales, such that gamma-band oscillations synchronize with phonemes, theta-band oscillations synchronize with syllables, and delta-band oscillations synchronize with intonational phrase boundaries.

References

Bhaya-Grossman, I., Chang, E.F., 2022. Speech computations of the human superior temporal gyrus. Annu. Rev. Psychol. 73, 79–102.

Bidelman, G.M., 2015. Induced neural beta oscillations predict categorical speech perception abilities. Brain Lang. 141, 62–69.

Bidelman, G.M., Mahmud, M.S., Yeasin, M., Shen, D., Arnott, S.R., Alain, C., 2019. Age-related hearing loss increases full-brain connectivity while reversing directed signaling within the dorsal–ventral pathway for speech. Brain Struct. Funct. 224, 2661–2676. https://doi.org/10.1007/s00429-019-01922-9.

Boatman, D., 2004. Cortical bases of speech perception: evidence from functional lesion studies. Cognition 92 (1–2), 47–65.

Catani, M., Jones, D.K., ffytche, D.H., 2005. Perisylvian language networks of the human brain. Ann. Neurol. 57, 8–16.

Della Rosa, P.A., Catricalà, E., Canini, M., Vigliocco, G., Cappa, S.F., 2018. The left inferior frontal gyrus: a neural crossroads between abstract and concrete knowledge. Neuroimage 175, 449–459.

Dietz, M.J., Friston, K.J., Mattingley, J.B., Roepstorff, A., Garrido, M.I., 2014. Effective connectivity reveals right-hemisphere dominance in audiospatial perception: implications for models of spatial neglect. J. Neurosci. 34 (14), 5003–5011.

Dietz, M.J., Nielsen, J.F., Roepstorff, A., Garrido, M.I., 2021. Reduced effective connectivity between right parietal and inferior frontal cortex during audiospatial perception in neglect patients with a right-hemisphere lesion. Hear. Res. 399, 108052.

Donhauser, P.W., Baillet, S., 2020. Two distinct neural timescales for predictive speech processing. Neuron 105, 385–393.

Edalati, M., Mahmoudzadeh, M., Safaie, J., Wallois, F., Moghimi, S., 2021. Violation of rhythmic expectancies can elicit late frontal gamma activity nested in theta oscillations. Psychophysiol 58, e13909. https://doi.org/10.1111/psyp.13909.

Falandays, J.B., Nguyen, B., Spivey, M.J., 2021. Is prediction nothing more than multi-scale pattern completion of the future? Brain Res. 1768, 147578.

Fei, N., Ge, J., Wang, Y., Gao, J.-H., 2020. Aging-related differences in the cortical network subserving intelligible speech. Brain Lang. 201, 104713.

Fujioka, T., Ross, B., Trainor, L.J., 2015. Beta-band oscillations represent auditory beat and its metrical hierarchy in perception and imagery. J. Neurosci. 35 (45), 15187–15198.

Gander, P.E., Kumar, S., Sedley, W., Nourski, K.V., Oya, H., Kovach, C.K., et al., 2019. Direct electrophysiological mapping of human pitch-related processing in auditory cortex. Neuroimage 202, 116076.

Garrido, M.I., Kilner, J.M., Stephan, K.E., Friston, K.J., 2009. The mismatch negativity: a review of underlying mechanisms. Clin. Neurophysiol. 120 (3), 453–463. https://doi.org/10.1016/j.clinph.2008.11.029.

Hamilton, L.S., Oganian, Y., Hall, J., Chang, E.F., 2021. Parallel and distributed encoding of speech across human auditory cortex. Cell 184, 4626–4639.

Heard, M., Lee, Y.S., 2020. Shared neural resources of rhythm and syntax: an ALE meta-analysis. Neuropsychologia 137, 107284.

Hickok, G., Poeppel, D., 2007. The cortical organization of speech processing. Nat. Rev. Neurosci. 8, 393–402.

Hovsepyan, S., Olasagasti, I., Giraud, A.-L., 2020. Combining predictive coding and neural oscillations enables online syllable recognition in natural speech. Nat. Commun. 11, 3117. https://doi.org/10.1038/s41467-020-16956-5.

Igelstrom, K.M., Graziano, M.S.A., 2017. The inferior parietal lobule and temporoparietal junction: a network perspective. Neuropsychologia 105, 70–83.

Jäncke, L., 2012. The relationship between music and language. Front. Psychol. 3, 123. https://doi.org/10.3389/fpsyg.2012.00123.

Koelsch, S., Gunter, T.C., von Cramon, D.Y., Zysset, S., Lohmann, G., Friederici, A.D., 2002. Bach speaks: a cortical "language-network" serves the processing of music. Neuroimage 17, 956–966. https://doi.org/10.1006/nimg.2002.1154.

Koelsch, S., Vuust, P., Friston, K., 2019. Predictive processes and the peculiar case of music. Trends Cogn. Sci. 23 (1), 63–77. https://doi.org/10.1016/j.tics.2018.10.006.

Lewis, A.G., Bastiaansen, M., 2015. A predictive coding framework for rapid neural dynamics during sentence-level language comprehension. Cortex 68, 155–168.

Lumaca, M., Dietz, M.J., Hansen, N.C., Quiroga-Martinez, D.R., Vuust, P., 2021. Perceptual learning of tone patterns changes the effective connectivity between Heschl's gyrus and planum temporale. Hum. Brain Mapp. 42, 941–952. https://doi.org/10.1002/hbm.25269.

Luthra, S., Li, M.Y.C., You, H., Brodbeck, C., Magnuson, J.S., 2021. Does signal reduction imply predictive coding in models of spoken word recognition? Psychon. Bull. Rev. 28, 1381–1389. https://doi.org/10.3758/s13423-021-01924-x.

McClelland, J.L., Elman, J.L., 1986. The TRACE model of speech perception. Cogn. Psychol. 18 (1), 1–86.

Meyer, L., 2018. The neural oscillations of speech processing and language comprehension: state of the art and emerging mechanisms. Eur. J. Neurosci. 48, 2609–2621.

Musso, M., Weiller, C., Horn, A., Glauche, V., Umarova, R., Hennig, J., et al., 2015. A single dual-stream framework for syntactic computations in music and language. Neuroimage 117, 267–283.

Oestreich, L.K.L., Randeniya, R., Garrido, M.I., 2019a. Auditory prediction errors and auditory white matter microstructure associated with psychotic-like experiences in healthy individuals. Brain Struct. Funct. 224, 3277–3289. https://doi.org/10.1007/s00429-019-01972-z.

Oestreich, L.K.L., Randeniya, R., Garrido, M.I., 2019b. Auditory white matter pathways are associated with effective connectivity of auditory prediction errors within a fronto-temporal network. Neuroimage 195, 454–462.

Oestreich, L.K.L., Whitford, T.J., Garrido, M.I., 2018. Prediction of speech sounds is facilitated by a functional fronto-temporal network. Front. Neural Circuits 12, 43. https://doi.org/10.3389/fncir.2018.00043.

Potes, C., Brunner, P., Gunduz, A., Knight, R.T., Schalk, G., 2014. Spatial and temporal relationships of electrocorticographic alpha and gamma activity during auditory processing. Neuroimage 97, 188–195.

Quiroga-Martinez, D.R., Hansen, N.C., Højlund, A., Pearce, M., Brattico, E., Holmes, E., et al., 2021. Musicianship and melodic predictability enhance neural gain in auditory cortex during pitch deviance detection. Hum. Brain Mapp. 42, 5595–5608. https://doi.org/10.1002/hbm.25638.

Rauschecker, J.P., Scott, S.K., 2009. Maps and streams in the auditory cortex: nonhuman primates illuminate human speech processing. Nat. Neurosci. 12 (6), 718–724.

Schlossmacher, I., Dilly, J., Protmann, I., Hofmann, D., Dellert, T., Roth-Paysen, M.-L., et al., 2022. Differential effects of prediction error and adaptation along the auditory cortical hierarchy during deviance processing. Neuroimage 259, 119445.

Schmidt, A., Diaconescu, A.O., Kometer, M., Friston, K.J., Stephan, K.E., Vollenweider, F.X., 2013. Modeling ketamine effects on synaptic plasticity during the mismatch negativity. Cereb. Cortex 23, 2394–2406. https://doi.org/10.1093/cercor/bhs238.

Seth, A.K., Friston, K.J., 2016. Active interoceptive inference and the emotional brain. Philos. Trans. R. Soc. Lond. B Biol. Sci. 371 (1708). https://doi.org/10.1098/rstb.2016.0007.

Siman-Tov, T., Granot, R.Y., Shany, O., Singer, N., Hendler, T., Gordon, C.R., 2019. Is there a prediction network? Meta-analytic evidence for a cortical-subcortical network likely subserving prediction. Neurosci. Biobehav. Rev. 105, 262–275.

Tsogli, V., Skouras, S., Koelsch, S., 2022. Brain-correlates of processing local dependencies within a statistical learning paradigm. Sci. Rep. 12, 15296. https://doi.org/10.1038/s41598-022-19203-7.

Vuust, P., Brattico, E., Seppänen, M., Näätänen, R., Tervaniemi, M., 2012. The sound of music: differentiating musicians using a fast, musical multi-feature mismatch negativity paradigm. Neuropsychologia 50, 1432–1443. https://doi.org/10.1016/j.neuropsychologia.2012.02.028.

Vuust, P., Ostergaard, L., Pallesen, K.J., Bailey, C., Roepstorff, A., 2009. Predictive coding of music—brain responses to rhythmic incongruity. Cortex 45, 80–92.

Wang, L., Schoot, L., Brothers, T., Alexander, E., Warnke, L., Kim, M., et al., 2022. Predictive coding across the left fronto-temporal hierarchy during language comprehension. Cereb. Cortex. https://doi.org/10.1093/cercor/bhac356. Epub.

Ylinen, S., Bosseler, A., Junttila, K., Huotilainen, M., 2017. Predictive coding accelerates word recognition and learning in the early stages of language development. Dev. Sci. 20, e12472. https://doi.org/10.1111/desc.12472.

Yvert, G., Perrone-Bertolotti, M., Baciu, M., David, O., 2012. Dynamic causal modeling of spatiotemporal integration of phonological and semantic processes: an electroencephalographic study. J. Neurosci. 32 (12), 4297–4306.

CHAPTER 12

Network changes underlying neural disorders: Relation to the MMN networks

Here I will review network connectivity changes underlying the disorders that were reviewed in Chapters 8 and 9. I will also endeavor to link these changes to the cognitive networks that underlie the mismatch negativity (MMN) used in predictive coding as applied to these disorders. I will first recall which resting-state networks (RSN) reliably communicate with each other in nondiseased people. Based on subdural intracranial recordings and mutual information, Chapeton et al. (2017) identified the connections between recording sites (Fig. 12.1) resulting in sparse, directed functional networks that are stable over minutes, hours, and days. Notably, the time delays associated with these connections are also highly preserved over multiple timescales. Note the highly reliable long-range connection between the angular gyrus (AG) and the precuneus, and the somewhat lower reliability between the inferior frontal gyrus (IFG) and superior temporal gyrus (STG) that take part in most of the modeling of the MMN network (Chapter 10).

12.1 Network changes in tinnitus

This section is partially based on previous reviews (Eggermont, 2021, 2022).

12.1.1 Structural changes

To separate anatomical substrates of tinnitus from the effects of comorbid hearing loss, Husain et al. (2011) used MRI and voxel-based morphometry (VBM) to identify cortical gray matter changes, and diffusion tensor imaging (DTI) to delineate changes in white matter (WM) tracts. The VBM data showed that hearing loss individuals without tinnitus had gray matter (GM) decreases in the anterior cingulate cortex (ACC) and in the superior and medial frontal gyri (SFG, MFG) compared to those with hearing loss and

Brain Responses to Auditory Mismatch and Novelty Detection
https://doi.org/10.1016/B978-0-443-15548-2.00012-0

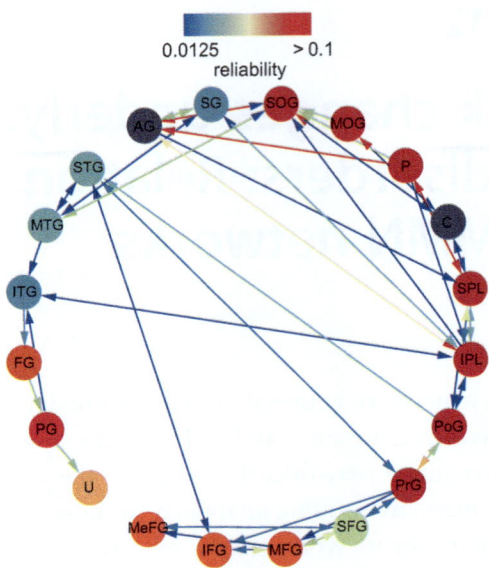

Fig. 12.1 Network based on reliable time-lag connectivity. Each node represents a distinct anatomical region containing electrode contacts from at least three participants, and each edge represents a connection between two regions. The reliability of a connection between two regions is represented by the *color of the edge*. We display edges only if the reliability of that connection was in the top 15% of all connections. Because most regions contain multiple electrodes, the reliability of connections within each region was calculated and is represented by the *color of each node*. Nodes that are colored *gray* did not have enough electrode coverage to compute the within-region reliability. AG, angular gyrus; C, cuneus; FG, fusiform gyrus; IFG, inferior frontal gyrus; IPL, inferior parietal lobule; ITG, inferior temporal gyrus; MeFG, medial frontal gyrus; MGF, middle frontal gyrus; MOG, middle occipital gyrus; MTG, middle temporal gyrus; P, precuneus; PG, parahippocampal gyrus; PoG, postcentral gyrus; PrG, precentral gyrus; SFG, superior frontal gyrus; SG, supramarginal gyrus; SOG, superior occipital gyrus; SPL, superior parietal lobule; STG, superior temporal gyrus; U, uncus. *(From Chapeton, J.I., Inati, S.K., Zaghloul, K.A., 2017. Stable functional networks exhibit consistent timing in the human brain. Brain 140, 628–640. ©Guarantors of Brain 2017, by permission of Oxford University Press.)*

tinnitus. Compared to normal–hearing controls, participants with only hearing loss showed gray matter decreases in SFG and MFG. In comparison to normal-hearing controls, DTI analysis in the hearing loss group showed decreases in fractional anisotropy (FA) values in the right superior and inferior longitudinal fasciculi, corticospinal tract, inferior fronto–occipital tract, superior occipital fasciculus, and anterior thalamic radiation. Husain et al. (2011) concluded that hearing loss rather than tinnitus had the greatest influence on gray and white matter alterations. Consequently, the interaction between hearing loss and tinnitus on structural changes in cortex, potentially affecting the ACC, is more complicated than an additive effect of tinnitus on a baseline hearing impaired condition.

Khan et al. (2021) examined white matter correlates of tinnitus and hearing loss. Diffusion imaging data were collected from 96 participants—43 with tinnitus and hearing loss (TIN_{HL}), 17 with tinnitus and normal–hearing thresholds (TIN_{NH}), 17 controls with hearing loss (CON_{HL}), and 19 controls with normal hearing (CON_{NH}). Fractional anisotropy (FA), mean diffusivity (MD), and probabilistic tractography analyses were conducted on the diffusion imaging data. They found differences in FA and structural connectivity specific to tinnitus, hearing loss, and both conditions when comorbid, suggesting the existence of tinnitus–specific neural networks (Fig. 12.2). These findings also suggest that age plays an important role in neural plasticity, and thus may account for some of the variability of results in the literature.

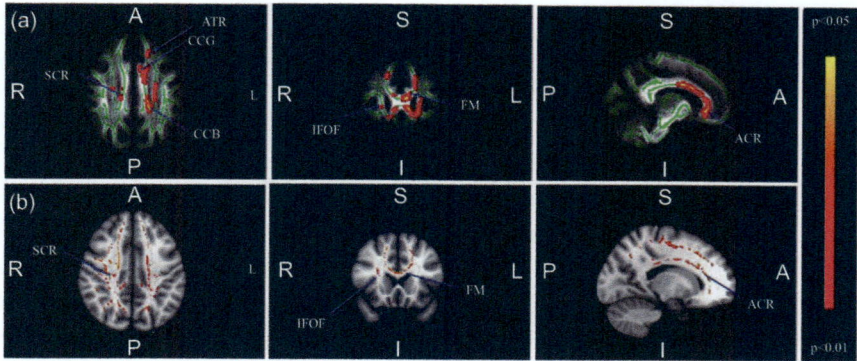

Fig. 12.2 Row (A) Regions of significant difference in group-level FA analysis (CON > TIN). *Green* represents the mean FA of all participants. The *red-yellow scale* represents regions of significant difference, with *red* representing regions of greatest difference. The regions showing significant differences between the groups included the right inferior fronto-occipital fasciculus (IFOF), right superior corona radiata (SCR), forceps minor (FM), genu (CCG) and body (CCB) of the corpus callosum, left anterior corona radiata (ACR), left anterior thalamic radiation (ATR), bilateral superior longitudinal fasciculus, and left inferior longitudinal fasciculus (not visible in this view). Row (B) Regions of significant difference in group-level MD analysis (CON < TIN). The *red-yellow scale* represents regions of significant difference, with *red* representing regions of greatest difference. *(From Khan, R.A., Sutton, B.P., Tai, Y., Schmidt, S.A., Shahsavarani, S., Husain, F.T., 2021. A large-scale diffusion imaging study of tinnitus and hearing loss. Sci. Rep. 11, 23395. https://doi.org/10.1038/s41598-021-02908-6. Open Access.)*

Importantly, Rauschecker et al. (2015) reported similar results in the internal capsule and parts of the corpus callosum and proposed a frontostriatal-gating system, in which the ventromedial prefrontal cortex (VMPFC) and nucleus accumbens (NAc) act as gatekeeping mechanisms for evaluation of sensory stimuli. The internal capsule, which showed reduced FA in the TIN_{HL} group compared to the CON_{HL} group in the study by Khan et al. (2021), is a nexus for ascending and descending fibers communicating with frontal regions of the brain and with the corpus callosum.

The general conclusion is that only sporadic structural changes (e.g., Khan et al., 2021) in gray and white matter are found that could be specifically attributed to tinnitus. Most changes, including those outside the auditory areas, have to be attributed to hearing loss.

12.1.2 Functional changes

12.1.2.1 Reviews

Several intrinsic resting-state networks (RSNs) have been implicated in tinnitus; these include the default mode (DMN; cf. Fig. 10.6), auditory, visual, and dorsal attention (DAN) networks. Studies reviewed in Husain and Schmidt (2014) suggest that tinnitus causes consistent modifications to brain networks, including greater connectivity between limbic areas and cortical networks not traditionally involved with emotion processing, and increased connectivity between attention and auditory processing brain regions. Hullfish et al. (2018) summarized "Ventrolateral PFC (including inferior frontal gyrus, IFG) has been argued to form part of a 'tinnitus core' network, which also includes auditory, inferior parietal, and parahippocampal cortex (De Ridder et al., 2014). Parahippocampal cortex has shown altered resting-state activity contralateral to the tinnitus ear (Vanneste et al., 2011) and resting-state fMRI correlation with auditory cortex (Maudoux et al., 2012a,b; Schmidt et al., 2017), and based on its prominent role in auditory memory is a potential source of persistent auditory predictions." Cheng et al. (2020) performed a metaanalysis on 14 studies in which the structural dataset comprised 242 tinnitus patients and 217 matched healthy controls, while the functional dataset included 130 tinnitus patients and 140 matched healthy controls. Compared to controls, they found gray matter alterations in the STG, middle temporal gyrus (MTG), angular gyrus, caudate nucleus, superior frontal gyrus, and supplementary motor area, as well as differences in spontaneous brain activity in the MTG, middle occipital gyrus,

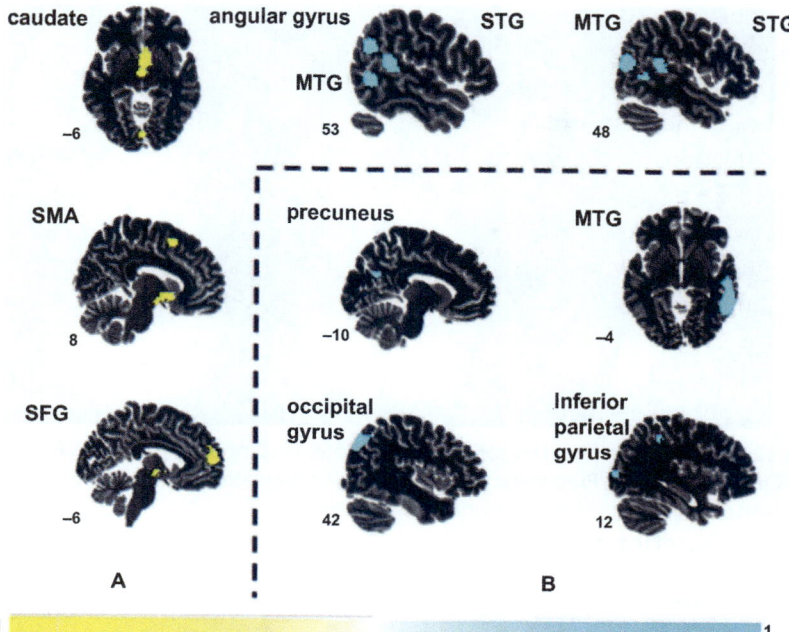

Fig. 12.3 Differences in the brain regions between tinnitus patients and HS. (A) Differences in gray matter (GM) volume between tinnitus patients and HS. (B) Changes in spontaneous brain activity of tinnitus patients compared with HS. *MTG*, middle temporal gyrus; *SFG*, superior frontal gyrus; *SMA*, supplementary motor area; *STG*, superior temporal gyrus. *(From Cheng, S., Xu, G., Zhou, J., Qu, Y., Li, Z., He, Z., et al., 2020. A multimodal meta-analysis of structural and functional changes in the brain of tinnitus. Front. Hum. Neurosci. 14, 28. https://doi.org/10.3389/fnhum.2020. 00028. Open access.)*

precuneus, and right inferior parietal (excluding supramarginal and angular) gyri (Fig. 12.3).

Moring et al. (2022) analyzed 17 resting-state voxel-based activity and 8 voxel-based morphology reports of tinnitus–associated regional alterations using activation likelihood estimation (ALE). ALEs were performed at two levels of statistical rigor: corrected for multiple comparisons and uncorrected. The corrected ALE applied cluster-level inference thresholding by intensity (z-score > 1.96; $P < .05$) followed by family-wise error correction for multiple comparisons ($P < .05$, 1000 permutations) and fail-safe correction for missing data. The corrected analysis identified one significant cluster comprising five foci in the posterior cingulate cortex (PCC) and precuneus, that is, not within the primary or secondary auditory cortices (Fig. 12.4). The uncorrected ALE identified additional regions within auditory and

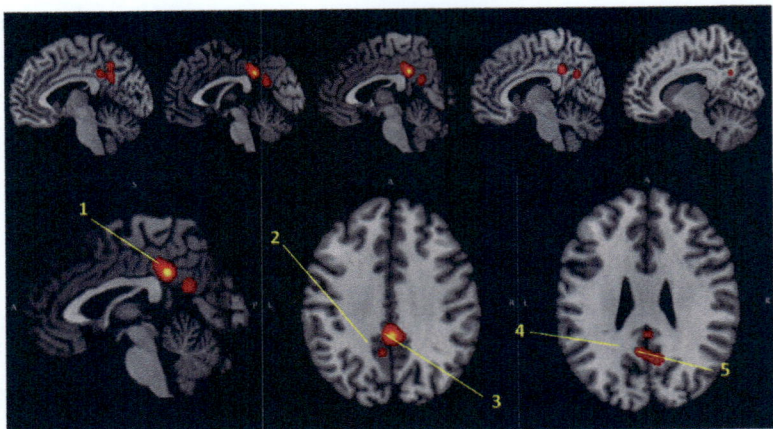

Fig. 12.4 Regions identified by corrected ALE. Note: 1 = Cingulate Gyrus; 2 = Precuneus; 3 = Cingulate Gyrus/Precuneus; 4 = Posterior Cingulate/Precuneus; 5 = Posterior Cingulate/Precuneus. *(From Moring, J.C., Husain, F.T., Gray, J., Franklin, C., Peterson, A.L., Resick, P.A., et al. 2022. Invariant structural and functional brain regions associated with tinnitus: a meta-analysis. PLoS One 17 (10), e0276140. https://doi.org/ 10.1371/journal.pone.0276140. Open Access.)*

cognitive processing networks. It is important to note that the cluster-level inference coordinate-based metaanalysis (CBMA) ALE with FWE did not detect any regions within the auditory RSN. Instead, the superior, middle, and inferior temporal gyri were detected by the single-threshold CBMA ALE. Moring et al. (2022) did find disease effects of tinnitus within the auditory system when the uncorrected ALE was implemented. Specifically, the MTG and subgyral (noncortical) regions within the auditory system demonstrated alterations. Individuals experiencing hearing loss with and without tinnitus demonstrate increased gray matter in the MTG when compared to individuals without tinnitus or hearing loss (Boyen et al., 2013). Therefore the MTG may be more related to hearing loss, rather than tinnitus. However, alterations within the right MTG have also been demonstrated among tinnitus sufferers both as structural differences and in spontaneous neural activity (Cheng et al., 2020). Dysregulation within the MTG may be a source of the tinnitus percept, while other regions of the brain may play a role in the loudness and emotional responses associated with tinnitus (Mohan et al., 2018). It needs to be noted that MRI measures are insensitive to neural synchrony changes that may accompany tinnitus (Noreña and Eggermont, 2003; Eggermont, 2022) so that the absence of auditory cortex involvement in the corrected results may also be due to insensitivity to neural synchrony changes. On the other hand, the BOLD response appears to be correlated with the local field potentials of which the ERPs are a compound response, implying that activation in an area is often likely to reflect the incoming input and the local processing in a given area rather than the spiking activity (Logothetis, 2003).

12.1.2.2 Data papers

Using fMRI, Trevis et al. (2017) investigated if impaired functioning of the cognitive control network that directs attentional focus is a mechanism erroneously maintaining the tinnitus sensation. People with chronic tinnitus and age and gender-matched healthy controls performed a cognitively demanding task. They showed reduced activation of a core node of the cognitive control network (the right middle frontal gyrus) and altered baseline connectivity between this node and nodes of the salience and autobiographical memory networks (Fig. 12.5).

Schmidt et al. (2017) examined two sources of variability in tinnitus subgroups: tinnitus severity and the length of time a person has had chronic tinnitus. Decreased correlations between two seed regions in the DMN—mPFC and PCC—and the precuneus were consistent across individuals who have had tinnitus for longer than one year, with more bothersome tinnitus demonstrating stronger decreases (Fig. 12.6 left). Schmidt et al. (2017) proposed DMN-precuneus decoupling as a potential marker of long-term tinnitus. Patients with moderately severe tinnitus showed increased correlations between seeds in the DAN—frontal eye fields (FEF) and posterior intraparietal sulci (pIPS)—and the precuneus. The precuneus is related to conscious and internal awareness, consequently coupling of the precuneus and DAN at rest is associated with bothersome tinnitus and is not always observed in patients with mild tinnitus and tinnitus duration between 6 and 12 months. Thus mild, recent-onset tinnitus may form a subgroup of tinnitus patients where resting-state functional connectivity (rs–FC) patterns have not yet changed from those seen in controls. The same group, Khan et al. (2021), propose that when an individual experiences auditory trauma, the internal capsule, which receives input from auditory fibers (such as the anterior thalamic radiation that showed reduced FA in the TIN_{HL} group compared to the TIN_{NH} group), relays the signal to frontal regions, where frontostriatal gating takes place. This circuit then determines whether persistent tinnitus starts. A frontostriatal circuit would consist of a closed–loop structure (Rauschecker et al., 2015), but the evaluation of a consistent negative stimulus may include the altered, unbalanced callosal excitation and

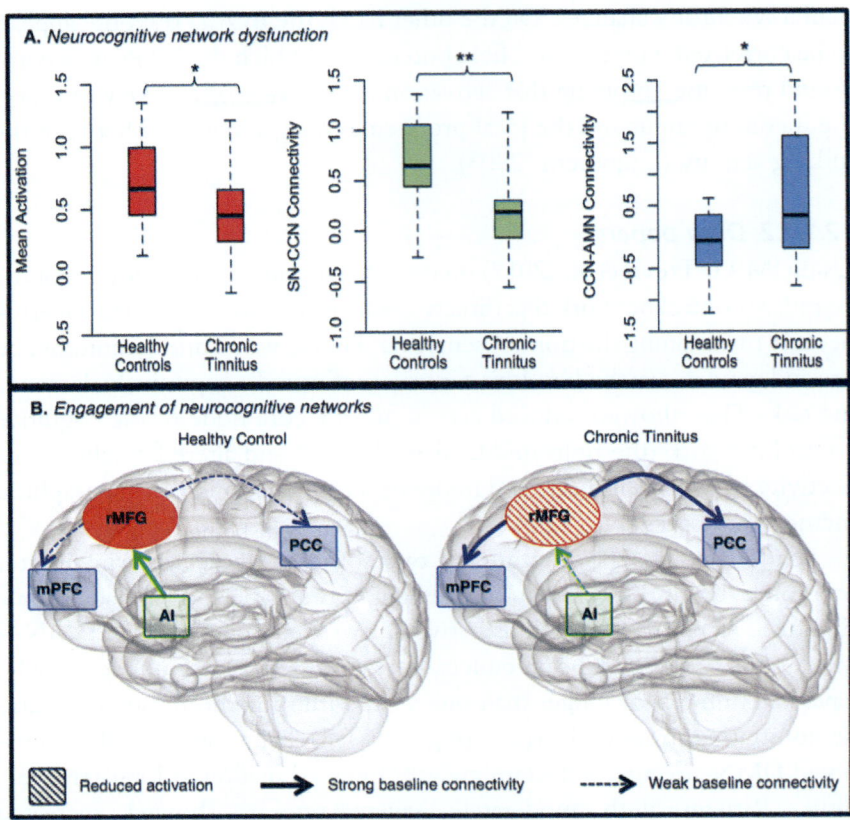

Fig. 12.5 Functional deficits associated with dysfunction of the cognitive control network (CCN) in chronic tinnitus ($n=15$) compared to healthy controls ($n=15$). (A) Neurocognitive network dysfunction: Boxplots highlight significantly attenuated activation of the posterior right middle frontal gyrus (rMFG) during the n-back task in chronic tinnitus; significantly lower baseline connectivity between the right anterior insula (rAI) node of the salience network and the affected CCN node (rMFG) in chronic tinnitus; and significantly higher baseline connectivity between the affected CCN node (rMFG) and nodes of autobiographical memory network (AMN), including the left posterior cingulate cortex (PCC, illustrated here) and the left medial prefrontal cortex (mPFC). (B) Engagement of neurocognitive networks. Illustration of large-scale neurocognitive network functioning in healthy controls (left) and chronic tinnitus (right) with nodes of the CCN *(red)*, SN *(green)*, and AMN *(blue)*. Here, while healthy controls show higher SN-CCN baseline connectivity associated with greater CCN activation, people experiencing chronic tinnitus show lower SN-CCN baseline connectivity and decreased CCN activation. This may underpin less proficient network switching, characterized by higher CCN-AMN baseline connectivity in chronic tinnitus, associated with difficulty switching attention away from the auditory environment (e.g., scanner noise), $**P<.01$; $*P<.05$. *AI*, anterior insula; *CCN-AMN*, cognitive control network-autobiographical network; *MFG*, medial frontal gurus; *mPFC*, medial prefrontal cortex; *SN*, salience network. *(From Trevis, K.J., Tailby, C., Grayden, D.B., McLachlan, N.M., Jackson, G.D., Wilson, S.J., 2017. Identification of a neurocognitive mechanism underpinning awareness of chronic tinnitus. Sci. Rep. 7, 15220. https://doi.org/10.1038/s41598-017-15574-4. Open access.)*

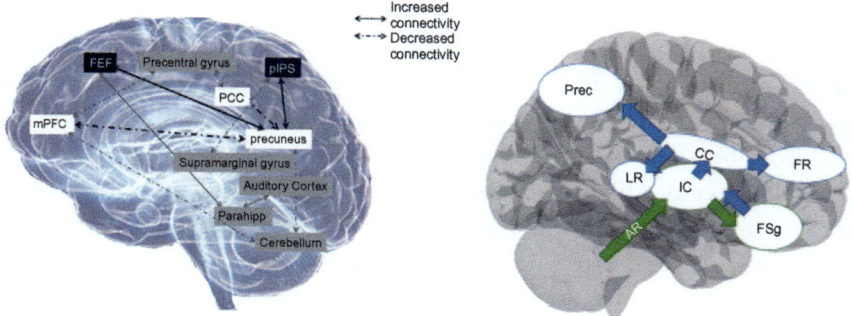

Fig. 12.6 Left. An illustration of the effect of tinnitus on resting-state functional connectivity, compared to controls. The *black boxes* correspond to regions of the dorsal attention network, while the *white boxes* are regions associated with the default mode network. In tinnitus patients, connectivity between the precuneus and the other regions of the default mode is decreased. This effect is more pronounced as tinnitus severity increases. Connectivity between the precuneus and regions in the dorsal attention network is increased in patients with bothersome tinnitus, but the effect is inconsistent in patients with mild tinnitus. The *gray boxes and lines* indicate findings that were described previously (Husain and Schmidt, 2014), but were not replicated in the current study. *FEF*, frontal eye fields; *mPFC*, medial prefrontal cortex; *parahipp*, parahippocampus; *PCC*, posterior cingulate cortex; *pIPS*, posterior intraparietal sulci. Right. Diagram of proposed mechanism for tinnitus persistence. In this model, sensory signals from auditory radiations are propagated to the internal capsule, from where they are projected to the ventromedial prefrontal cortex and nucleus accumbens. There, frontostriatal gating as described by Rauschecker et al. (2015) takes place. Following evaluation of the tinnitus signal, frontal regions propagate signal back to the internal capsule, and the perception of a negative stimulus has a wider impact on limbic and frontal regions. *Green arrows* represent signal propagation prior to frontostriatal gating, while *blue arrows* represent the signal following frontostriatal gating. *AR*, acoustic radiations; *CC*, corpus callosum; *FR*, frontal regions; *FSg*, frontostriatal gating (consisting of the ventromedial prefrontal cortex and nucleus accumbens); *IC*, internal capsule; *LR*, limbic regions; *Prec*, precuneus. *(Left Panel: From Schmidt, S.A., Carpenter-Thompson, J., Husain, F.T., 2017. Connectivity of precuneus to the default mode and dorsal attention networks: a possible invariant marker of long-term tinnitus. Neuroimage Clin. 16, 196–204. Open Access. Right Panel: From Khan, R.A., Sutton, B.P., Tai, Y., Schmidt, S.A., Shahsavarani, S., Husain, F.T., 2021. A large-scale diffusion imaging study of tinnitus and hearing loss. Sci. Rep. 11, 23395. https://doi.org/10.1038/s41598-021-02908-6. Open Access.)*

inhibition. Because the corpus callosum is one of the most widely connected neural structures, Khan et al. (2021) believe that tinnitus-related alterations in other white matter structures (such as the superior longitudinal fasciculi, and interior fronto-occipital fasciculi) and brain networks (such as the DMN) are likely a result of the persistence of tinnitus (Fig. 12.6 right).

The precuneus plays an important role in the DMN, and the tinnitus signal would be a constant disturbance to the "rest" state, which may stimulate structural changes in the precuneus. This study suggests that when tinnitus is comorbid with hearing loss, associated neural plasticity can be differentiated from plasticity associated with tinnitus without hearing loss.

Hullfish et al. (2018) performed fMRI on a large clinical sample with chronic tinnitus and moderate high-frequency hearing loss (>2 kHz). Subjects were presented with sound stimuli matched to their tinnitus frequency (TF) as well as similar stimuli presented at a control frequency (CF) >1 octave higher or lower than the TF, and typically in the normal-hearing frequency region. Whole-brain subtraction analysis showed that, during TF stimulation, subjects had greater BOLD activity in several regions, mainly in frontal cortex, parietal cortex, and the cerebellum in comparison to CF stimulation. No regions showed greater BOLD activity during CF. Many of these TF-activated regions resemble an extended cortical network for semantic cognition, which involves the representation and control of semantic knowledge. Hullfish et al. (2018) "hypothesized that TF stimuli will elicit greater activity and/or functional connectivity in areas related to the cognitive and emotional aspects of tinnitus, i.e. tinnitus-related distress. Conversely, CF stimuli might elicit greater activity/connectivity in areas related to auditory perception and attention." Rosemann and Rauschecker (2022) investigated alterations related to tinnitus perception along with tinnitus distress and cognitive abilities. Tinnitus patients and age, sex, and hearing loss matched controls underwent MRI and audiometric as well as cognitive assessments. Tinnitus patients had increased gray matter volume in the middle frontal gyrus and frontal pole compared to controls. Moreover, there was increased cortical thickness in the precuneus associated with tinnitus distress. Chen et al. (2022) showed that the FC directed strength from the A1 to the parahippocampal gyrus (PHG) and IFG and from the PHG to the IFG was lower on the affected side in tinnitus patients than that in control subjects. However, the FC from the IFG to A1 was stronger in tinnitus patients than that in the control subjects. In patients with on average 40-dB hearing loss, and with or without tinnitus, these changes were enhanced.

12.2 Functional connectivity in dyslexia

Shaywitz et al. (1998) compared fMRI activation patterns in dyslexic and nonimpaired subjects as they performed tasks that made progressively greater demands on phonologic analysis. These patterns differed significantly between the groups; dyslexic readers showed hypoactivation in posterior regions (Wernicke's area, the angular gyrus, and striate cortex) and hyperactivation in the IFG. Cao et al. (2008) examined effective connectivity between three left hemisphere brain regions (IFG, inferior parietal lobule (IPL), fusiform gyrus) and bilateral MFG in children with reading difficulties and age-matched control children during rhyming judgments to visually presented words. Only for conflicting trials they found that the modulatory effect from left fusiform gyrus to left IPL was stronger in controls than in children with reading difficulties. Only in control children modulatory effects from left fusiform gyrus and left IPL to left IFG were stronger for conflicting trials than for nonconflicting trials. Only in control children modulatory effects from left IFG to IPL, from medial frontal gyrus to left IPL, and from left IPL to medial frontal gyrus were positively correlated with reading skills (Cao et al., 2008). This was confirmed by Quaglino et al. (2008) who compared the effective connectivity network between three groups—dyslexic, age-matched, and reading level-matched children—using a structural equation model including the supramarginal cortex, fusiform cortex, and IFG areas of the left hemisphere. In dyslexic patients, in contrast with chronological age- and reading level-matched groups, no causal relationship was demonstrated between supramarginal cortex and IFG. However, a significant causal relationship was demonstrated between fusiform cortex and IFG both in dyslexic children and in the reading level-matched group.

Díaz et al. (2012) noted that "neuroimaging studies on reading and phonological tasks in dyslexic participants relative to controls, have reported higher activity in IFG (Shaywitz et al., 1998; Georgiewa et al., 2002) and less activation mostly in the left parietotemporal, left occipitotemporal, and left frontal areas (Shaywitz et al., 2002; Paulesu et al., 2001; Raschle et al., 2012), although this is not the case in all studies (e.g., Temple et al., 2000; Steinbrink et al., 2009)." Díaz et al. (2012) corroborated the findings of higher activity in dyslexics compared with controls in the right IFG, "potentially reflecting a compensatory mechanism in dyslexia (Hoeft et al., 2011; Shaywitz et al., 1998; Georgiewa et al., 2002)." Intact phonetic representations bilaterally in primary and secondary auditory cortical areas were found in a study that combined MRI with voxel-based morphometry (VBM) pattern analysis and functional connectivity (FC) analysis. However, the functional and structural connectivity between the bilateral auditory cortices and the left IFG was significantly impaired in dyslexics (Boets et al., 2013). Olulade et al. (2015) reported absence of an IFG gradient and connectivity between the lateral aspect of the IFG and the anterior occipito-temporal cortex in the dyslexic children. Together, these results provide insights into the source of the anomalies reported in previous studies of dyslexia and add evidence to an orthographic role of IFG in reading. Schurz et al. (2015) found reduced connectivity in dyslexic readers between left posterior temporal areas (fusiform, inferior temporal, middle temporal, superior temporal) and the left IFG. They also found that connectivity between multiple reading-related areas and areas of the DMN, in particular the precuneus, was stronger in dyslexic compared with nonimpaired readers. Furthermore, stronger connectivity in dyslexic readers was found for IFG pars opercularis–left IPL/angular gyrus and IFG pars opercularis–right angular gyrus.

Molinaro et al. (2016) recorded MEGs from dyslexic readers and age-matched controls while they were listening to ~10-s long sentences. Compared to controls, dyslexic readers had (1) an impaired neural entrainment to speech in the delta band (0.5–1 Hz), (2) a reduced delta synchronization in both the right auditory cortex and the left IFG, and (3) an impaired feedforward functional coupling between neural oscillations in the right auditory cortex and the left IFG (Fig. 12.7). Molinaro et al. (2016) confirmed the findings by Boets et al. (2013) and also found that the reduced speech–brain synchronization in dyslexic readers compared to normal readers appears preserved through the developmental period from childhood to adulthood.

Morken et al. (2017) took a longitudinal approach to cortical connectivity based on fMRI data in reading tasks in children with dyslexia and children with typical reading development. The participants were followed with repeated measurements through preliteracy (6 years old), emergent literacy (8 years old), and literacy (12 years old) stages. Using Dynamic Causal Modeling (DCM), they found that the difference between groups centered on connections going to and from the IFG (two connections) and the occipito-temporal cortex (three connections). For all five connections, the typical group showed stable or decreasing connectivity measures with age. The dyslexia group, on the other hand, showed a marked upregulation (occipito-temporal connections) or down-regulation (IFG connections) from testing at 6 years to 8 years, followed by normalization from 8 years to 12 years (Fig. 12.8). Morken et al. (2017) interpreted this as a delay in the dyslexia group to develop into the preliteracy and emergent-literacy stages. By age 12, there was no statistically significant difference in connectivity between the groups, but differences in literacy skills were still present and were in fact larger than when measured at younger ages. Note that this

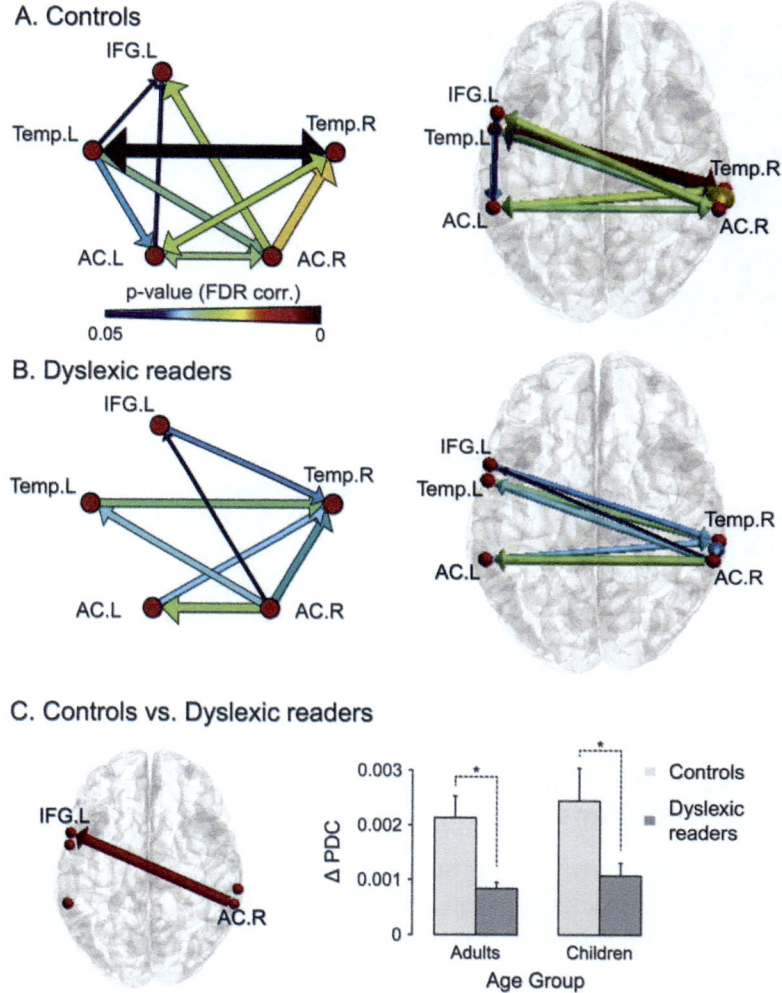

Fig. 12.7 Partial direct coherence analysis. Network dynamics for control (panel A) and dyslexic participants (B) among the five seeds in the 0.5–1 Hz frequency band (during speech perception compared to baseline) plotted on both connectivity graphs and dorsal views of the brain renderings. *Arrow orientation* represents the causal direction of the observed coupling; *arrow color and thickness* represent the statistical strength of the connection (*P*-values). (C) Left panel: Differential connection strength between control and dyslexic readers ($p_{FDR} < 0.05$, age and IQ corrected). Right panel: Strength of AC.R→IFG.L connection (for dyslexic readers and their control peers) plotted separately for adults and children. *AC*, auditory cortex; *IFG*, inferior frontal gyrus; *Temp*, temporal gyrus. (From Molinaro, N., Lizarazu, M., Lallier, M., Bourguignon, M., Carreiras, M., 2016. Out-of-synchrony speech entrainment in developmental dyslexia. Hum. Brain Mapp. 37, 2767–2783. VC 2016 Wiley Periodicals, Inc., with permission of John Wiley and Sons.)

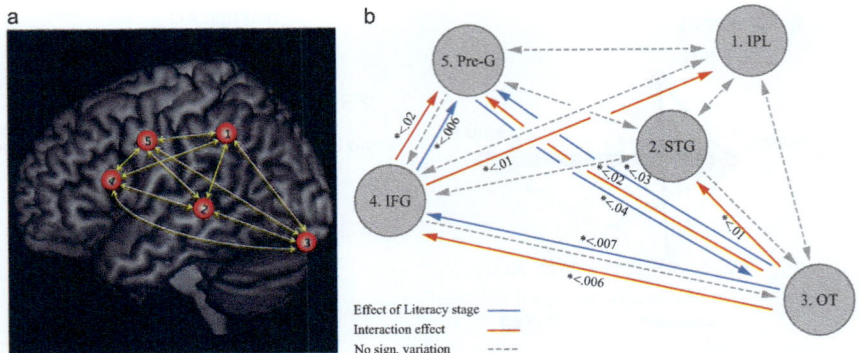

Fig. 12.8 Effective connectivity model. (A) Anatomical localization of the five areas. (B) DCM result with all connections. *Asterisks* mark significant effects with *P*-values (≤.05). IFG (4), inferior frontal gyrus; IPL (1), inferior parietal lobe; OT (3), occipito-temporal cortex; *Pre-G (5)*, precentral gyrus; *STG (2)*, superior temporal gyrus. *(From Morken, F., Helland, T., Hugdahl, K., Specht, K., 2017. Reading in dyslexia across literacy development: a longitudinal study of effective connectivity. Neuroimage 144, 92–100. Open Access.)*

analysis found a different connectivity set than Molinaro et al. (2016); among others an involvement of the visual areas, reflecting the reading tasks.

12.3 Network changes across aging and neurodegenerative disorders

The prefrontal cortex comprises the lateral, medial, and orbitofrontal regions. Jobson et al. (2021) write: "Collectively the mPFC possesses a range of corresponding couplings that vary across ageing and disorders in terms of directional intensity. In terms of network disturbances, the DMN has particularly arisen as a recurrently affected large-scale circuit involving the mPFC across ageing-related disorders (Fig. 12.9A), perhaps due to its common role in memory consolidation or autobiographical processes. Specifically, decreased FC is consistently found between long-distance anterior and posterior subsystems, which has been confirmed by network-based 'effective connectivity' measures in MCI, Alzheimer's disease and APOE4 elderly carriers (Fig. 12.9B). Additional mPFC connections within networks that were found to be disturbed include the salience network within healthy ageing and fronto-temporal dementia (FTD) patients, implicating that a range of critical circuits for optimal mPFC function are affected within health and disease alike. However, each disorder has also demonstrated distinct disease-specific patterns, as ageing acts seemingly on a continuum of

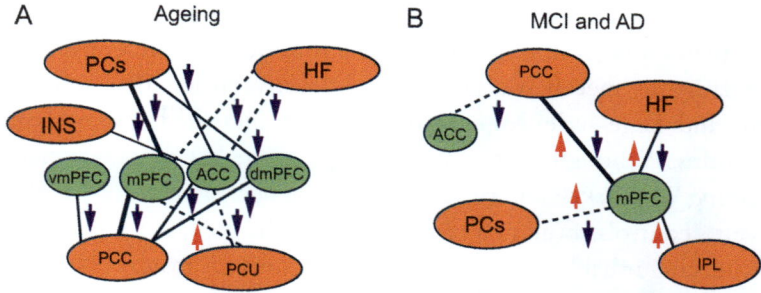

Fig. 12.9 Summary of altered mPFC functional connectivity (FC) trends across aging, MCI, and AD. (A) Pairwise FC changes (*upwards red arrow* indicates an increase and *downwards blue arrow* a decrease) in healthy *(thin line)*, Alzheimer's disease (AD) susceptible *(dashed line)* or both *(thick line)* aged subjects between mPFC subregions *(green circles)* and parietal cortices (PCs), insula (INS), hippocampal formation (HF), posterior cingulate cortex (PCC), and precuneus (PCU) brain regions *(orange circles)*. (B) Pairwise FC changes in mild cognitive impairment (MCI; *thin line*), Alzheimer's disease *(dashed line)* or both *(thick line)* between mPFC subregions and PCC, HF, PCs, or inferior parietal lobule (IPL). *(From Jobson, D.D., Hase, Y., Clarkson, A.N., Kalaria, R.N., 2021. The role of the medial prefrontal cortex in cognition, ageing and dementia. Brain Commun. 3 (3), fcab125. https://doi.org/10.1093/braincomms/fcab125. Open Access.)*

declining FC to mPFC, which continues into MCI and then Alzheimer's disease with progressively exaggerated decline as the individuals worsen in cognitive state. Moreover, in Alzheimer's disease, the mPFC circuits uniquely disconnect from the hippocampus and may impact upon the symptomatic memory deterioration within individuals."

Sharma et al. (2021) examined the association between subjective cognitive decline and posterior DMN (pDMN) network connectivity with medial temporal lobe (MTL, including hippocampus and parahippocampal gyrus, PHG) regions using rs-fMRI in older participants (~70 yrs of age). They found significant connectivity between the posterior cingulate cortex (PCC) and MTL in the cognitive unimpaired group but this was absent in subjective cognitive decline. Across all participants, self-perception of frequency of forgetting, but not objective memory, was strongly correlated with connectivity between the PCC and left PHG.

12.4 Autism spectrum disorder

12.4.1 Auditory system involvement

Lai et al. (2012) compared speech and song neural systems in low–functioning autistic and age-matched control children using passive auditory stimulation during fMRI and DTI. Activation in left IFG was reduced in autistic children relative to controls during speech stimulation, but was greater than controls during song stimulation. Functional connectivity for song relative to speech was also increased between left IFG and STG in autism, and large-scale connectivity showed increased frontal–posterior connections. Although fractional anisotropy of the left arcuate fasciculus—connecting Wernicke's and Broca's areas—was decreased in autistic children relative to controls, structural terminations of the arcuate fasciculus in IFG were indistinguishable between autistic and control groups. Wilson et al. (2022) used resting-state fMRI data to examine functional connectivity of the primary auditory cortex in ASD and healthy controls. The ASD group showed reduced functional connectivity between the A1 and four regions: the medial occipital cortex, primary motor cortex, insular cortex, and Wernicke's area. Schelinski et al. (2022) assessed the functional integrity of the auditory midbrain in ASD in three independent fMRI experiments. They focused on voice identity perception and recognizing speech-in-noise. They found: (1) The right inferior colliculus (IC) responded less in the ASD than in the control group for voice identity, in contrast to speech recognition. (2) The right IC also responded less in ASD than in controls when passively listening to vocal sounds in contrast to nonvocal sounds. (3) While in controls the left and right IC response was higher for recognizing speech in background noise in contrast to clear speech, in ASD this was only the case for the left, but not the right IC. Together, the results reveal that impaired processing of communication signals in ASD is associated with altered responses in a specific subcortical auditory sensory pathway structure—the right IC.

12.4.2 Nonauditory areas

Assaf et al. (2010) investigated the role of altered functional connectivity of DMN subnetworks in patients with high-functioning ASD compared to matched healthy controls, using resting-state fMRI and independent component analysis (ICA). Compared to controls, patients showed decreased FC between the precuneus and mPFC/ACC, and DMN core areas with other DMN subnetwork areas. These results suggest that DMN subnetwork underconnectivity contributes to the core deficits seen in ASD. Using seed-based connectivity and ICA approaches, von dem Hagen et al. (2013) looked at resting FC in adult high-functioning ASD within and between specific "social" brain regions. They found reduced FC within the DMN in individuals with ASD. Two further networks identified by

ICA, the SAL incorporating the insula, and an MTL network incorporating the amygdala, showed reduced internetwork connectivity. This was corroborated by reduced seed-based connectivity between the insula and amygdala. Supekar et al. (2013) showed greater functional connectivity in the brains of children with ASD in comparison to those of typically developing children. This hyperconnectivity in ASD was observed at the whole-brain and subsystems levels, across long- and short-range connections, and was associated with higher levels of fluctuations in regional brain signals. Brain hyperconnectivity predicted symptom severity in ASD, such that children with greater functional connectivity had more severe social deficits. The same group, Uddin et al. (2013), proposed that discrepancies between findings of autism-related hypoconnectivity and hyperconnectivity might be reconciled by taking developmental changes into account. They reviewed neuroimaging studies of autism, with an emphasis on fMRI studies of resting-state FC (rs-FC) in children, adolescents, and adults. Across several studies, a consistent pattern emerged that rs-FC in adolescents and adults with autism is generally reduced compared with age-matched controls, whereas rs-FC in younger autistic children appears to be increased. Uddin et al. (2015) then acquired functional neuroimaging data from 2 cohorts, each consisting of 17 children with ASDs and 17 age- and IQ-matched typically developing (TD) children, during stimulus-evoked brain states involving performance of social attention and numerical problem-solving tasks, as well as during intrinsic, resting brain states. Effective connectivity between key nodes of the salience network, default mode network, and central executive network was used to obtain indices of functional organization across evoked and intrinsic brain states (Fig. 12.10). This showed evidence for reduced discriminability between evoked and resting-state brain states in children with ASD. Specifically, children with ASD do not exhibit as drastic changes in connectivity patterns between task-evoked processing and resting states as do TD children.

Yerys et al. (2015) measured rs-FC in 8–13 year old nonmedicated children with ASD and typically developing controls using seed-based and network segregation FC methods. Relative to controls the ASD group showed both hypo (within DMN) and hyper FC (non-DMN) regions. ASD symptoms correlated negatively with the connection strength of the DMN midline core (mPFC-PCC). Network segregation analysis based on the participation coefficient showed more segregation for the ASD group. Bi et al. (2018) showed abnormal FC of resting-state networks (RSNs) in ASD patients (aged 8–18 years) compared to healthy controls. The RSNs

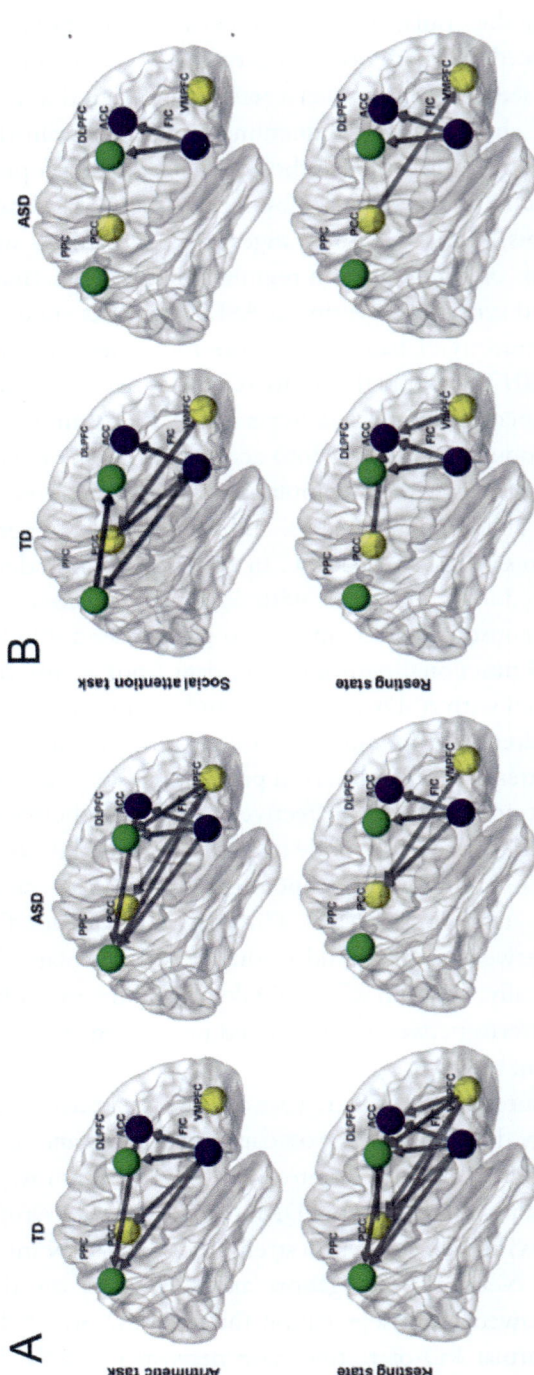

Fig. 12.10 Effective connectivity between network nodes. (A) Social attention and arithmetic verification tasks were used to probe task-evoked brain states. Granger causality analysis of the six key nodes of the salience (*blue*), central executive (*green*), and default mode (*yellow*) networks in TD children and children with ASD during the arithmetic task (B. Left), social attention task (B. Right), and resting state. ROIs within the SN, right CEN, and DMN were based on a previous publication which identified these nodes using ICA of resting-state fMRI data (Uddin et al., 2011). Network nodes were based on 8-mm radius spheres created around coordinates from the previous study. *ACC*, anterior cingulate cortex; *DLPFC*, dorsolateral prefrontal cortex; *FIC*, frontoinsular cortex; *PCC*, posterior cingulate cortex; *PPC*, posterior parietal cortex; *VMPFC*, ventromedial prefrontal cortex. (*From Uddin, L.Q., Supekar, K., Lynch, C.J., Cheng, K.M., Odriozola, P., Barth, M.E., et al., 2015. Brain state differentiation and behavioral inflexibility in autism. Cer. Cortex 25, 4740–4747. ©2014 The author, by permission of Oxford University Press.*)

with decreased FC in ASD patients included the DAN, DMN, executive control network (ECN), core network, visual network, and self-referential network. The RSNs with increased FC in ASD patients included the auditory network and the somatomotor network. Pereira et al. (2018) examined the fMRI-based FC of the DMN to investigate system-level alterations in adolescents and young adults with high-functioning autism compared to age- and intelligence quotient-matched healthy controls. The main findings were that patients with ASD had decreased connectivity between the PCC and areas of the executive control component of the DMN and increased FC between the anteromedial PFC and areas of the sensorimotor component of the DMN. Overall, these results are similar to those observed about brain hyperconnectivity predicting symptom severity in ASD (Supekar et al., 2013). Individuals with greater FC showed more severe social deficits, and they argue that this brain–behavior relationship suggests that aberrant FC may underlie social deficits, which are some of the hallmarks of ASD. From a rs-fMRI study, Rolls et al. (2020) found (Fig. 12.11) that directed connectivity from the MTG, implicated in face expression processing and theory of mind, to the precuneus and cuneus, implicated in processing about the self, was reduced in autism, as well as connectivity from the MTG to the VMPFC, involved in emotion and emotion-related decision making. In contrast, effective connectivity from memory (hippocampus) and emotion-related areas (amygdala) to the MTGs was increased in autism as was connectivity from the basal ganglia to the precentral and postcentral gyrus and mid-cingulate cortex.

Nair et al. (2020) reviewed functional connectivity studies of the DMN in adolescents with ASD (mean age range = 14.2–24.3 years). Of these, 15 of 29 studies in ASD adolescents reported predominant hypoconnectivity in the DMN. Fig. 12.12 provides a schematic representation of findings from all the studies for ASD, with yellow dots representing the DMN hub regions, blue (hypo-connected) or red (hyperconnected) dots representing connectivity with other brain regions, and thickness of lines connecting the dots representing frequency of findings across studies in each group.

12.5 ADHD

In individuals with ADHD (7.2–21.8 yr old), Sripada et al. (2014) found significant and specific maturational delays in connections within the DMN and in DMN interconnections with two task positive networks (TPNs): frontoparietal network (FPN) and ventral attention network (VAN).

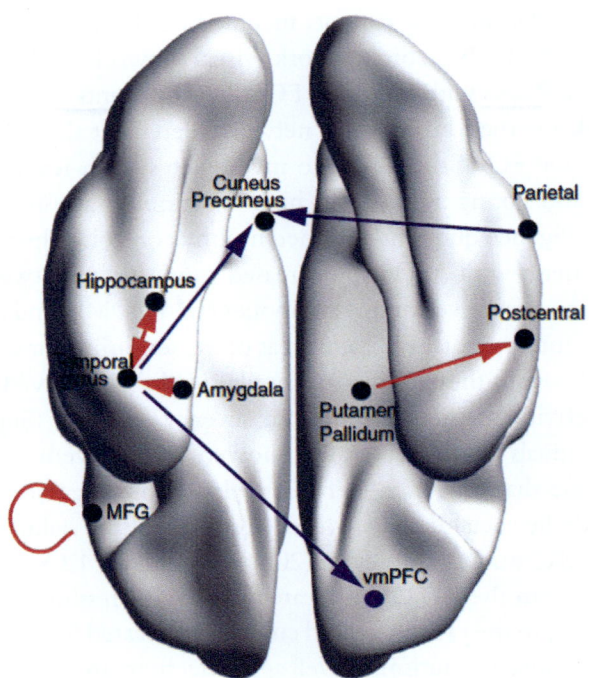

Fig. 12.11 Summary of some of the main changes in effective connectivity in autism. The increased effective connectivities in autism are shown in *red*, and decreased in *blue*. A *blue circle* indicates a decrease in activity in autism. *(From Rolls, E.T., Zhou, Y., Cheng, W., Gilson, M., Deco, G., Feng, J., 2020. Effective connectivity in autism. Autism Res. 13 (1), 32–44. © 2019 International Society for Autism Research, with permission of John Wiley and Sons.)*

In particular, a maturational lag was observed within the midline core of the DMN, as well as in DMN connections with right lateralized prefrontal regions (in FTN) and anterior insula (in VAN). Altered DMN–TPN interactions may be underlying attention dysfunction in ADHD. Sidlauskaite et al. (2016) compared connectivity within and between the DAN and VAN, the DMN and the SAL networks in ADHD and controls. The ADHD group displayed hyperconnectivity between the two attention networks and within the DMN and VAN. The SAL was hypo–connected to the DAN. There were trends toward hyperconnectivity within the DAN and between the SAL and VAN in ADHD. Pruim et al. (2019) identified functional connectivity abnormalities related to ADHD across 14 networks within a large rs-fMRI dataset ($n = 409$; age $= 17.5 \pm 3.3$ years). They found aberrant FC of the PCC and frontal areas within the DMN. They found that

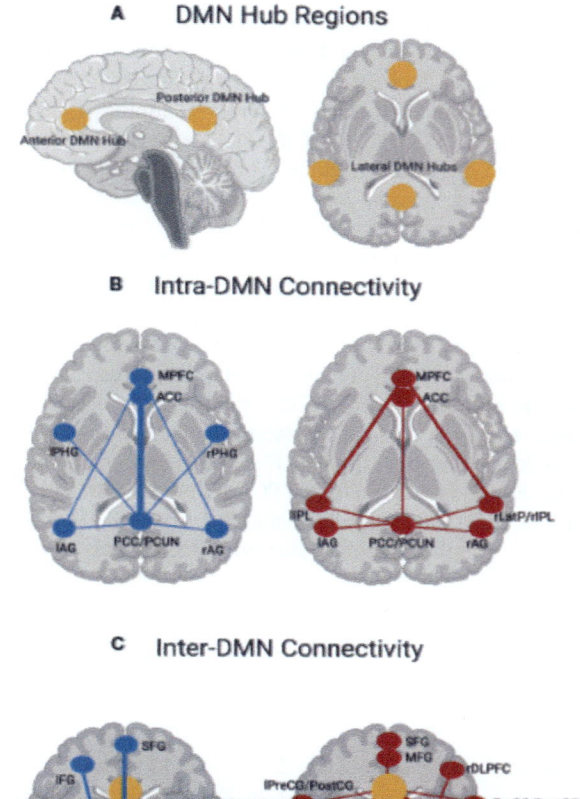

Fig. 12.12 (A) DMN hub regions included in analyses across most studies in the present review denoted in *orange circles*—anterior hubs include medial prefrontal regions (mPFC) and anterior cingulate cortex (ACC), posterior hubs include the posterior cingulate cortex (PCC) and precuneus (PCUN), lateral hubs include inferior parietal lobule (IPL), angular gyrus (AG), and medial temporal regions such as parahippocampal gyrus (PHG). (B) Intra-DMN connectivity findings across ASD adolescent studies (left panel)—underconnectivity findings (denoted by *blue dots*) mostly involve the posterior hubs of the DMN. *Thicker lines* such as between the PCC/PCUN and the anterior hubs of the DMN (mPFC, ACC) denote overlapping findings across multiple studies, with *thinner lines* indicating underconnectivity within DMN regions found in fewer studies. Overconnectivity findings (denoted by *red dots*) for the ASD group involve the anterior hubs of the DMN slightly more prominently than the posterior hubs on the DMN. *Thicker lines* such as between the

(Continued)

FIG. 12.12, Cont'd mPFC and left and right IPL denote overlapping findings across multiple studies, with *thinner lines* indicating overconnectivity within DMN regions found in fewer studies. (C) Inter-DMN connectivity findings across ASD adolescent studies (left panel)—underconnectivity findings (denoted by *blue dots*) mostly involve the posterior hub (PCC/PCUN) of the DMN especially in its connectivity with prefrontal regions and the right anterior insula (rAI) hub of the salience network denoted by *thicker lines* with additional findings of underconnectivity between DMN and other brain regions denoted by *thinner lines*. Overconnectivity findings (denoted by *red dots*) for the ASD group also involve the posterior hubs of the DMN mostly with somatomotor regions as well as anterior hubs of the DMN with the salience network hub (rAI) denoted by *thicker lines*. Other regions demonstrating overconnectivity with the DMN for the ASD group are denoted by *thinner lines*. Additional abbreviations used in figure: *AG*, angular gyrus; *CING*, cingulate gyrus; *DLPFC*, dorsolateral prefrontal gyrus; *FFA*, fusiform face area; *IFG*, inferior frontal gyrus; *ITG*, inferior temporal gyrus; *LatP*, lateral parietal lobule; *MFG*, middle frontal gyrus; *MTG*, middle temporal gyrus; *PHG*, parahippocampal gyrus; *POSTCG*, postcentral gyrus; *PreCG*, precentral gyrus; *PUT*, putamen; *SFG*, superior frontal gyrus; *SMG*, supramarginal gyrus; *SPG*, superior parietal gyrus; *STG*, superior temporal gyrus; *Thal*, thalamus; *TPJ*, temporo-parietal junction. *(Modified from Nair, A., Jolliffe, M., Lograsso, Y.S.S., Bearden, C.E., 2020. A review of default mode network connectivity and its association with social cognition in adolescents with autism spectrum disorder and early-onset psychosis. Front. Psychiatry 11, 614. https://doi.org/10.3389/fpsyt.2020.00614. Open Access.)*

"the frontal FC marker within the DMN was associated with multiple neurocognitive measures, whereas two PCC markers were unrelated to the neurocognitive measures, suggesting that subdivisions within the DMN play a dissociable role in the etiology of ADHD. Specifically, regarding PCC, its absent association with neurocognitive measures and the myriad of findings described in literature on PCC abnormalities across psychiatric disorders support the idea that abnormalities of this region 1) are directly related to ADHD, 2) are not influenced by cognitive risk factors for ADHD, or 3) are unspecific to ADHD and a consequence of accumulating remote pathological effects."

Based on data in 21 studies including 700 ADHD patients and 580 controls Gao et al. (2021) conducted metaanalyses. The results showed that ADHD was characterized by hyperconnectivity between the FPN and regions of the DMN and AN as well as hypoconnectivity between the FPN and regions of the VAN and the somatosensory network (SSN; Fig. 12.13). Based on previous literature, Damiani et al. (2021) selected five core subcortical regions (amygdala, caudate, putamen, pallidum, and hippocampus) as regions of interest, measuring their whole-brain voxel-wise rs-FC in a sample of 95 ADHD and 90 neurotypical children and adolescents

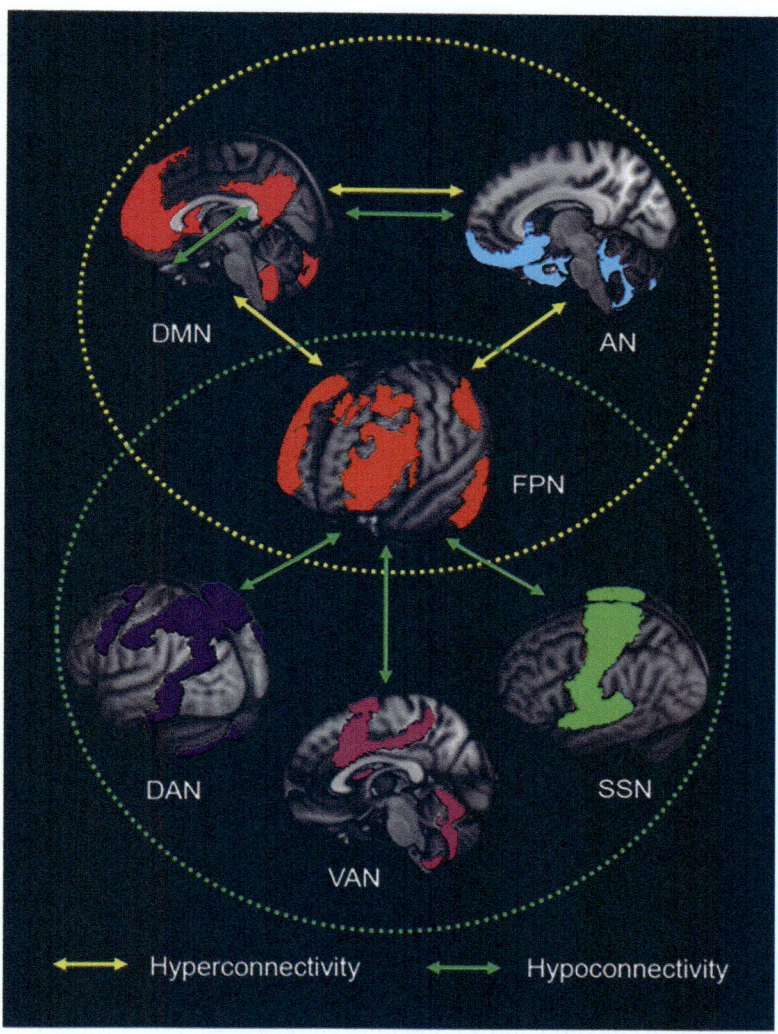

Fig. 12.13 Neurocognitive network dysfunction patterns of attention-deficit/ hyperactivity disorder (ADHD). The *yellow circle* represents the aberrant interplay among the default mode network (DMN), frontoparietal network (FPN), and affective network (AN) in ADHD. The *green circle* represents ADHD-associated hypoconnectivity between the FPN and the dorsal attention network (DAN), ventral attention network (VAN), and somatosensory network (SSN). The overlap of the patterns suggests that the FPN is a core intrinsic network of ADHD pathophysiology. *(From Gao, Y., Shuai, D., Bu, X., Hu, X., Tang, S., Zhang, L., et al., 2019. Impairments of large-scale functional networks in attention-deficit/hyperactivity disorder: a meta-analysis of resting-state functional connectivity. Psychol. Med. 49 (15), 2475–2485, by permission of Cambridge University Press.)*

aged from 7 to 18. The only subcortical structure showing significant differences in rs–FC was the caudate nucleus. Specifically, they found increased rs–FC with ACC and right insula, two mesolimbic regions in the SAL. The degree of hyper rs–FC was positively correlated with ADHD symptomatology. Marcos-Vidal et al. (2022) compared local and distant FC between 31 medication-naïve adults with ADHD and 31 healthy controls and tested whether this pattern was associated with symptoms severity scores. "The ADHD group showed increased local FC in the dACC and the SFG and decreased local connectivity in the PCC. Results parallel those obtained in children samples suggesting a deficient integration within the DMN and segregation between DMN, FPN, and VAN." In post hoc analyses, with seeds in the dACC/SFG, adults with ADHD showed increased FC between the dACC/SFG and the motor cortex. However, for the PCC as a seed they found decreased FC between that cluster and the mPFC in patients with ADHD. As previously stated, the mPFC and the PCC are the two principal hubs of the DMN, and thus, show high FC in control populations.

12.6 MCI and Alzheimer's disease

12.6.1 General network changes

12.6.1.1 Reviews

In an exhaustive review, Teipel et al. (2016) found that "hundreds of peer-reviewed (cross-sectional and longitudinal) papers have shown in patients with MCI and mild AD compared to Controls (1) impairment of callosal (splenium), thalamic, and anterior–posterior white matter bundles; (2) reduced correlation of resting state blood oxygen level-dependent activity across several intrinsic brain circuits including default mode and attention-related networks; and (3) abnormal power and functional coupling of resting state cortical EEG rhythms." Ibrahim et al. (2021) conducted a systematic review aimed at determining the diagnostic power of rs-fMRI to identify FC abnormalities in the DMN of patients with AD or MCI compared with healthy controls using machine learning methods. Among AD patients, the PCC/precuneus was noted to be a highly affected hub of the DMN that demonstrated overall reduced FC, whereas reduced DMN FC between the PCC and anterior cingulate cortex (ACC) was observed in MCI patients. Evidence indicates that the nodes of the DMN can offer moderate to high diagnostic power to distinguish AD and MCI patients.

12.6.1.2 Data papers

Wang et al. (2016) compared the differences between white matter connectivity characteristics at global, regional, and local levels in patients with probable AD and normal elderly subjects, using connectivity networks constructed from DTI. They found that the AD patients had significantly weaker connections in multiple local cortical and subcortical regions, such as precuneus, temporal lobe, hippocampus, and thalamus (Fig. 12.14).

Liu et al. (2022) combined voxel-based morphometry (VBM), amplitude of low-frequency fluctuations (ALFFs), regional homogeneity (Reho), and rs-FC approaches to explore concurrent structural and functional alterations in patients with amnestic MCI (aMCI). They found that, in patients with aMCI compared with healthy controls, both ALFF and Reho were decreased in the right superior frontal gyrus (rSFG) and right medial frontal gyrus (rMFG). In addition, both gray matter volume and Reho were decreased in the left IFG (LIFG) of patients with aMCI. Furthermore, these overlapping clusters from VBM, ALFF, and Reho analyses were used as seed regions to analyze rs-FC. Liu et al. (2022) thus found that, compared with healthy controls, patients with aMCI had decreased rs-FC

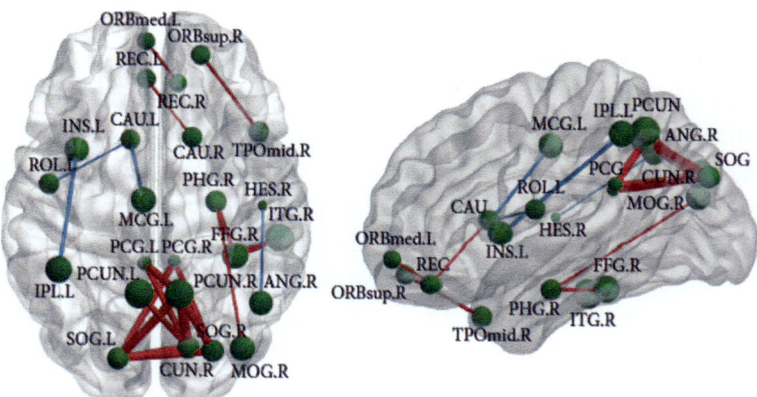

Fig. 12.14 The axial and the sagittal views of the significantly different connections (*P* <.05) based on the fiber counts between the AD group and the NC group. The stronger connections (higher fiber counts between a pair of ROIs) in the AD group are shown in *blue*, while the weaker connections (lower fiber counts between a pair of ROIs) are in *red*. Refer to Table 12.2 for the label of each ROI. *(From Wang, T., Shi, F., Jin, Y., Yap, P.-T., Wee, C.-Y., Zhang, J., et al., 2016. Multilevel deficiency of white matter connectivity networks in Alzheimer's disease: a diffusion MRI study with DTI and HARDI models. Neural Plast. 2947136. https://doi.org/10.1155/2016/2947136. Open Access.)*

between rSFG and the right temporal lobe (subgyral); the rMFG seed and LSTG, LIPL, and rACC; the LIFG seed and left precentral gyrus, left cingulate gyrus, and LIPL. These findings highlighted shared-imaging features in structural and functional MRI, suggesting that rSFG, rMFG, and LIFG may play a major role in the pathophysiology of aMCI, which might be useful to better understand the underlying neural mechanisms of aMCI and AD (Liu et al., 2022).

Khan et al. (2020) examined functional connectivity differences in AD patients and cognitively normal participants, as well as the entire AD pathological spectrum. They "revealed decreased FC in the anterior and central precuneus, and dorsal PCC in AD patients compared to cognitively normal participants. Functional abnormalities in the dorsal PCC and central precuneus were also related to amyloid burden and volumetric hippocampal loss. Across the entire AD spectrum, functional connectivity of the central precuneus was associated with disease severity and specific deficits in memory and executive function." This evidence shows that "the posteromedial cortex is selectively impacted in AD, with prominent network failures of the dorsal PCC and central precuneus underpinning the neurodegenerative and cognitive dysfunctions associated with the disease."

12.6.2 Connectivity with hippocampus

In relation to Alzheimer's disease progression, Sohn et al. (2014) explored the changes in FC of the left and right hippocampus by comparing the resting-state ALFF from these regions. They found that the total functional connectivity of both the right and left hippocampus was maintained during aMCI and the early stages of AD and that it decreased in the later stages of AD. Seeding the left hippocampus revealed a significant decrease in the functional connectivity with the PCC region and lateral parietal areas, and an increase in FC in the ACC starting within aMCI patients. Gardini et al. (2015) investigated the relationship between semantic memory impairment and DMN intrinsic connectivity in MCI. Patients presented increased DMN connectivity between the medial prefrontal regions and the posterior cingulate and between the PCC and the parahippocampal gyrus (PHG) and anterior hippocampus. MCI patients also showed a significant negative correlation of medial prefrontal gyrus connectivity with PHG and posterior hippocampus and visual naming performance. This suggested to Gardini et al. (2015) that increasing DMN connectivity may contribute to semantic memory deficits in MCI, specifically in visual naming. Liu et al. (2016) found that several regions of the DMN showed reduced connectivity with

PHG in the AD patients, associated with disease severity in the MCI and AD subjects. They also found positive correlations between the connectivity strengths of the left PHG-PCC/precuneus (Pcu) and left PHG-left MTG and the Mini-Mental State Examination. This indicated that with progression from MCI to severe AD, reduction in the FC of the PHG increases severely. Furthermore, Liu et al. (2016) found that the severe AD and MCI groups exhibited significantly decreased connectivity with the right PHG in all of the identified brain regions. In the identified brain regions, the PCC/Pcu/Cuneus, MPFC, and anterior MTG showed significantly stronger connectivity in the normal control participants compared with those in the AD subjects. Only the posterior MTG showed weaker connectivity in the severe AD subjects compared with the MCI subjects; in addition, the posterior/anterior MTG and ANG/STG exhibited weaker connectivity in the subjects with severe AD compared with those with moderate AD.

Zhang et al. (2016) computed resting-state FC between 18 metaanalytically determined peak coordinates of brain activation during successful memory retrieval. They observed higher FC between the PHG, parietal cortex, and the middle frontal gyrus was observed in both AD and MCI compared to HC. The increase in FC between the PHG and middle frontal gyrus (MFG) was associated with reduced episodic memory in aMCI (Fig. 12.15), independent of amyloid-β binding and apolipoprotein E ϵ4-carrier status. Zhang et al. (2016) concluded that "increased FC of PHG-prefrontal is predictive of impaired episodic memory in aMCI and may reflect a dysfunctional change within the episodic memory-related neural network."

Zhou et al. (2022) selected regions of interest with a high level of resting-state functional connectivity with hippocampus as seeds to track fibers based on DTI. In this way, hippocampus-temporal and thalamus-related fibers were selected, and each fiber's DTI parameters were extracted. Compared with normal fibers, the degenerated fibers detected by the DTI indexes, especially for hippocampus-temporal fibers, show significantly higher correlations with cognitive scores (Fig. 12.16). Furthermore, compared with the hippocampus-temporal fibers, thalamus-related fibers have shown significantly higher correlations with depression scores within MCI.

12.6.3 Network changes correlate with amyloid and tau deposits

Myers et al. (2014) compared spatial patterns of amyloid-β-plaques (measured by Pittsburgh compound B PET) and intrinsic functional connectivity

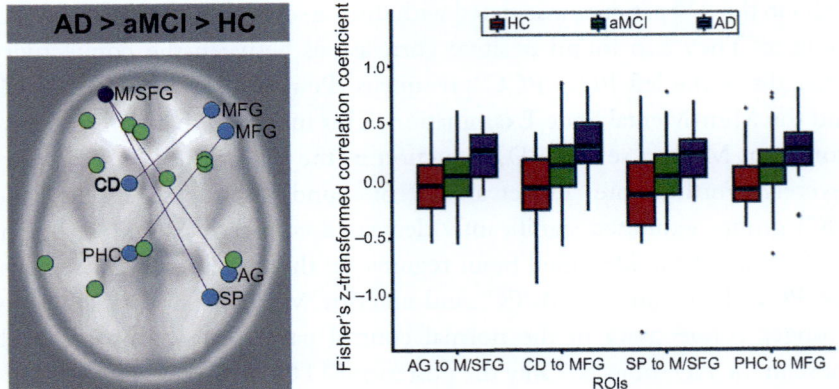

Fig. 12.15 Left. Spatial projection of ROI-to-ROI connections onto an axial brain slice. The strength of FC was increased in AD dementia compared to the strength of FC in MCI which in turn was larger than the strength of FC in HC. *ROI color* corresponds to the number of connections to (or from) each ROI. Abbreviations: *AD*, Alzheimer's disease with dementia; *AG*, angular gyrus; *aMCI*, amnestic mild cognitive impairment; *CD*, caudate; *FC*, functional connectivity; *HC*, cognitively healthy; *M/SFG*, middle/ superior frontal gyrus; *MFG*, middle frontal gyrus; *PHC*, parahippocampal gyrus; *ROI*, region of interest; *SP*, superior parietal. Right. Boxplot of functional connectivity values as a function of diagnosis on the four connections which significantly increased between HC and AD. Abbreviations: *AD*, Alzheimer's disease with dementia; *AG*, angular gyrus; *aMCI*, amnestic mild cognitive impairment; *CD*, caudate; *HC*, cognitively healthy; *MFG*, middle frontal gyrus; *M/SFG*, middle/superior frontal gyrus; *PHC*, parahippocampal; *ROIs*, regions of interest; *SP*, superior parietal. *(Reprinted from Zhang, Y., Simon-Vermot, L., Araque Caballero, M.Á., Gesierich, B., Taylor, A.N.W., Duering, M., et al., 2016. Enhanced resting-state functional connectivity between core memory-task activation peaks is associated with memory impairment in MCI. Neurobiol. Aging 45, 43–49. ©2016, with permission from Elsevier.)*

(measured by rs-fMRI) in patients with prodromal Alzheimer's disease via spatial correlations in intrinsic networks covering frontoparietal heteromodal cortices. They found that amyloid-β and intrinsic connectivity patterns were positively correlated in the DMN and several attention networks, confirming that amyloid-β aggregates in areas of high intrinsic connectivity on a within-network basis. The internetwork gradient of the correlation strength depended on network plaque load. Using rs-fMRI and graph-theory approaches, Wang et al. (2015) investigated the topological organization of whole-brain functional networks in 16 APOE ε4 carriers and 26 matched noncarriers with AD at the global whole-brain, intermediate module, and regional node/connection level. They showed that the APOE ε4 carriers performed worse on delayed memory but better on a late item generation of a verbal fluency task than noncarriers. They also found that APOE ε4

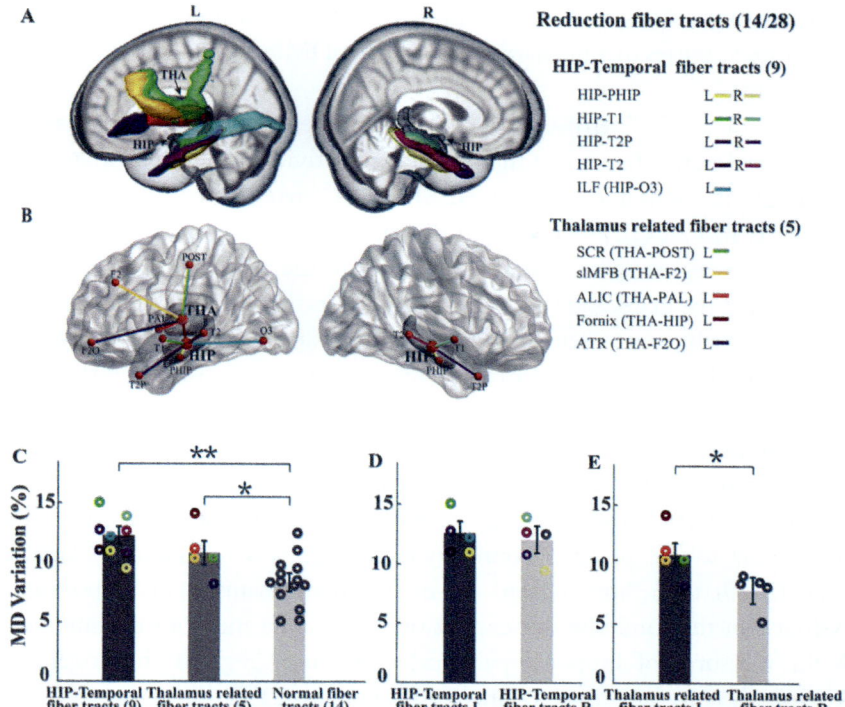

Fig. 12.16 Spatial distribution and statistical comparison of degenerated fiber tracts in MCI. (A) Reduction fiber tracts in MCI were shown in glass brain representations in standard space. (B) Schematic illustration of degenerated white matter in MCI. (C) Statistical comparisons of MD variation for hippocampus-temporal fiber tracts ($n = 9 \times 42$), thalamus-related fiber tracts ($n = 5 \times 42$), and normal fiber tracts ($n = 14 \times 42$). (D) Statistical comparisons of MD variation between two hemispheres for hippocampus-temporal fiber tracts (L: $n = 9 \times 42$, R: $n = 9 \times 42$). (E) Statistical difference of MD variation between two hemispheres for thalamus-related fiber tracts (L: $n = 5 \times 42$, R: $n = 5 \times 42$). Notes: (1) All tracked fibers ($n = 28$) were classified into degenerated hippocampus-temporal fiber tracts ($n = 9$), degenerated thalamus-related fiber tracts ($n = 5$), and normal fiber tracts ($n = 14$). (2) Nodes represented ROIs from AAL templates, and edges represented connected fibers, in which the hippocampus and the thalamus were two important nodes. (3) The MD variation was showed by vertical bar charts (mean ± standard error of the mean [SEM]; *$P < .05$ and **$P < .005$; rank-sum test with FDR correction), in which each dot represented the mean MD variation for each fiber tract. *(From Zhou, Y., Si, X., Chen, Y., Chao, Y., Lin, C.-P., Li, S., et al., 2022. Hippocampus-and thalamus-related fiber specific white matter reductions in mild cognitive impairment. Cereb. Cortex 32, 3159–3174, ©2021, by permission of Oxford University Press.)*

significantly disrupted whole-brain topological organization as reflected in "(1) reduced parallel information transformation efficiency; (2) decreased intramodular connectivity within the posterior default mode network (pDMN) and intermodular connectivity of the pDMN and ECN with other neuroanatomical systems; and (3) impaired functional hubs and their rich-club connectivities that primarily involve the pDMN, ECN, and sensorimotor systems."

Harrison et al. (2016) used a paired-associates memory task to examine differences in task-dependent functional connectivity of the anterior and posterior hippocampus in nondemented APOE ε4 carriers and noncarriers to test the theory that APOE ε4-mediated differences would be more pronounced in the anterior region, which is affected earlier in the AD course. They found that "during the encoding task, APOE ε4 carriers had lower FC change compared to baseline between the anterior hippocampus and right precuneus, anterior insula and cingulate cortex. During the retrieval task, bilateral supramarginal gyrus and right precuneus showed lower FC change with anterior hippocampus in carriers. Also, during retrieval, carriers showed lower connectivity change in the posterior hippocampus with auditory cortex. In each case, APOE ε4 carriers showed strong negative connectivity changes compared to noncarriers where positive connectivity change was measured." Harrison et al. (2019) then "investigated tau-mediated mechanisms of hippocampal dysfunction that underlie the expression of episodic memory decline using fMRI measures of hippocampal local coherence (regional homogeneity; ReHo), distant functional connectivity and tau-PET. They showed that age and tau pathology are related to higher hippocampal ReHo. Functional disconnection between the hippocampus and other components of the MTL memory system, particularly an anterior-temporal network specialized for object memory, is also associated with higher hippocampal ReHo and greater tau burden in anterior-temporal regions."

Shigemoto et al. (2018) investigated 18 patients with amnestic mild cognitive impairment/mild AD (AD-spectrum group) and 35 cognitively normal older adults using diffusion MRI, amyloid, and tau PET imaging. DTI was performed to identify white matter pathways correlated with each of the six variables of tau deposition in the bilateral hippocampi, temporal lobes, posterior and anterior cingulate cortices, precunei, orbitofrontal lobes, and entire cerebrum. They found that the normal older adults group had "increased connectivity along with an increased tau deposition in the bilateral hippocampi, temporal lobes, and entire cerebrum, whereas the AD-spectrum group showed decreased connectivity in the bilateral hippocampi, temporal lobes, anterior and posterior cingulate cortices, precunei, and entire cerebrum" (Fig. 12.17).

Pereira et al. (2019) "applied a joint independent component analysis to ^{18}F–Flutemetamol (amyloid-β) and ^{18}F–Flortaucipir (tau) PET images to

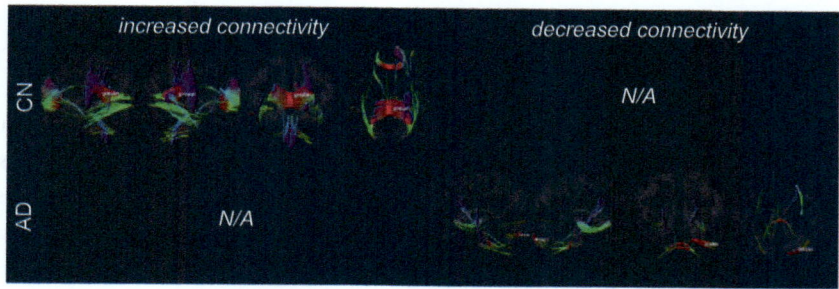

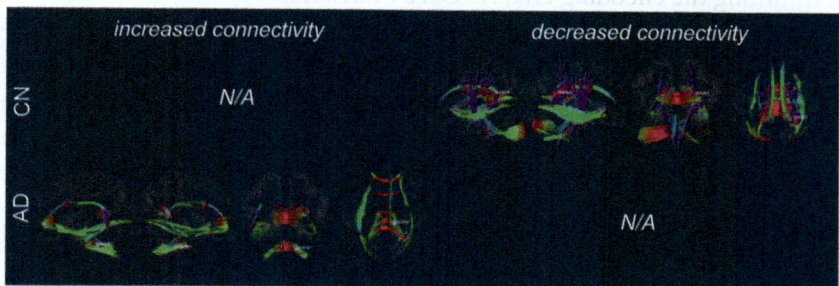

Fig. 12.17 Top. Structural connectivity correlated with tau deposition in temporal lobe. The CN group shows increased connectivity (FDR = 0.0023), whereas the AD-spectrum group shows decreased connectivity (FDR = 0.015). Bottom. Structural connectivity correlated with tau deposition in orbitofrontal lobe. The CN group shows decreased connectivity (FDR = 0.0064), whereas the AD-spectrum group shows increased connectivity (FDR = 0.0017). *Red*: left-right, *green*: anterior-posterior, *blue*: superior-inferior. *(From Shigemoto, Y., Sone, D., Maikusa, N., Okamura, N., Furumoto, S., Kudo, Y., et al., 2018. Association of deposition of tau and amyloid-β proteins with structural connectivity changes in cognitively normal older adults and Alzheimer's disease spectrum patients. Brain Behav. 8, e01145. https://doi.org/10.1002/brb3.1145. Open Access.)*

identify amyloid-β and tau networks across different stages of Alzheimer's disease. They then assessed whether these patterns were associated with resting-state functional networks and white matter tracts. Analyses revealed nine patterns that were linked across tau and amyloid-β data. The amyloid-β and tau patterns showed a fair to moderate overlap with distinct functional networks but only tau was associated with white matter integrity loss and multiple cognitive functions."

12.6.4 From mild cognitive impairment to Alzheimer's disease

Alzheimer's disease is associated with abnormal resting-state network architecture of the DMN, the dorsal-attention network, the executive control network, the salience network (SAL), and the sensorimotor network. Zhan et al. (2016) found longitudinal alterations of functional connectivity within the DMN, where they were correlated with variation in cognitive ability. In mild cognitive impairment, the SAL as well as the interaction between the DMN and the SAL was disrupted. They also found that longitudinal alterations of functional connectivity are more profound in earlier stages than in later stages of the disease.

Dillen et al. (2017) used rs-fMRI to investigate the functional connectivity strength across five DMN nodes in healthy controls, subjective cognitive decline participants, and prodromal Alzheimer's disease patients. Functional connectivity of the ventral medial prefrontal cortex (VMPFC) and PCC in prodromal AD patients was disrupted. The strength of the VMPFC-PCC connectivity correlated positively with memory performance in prodromal AD. In subjective cognitive decline and prodromal AD patients, the hippocampus decoupled from posterior DMN nodes. There was no connectivity between the hippocampus and the anterior DMN. Only in the healthy control group there was communication between the hippocampus and DMN regions but not in the other two groups.

Berron et al. (2020) analyzed data from 256 amyloid-β-negative cognitively unimpaired, 103 amyloid-β-positive cognitively unimpaired, and 83 amyloid-β-positive individuals with MCI. Amyloid-β and tau pathology were measured using the cerebrospinal fluid (CSF) amyloid-$\beta_{42/40}$ ratio and phosphorylated tau, respectively. Amyloid-β-positive cognitively unimpaired individuals showed decreased FC between the medial temporal lobe and regions in the anterior-temporal system, most prominently between left perirhinal/entorhinal cortices and medial prefrontal cortex. Correlation analysis in this group revealed decreasing FC between bilateral perirhinal/entorhinal cortices, anterior hippocampus, and posterior-medial regions with increasing levels of phosphorylated tau. The amyloid-β-positive individuals with MCI showed reduced FC between the medial temporal lobe and posterior-medial regions, predominantly between the anterior hippocampus and posterior cingulate cortex. Berron et al. (2020) also found hyperconnectivity within the medial temporal lobe and its immediate proximity. Lower medial temporal-cortical FC networks of cognitively unimpaired individuals were associated with reduced memory performance and more rapid longitudinal memory decline. In conclusion, Berron et al. (2020) show "that the earliest changes in preclinical Alzheimer's disease might involve decreased connectivity within the anterior-temporal system, and early changes in connectivity might be related to memory impairment, but not to structural changes."

12.7 Integration of the MMN network and disorder-specific changes

Are there disorder-specific network changes that interact with the "universal" auditory MMN network? Not likely, as Näätänen et al. (2014) already emphasized; "while not serving as a specific marker to any particular disorder, MMN may be useful for understanding factors of cognition in various disorders."

In Table 12.1 we show disorder-specific network changes in developmental disorders: dyslexia, ASD, and ADHD. The diagonal shows intranetwork changes in connectivity, the off-diagonal entries are internetwork connectivity. With the exception of the IFG in dyslexia, intranetwork FCs are decreased. It is noted that dyslexia and ASD share common decreases in connectivity between the STG and left IFG, as well as increased FC between pCUN/PCC and IPL/IPC. ADHD and ASD share only a common decreased FC between mPFC and ACC.

In Table 12.2, I compiled disorder-specific network changes in aging, tinnitus, and MCI/AD. With the exception of tinnitus in PCC, intranetwork FCs—including the IFG—are decreased. It is noted that the connectivity changes between the STG and IFG, which play a role in most auditory MMN network changes, are increased in aging and MCI/AD despite the fact that the connectivity within the IFG is decreased. We also note that there are no clear examples of modulatory effects of the other DMN network nodes—PCC/pCUN, IPL/IPC—on the IFG. Overall, with the exception of tinnitus, the number of decreased connectivities outweighs the increases.

There appears to be limited connections between disorder-specific network changes in the DMN and those in the IFG and STG that play such a dominant role in the MMN network. Consequently, the use of predictive coding networks based on auditory MMN as error signal only emphasizes the communality of cognitive decrease—potentially differentiated between temporal and frequency mismatches—however does not discriminate nor is of further diagnostic value. On the other hand, dynamic causal modeling elucidates the underlying intrinsic, particularly inhibitory changes that are affected by neurological disorders, in the various network nodes that affect the intranode FCs. This may be the most relevant, and so far underused,

Table 12.1 Connectivity changes in developmental disorders.

	DMN	DLPFC	mPFC	pCUN	AG	CEN	PCC	ACC	LIFG	STG	SAL	DAN	pHIP	IPL/IPC	MFG
DMN	↓[7,8] ↑[12]	↑[12]													
DLPFC															
mPFC			↑[12]	↓[14]			↓[9]	↑[8] ↓[17]	↓[12]	↓[12]				↓[13]	
pCUN			↓[12]		↑[1] ↓[12]		↑[9,14]	↓[7,13]		↓[13]				↑[4] ↑[13]	
AG									↑[4]						
CEN						↓[10]	↓[11]	↑[7,12]							
PCC		↓[9]	↑[9,14]			↑[11]	↓[13]	↓[9]						↑[4] ↑[13]	↑[11]
ACC			↓[8] ↓[17]			↓[7,12]	↓[9]	↓[12]			↑[16]	↑[16,19]			↑[11]
LIFG	↓[12]			↑[18]	↑[4]				↑[1,3]	↓[4,5,6] ↓[12,15]				↓[2,3,4]	
STG	↓[12]		↓[13]						↓[4,5,6] ↓[12,15]						
SAL								↑[16]				↓[20]			
DAN								↑[16,19]			↓[20]				
pHIP															
IPL/IPC			↓[13]	↑↑[13]			↑[4] ↑[13]		↓[2,3,4]				↑[12]		
MFG							↑[11]	↓[11]							

↑↓ Dyslexia; ↑↓ ASD; ↑↓ ADHD

[1] Shaywitz et al. (1998); [2] Cao et al. (2008); [3] Georgiewa et al. (2002); [4] Schurz et al. (2015); [5] Boets et al. (2013); [6] Molinaro et al. (2016); [7] Assaf et al. (2010); [8] von dem Hagen et al. (2013); [9] Uddin et al. (2015); [10] Bi et al. (2018); [11] Pereira et al. (2018); [12] Nair et al. (2020); [13] Rolls et al. (2020); [14] Lai et al. (2012); [15] Yerys et al. (2015); [16] Damiani et al. (2021); [17] Marcos-Vidal et al. (2022); [18] Chiang et al. (2022); [19] Lin et al. (2021); [20] Sidlauskaite et al. (2016).

Abbreviations: *ACC*, anterior cingulate cortex; *AG*, angular gyrus; *CEN*, central executive network; *DAN*, dorsal attention network; *DLPFC*, dorsolateral prefrontal cortex; *DMN*, default mode network; *IPl/IPC*, inferior parietal lobe/cortex; *LIFG*, left inferior frontal gyrus; *MFG*, middle frontal gyrus; *mPFC*, medial prefrontal cortex; *PCC*, posterior cingulate cortex; *pHIP*, parahippocampus; *SAL*, salience network; *STG*, superior temporal gyrus.

Table 12.2 Connectivity changes in age-related disorders.

	DMN	DLPFC	mPFC	pCUN	AG	CEN	PCC	ACC	LIFG	STG	SAL	DAN	pHIP	IPL/IPC	MFG
DMN	↓[1,7,12]														
DLPFC		↓[2]													
mPFC			↑[16]↓[16]↓[18]	↑[16]↓[16]↓[18] ↑[2,4]↓[19]			↑[10]↓[15]↓[16]								
pCUN					↓[2,6]		↑[18]					↑[18]			
AG						↓[3,7]									
CEN	↓[12]														
PCC	↓[4]						↓[6]↑[19]					↑[18]	↑[8,9]		
ACC	↓[4]			↓[4,16]↓[16]											
LIFG									↓[5]						
STG									↑[20]↑[21]	↑[20]↑[21]					
SAL	↓[14]										↓[7,14]				↑[13]
DAN															↑[5]
pHIP	↓[15]		↓[11]	↓[9,11]			↑[8,9]								↑[5]
IPL/IPC			↑[16]	↓[17]											
MFG								↓[5]	↓[5]				↑[13]		

↑↓ MCI/AD; ↑↓ aging; ↑↓ tinnitus

[1] Gour et al. (2014); [2] Wang et al. (2016); [3] Zhao et al. (2016); [4] Ibrahim et al. (2021); [5] Liu et al. (2022); [6] Khan et al. (2020); [7] Ripp et al. (2020); [8] Sohn et al. (2014); [9] Tahmasian et al. (2015); [10] Gardini et al. (2015); [11] Liu et al. (2016); [12] Wang et al. (2015); [13] Zhang et al. (2016); [14] Zhan et al. (2016); [15] Dillen et al. (2017); [16] Jobson et al. (2021); [17] Alderson et al. (2017); [18] Schmidt et al. (2017); [19] Moring et al. (2022); [20] Cooray et al. (2014); [21] Cope et al. (2022).

Abbreviations: ACC, anterior cingulate cortex; AG, angular gyrus; CEN, central executive network; DAN, dorsal attention network; DLPFC, dorsolateral prefrontal cortex; DMN, default mode network; IPL/IPC, inferior parietal lobe/cortex; LIFG, left inferior frontal gyrus; MFG, middle frontal gyrus; mPFC, medial prefrontal cortex; PCC, posterior cingulate cortex; pHIP, parahippocampus; SAL, salience network; STG, superior temporal gyrus.

contribution of predictive coding for therapeutics. The limitation of Tables 12.1 and 12.2 is that no differentiation is made here between left and right structures, and that the mPFC is not further specified as to its parts. The same can be said of not differentiating between the various parts of the IFG (see Chapter 10) and STG.

12.8 Summary

In tinnitus, only sporadic structural changes in gray and white matter are found that could be specifically attributed to tinnitus. Most changes, including those outside the auditory areas, result from hearing loss. Functional changes relevant for tinnitus, according to and activation likelihood estimation from fMRI publications, were identified only in posterior cingulate cortex and precuneus. However, EEG/MEG studies typically identify auditory cortical and cognitive areas as well. It needs to be noted that fMRI measures are not very sensitive to neural synchrony changes so that the absence of auditory cortex involvement in the ALE results is likely due to insensitivity to temporal synchrony changes of fMRI.

In dyslexia, studies find less activity in dyslexic participants in auditory cortical areas; however, higher activity in dyslexics compared with controls is found in the right IFG. In addition, there is reduced connectivity in dyslexic readers between left posterior temporal areas and the left IFG. Furthermore, connectivity between multiple reading-related areas and areas of the DMN, in particular the precuneus, was stronger in dyslexic compared with nonimpaired readers. Another finding of stronger connectivity in dyslexic readers was linked to IFG pars opercularis–left IPL/angular gyrus and also with right angular gyrus.

In ASD, resting-state fMRI data show reduced FC between the A1 and four regions: the medial occipital cortex, primary motor cortex, insular cortex, and Wernicke's area. Nonauditory areas show general hyperconnectivity across long- and short-range connections. Contrasting with findings in TD children, in ASD there are no drastic changes in FC between task-evoked processing and resting states. In young adults, the main finding is that ASD had *decreased* connectivity between the PCC and areas of the executive control component of the DMN and *increased* FC between the anteromedial PFC and areas of the sensorimotor component of the DMN. The hypoconnectivity within and between DMN and other brain regions is a dominant finding. ADHD is characterized by hyperconnectivity between the FPN and regions of the DMN and the limbic network as well as hypoconnectivity between the FPN and regions of the VAN and the somatosensory network. The only subcortical structure showing significant differences in rs–FC was the caudate nucleus.

Aging acts seemingly on a continuum of declining FC from and to medial PFC, which continues into MCI and then Alzheimer's disease with progressively exaggerated decline as the individuals worsen in cognitive state. Moreover, in Alzheimer's disease, the mPFC circuits uniquely disconnect from the hippocampus and may impact upon the symptomatic memory deterioration within individuals. Specifically, normal older adults show increased connectivity along with an increased tau deposition in the entire cerebrum, whereas the AD-spectrum group showed decreased connectivity in the entire cerebrum. Compared with healthy elderly, MCI patients show an extensive semantic memory decline in category fluency, visual naming, naming from definition, words association, and reading tasks. These patients presented increased DMN connectivity between the medial prefrontal regions and the posterior cingulate and between the PCC and the parahippocampal gyrus (PHG) and anterior hippocampus.

Among AD patients, the PCC/precuneus is a highly affected hub of the DMN that demonstrated overall reduced FC, whereas reduced FC in the DMN between the PCC and anterior cingulate cortex (ACC) was observed in MCI patients. Furthermore, decreased FC in the anterior and central precuneus, and dorsal PCC is present in AD patients compared to cognitively normal participants, and is related to amyloid burden and volumetric hippocampal loss. At the global network level, amyloid-β and intrinsic connectivity patterns were positively correlated in the DMN and several frontoparietal attention networks, confirming that amyloid-β aggregates in areas of high intrinsic connectivity on a within-network basis.

Including all disorders covered here and relating the findings to predictive coding, there appears to be limited connections between disorder-specific network changes in the DMN and those in the IFG and STG that play such a dominant role in the MMN network. Consequently, the use of predictive coding networks based on auditory MMN as error signal only emphasizes the communality of cognitive decrease—potentially differentiated between temporal and frequency mismatches—however does not discriminate nor is of further diagnostic value. On the other hand, dynamic causal modeling elucidates the underlying intrinsic, particularly inhibitory changes that are affected by neurological disorders, in the various network nodes that affect the intranode FCs.

References

Alderson, T., Kehoe, E., Maguire, L., Farrell, D., Lawlor, B., Kenny, R.A., et al., 2017. Disrupted thalamus white matter anatomy and posterior default mode network effective connectivity in amnestic mild cognitive impairment. Front. Aging Neurosci. 9, 370. https://doi.org/10.3389/fnagi.2017.00370.

Assaf, M., Jagannathan, K., Calhoun, V.D., Miller, L., Stevens, M.C., Sahl, R., et al., 2010. Abnormal functional connectivity of default mode sub-networks in autism spectrum disorder patients. Neuroimage 53, 247–256.

Berron, D., van Westen, D., Ossenkoppele, R., Strandberg, O., Hansson, O., 2020. Medial temporal lobe connectivity and its associations with cognition in early Alzheimer's disease. Brain 143, 1233–1248.

Bi, X., Zhao, J., Xu, Q., Sun, Q., Wang, Z., 2018. Abnormal functional connectivity of resting state network detection based on linear ICA analysis in autism spectrum disorder. Front. Physiol. 9, 475. https://doi.org/10.3389/fphys.2018.00475.

Boets, B., Op de Beeck, H.P., Vandermosten, M., Scott, S.K., Gillebert, C.R., Mantini, D., et al., 2013. Intact but less accessible phonetic representations in adults with dyslexia. Science 342, 1251–1254.

Boyen, K., Langers, D.R.M., de Kleine, E., van Dijk, P., 2013. Gray matter in the brain: differences associated with tinnitus and hearing loss. Hear. Res. 295, 67–78.

Cao, F., Bitan, T., Booth, J.R., 2008. Effective brain connectivity in children with reading difficulties during phonological processing. Brain Lang. 107, 91–101.

Chapeton, J.I., Inati, S.K., Zaghloul, K.A., 2017. Stable functional networks exhibit consistent timing in the human brain. Brain 140, 628–640.

Chen, J., Zhao, Y., Zou, T., Wen, X., Zhou, X., Yu, Y., et al., 2022. Sensorineural hearing loss affects functional connectivity of the auditory cortex, parahippocampal gyrus and inferior prefrontal gyrus in tinnitus patients. Front. Neurosci. 16, 816712. https://doi.org/10.3389/fnins.2022.816712.

Cheng, S., Xu, G., Zhou, J., Qu, Y., Li, Z., He, Z., et al., 2020. A multimodal meta-analysis of structural and functional changes in the brain of tinnitus. Front. Hum. Neurosci. 14, 28. https://doi.org/10.3389/fnhum.2020.00028.

Chiang, H.-L., Tseng, W.-Y., Wey, H.-Y., Gau, S., 2022. Shared intrinsic functional connectivity alterations as a familial risk marker for ADHD: a resting-state functional magnetic resonance imaging study with sibling design. Psychol. Med. 52, 1736–1745.

Cooray, G., Garrido, M., Hyllienmark, L., Brismar, T., 2014. A mechanistic model of mismatch negativity in the ageing brain. Clin. Neurophysiol. 125, 1774–1782. https://doi.org/10.1016/j.clinph.2014.01.015.

Cope, T., Hughes, L., Phillips, H., Adams, N., Jafarian, A., Nesbitt, D., et al., 2022. Causal evidence for the multiple demand network in change detection: auditory mismatch magnetoencephalography across focal neurodegenerative disease. J. Neurosci. 42 (15), 3197–3215. https://doi.org/10.1523/JNEUROSCI.1622-21.2022.

Damiani, S., Tarchi, L., Scalabrini, A., Marini, S., Provenzani, U., Rocchetti, M., et al., 2021. Beneath the surface: hyper-connectivity between caudate and salience regions in ADHD fMRI at rest. Eur. Child Adolesc. Psychiatry 30, 619–631. https://doi.org/10.1007/s00787-020-01545-0.

De Ridder, D., Vanneste, S., Weisz, N., Londero, A., Schlee, W., Elgoyhen, A.B., et al., 2014. An integrative model of auditory phantom perception: tinnitus as a unified percept of interacting separable subnetworks. Neurosci. Biobehav. Rev. 44, 16–32.

Díaz, B., Hintz, F., Kiebel, S.J., von Kriegstein, K., 2012. Dysfunction of the auditory thalamus in developmental dyslexia. Proc. Natl. Acad. Sci. U. S. A. 109 (34), 13841–13846.

Dillen, K.N.H., Jacobs, H.I.L., Kukolja, J., Richter, N., von Reutern, B., Onur, Ö.A., et al., 2017. Functional disintegration of the default mode network in prodromal Alzheimer's Disease. J. Alzheimers Dis. 59, 169–187. https://doi.org/10.3233/JAD-161120.

Eggermont, J.J., 2021. Brain oscillations, synchrony and plasticity. In: Basic Principles and Application to Auditory-Related Disorders. Academic Press, London, ISBN: 978-0-12-819818-6, pp. 1–250.

Eggermont, J.J., 2022. Tinnitus and hyperacusis. In: Facts, Theories and Clinical Implications. Academic Press, UK.

Gao, Y., Shuai, D., Bu, X., Hu, X., Tang, S., Zhang, L., et al., 2021. Impairments of large-scale functional networks in attention-deficit/hyperactivity disorder: a meta-analysis of resting-state functional connectivity. Psychol. Med. 49, 2475–2485. https://doi.org/10.1017/S003329171900237X.

Gardini, S., Venneri, A., Sambataro, F., Cuetos, F., Fasano, F., Marchi, M., et al., 2015. Increased functional connectivity in the default mode network in mild cognitive impairment: a maladaptive compensatory mechanism associated with poor semantic memory performance. J. Alzheimers Dis. 45, 457–470. https://doi.org/10.3233/JAD-142547.

Georgiewa, P., Rzanny, R., Gaser, C., Gerhard, U.J., Vieweg, U., Freesmeyer, D., et al., 2002. Phonological processing in dyslexic children: a study combining functional imaging and event related potentials. Neurosci. Lett. 318 (1), 5–8.

Gour, N., Felician, O., Didic, M., Koric, L., Gueriot, C., Chanoine, V., et al., 2014. Functional connectivity changes differ in early and late-onset Alzheimer's disease. Hum. Brain Mapp. 35, 2978–2994.

Harrison, T.M., Burggren, A.C., Small, G.W., Bookheimer, S.Y., 2016. Altered memory-related functional connectivity of the anterior and posterior hippocampus in older adults at increased genetic risk for Alzheimer's disease. Hum. Brain Mapp. 37, 366–380. https://doi.org/10.1002/hbm.23036.

Harrison, T.M., Maass, A., Adams, J.N., Du, R., Baker, S.L., Jagust, W.J., 2019. Tau deposition is associated with functional isolation of the hippocampus in aging. Nat. Commun. 10, 4900. https://doi.org/10.1038/s41467-019-12921-z.

Hoeft, F., McCandliss, B.D., Black, J.M., Gantman, A., Zakerani, N., Hulme, C., et al., 2011. Neural systems predicting long-term outcome in dyslexia. Proc. Natl. Acad. Sci. U. S. A. 108 (1), 361–366.

Hullfish, J., Abenes, I., Kovacs, S., Sunaert, S., De Ridder, D., Vanneste, S., 2018. Functional brain changes in auditory phantom perception evoked by different stimulus frequencies. Neurosci. Lett. 683, 160–167.

Husain, F.T., Schmidt, S.A., 2014. Using resting state functional connectivity to unravel networks of tinnitus. Hear. Res. 307, 153–162.

Husain, F.T., Medina, R.E., Davis, C.W., Szymko-Bennett, Y., Simonyan, K., Pajor, N.M., et al., 2011. Neuroanatomical changes due to hearing loss and chronic tinnitus: a combined VBM and DTI study. Brain Res. 1369, 74–88.

Ibrahim, B., Suppiah, S., Ibrahim, N., Mohamad, M., Hassan, H.A., Nasser, N.S., et al., 2021. Diagnostic power of resting-state fMRI for detection of network connectivity in Alzheimer's disease and mild cognitive impairment: a systematic review. Hum. Brain Mapp. 42, 2941–2968. https://doi.org/10.1002/hbm.25369.

Jobson, D.D., Hase, Y., Clarkson, A.N., Kalaria, R.N., 2021. The role of the medial prefrontal cortex in cognition, ageing and dementia. Brain Commun. 3 (3), fcab125. https://doi.org/10.1093/braincomms/fcab125.

Khan, W., Amad, A., Giampietro, V., Werden, E., De Simoni, S., O'Muircheartaigh, J., et al., 2020. The heterogeneous functional architecture of the posteromedial cortex is associated with selective functional connectivity differences in Alzheimer's disease. Hum. Brain Mapp. 41, 1557–1572. https://doi.org/10.1002/hbm.24894.

Khan, R.A., Sutton, B.P., Tai, Y., Schmidt, S.A., Shahsavarani, S., Husain, F.T., 2021. A large-scale diffusion imaging study of tinnitus and hearing loss. Sci. Rep. 11, 23395. https://doi.org/10.1038/s41598-021-02908-6.

Lai, G., Pantazatos, S.P., Schneider, H., Hirsch, J., 2012. Neural systems for speech and song in autism. Brain 135, 961–975.

Lin, H., Lin, Q., Li, H., Wang, M., Chen, H., Liang, Y., et al., 2021. Functional connectivity of attention-related networks in drug-naïve children with ADHD. J. Atten. Disord. 25 (3), 377–388. https://doi.org/10.1177/1087054718802017.

Liu, J., Zhanga, X., Yu, C., Duan, Y., Zhuo, J., Cui, Y., et al., 2016. Impaired parahippocampus connectivity in mild cognitive impairment and Alzheimer's disease. J. Alzheimer's Dis. 49, 1051–1064. https://doi.org/10.3233/JAD-150727.

Liu, L., Wang, T., Du, X., Zhang, X., Xue, C., Ma, Y., Wang, D., 2022. Concurrent structural and functional patterns in patients with amnestic mild cognitive impairment. Front. Aging Neurosci. 14, 838161. https://doi.org/10.3389/fnagi.2022.838161.

Logothetis, N.K., 2003. The underpinnings of the BOLD functional magnetic resonance imaging signal. J. Neurosci. 23 (10), 3963–3971.

Marcos-Vidal, L., Martínez-García, M., Martín-de Blas, D., Navas-Sánchez, F.J., Pretus, C., Ramos-Quiroga, J.A., et al., 2022. Local functional connectivity as a parsimonious explanation of the main frameworks for ADHD in medication-naïve adults. J. Atten. Disord. 26 (12), 1563–1575. https://doi.org/10.1177/10870547211031998.

Maudoux, A., Lefebvre, P., Cabay, J.E., Demertzi, A., Vanhaudenhuyse, A., Laureys, S., et al., 2012a. Auditory resting-state network connectivity in tinnitus: a functional MRI study. PLoS One 7, e36222.

Maudoux, A., Lefebvre, P., Cabay, J.E., Demertzi, A., Vanhaudenhuyse, A., Laureys, S., et al., 2012b. Connectivity graph analysis of the auditory resting state network in tinnitus. Brain Res. 1485, 10–21.

Mohan, A., De Ridder, D., Idiculla, R., DSouza, C., Vanneste, S., 2018. Distress-dependent temporal variability of regions encoding domain-specific and domain-general behavioral manifestations of phantom percepts. Eur. J. Neurosci. 48, 1743–1764.

Molinaro, N., Lizarazu, M., Lallier, M., Bourguignon, M., Carreiras, M., 2016. Out-of-synchrony speech entrainment in developmental dyslexia. Hum. Brain Mapp. 37, 2767–2783.

Moring, J.C., Husain, F.T., Gray, J., Franklin, C., Peterson, A.L., Resick, P.A., et al., 2022. Invariant structural and functional brain regions associated with tinnitus: a meta-analysis. PLoS One 17 (10), e0276140. https://doi.org/10.1371/journal.pone.0276140.

Morken, F., Helland, T., Hugdahl, K., Specht, K., 2017. Reading in dyslexia across literacy development: a longitudinal study of effective connectivity. Neuroimage 144, 92–100.

Myers, N., Pasquini, L., Göttler, J., Grimmer, T., Koch, K., Ortner, M., et al., 2014. Within-patient correspondence of amyloid-b and intrinsic network connectivity in Alzheimer's disease. Brain 137, 2052–2064.

Näätänen, R., Sussman, E.S., Salisbury, D., Shafer, V.L., 2014. Mismatch negativity (MMN) as an index of cognitive dysfunction. Brain Topogr. 27, 451–466.

Nair, A., Jolliffe, M., Lograsso, Y.S.S., Bearden, C.E., 2020. A review of default mode network connectivity and its association with social cognition in adolescents with autism spectrum disorder and early-onset psychosis. Front. Psychiatry 11, 614. https://doi.org/10.3389/fpsyt.2020.00614.

Noreña, A.J., Eggermont, J.J., 2003. Changes in spontaneous neural activity immediately after an acoustic trauma: implications for neural correlates of tinnitus. Hear. Res. 183, 137–153.

Olulade, O.A., Flowers, D.L., Napoliello, E.M., Eden, G.F., 2015. Dyslexic children lack word selectivity gradients in occipito-temporal and inferior frontal cortex. Neuroimage Clin. 7, 742–754.

Paulesu, E., Démonet, J.F., Fazio, F., McCrory, E., Chanoine, V., Brunswick, N., et al., 2001. Dyslexia: cultural diversity and biological unity. Science 291 (5511), 2165–2167.

Pereira, A.M., Campos, B.M., Coan, A.C., Pegoraro, L.F., de Rezende, T.J.R., Obeso, I., et al., 2018. Differences in cortical structure and functional MRI connectivity in high functioning autism. Front. Neurol. 9, 539. https://doi.org/10.3389/fneur.2018.00539.

Pereira, J.B., Ossenkoppele, R., Palmqvist, S., Strandberg, T.O., Smith, R., Westman, E., Hansson, O., 2019. Amyloid and tau accumulate across distinct spatial networks and are differentially associated with brain connectivity. eLife 8, e50830. https://doi.org/10.7554/eLife.50830.

Pruim, R.H.R., Beckmann, C.F., Oldehinkel, M., Oosterlaan, J., Heslenfeld, D., Hartman, C.A., 2019. An integrated analysis of neural network correlates of categorical and dimensional models of Attention-Deficit/Hyperactivity Disorder. Biol. Psychiatry Cogn. Neurosci. Neuroimag. 4, 472–483. www.sobp.org/BPCNNI.

Quaglino, V., Bourdin, B., Czternasty, G., Vrignaud, P., Fall, S., Meyer, M.E., et al., 2008. Differences in effective connectivity between dyslexic children and normal readers during a pseudoword reading task: an fMRI study. Neurophysiol. Clin. 38, 73–82.

Raschle, N.M., Zuk, J., Gaab, N., 2012. Functional characteristics of developmental dyslexia in left-hemispheric posterior brain regions predate reading onset. Proc. Natl. Acad. Sci. U. S. A. 109 (6), 2156–2161.

Rauschecker, J.P., May, E.S., Maudoux, A., Ploner, M., 2015. Frontostriatal gating of tinnitus and chronic pain. Trends Cogn. Sci. 19 (10), 567–578.

Ripp, I., Stadhouders, T., Savio, A., Goldhardt, O., Cabello, J., Calhoun, V., et al., 2020. Integrity of neurocognitive networks in dementing disorders as measured with simultaneous PET/fMRI. J. Nucl. Med. 61, 1341–1347. https://doi.org/10.2967/jnumed.119.234930.

Rolls, E.T., Zhou, Y., Cheng, W., Gilson, M., Deco, G., Feng, J., 2020. Effective connectivity in autism. Autism Res. 13 (1), 32–44.

Rosemann, S., Rauschecker, J.P., 2022. Neuroanatomical alterations in middle frontal gyrus and the precuneus related to tinnitus and tinnitus distress. Hear. Res. 424, 108595.

Schelinski, S., Tabas, A., Kriegstein, K., 2022. Altered processing of communication signals in the subcortical auditory sensory pathway in autism. Hum. Brain Mapp. 43, 1955–1972. https://doi.org/10.1002/hbm.25766.

Schmidt, S.A., Carpenter-Thompson, J., Husain, F.T., 2017. Connectivity of precuneus to the default mode and dorsal attention networks: a possible invariant marker of long-term tinnitus. Neuroimage Clin. 16, 196–204.

Schurz, M., Wimmer, H., Richlan, F., Ludersdorfer, P., Klackl, J., Kronbichler, M., 2015. Resting-state and task-based functional brain connectivity in developmental dyslexia. Cereb. Cortex 25 (10), 3502–3514. https://doi.org/10.1093/cercor/bhu184.

Sharma, N., Murari, G., Vandermorris, S., Verhoeff, N.P.L.G., Herrmann, N., Chen, J.J., Mah, L., 2021. Functional connectivity between the posterior default mode network and parahippocampal gyrus is disrupted in older adults with subjective cognitive decline and correlates with subjective memory ability. J. Alzheimers Dis. 82, 435–445. https://doi.org/10.3233/JAD-201579.

Shaywitz, S.E., Shaywitz, B.A., Pugh, K.R., Fulbright, R.K., Constable, R.T., Mencl, W.E., et al., 1998. Functional disruption in the organization of the brain for reading in dyslexia. Proc. Natl. Acad. Sci. U. S. A. 95, 2636–2641.

Shaywitz, B.A., Shaywitz, S.E., Pugh, K.R., Mencl, W.E., Fulbright, R.K., Skudlarski, P., et al., 2002. Disruption of posterior brain systems for reading in children with developmental dyslexia. Biol. Psychiatry 52, 101–110.

Shigemoto, Y., Sone, D., Maikusa, N., Okamura, N., Furumoto, S., Kudo, Y., et al., 2018. Association of deposition of tau and amyloid-β proteins with structural connectivity changes in cognitively normal older adults and Alzheimer's disease spectrum patients. Brain Behav. 8, e01145. https://doi.org/10.1002/brb3.1145.

Sidlauskaite, J., Sonuga-Barke, E., Roeyers, H., Wiersema, J.R., 2016. Altered intrinsic organisation of brain networks implicated in attentional processes in adult attention deficit/hyperactivity disorder: a resting-state study of attention, default mode and salience network connectivity. Eur. Arch. Psychiatry Clin. Neurosci. 266, 349–357. https://doi.org/10.1007/s00406-015-0630-0.

Sohn, W.S., Yoo, K., Na, D.L., Jeong, Y., 2014. Progressive changes in hippocampal resting-state connectivity across cognitive impairment. Alzheimer Dis. Assoc. Disord. 28, 239–246.

Sripada, C.S., Kessler, D., Angstadt, M., 2014. Lag in maturation of the brain's intrinsic functional architecture in attention-deficit/hyperactivity disorder. Proc. Natl. Acad. Sci. U. S. A. 111 (39), 14259–14264.

Steinbrink, C., Ackermann, H., Lachmann, T., Riecker, A., 2009. Contribution of the anterior insula to temporal auditory processing deficits in developmental dyslexia. Hum. Brain Mapp. 30, 2401–2411.

Supekar, K., Uddin, L.Q., Khouzam, A., Phillips, J., Gaillard, W.D., Kenworthy, L.E., et al., 2013. Brain hyperconnectivity in children with autism and its links to social deficits. Cell Rep. 5, 738–747.

Tahmasian, M., Pasquini, L., Scherr, M., Meng, C., Förster, S., Bratec, S.M., et al., 2015. The lower hippocampus global connectivity, the higher its local metabolism in Alzheimer disease. Neurology 84, 1956–1963.

Teipel, S., Grothe, M.J., Zhou, J., Sepulcre, J., Dyrba, M., Sorg, C., et al., 2016. Measuring cortical connectivity in Alzheimer's disease as a brain neural network pathology: toward clinical applications. J. Int. Neuropsychol. Soc. 22, 138–163. https://doi.org/10.1017/S1355617715000995.

Temple, E., Poldrack, R.A., Protopapas, A., Nagarajan, S., Salz, T., Tallal, P., et al., 2000. Disruption of the neural response to rapid acoustic stimuli in dyslexia: evidence from functional MRI. Proc. Natl. Acad. Sci. U. S. A. 97 (25), 13907–13912.

Trevis, K.J., Tailby, C., Grayden, D.B., McLachlan, N.M., Jackson, G.D., Wilson, S.J., 2017. Identification of a neurocognitive mechanism underpinning awareness of chronic tinnitus. Sci. Rep. 7, 15220. https://doi.org/10.1038/s41598-017-15574-4.

Uddin, L.Q., Supekar, K.S., Ryali, S., Menon, V., 2011. Dynamic reconfiguration of structural and functional connectivity across core neurocognitive brain networks with development. J. Neurosci. 31 (50), 18578–18589.

Uddin, L.Q., Supekar, K., Menon, V., 2013. Reconceptualizing functional brain connectivity in autism from a developmental perspective. Front. Hum. Neurosci. 7, 458. https://doi.org/10.3389/fnhum.2013.00458.

Uddin, L.Q., Supekar, K., Lynch, C.J., Cheng, K.M., Odriozola, P., Barth, M.E., et al., 2015. Brain state differentiation and behavioral inflexibility in autism. Cereb. Cortex 25, 4740–4747. https://doi.org/10.1093/cercor/bhu161.

Vanneste, S., Heyning, P.V., De Ridder, D., 2011. Contralateral parahippocampal gamma-band activity determines noise-like tinnitus laterality: a region of interest analysis. Neuroscience 199, 481–490.

von dem Hagen, E.A.H., Stoyanova, R.S., Baron-Cohen, S., Calder, A.J., 2013. Reduced functional connectivity within and between social resting state networks in autism spectrum conditions. Soc. Cogn. Affect. Neurosci. 8, 694–701. https://doi.org/10.1093/scan/nss053.

Wang, J., Wang, X., He, Y., Yu, X., Wang, H., He, Y., 2015. Apolipoprotein E ε4 modulates functional brain connectome in Alzheimer's disease. Hum. Brain Mapp. 36, 1828–1846. https://doi.org/10.1002/hbm.22740.

Wang, T., Shi, F., Jin, Y., Yap, P.-T., Wee, C.-Y., Zhang, J., et al., 2016. Multilevel deficiency of white matter connectivity networks in Alzheimer's disease: a diffusion MRI study with DTI and HARDI models. Neural Plast., 2947136. https://doi.org/10.1155/2016/2947136.

Wilson, K.C., Kornisch, M., Ikuta, T., 2022. Disrupted functional connectivity of the primary auditory cortex in autism. Psychiatry Res. Neuroimaging 324, 111490.

Yerys, B.E., Gordon, E.M., Abrams, D.N., Satterthwaite, T.D., Weinblatt, R., Jankowski, K.F., et al., 2015. Default mode network segregation and social deficits in autism spectrum disorder: evidence from non-medicated children DMN in children with ASD. Neuroimage Clin. 9, 223–232.

Zhan, Y., Ma, J., Alexander-Bloch, A.F., Xu, K., Cui, Y., Feng, Q., et al., 2016. Longitudinal study of impaired intra- and inter-network brain connectivity in subjects at high risk for Alzheimer's disease. J. Alzheimers Dis. 52, 913–927. https://doi.org/10.3233/JAD-160008.

Zhang, Y., Simon-Vermot, L., Araque Caballero, M.Á., Gesierich, B., Taylor, A.N.W., Duering, M., et al., 2016. Enhanced resting-state functional connectivity between core memory-task activation peaks is associated with memory impairment in MCI. Neurobiol. Aging 45, 43–49.

Zhao, Y., Song, Q., Li, X., Li, C., 2016. Neural hyperactivity of the central auditory system in response to peripheral damage. Neural Plast., 2162105. https://doi.org/10.1155/2016/2162105.

Zhou, Y., Si, X., Chen, Y., Chao, Y., Lin, C.-P., Li, S., et al., 2022. Hippocampus-and thalamus-related fiber specific white matter reductions in mild cognitive impairment. Cereb. Cortex 32, 3159–3174. https://doi.org/10.1093/cercor/bhab407.

A primer on cortical auditory evoked potentials and magnetic fields

Cortical auditory evoked potentials and magnetic fields comprise obligatory ones responding to external stimuli, and endogenous event-related potentials (ERPs). An ERP is an electrophysiological brain response that is the direct result of a specific sensory, cognitive, or motor event. ERPs encompass the obligatory cortical auditory potentials (CAEPs); however, its use is typically reserved for responses to cognitive or motor events (tasks). The dominant cognitive ERPs are the mismatch negativity (MMN) and the P3(00) complex; however, we will also describe the contingent negative variation (CNV), the readiness potential (RP), and a few more.

A.1 Obligatory—Exogenous—Cortical auditory evoked potentials

A.1.1 The auditory steady-state response

Auditory steady-state responses (ASSRs) can be recorded from the human scalp when a sound, e.g., a click, is presented at a rate sufficiently fast that the response to any one sound overlaps the responses to preceding sounds. The most widely studied auditory steady-state response is evoked by sounds presented at rates near 40 Hz. The sinusoidal modulation of the amplitude (SAM) of a continuous tone also evokes an ASSR, particularly at modulation rates between 40 and 100 Hz (Lins and Picton, 1995). The cortical topographic organization of the ASSR depended on AM rate. Overall, faster AM rates lead to ASSRs located in more posterior and medial parts of the supratemporal gyrus (STG), whereas slower AM rates peaked at more anterior and lateral locations (Weisz and Lithari, 2017).

A.1.2 Thalamocortical onset responses

The middle-latency response (MLR) components are typically indicated by P_0, N_a, P_a, N_b, and P_b, and the longer latency components, or CAEPs, by P_1, N_1, P_2, and N_2 (Fig. A.1).

The MLRs have latencies from 10 to 50 ms and comprise components starting with the longest latency one from the auditory brainstem response (ABR; P_0) to the earliest responses from the auditory cortex (P_a; Fig. A.1; Eggermont, 2014). The scalp potential field configuration observed for N_a (Deiber et al., 1988) suggested a deep generator, possibly located at the midbrain or thalamus level. The latter is unlikely since the dipoles in the thalamus are not well aligned reducing its contribution to the scalp recording (Eggermont, 2007) so the thalamocortical radiation is

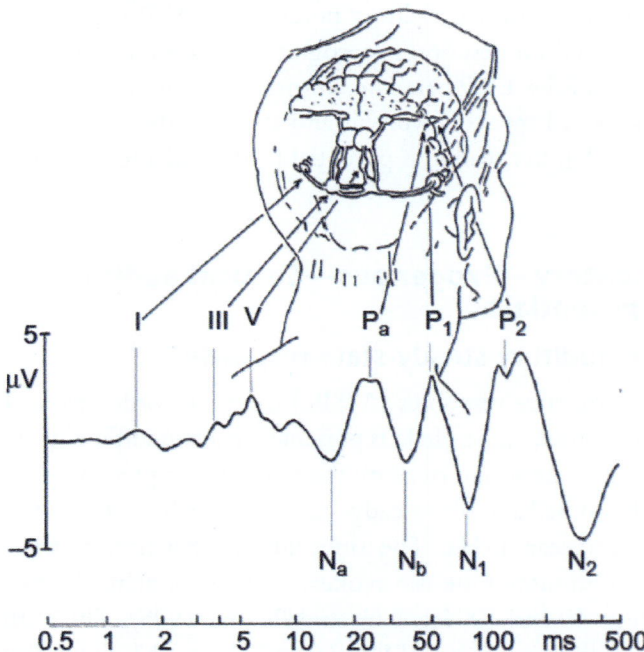

Fig. A.1 Waveform and nomenclature of ABR, middle- and long-latency obligatory components as recorded from Cz in response to clicks at a repetition rate of 1.3/s. The ABR components ("waves") are labeled I, III, and V. The middle-latency components are indicated with N_a, P_a, N_b, P_b (P_1). The long-latency components are indicated with P_1, N_1, P_2, and N_2. Note that P_b typically overlaps with P_1. *(Reprinted from Eggermont, J.J., 2014. Noise and the brain. In: Effects of non-damaging sound on the developing brain, pp. 84–119 (Chapter 4). ©2014, with permission from Elsevier.)*

a more likely source. According to Deiber et al. (1988), the potential field for P_a is different from that for N_a, suggesting that distinct generators are responsible for these two components. This contrasts with the dipole source analysis of Scherg and Von Cramon (1985), which suggested a single source for N_a-P_a in the primary auditory cortex. By recording cortical auditory evoked magnetic fields (CAEFs) Reite et al. (1988) identified an approximately 50-ms latency component (P_b) called the P50m. This longest latency component of the MLR (P_b) overlaps in time with the shortest latency response of the late components (P_1), but the origin of these two responses is different (Eggermont and Ponton, 2002). The P_b is driven by the thalamocortical afferents whereas the P_1 results from a drive by the mesencephalic reticular activating system (mRAS).

The CAEPs comprise every component with latencies ≥ 50 ms, although there are further subdivisions as to whether the potential is generated by the stimulus (obligatory components) or by a stimulus-related event (e.g., a change detection or task and then they are generally called event-related potentials; ERPs). The time boundary between MLRs and CAEPs at approximately 50 ms separates those AEP components that are strongly affected by attention (latencies > 50 ms) and those that are not (latencies < 50 ms).

A.1.3 Evoked magnetic fields

Postsynaptic potentials (PSPs), which underlie the CAEPs, are monophasic changes and are either reflecting depolarization (the cell interior becomes less negative than its resting value) or hyperpolarization (the cell interior becomes more negative than its resting value). These changes result in a current flow across the membrane, so if the interior of the cell becomes more positive (depolarization), then the cell exterior at that location has to become more negative to counterbalance the charge across the membrane. Depolarization of the neuron produces an inflow of Na^+ ions into the dendrite through specialized ion channels in the postsynaptic membrane and the location where this happens is called a "sink." As a result of the local current flow into the cell, the other end of the dendrite will become relatively negative and this site is called a (passive) "source." The sink–source configuration forms a dipole.

The dipole formed by the sink–source configuration in auditory cortex is a dynamic entity because the PSPs last only ~ 15 ms. Thus the currents flowing into the cell from the sink to the source and outside from the source to the sink are time dependent. Time-varying currents generate time-varying magnetic fields and these, if close enough to the scalp, can be recorded using magnetic field sensors. The voltage changes produced by changing currents have the same time course as the changes in magnetic fields and thus the waveforms of potentials and fields are generally the same. The differences between potentials and fields arise from the fact that potentials can also be recorded from radially (i.e., perpendicular to the scalp) oriented dipoles. In addition, they can be easily recorded from sources much deeper below the surface, and from sources in cortex, midbrain, and brainstem than in case of the magnetic fields (Eggermont, 2007).

A.2 Description of CAEP components

A.2.1 The P_1

The P_1 is the first vertex positive peak of the P_1-N_1-P_2 complex (Fig. A.1). This component typically occurs approximately 50 ms after stimulus onset in normal-hearing young adults. As mentioned before, P_1 and P_b are distinct components (Ponton et al., 2002). Neural generators of P_1 include primary auditory cortex (A1, Heschl's gyrus), hippocampus, planum temporale, and lateral temporal regions and possibly subcortical regions (Miyazato et al., 1996, 1999; Martin et al., 2008). In adults, the amplitude of P_1 is small and N_1 and P_2 dominate the response. This is largely the result of the temporal overlap of P_1 with N_1, which partially cancels the P_1 (Ponton and Eggermont, 2001).

A.2.2 The N_1

The N_1 peak (Fig. A.1) is composed of multiple, partially temporally overlapping, independent components (Näätänen and Picton, 1987; Picton et al., 1999). One negative component with a peak latency of 100 ms is generated in the auditory cortex on the supratemporal plane, as originally proposed by Vaughan and Ritter (1970). This component is probably generated over a much wider region of the supratemporal plane than that occupied by the A1 on Heschl's gyrus (HG). Another component of N_1 is found in the T complex that consists of a positive wave around 100 ms (Ta) and a negative wave (Tb) at approximately 150 ms. The T complex is generated on the lateral aspect of the temporal and parietal cortex, as originally proposed by Wolpaw and Penry (1975). This component has a radially oriented dipole and therefore is not recorded magnetically. Ta originates in the auditory association areas, activated by connections from the primary auditory cortex and also possibly directly from the thalamus. A third, vertex-negative component with a latency of approximately 100 ms is generated in the motor and premotor cortices under the influence of the reticular formation (as for P_2) and the ventrolateral nucleus of the thalamus. This latter component of N_1 may be the most likely component of N_1 to reflect attention switching to the eliciting sound (Alcaini et al., 1994).

A.2.3 The P_2

The P_2 follows N_1 and is a positive waveform elicited in normal-hearing adults approximately 180 ms after the onset of a sound stimulus (Fig. A.1). Like the previously described components, P_2 has multiple generators located in multiple auditory areas, including primary auditory cortex, secondary cortex, and driven by the caudal mesencephalic reticular activating system (mRAS; Velasco et al., 1989; Özkan et al., 2022). Because P_2 morphology often covaries with N_1 latency and amplitude, it is erroneously assumed that the N_1 and P_2 reflect the same neural mechanisms. Especially, as we have shown that P_2 matures much earlier (reaching adult values by as early as 2–3 years of age) than the N_1, which has a developmental time course extending into adolescence (Ponton et al., 2000). Eggermont (1988) has earlier shown that the short maturation time constant for P_2 latency is similar to that of the auditory brain stem response I-V interpeak interval, which may suggest that the maturation of the P_2 latency is only limited by maturational processes affecting the brain stem.

Shahin et al. (2003) found that equivalent current dipoles fitted to the N_1 and P_2 field patterns localized to spatially differentiable regions of the secondary auditory cortex. Sheehan et al. (2005) noted that P_2 reflects activity from several distinct generators, including at least two locations in the supratemporal auditory cortex, bilateral temporo-parietal cortex, and the mRAS. Ross and Tremblay (2009) showed N_1 and P_2 to originate from different anatomical structures that likely serve different functions. N_1 sources lay in the posterior part of auditory cortex, the planum temporale, whereas the center of activity for P_2 lay in anterior auditory cortex, the lateral part of Heschl's gyrus (Fig. A.2). P_2 sources have also been identified in planum temporale, Brodmann's area 22, and auditory association cortices (Crowley and Colrain, 2004).

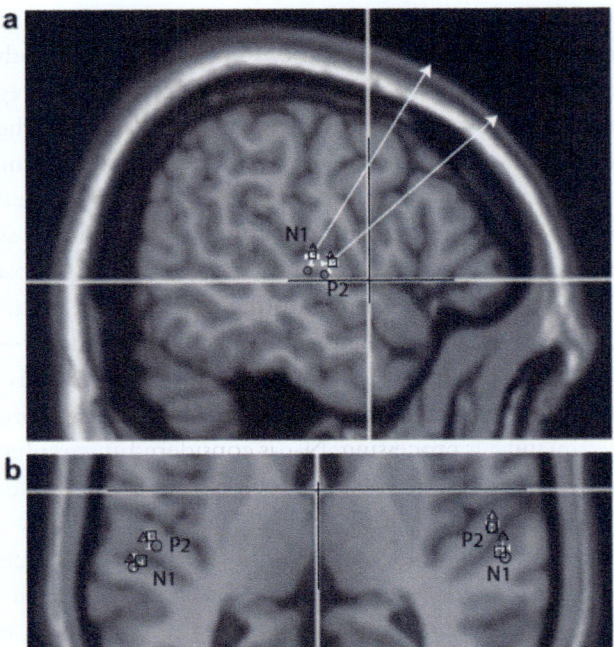

Fig. A.2 Overlay of group-averaged dipole locations onto a MRI atlas brain. A section of a coronal slice at Talairach z-coordinate $z_{tal} = 8$. The *white cursor lines* correspond to $x_{tal} = 0$ and $z_{tal} = 0$. The *white error bars* indicate the 95% confidence limits for the group mean location in x and y direction for N_1 and P_2 sources in the left and right hemisphere. The open symbols indicate the group mean dipole locations for the young *(circle)*, middle-aged *(square)*, and older *(triangle)* groups. P_2 and N_1 sources are separated with more medial and anterior locations of the P_2 sources. Also the hemispherical asymmetry is obvious for both the anatomical structure and dipole locations. The upper figure shows a view from right onto a sagittal slice at $x_{tal} = -54$. The *white arrows* represent a projection of mean dipole orientation onto the y-z plane. The orientation of the P_2 dipole is rotated in anterior direction compared to the N_1 orientation. *(Reprinted from Ross, B., Tremblay, K., 2009. Stimulus experience modifies auditory neuromagnetic responses in young and older listeners. Hear. Res. 248, 48–59. ©2008, with permission from Elsevier.)*

A.2.4 The N_2 and N_{2b}

The N_2 (Fig. A.1) is generated in medial frontal areas, including anterior cingulate cortex (ACC; Bekker et al., 2005; Jonkman et al., 2007), and the right ventral and dorsolateral prefrontal cortex (vPFC, DLPFC; Lavric et al., 2007). Karch et al. (2010) conducted a simultaneous EEG and fMRI study and reported that the N_2 amplitude was associated with BOLD responses in medial frontal brain regions and the putamen. These studies combined emphasize the importance of frontal attention-related brain regions in the generation of the N_2. The large age-related differences in N_2 amplitude or absence of the N_2 in older adults, and the increased variability in attention modulation of N_2 amplitudes in older adults, may represent an age-related change in the structure and function of these frontal attention networks. The N_2 appears frequently in combination with the P_3 in active listening conditions. As Amenado and Diaz (1998) put it: "Among the proposed different kinds of N_2 components that can be distinguished, N_{2b} is a sharp negative component with a central modality-non-specific topography—also called late MMN, often preceding P_3 (Novak et al., 1990). N_{2b} is elicited by attended infrequent (target) stimuli when they have to be actively selected by the subject to further processing. N_{2b} is considered an index of controlled orienting to and detection of deviant stimuli occurring in the attended auditory input (Näätänen, 1988, 1990; Novak et al., 1990)."

A.2.5 The T-complex

The T-complex represents a series of peaks in the latency range of 70–160 ms, recorded at temporal scalp locations. First described by Wolpaw and Penry (1975), the T-complex consists of a small negativity that peaks at about 70–80 ms (referred to as N_a), a positivity at approximately 100 ms (Ta), followed by a larger negativity at approximately 140–160 ms (Tb). The maturation of the obligatory CAEPs components, P_1, N_{1b}, and P_2, has been described by many investigators. We (Tonnquist-Uhlen et al., 2003) provided a comprehensive description of T-complex maturation from mid-childhood (5–6 yrs) to young adulthood (Fig. A.3).

The gradual nature of the maturational changes is evident in the surface plot reconstructions of the grand-mean waveforms shown in Fig. A.3. Distinct differences in T-complex morphology are apparent for the CAEPs recorded at electrodes T3 and T4 versus T5 and T6. At electrode T4, all the components of the T-complex (Na, Ta, and Tb) are identifiable in the grand-mean data, beginning with the 5- and 6-year-olds. The peaks of the T-complex are less distinct at electrode T3, but can be identified in the majority of subjects, including the 5- and 6-year-olds. The grand-mean waveforms from T5 and T6 are dominated by a large negativity (70–80 ms) in the age groups 5–6, 7, and 8 years. At 9 years of age, a positivity (100 ms) and a later negativity (150 ms) emerge. Overall, T-complex

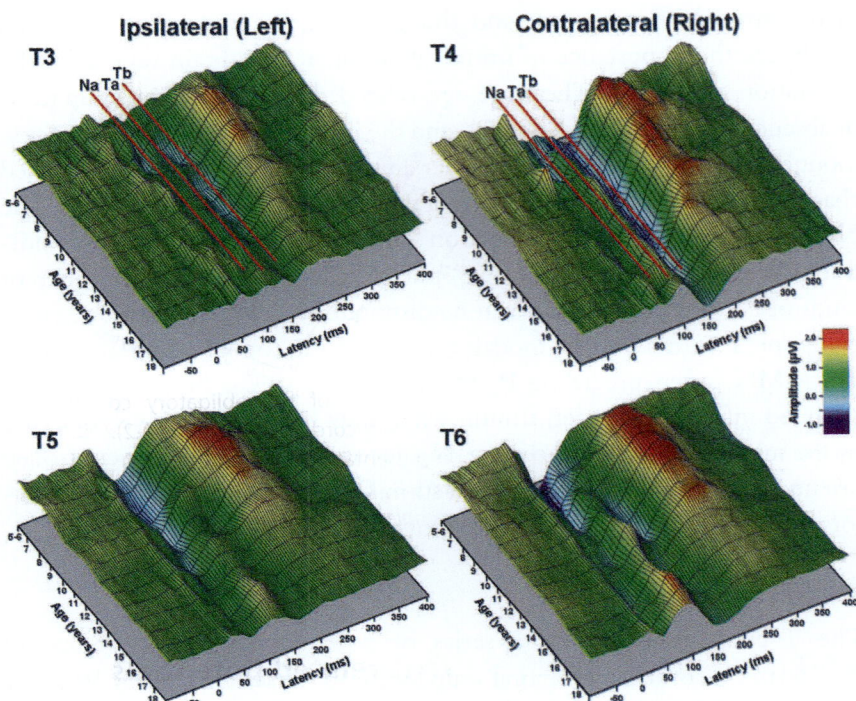

Fig. A.3 Surface plot representations of the same data as in Fig. A.1 show the age-related changes in waveform pattern at the four different electrode positions. The youngest age groups are in the back and the adult group at the front. At T5 and T6, the gradual emergence of wave Ta around 10–11 years of age is clearly visible. At T3 and T4, the pattern does not change much over time. *(Reprinted from Tonnquist-Uhlen, I., Ponton C.W., Eggermont, J.J. Kwong, B., Don, M., 2003. Maturation of human central auditory system activity: the T-complex. Clin. Neurophysiol. 114, 685–701. ©2003, with permission from Elsevier.)*

amplitude decreases as age increases and remains larger over the contralateral right hemisphere. There is evidence that the AEPs recorded at T5 and T6 do not represent independent T-complex activity; rather, they largely represent the inverted P_1, N_{1b}, and P_2 potentials measured at C3, C4, and other central and more frontal scalp locations. This is clearly illustrated in the surface plots shown in Fig. A.4. The surface plot for the grand-mean AEPs recorded at electrode C4 is shown on the left while the mathematically inverted AEPs recorded at electrode T6 are shown on right.

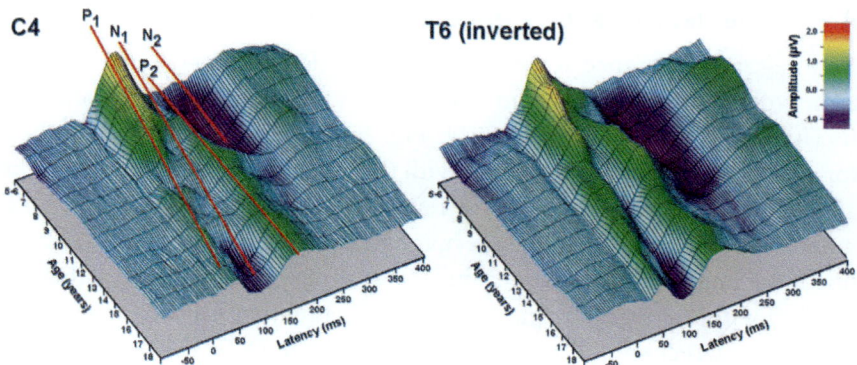

Fig. A.4 Comparison of the age-related changes of the obligatory components $(P_1-N_{1b}-P_2)$ at C4 and the inverted T-complex recorded at T6 (Fig. A.2). Note the striking resemblance between the two representations. *(Reprinted from Tonnquist-Uhlen, I., Ponton C.W., Eggermont, J.J. Kwong, B., Don, M., 2003. Maturation of human central auditory system activity: the T-complex. Clin. Neurophysiol. 114, 685–701. ©2003, with permission from Elsevier.)*

A.3 Endogenous auditory event-related potentials

Because of the large number of negative polarity ERPs, the observed latency ranges are compiled in Fig. A.5 and suggests that a number of different names for similar latency ERPs are used. This will be elucidated in the following sections.

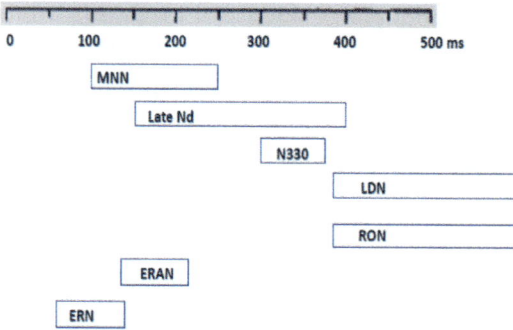

Fig. A.5 Survey of negative ERP latencies. *ERAN*, early right anterior negativity; *ERN*, error-related negativity; *LDN*, late negative difference, late MMN; *MMN*, mismatch negativity; *N330*, a mismatch response reflecting speech versus nonspeech detection; *Nd*, negative difference potential; *RON*, reorienting negativity requires attention.

A.3.1 The negative difference wave, Nd

The negative difference wave (Nd), N_{2b}, and P300 reflect selective attention, voluntary attention, and cognitive context updating, respectively (Näätänen, 1982; Itagaki et al., 2011; Fig. A.6). Nd is considered an indicator of selective attention associated with the relationship between a stimulus and a task. Early Nd is believed to reflect selection of information source, whereas late Nd is believed to reflect voluntary attention. Late Nd was identified as a negative deflection in the 150–400 ms period by subtracting the results for the nontarget stimuli on the nonattentive ear side from those for the nontarget stimuli on the attentive ear side (Näätänen, 1982). "Nd onset latency increases as the physical separation between the attended and unattended stimuli decreases, suggesting that Nd onset reflects the time required to determine whether a stimulus belongs to the unattended or the attended channel (i.e., whether or not it should receive extended processing). Whilst developmental studies reveal that Nd onset decreases into the teen years and beyond, Nd responses are clearly apparent in pediatric populations and can serve as a useful metric of the integrity of brain processes involved in selective attention in clinical groups of children" (Gomes et al., 2012).

A.3.2 The N330

The N330 match/mismatch responses measured from temporal electrodes, reflecting speech versus nonspeech detection, bilaterally. The ERP components recorded the temporal region (Galilee et al., 2017; Fig. A.7), included a negative going component peaking at 150 ms (N150), a positive going component peaking at approximately 280 ms (P250), a negative going component peaking at 330 ms (N330), and a positive going LSW (P600).

A.3.3 The mismatch negativity

A specific class of ERPs is formed by those responses that can only be obtained following an unexpected sound, such as an infrequent (10%–15% of the time) tone of 1000 Hz among a series of more frequent (85%–90%) tones of 1100 Hz. Under passive listening conditions the difference between the responses to the standard and the unexpected, also called deviant or oddball, sound obtained is the mismatch negativity. As originally described by Näätänen et al. (1978), the MMN is elicited by a discriminable change in a repeated sound, even when the sound is not attended to. This is illustrated in Fig. A.8.

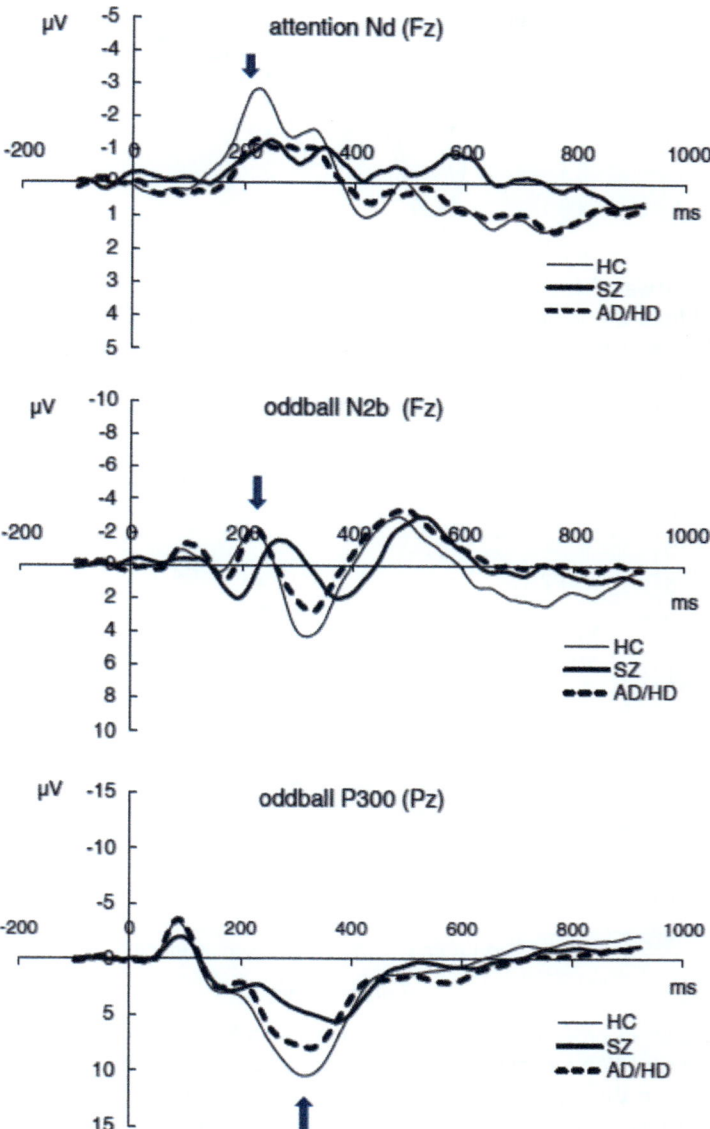

Fig. A.6 Grand-averaged waveforms for attentional Nd (Fz), oddball N_{2b} (Fz), and oddball P300 (Pz). The top of this figure shows the waveforms obtained by subtracting the responses to the nontarget stimuli to the nonattentive ear side from those to the nontarget stimuli to the attentive ear side in the attentional task. The *arrow* indicates late Nd. The middle part of this figure shows overall averaged waveforms obtained by subtracting the responses to the high-frequency nontarget stimuli from those to low-frequency stimuli in the oddball task. The *arrow* indicates N_{2b}. The bottom of this figure shows overall averaged waveforms for the responses to low-frequency stimuli in the oddball task. The *arrow* indicates P300. *(Reprinted from Itagaki, S., Yabe, H., Mori, Y., Ishikawa, H., Takanashi, Y., Shin-ichi Niwa, S.-i., 2011. Event-related potentials in patients with adult attention-deficit/hyperactivity disorder versus schizophrenia. Psychiatry Res. 189, 288–291. ©2011, with permission from Elsevier.)*

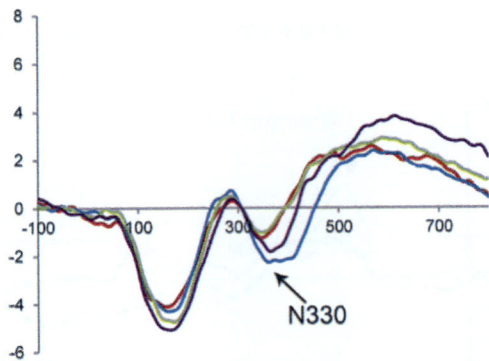

Fig. A.7 ERP waveforms in the temporal area. The figure represents the ERP waveforms recorded from the temporal left. *(From Galilee, A., Stefanidou, C., McCleery, J.P., 2017. Atypical speech versus non-speech detection and discrimination in 4- to 6-yr old children with autism spectrum disorder: an ERP study. PLoS One 12 (7), e0181354. https://doi. org/10.1371/journal.pone.018135. Figure 4. Open Access.)*

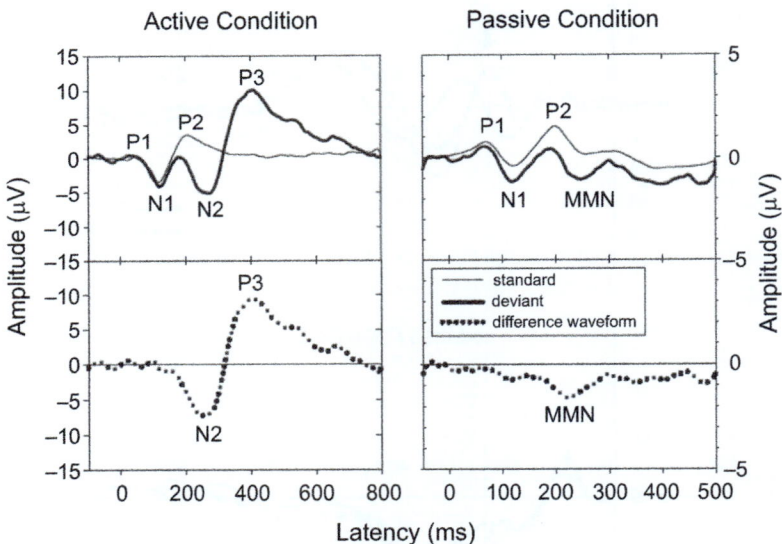

Fig. A.8 ERPs recorded while subjects actively discriminate auditory stimuli are shown on the left, and ERPs recorded while subjects ignore auditory stimuli (passive condition) are shown on the right. Averaged waveforms recorded to the frequently occurring standard stimulus are indicated by *thin lines*, whereas averaged waveforms recorded to infrequently occurring deviant stimuli are indicated by *thick lines*. In both the active and passive conditions, the averaged waveform in response to standards shows a clear P_1-N_1-P_2 pattern. In the active condition, the averaged waveform in response to deviants shows two additional components, N_2 and P_3. In the passive condition, there is increased negativity in response to the deviant stimuli. This increased negativity is the MMN. In the lower graphs, these discriminative components have been isolated by subtracting the response to standards from the response to deviants. These difference waveforms clearly show the N_2 and P_3 components in the active condition and the MMN in the passive condition. *(From Martin, B.A., Tremblay, K.L., Korczak, P., 2008. Speech evoked potentials: from the laboratory to the clinic. Ear Hear. 29, 285–313, with permission from Wolters-Kluwer Health, Inc.)*

The auditory MMN is obtained in adults with normal hearing as a frontocentral negativity, which occurs at approximately 100–250 ms after the onset of the deviant stimulus (for review, see Näätänen, 1990). It is maximally recorded from electrodes in the vicinity of Fz. The MMN is generated in primary and secondary auditory cortices and may have an additional generator in frontal cortex as well (Martin et al., 2008). The auditory cortex generators are believed to index brain activity related to change detection, whereas the frontal generator may be related to attention switching or orienting that is triggered by the change detection mechanism (Rinne et al., 2000). The MMN provides a neurophysiological correlate of the accuracy of the sound representation in the brain. The accuracy of this representation at least partly determines the individual's sound discrimination accuracy (Näätänen and Alho, 1997; Picton et al., 2000). The shape or size of the MMN is generally not indicative of what feature, if more than one is present, differentiates the deviant from the standard stimulus, only that a stimulus contrast exists and how big the difference might be (Martin et al., 2008).

A.3.4 The P3(00)

The P_3 (also known as P300) was first described by Sutton et al. (1965). The generators of auditory P_3 included auditory cortex, centroparietal cortex, hippocampus, and frontal cortex (Martin et al., 2008). There are two types of P_3 responses (P_{3a} and P_{3b}). P_{3a} has a more frontal scalp distribution and is less affected by attention (Squires et al., 1975). It often occurs following the MMN (Fig. A.8). It may signal the obtrusiveness of a deviant stimulus placed within a train of homogeneous stimuli or a preconscious, rapid attention switch to a deviant (Martin et al., 2008). In contrast, P_{3b} has a more parietal scalp location, is task related, and is dependent on attention (Picton, 1992). P_{3b} is a large centroparietal (i.e., at Pz) positivity seen, in normal-hearing adults, approximately 300 ms after deviant stimulus onset.

A.3.5 The ERAN

The early right anterior negativity (ERAN) is an early response to music-syntactic irregularities. In contrast to the classical MMN, which depends on regularities in ongoing auditory input, the ERAN depends on syntactic knowledge that transcends current auditory sensations. This is because music-syntactic regularities are represented in long-term memory. As with

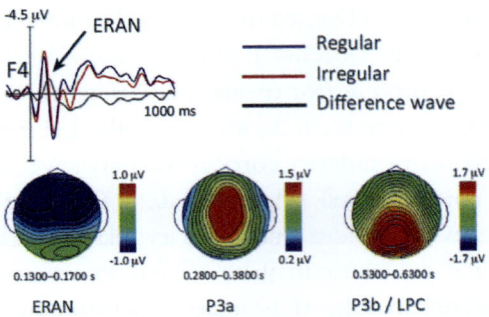

Fig. A.9 Influence of predictive processes on music-syntactic processing. When participants listen to musical sequences with irregular, thus unexpected, harmonies (as composed by a composer) elicits an early right anterior negativity (ERAN), even though participants are presented repeatedly with the same sequences and are told whether the sequence will be regular or irregular (*red line*: electric brain responses to irregular harmonies; *blue line*: brain responses to regular harmonies; the *black line* indicates the difference wave: regular subtracted from irregular harmonies). The isopotential maps in the lower panel show the frontal scalp distribution of the ERAN, and that the ERAN was followed by a P_{3a} with frontal preponderance and a P_{3b}/late positive component (LPC) with parietal preponderance (isopotential maps indicate difference potentials: regular subtracted from irregular harmonies). *(Reprinted from Koelsch, S., Vuust, P., Friston, K., 2019. Predictive processes and the peculiar case of music. Trends Cogn. Sci. 23 (1), 63–77. ©2018 with permission from Elsevier.)*

the MMN, the amplitude of the ERAN does not change when individuals know that a syntactic irregularity is pending. The ERAN was followed by a P_{3a} and a P_{3b} (Fig. A.9). Whereas P_{3a} and P_{3b} declined systematically across presentations, there was no systematic attenuation of the ERAN amplitude (Koelsch et al., 2019).

A.3.6 The late discriminative negativity, LDN

A set of ERP components, such as N_1, P_2, MMN, P_{3a}, and late discriminative negativity (LDN, or late MMN) or reorienting negativity (RON), have been identified as reflecting various stages of sensory and attentional processing (Yang et al., 2015). In addition to MMN and P_{3a}, LDN, which can also be elicited by deviant stimuli in the passive oddball paradigm in late time window, has different characteristics from the classic MMN. LDN, usually occurring between 400 and 700 ms after change onset with a frontocentral distribution, is often found in young children, and its amplitude tends to decrease with age. Bishop et al. (2010) found that LDN was larger for small deviants than for large deviants, and suggested on this basis that the LDN reflects the further processing of auditory stimuli when the salient features of the stimulus are difficult to discriminate. Mueller et al. (2008) showed that LDN was mainly found in the attended condition and was considerably reduced in children and absent in adults under the unattended condition. The LDN may also reflect the attention reorienting back to the original task, similar to the reorienting negativity (RON), which reflects attentional allocation after distraction and general reorientation of attention (Schröger et al., 2000). To summarize, at least three ERP components—MMN (or P-MMR), P_{3a}, and LDN—could be identified to index various stages of attentional processes during the passive auditory discrimination task, including preattentive change detection, involuntary attention switching, and attention reorienting (Yang et al., 2015).

Bishop et al. (2011) further "considered the characteristics of the LDN, which has received less research attention than the MMN. There are two characteristics of the LDN that have been described in the literature: it is greater in children than in adults, and more pronounced for speech than for non-speech signals. Both findings were obtained in the current dataset. Nevertheless, the speech–non-speech comparison needs to be treated with caution, as the stimuli contrasted in important respects other than 'speechness'. … One intriguing and novel observation about the LDN is that it was reliably larger for small than for large deviants. This is the opposite pattern to that seen for the MMN, and agrees with other sources of evidence (Ceponiene et al., 2004) that this component should not be regarded as a late manifestation of the MMN." Thus LDN is observed in response to complex stimuli, especially linguistic stimuli, in both children and adults (David et al., 2020). The amplitude of the LDN tends to diminish with increasing age, with larger LDN response in children than in adults (Fig. A.10).

A.3.7 The reorienting negativity (RON)

Auditory ERPs comprise the "distraction potential"—which includes the MMN associated with automatic auditory deviance detection, P_{3a}, and the Reorientation Negativity (RON), related to attentional reorientation after distraction, and has been associated with novelty detection, orientation response, and involuntary attention (Solís-Vivanco et al., 2015). Task-irrelevant frequency deviations prolong reaction times in the duration discrimination task by more than 35 ms and elicited the MMN and P_{3a} components of the event-related potential. The task-related P_{3a} was followed by

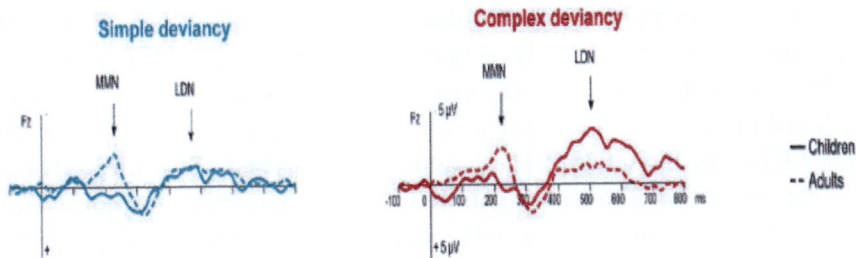

Fig. A.10 Cerebral responses to simple and complex deviancy in children and in adults. Figure displays the direct group comparison for each condition, brain responses are represented in *solid line* for children and in *dashed line* for adults. Figures show the difference waveforms in response to the simple *(blue)* and complex *(red)* phonological change and scalp potential and current density distributions of the MMN and the LDN. Statistical maps present the topographical differences calculated using permutation analyses. *(From David, C., Roux, S., Bonnet-Brilhault, F., Ferré, S., Gomot, M., 2020. Brain responses to change in phonological structures of varying complexity in children and adults. Psychophysiology 57 (9), e13621. © 2020 Society for Psychophysiological Research, with permission of John Wiley and Sons.)*

a negative deflection called RON (reorienting negativity). P_{3a} and RON are absent in the ignore condition (Fig. A.11). All effects are highly stable between sessions (product–moment correlations between 0.76 and 0.90). Scalp current density analysis suggested frontal generators for P_{3a} and for RON (Schröger et al., 2000).

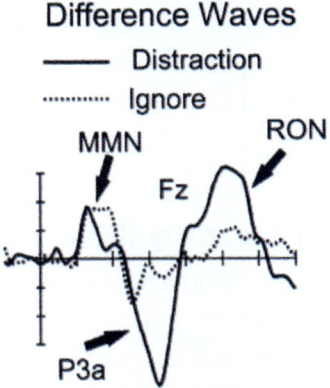

Fig. A.11 Grand-average difference waveforms obtained by subtracting the ERPs elicited by the frequent standard tones from the ERPs elicited by the infrequent deviant tones, separately for the distraction *(continuous lines)* and the ignore *(dotted lines)* condition. *(Reprinted from Schröger, E., Giard, M.-H., Wolff, Ch., 2000. Auditory distraction: event-related potential and behavioral indices. Clin. Neurophysiol. 111, 1450–1460. © 2000, with permission from Elsevier.)*

A.3.8 The error-related negativity

The ERN is a frontocentrally maximal response-locked negative deflection in the ERP that peaks approximately 50 ms following errors and has been observed across tasks that employ a variety of stimulus and response modalities and levels of difficulty. MRI source localization, MEG, and intracerebral recording indicate that the ERN is generated in the anterior cingulate cortex (ACC), a region of the medial prefrontal cortex (mPFC) implicated in the pathophysiology of a number of affective disorders (Weinberg et al., 2010). An unpredictable context potentiates amygdala activity. Fewer errors are made during unpredictable relative to predictable tone sequences as reflected in the ERN (Fig. A.12; Speed et al., 2017).

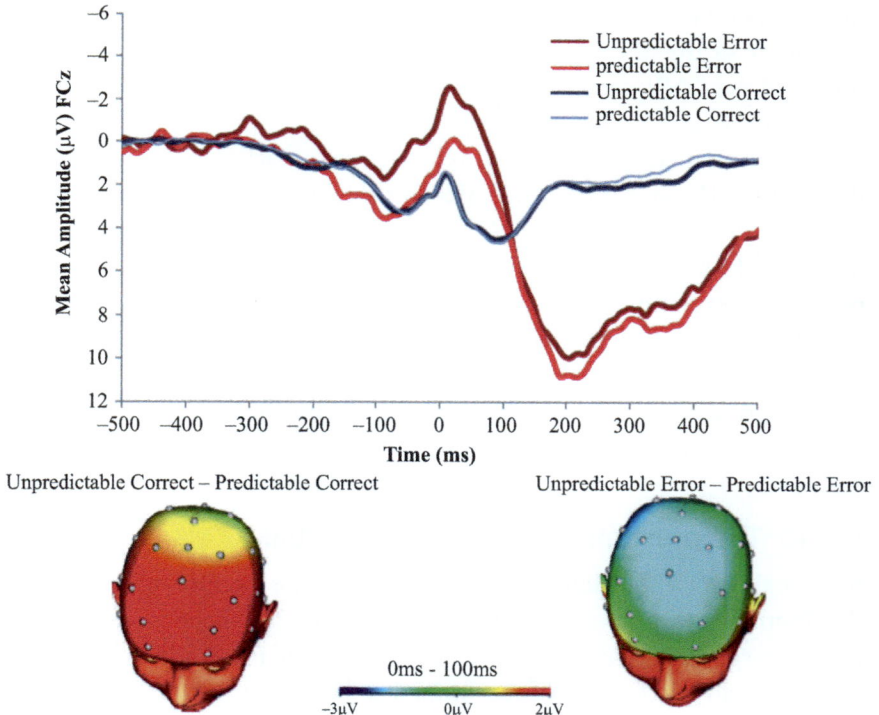

Fig. A.12 ERP waveforms (top) display the average electrocortical response to error (ERN) and correct (CRN) trials while participants were exposed to predictable versus unpredictable tone sequences. The ERN was scored between 0 and 100 ms postresponse at FCz. Head maps (bottom) show the scalp topography of the difference between unpredictable and predictable (unpredictable minus predictable) tone sequences on the CRN (left) and ERN (right). *(Reprinted from Speed, B.C., Jackson, F., Nelson, B.D., Infantolino, Z.P., Hajcak, G., 2017. Unpredictability increases the error-related negativity in children and adolescents. Brain Cogn. 119, 25–31. ©2017, with permission from Elsevier.)*

A.3.9 The contingent negative variation

The contingent negative variation (CNV) is a surface–negative brain potential generated during delay periods between stimuli and likely measure prefrontal cortex (PFC) activity (Rosahl and Knight, 1995). The CNV has cognitive and motor components and is observed during response anticipation. Berchicci et al. (2020) recorded anticipatory brain activities and evaluated whether temporal orienting processes are reflected by the novel prefrontal negative (pN) component, as already shown for the contingent negative variation. Fourteen young healthy participants underwent EEG and fMRI recordings in separate sessions; they were asked to perform a Go/NoGo task in which temporal orienting was manipulated: the external condition (a visual display indicating the time of stimulus onset) and the internal condition (time information not provided). In both conditions, the source of the pN was localized in the pars opercularis of the IFG; the source of the CNV was localized in the supplementary motor area and cingulate motor area, as expected (Fig. A.13).

EEG-informed fMRI analysis relating single-trial CNV amplitudes to the BOLD MRI signal addressed trial-to-trial fluctuations in prior probability processing. Scheibe et al. (2010) identified a set of regions mainly consisting of frontal, parietal, and striatal regions that represents unspecific response preparation on a trial-to-trial basis. A subset of these regions, namely, the dorsolateral prefrontal cortex (dlPFC), the inferior frontal gyrus (IFG), and the inferior parietal lobule (IPL), showed activations that exclusively represented the contributions of prior probability to the trial-to-trial fluctuations of the CNV (Fig. A.14).

A.3.10 The prestimulus readiness potential

Starr et al. (1995) instructed participants (1) to press a response button with the thumb of the dominant hand to each target or (2) to keep a mental count of each target. A prestimulus slow negative potential was identified before every stimulus except nontargets immediately after targets. The amplitude of the prestimulus negativity was significantly affected by task instructions and was up to 4 times larger during the button press than the mental count condition. In contrast, the amplitudes and latencies of the event-related components (N100, P200, N200, and P300), when slow potentials were removed by filtering, were not different as a function of press or count instructions (Fig. A.15). The major portion of the prestimulus negative potential is considered a readiness potential (RP) reflecting preparations to make a motor response.

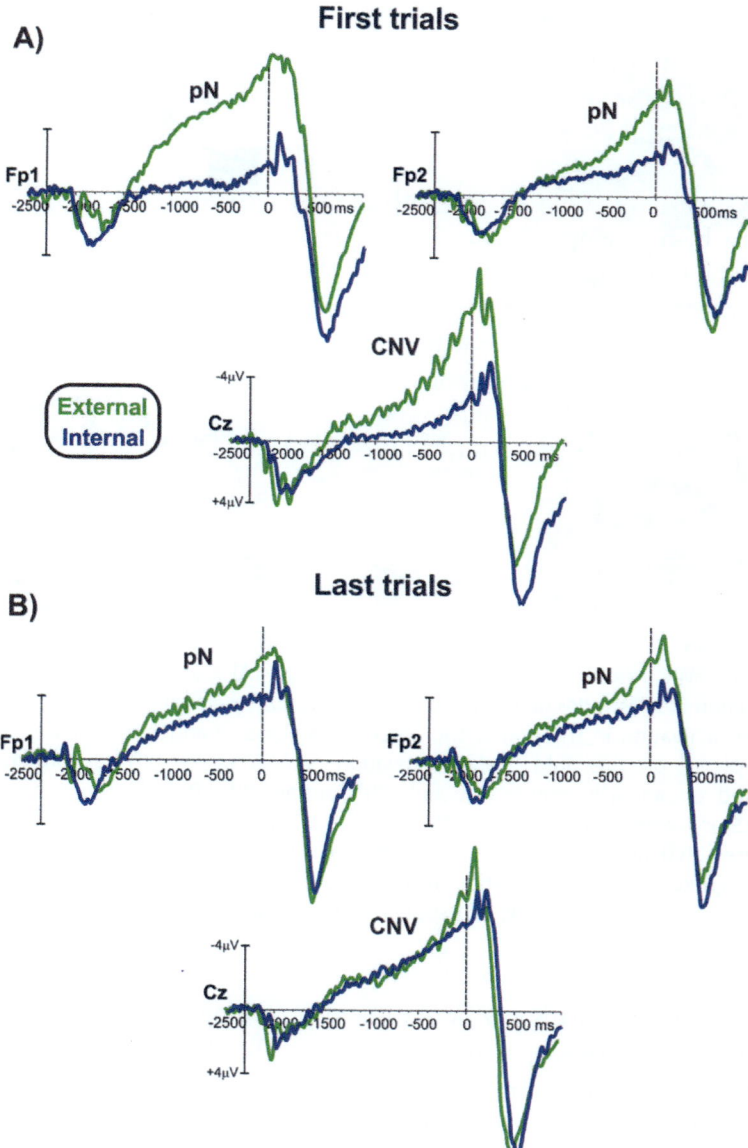

Fig. A.13 Time on task effect: Different effect for different conditions. (A) Before learning: overlap of ERP waveforms in the external *(green lines)* and internal *(blue lines)* conditions. (B) After learning: overlap of ERP waveforms in the external *(green lines)* and internal *(blue lines)* conditions. *(Reprinted from Berchicci, M., Sulpizio, V., Mento, G., Lucci, G., Civale, N., Galati, G., et al., 2020. Prompting future events: effects of temporal cueing and time on task on brain preparation to action. Brain Cogn. 141, 105565. ©2020, with permission from Elsevier.)*

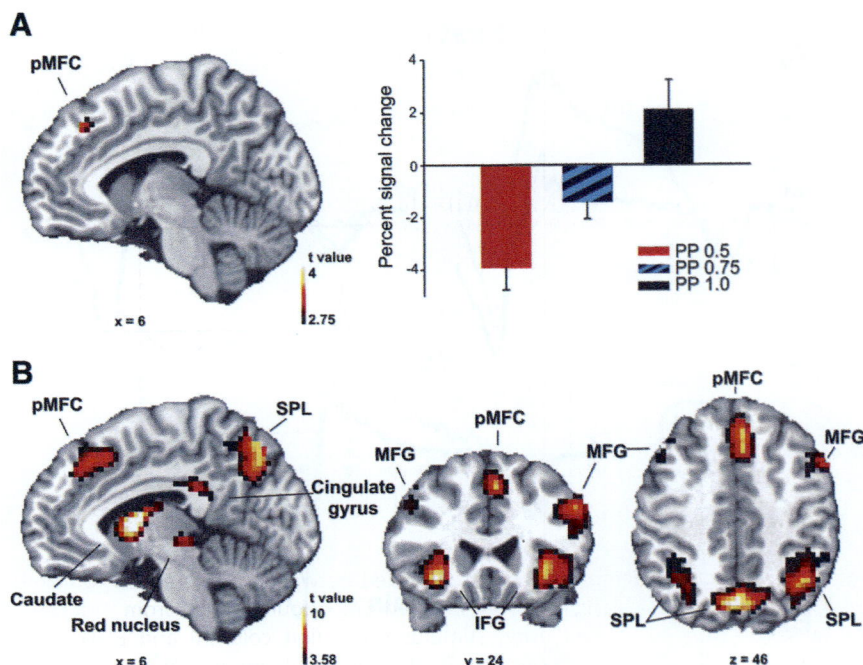

Fig. A.14 fMRI results of the conjunction analysis. (A) Conjunction analysis. The conjunction analysis designed to identify voxels parametrically modulated by PP revealed activation in the pMFC. The percentage signal changes for each condition are depicted on the right. Neural activation (shown at $P < .001$, uncorrected) is projected on an MNI template (Colin). (B) Contrast PP $1.0 > 0.5$. The contrast PP $1.0 > 0.5$ activated regions in the pMFC, MFG, IFG, SPL, cingulate gyrus, and midbrain structures. Activation clusters (shown at p_0.001, uncorrected) are projected on an MNI template (Colin). *IFG*, inferior frontal gyrus; *MFG*, medial frontal gyrus; *pMFC*, posterior medial frontal cortex; *SPL*, superior parietal lobe. *(From Scheibe et al. (2010).)*

Readiness potentials can be measured preceding auditory, visual, and somatosensory evoked potentials (Bianco et al., 2020). They confirmed the presence of a visual negativity (vN) component for the visual modality starting about 800 ms before stimulus with source in extrastriate areas and found novel modality-specific sensory readiness components for the auditory and somatosensory modalities. Surprisingly, an auditory positivity (aP) started about 800 ms before stimulus with source in bilateral auditory cortices and the somatosensory negativity (sN) started about 500 ms before stimulus with source in the somatosensory secondary cortex, contralateral to the stimulated hand (Fig. A.16). The scalp topography and intracranial sources of these three slow preparatory activities were mirrored with

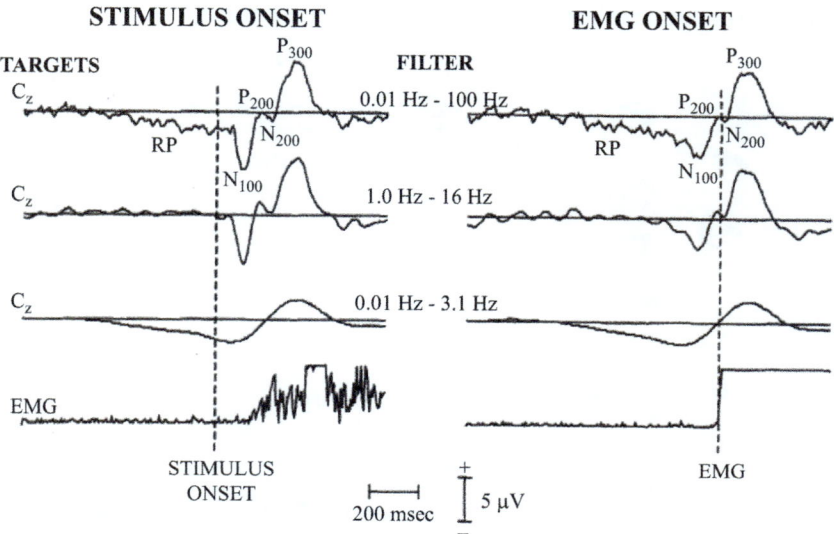

Fig. A.15 Averaged potentials to targets recorded between Cz referenced to linked earlobes (first 3 traces) along with the rectified EMG (fourth trace) from 1 subject. The averages were computed using stimulus onset (left column) and EMG onset (right column). The components are identified by their polarity (P or N for positive or negative, respectively) and approximate latency in ms. RP is a slow negative shift preceding both stimulus and EMG onset. The raw averages in the top line (0.01–100 Hz) were band-pass filtered: first, from 1 to 16 Hz (traces in second line) to attenuate the slow RP for measurement of the amplitudes of the event-related peaks; second, from 0.01 to 3.1 Hz (traces in third line) to attenuate the event-related components for measurement of the amplitude of the RP. In this and all subsequent figures, positivity at the "active" electrode (in this case Cz) is plotted up, a 5-µV calibration is provided for reference for evoked potentials, and a horizontal line is drawn through the average voltage in the baseline period. *(Reprinted from Starr, A., Sandroni, P., Michalewski, H.J., 1995. Readiness to respond in a target detection task: pre- and post-stimulus event-related potentials in normal subjects. Electroencephalogr. Clin. Neurophysiol. 96, 76–92. ©1995, with permission from Elsevier.)*

inverted polarity at early poststimulus stage evoking the well-known visual P_1, auditory N_1, and somatosensory P100 components. Present findings contribute to widening the family of slow wave preparatory components, providing evidence about the relationship between top–down and bottom–up processing in sensory perception.

The polarity shift between prestimulus and poststimulus activities in all three modalities (Fig. A.16) might be explained by different laminar activities in the two time intervals. Present results are in line with the proposal that poststimulus evoked cortical activity can be considered as a part of a

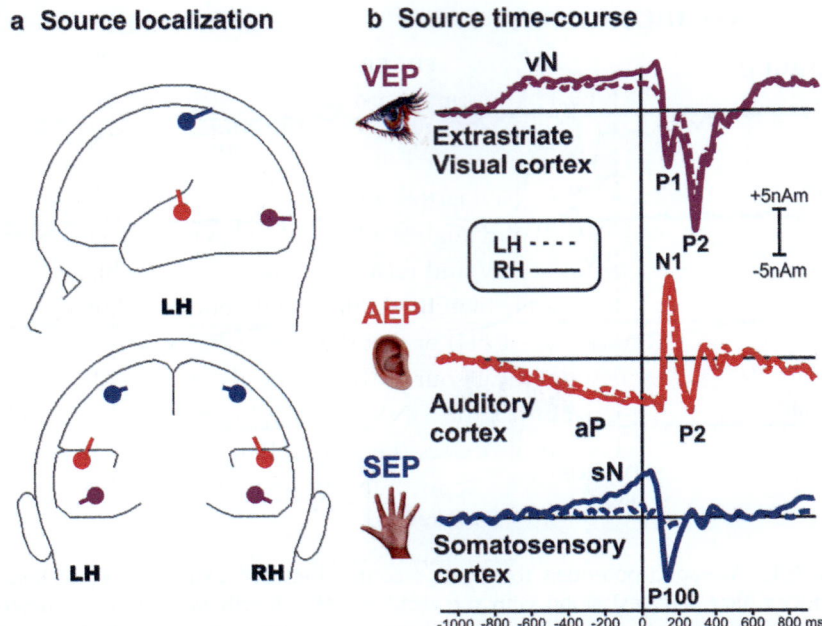

Fig. A.16 Source analysis of the prestimulus interval. (A) Source localization and orientation, (B) source time course (dipole moment). *LH*, left hemisphere; *RH*, right hemisphere. *(From Bianco, V., Perri, R.L., Berchicci, M., Quinzi, F., Spinelli, D., Di Russo, F., 2020. Modality-specific sensory readiness for upcoming events revealed by slow cortical potentials. Brain Struct. Funct. 225, 149–159. With permission from Springer Nature.)*

general recognition process, where prior expectations and sensory information interact according to a predictive model. At cortical level, present findings may support the predictive coding theory (Friston, 2005) proposing that neurons in the superficial layers represent top-down prediction sensory inputs, while neurons in deep layers encode bottom-up sensory inputs (Bianco et al., 2020).

A.3.11 Relating CNV and RP?

The readiness potential, a slow buildup of electrical potential recorded at the scalp using EEG, has been associated with neural activity involved in movement preparation. It became famous thanks to Libet et al. (1983), who used the time difference between the RP and self-reported time of conscious intention to argue that we lack free will. The RP's informativeness about self-generated action and derivatively about free will has prompted continued research on this neural phenomenon. Schurger et al. (2021) argue that recent advances in our understanding of the RP, including computational modeling of the phenomenon, call for a reassessment of its relevance for understanding volition and the philosophical problem of free will.

Schurger et al. (2021) also argued that the relationship between CNV and RP is contested. Some have posited that the RP is identical to the late component of the CNV (Rohrbaugh et al., 1976). It has been suggested that the RP is identical to the CNV and related to anticipation (Schlegel et al., 2013). It also remains possible that the RP has a different neural mechanism from the CNV (Ikeda et al., 1994) or that they share the same cortical generator, yet differ in subcortical sources (Ikeda et al., 1997). Meanwhile, others have hypothesized that the CNV is a combination of both motor and nonmotor preparatory processes (i.e., the RP and stimulus-preceding negativity (SPN), respectively; Damen and Brunia, 1987). More recently, it was observed that the amplitudes of the waveform obtained by superimposing the SPN on the RP were similar to that of the CNV at central and parietal sites but were smaller at frontal sites (Kotani et al., 2011). They suggest that the CNV and SPN may both be related to attention, sharing underlying neural mechanisms in the parietal area but differing in frontal regions.

In sum, the relationship between the RP and other slow cortical waves remains murky. The commonalities among them make it difficult to determine which aspects of the RP, other than the lateralized readiness potential (LRP), are specifically linked to motor preparation (Schurger et al., 2021).

References

Alcaini, M., Giard, M.H., Thevenet, M., Pernier, J., 1994. Two separate frontal components in the N1 wave of the human auditory evoked response. Psychophysiology 31 (6), 611–615.

Amenado, E., Diaz, F., 1998. Automatic and effortful processes in auditory memory reflected by event-related potentials. Age-related findings. Electroencephalogr. Clin. Neurophysiol. 108 (4), 361–369.

Bekker, E.M., Kenemans, J.L., Verbaten, M.N., 2005. Source analysis of the N2 in a cued go/Nogo task. Brain Res. Cogn. Brain Res. 22, 221–231.

Berchicci, M., Sulpizio, V., Mento, G., Lucci, G., Civale, N., Galati, G., et al., 2020. Prompting future events: effects of temporal cueing and time on task on brain preparation to action. Brain Cogn. 141, 105565.

Bianco, V., Perri, R.L., Berchicci, M., Quinzi, F., Spinelli, D., Di Russo, F., 2020. Modality-specific sensory readiness for upcoming events revealed by slow cortical potentials. Brain Struct. Funct. 225, 149–159. https://doi.org/10.1007/s00429-019-01993-8.

Bishop, D.V., Hardiman, M.J., Barry, J.G., 2010. Lower-frequency event- related desynchronization: a signature of late mismatch responses to sounds, which is reduced or absent in children with specific language impairment. J. Neurosci. 30, 15578–15584. https://doi.org/10.1523/jneurosci.2217-10.2010.

Bishop, D.V.M., Hardiman, M.J., Barry, J.G., 2011. Is auditory discrimination mature by middle childhood? A study using time-frequency analysis of mismatch responses from 7 years to adulthood. Dev. Sci. 14 (2), 402–416. https://doi.org/10.1111/j.1467-7687.2010.00990.x.

Ceponiene, R., Lepist, T., Soininen, M., Aronen, E., Alku, P., Näätänen, R., 2004. Event-related potentials associated with sound discrimination versus novelty detection in children. Psychophysiology 41 (1), 130–141. https://doi.org/10.1111/j.1469-8986.2003.00138.x.

Crowley, K.E., Colrain, I.M., 2004. A review of the evidence for P2 being an independent component process: age, sleep and modality. Clin. Neurophysiol. 115, 732–744.

Damen, E.J.P., Brunia, C.H.M., 1987. Changes in heart rate and slow brain potentials related to motor preparation and stimulus anticipation in a time estimation task. Psychophysiology 24 (6), 700–713.

David, C., Roux, S., Bonnet-Brilhault, F., Ferré, S., Gomot, M., 2020. Brain responses to change in phonological structures of varying complexity in children and adults. Psychophysiology 57 (9), e13621.

Deiber, M.P., Ibañez, V., Fischer, C., Perrin, F., Maugière, P., 1988. Sequential mapping favours the hypothesis of distinct generators for Na and Pa middle latency auditory evoked potentials. Electroencephalogr. Clin. Neurophysiol. 71 (3), 187–197.

Eggermont, J.J., 1988. On the rate of maturation of sensory evoked potentials. Electroencephalogr. Clin. Neurophysiol. 70, 293–305.

Eggermont, J.J., 2007. Electric and magnetic fields of synchronous neural activity propagated to the surface of the head: peripheral and central origins of AEPs. In: Burkard, R.R., Don, M., Eggermont, J.J. (Eds.), Auditory Evoked Potentials. Lippincott Williams & Wilkins, Baltimore, MD, pp. 2–21.

Eggermont, J.J., 2014. Noise and the brain. In: Experience Dependent Developmental and Adult Plasticity. Academic Press, London.

Eggermont, J.J., Ponton, C.W., 2002. The neurophysiology of auditory perception: from single-units to evoked potentials. Audiol. Neurootol. 7, 71–99.

Friston, K., 2005. A theory of cortical responses. Philos. Trans. R. Soc. B 360, 815–836. https://doi.org/10.1098/rstb.2005.1622.

Galilee, A., Stefanidou, C., McCleery, J.P., 2017. Atypical speech versus non-speech detection and discrimination in 4- to 6-yr old children with autism spectrum disorder: an ERP study. PLoS One 12 (7), e0181354. https://doi.org/10.1371/journal.pone.018135.

Gomes, H., Duff, M., Ramos, M., Molholm, S., Foxe, J.J., Halperin, J., 2012. Auditory selective attention and processing in children with attention-deficit/hyperactivity disorder. Clin. Neurophysiol. 123, 293–302.

Ikeda, A., Shibasaki, H., Nagamine, T., Terada, K., Kaji, R., Fukuyama, H., et al., 1994. Dissociation between contingent negative variation and Bereitschaftspotential in a patient with cerebellar efferent lesion. Electroencephalogr. Clin. Neurophysiol. 90, 359–364.

Ikeda, A., Shibasaki, H., Kaji, R., Terada, K., Nagamine, T., Honda, M., et al., 1997. Dissociation between contingent negative variation (CNV) and Bereitschaftspotential (BP) in patients with parkinsonism. Electroencephalogr. Clin. Neurophysiol. 102, 142–151.

Itagaki, S., Yabe, H., Mori, Y., Ishikawa, H., Takanashi, Y., Shin-ichi Niwa, S.-I., 2011. Event-related potentials in patients with adult attention-deficit/hyperactivity disorder versus schizophrenia. Psychiatry Res. 189, 288–291.

Jonkman, L.M., Sniedt, F.L., Kemner, C., 2007. Source localization of the Nogo-N2: a developmental study. Clin. Neurophysiol. 118, 1069–1077.

Karch, S., Feuerecker, R., Leicht, G., Meindl, T., Hantschk, I., Kirsch, V., et al., 2010. Separating distinct aspects of the voluntary selection between response alternatives: N2- and P3-related BOLD responses. Neuroimage 51, 356–364.

Koelsch, S., Vuust, P., Friston, K., 2019. Predictive processes and the peculiar case of music. Trends Cogn. Sci. 23 (1), 63–77. https://doi.org/10.1016/j.tics.2018.10.006.

Kotani, Y., Ohgami, Y., Arai, J.-I., Kiryu, S., Yusuke Inoue, Y., 2011. Motor and nonmotor components of event-brain potential in preparation of motor response. J. Behav. Brain Sci. 1, 234–241.

Lavric, A., Clapp, A., Rastle, K., 2007. ERP evidence of morphological analysis from orthography: a masked priming study. J. Cogn. Neurosci. 19, 866–877.

Libet, B., Gleason, C.A., Wright, E.W., Pearl, D.K., 1983. Time of conscious intention to act in relation to onset of cerebral activity (readiness-potential). The unconscious initiation of a freely voluntary act. Brain 106, 623–642.

Lins, O.G., Picton, T.W., 1995. Auditory steady-state responses to multiple simultaneous stimuli. Electroencephalogr. Clin. Neurophysiol. 96, 420–432.

Martin, B.A., Tremblay, K.L., Korczak, P., 2008. Speech evoked potentials: from the laboratory to the clinic. Ear Hear. 29, 285–313.

Miyazato, H., Skinner, R.D., Reese, N.B., Mukawa, J., Garcia-Rill, E., 1996. Midlatency auditory evoked potentials and the startle response in the rat. Neuroscience 75 (1), 289–300.

Miyazato, H., Skinner, R.D., Cobb, M., Andersen, B., Garcia-Rill, E., 1999. Midlatency auditory-evoked potentials in the rat: effects of interventions that modulate arousal. Brain Res. Bull. 48 (5), 545–553.

Mueller, V., Brehmer, Y., von Oertzen, T., Li, S.C., Lindenberger, U., 2008. Electrophysiological correlates of selective attention: a lifespan comparison. BMC Neurosci. 9, 18. https://doi.org/10.1186/1471-2202-9-18.

Näätänen, R., 1982. Processing negativity: an evoked-potential reflection of selective attention. Psychol. Bull. 92, 605–640.

Näätänen, R., 1988. Implications of ERP data for psychological theories of attention. Biol. Psychol. 26, 117–163.

Näätänen, R., 1990. The role of attention in auditory information processing as revealed by event-related potentials and other brain measures of cognitive function. Behav. Brain Sci. 13, 201–288.

Näätänen, R., Alho, K., 1997. Mismatch negativity—the measure for central sound representation accuracy. Audiol. Neurootol. 2, 341–353.

Näätänen, R., Picton, T., 1987. The N1 wave of the human electric and magnetic response to sound: a review and an analysis of the component structure. Psychophysiology 24, 375–425.

Näätänen, R., Gaillard, A.W., Mantysalo, S., 1978. Early selective-attention effect on evoked potential reinterpreted. Acta Psychol. (Amst) 42, 313–329.

Novak, G.P., Ritter, W., Vaughan Jr., H.G., Wiznitzer, M.L., 1990. Differentiation of negative event-related potentials in an auditory discrimination task. Electroencephalogr. Clin. Neurophysiol. 75, 255–275.

Özkan, M., Köse, B., Algın, O., Oguz, S., Erden, M.E., Çavdar, S., 2022. Non-motor connections of the pedunculopontine nucleus of the rat and human brain. Neurosci. Lett. 767, 136308.

Picton, T.W., 1992. The P300 wave of the human event-related potential. J. Clin. Neurophysiol. 9, 456–479.

Picton, T.W., Alain, C., Woods, D.L., John, M.S., Scherg, M., Valdes-Sosa, P., et al., 1999. Intracerebral sources of human auditory-evoked potentials. Audiol. Neurootol. 4, 64–79.

Picton, T.W., Alain, C., Otten, L., Ritter, W., Achim, A., 2000. Mismatch negativity: different water in the same river. Audiol. Neuro Otol. 5, 111–139.

Ponton, C.W., Eggermont, J.J., 2001. Of kittens and kids. Altered cortical maturation following profound deafness and cochlear implant use. Audiol. Neurootol. 6, 363–380.

Ponton, C.W., Eggermont, J.J., Kwong, B., Don, M., 2000. Maturation of human central auditory system activity: evidence from multi-channel evoked potentials. Clin. Neurophysiol. 111, 220–236.

Ponton, C.W., Eggermont, J.J., Khosla, D., Kwong, B., Don, M., 2002. Maturation of human central auditory system activity: separating auditory evoked potentials by dipole source modeling. Clin. Neurophysiol. 113, 407–420.

Reite, M., Teale, P., Zimmerman, J., Davis, K., Whalen, J., 1988. Source location of a 50 msec latency auditory evoked field component. Electroencephalogr. Clin. Neurophysiol. 70 (6), 490–498.

Rinne, T., Alho, K., Ilmoniemi, R.J., Virtanen, J., Näätänen, R., 2000. Separate time behaviors of the temporal and frontal MMN sources. Neuroimage 12, 14–19.

Rohrbaugh, J.W., Syndulko, K., Lindsley, D.B., 1976. Brain wave components of the contingent negative variation in humans. Science 191 (4231), 1055–1057.

Rosahl, S.K., Knight, R.T., 1995. Role of prefrontal cortex in generation of the contingent negative variation. Cereb. Cortex 5 (2), 123–134.

Ross, B., Tremblay, K., 2009. Stimulus experience modifies auditory neuromagnetic responses in young and older listeners. Hear. Res. 248, 48–59.

Scheibe, C., Ullsperger, M., Sommer, W., Heekeren, H.R., 2010. Effects of parametrical and trial-to-trial variation in prior probability processing revealed by simultaneous electroencephalogram/functional magnetic resonance imaging. J. Neurosci. 30 (49), 16709–16717.

Scherg, M., Von Cramon, D., 1985. Two bilateral sources of the late AEP as identified by a spatio-temporal dipole model. Electroencephalogr. Clin. Neurophysiol. 62, 32–44.

Schlegel, A., Alexander, P., Sinnott-Armstrong, W., Roskies, A., Tse, P.U., Wheatley, T., 2013. Barking up the wrong free: readiness potentials reflect processes independent of conscious will. Exp. Brain Res. 229, 329–335. https://doi.org/10.1007/s00221-013-3479-3.

Schröger, E., Giard, M.-H., Wolff, C., 2000. Auditory distraction: event-related potential and behavioral indices. Clin. Neurophysiol. 111, 1450–1460.

Schurger, A., Hu, P.B., Pak, J., Roskies, A.L., 2021. What is the readiness potential? Trends Cogn. Sci. 25 (7), 558–570. https://doi.org/10.1016/j.tics.2021.04.001.

Shahin, A., Bosnyak, D.J., Trainor, L.J., Roberts, L.E., 2003. Enhancement of neuroplastic P2 and N1c auditory evoked potentials in musicians. J. Neurosci. 23 (12), 5545–5552.

Sheehan, K.A., McArthur, G.M., Bishop, D.V.M., 2005. Is discrimination training necessary to cause changes in the P2 auditory event-related brain potential to speech sounds? Cogn. Brain Res. 25, 547–553.

Solís-Vivanco, R., Rodríguez-Violante, M., Rodríguez-Agudelo, Y., Schilmann, A., Rodríguez-Ortiz, U., Ricardo-Garcell, J., 2015. The P3a wave: a reliable neurophysiological measure of Parkinson's disease duration and severity. Clin. Neurophysiol. 126, 2142–2149.

Speed, B.C., Jackson, F., Nelson, B.D., Infantolino, Z.P., Hajcak, G., 2017. Unpredictability increases the error-related negativity in children and adolescents. Brain Cogn. 119, 25–31.

Squires, N.K., Squires, K., Hillyard, S.A., 1975. Two varieties of long-latency positive waves evoked by unpredictable stimuli in man. Electroencephalogr. Clin. Neurophysiol. 38, 387–401.

Starr, A., Sandroni, P., Michalewski, H.J., 1995. Readiness to respond in a target detection task: pre- and post-stimulus event-related potentials in normal subjects. Electroencephalogr. Clin. Neurophysiol. 96, 79–92.

Sutton, S., Braren, M., Zubin, J., John, E.R., 1965. Evoked potential correlates of stimulus uncertainty. Science 150 (3700), 1187–1188.

Tonnquist-Uhlen, I., Ponton, C.W., Eggermont, J.J., Kwong, B., Don, M., 2003. Maturation of human central auditory system activity: the T-complex. Clin. Neurophysiol. 114, 685–701.

Vaughan, H.G., Ritter, W., 1970. The sources of auditory evoked responses recorded from the human scalp. Electroencephalogr. Clin. Neurophysiol. Suppl. 28, 360–367.

Velasco, M., Velasco, F., Velasco, A.L., 1989. Intracranial studies on potential generators of some vertex auditory evoked potentials in man. Stereotact. Funct. Neurosurg. 53 (1), 49–73. https://doi.org/10.1159/000099517.

Weinberg, A., Olvet, D.M., Hajcak, G., 2010. Increased error-related brain activity in generalized anxiety disorder. Biol. Psychol. 85, 472–480.

Weisz, N., Lithari, C., 2017. Amplitude modulation rate dependent topographic organization of the auditory steady-state response in human auditory cortex. Hear. Res. 354, 102–108.

Wolpaw, J.R., Penry, J.K., 1975. A temporal component of the auditory evoked response. Electroencephalogr. Clin. Neurophysiol. 39 (6), 609–620.

Yang, M.-T., Hsu, C.-H., Yeh, P.-W., Lee, W.-T., Liang, J.-S., Fu, W.-M., Lee, C.-Y., 2015. Attention deficits revealed by passive auditory change detection for pure tones and lexical tones in ADHD children. Front. Hum. Neurosci. 9, 470. https://doi.org/10.3389/fnhum.2015.00470.

Index

Note: Page numbers followed by *f* indicate figures and *t* indicate tables.

CPI Antony Rowe
Eastbourne, UK
July 14, 2023